Mathematical Models with Applications

COMAP'S

Mathematical Models with Applications

Developed by

COMAP, Inc.

The Consortium for Mathematics
and Its Applications

57 Bedford Street, Suite 210

Lexington, Massachusetts 02420

Project Leadership

Solomon Garfunkel
COMAP, Inc., Lexington, MA

Roland Cheyney
COMAP, Inc., Lexington, MA

Lead Author

Jerry Lege
COMAP, Inc., Lexington, MA

Authors

Ronald E. Bell, II
University of Texas
at Austin, TX

Marsha Davis
E. CT State University,
Williamantic, CT

Sandra Nite
Sweeny High School,
Sweeny TX

W. H. Freeman and Company
New York

www.whfreeman.com

The Consortium for Mathematics and Its Applications (COMAP)

57 Bedford Street, Suite 210
Lexington, MA 02420

Published and distributed by

W. H. Freeman and Company

41 Madison Avenue, New York, NY 10010

ISBN 0-7167-4458-9 (EAN: 9780716744580)

Printed in the United States of America.
Fourth printing, 2004

Project Leaders: Solomon Garfunkel, COMAP, Inc., Lexington, MA
Roland Cheyney, COMAP, Inc., Lexington, MA

Lead Author: Jerry Lege, COMAP, Inc., Lexington, MA

Authors: Ronald E. Bell, II, University of Texas at Austin, TX;
Marsha Davis, Eastern Connecticut State University, Willimantic, CT;
Sandra Nite, Sweeny High School, Sweeny, TX

Publisher: Michelle Julet
Acquisitions Editor: Craig Bleyer
Marketing manager: Mike Saltzman
Supplements and New Media Editor: Mark Santee
Cover and Text Design: Daiva Kiliulis
Front Cover: Corbis Images
Photo Research: Michele Doherty, Susan Van Etten
Illustrations: David Barber, Lianne Dunn, Ashley Van Etten
Production Department: Michele Doherty, Daiva Kiliulis, George Ward, Gail Wessell, Pauline Wright
Composition: Daiva Kiliulis
Manufacturing: R.R. Donnelly and Sons Co.

Dear Student,

Mathematical Models with Applications is a different kind of math book than you may have used, for a different kind of math course than you may have taken. In addition to presenting mathematics for you to learn, we have tried to present mathematics for you to use. The word "models" in the title is the key. Real problems do not come in the form of problems at the end of chapters in a math book. Real problems don't look like math problems. Real problems ask questions such as: How do we know how much rain falls in a state? What kind of car should I buy? How do we effectively control animal population? Real problems can be messy, but are often more interesting than the types of math problems you have worked on before.

Mathematical models are tools to explain, clarify, and solve these types of real world problems. Sometimes the actual models you develop will be useful for looking at other problems, but the key here is the process of modeling: forming a theory, testing it, and then revisiting it based on the results of the test. You will probably not know in advance what mathematics to use. The mathematics you settle on may be a mix of ideas in geometry, algebra, and data analysis. You may need to use graphing calculators to help sort through data, or speed up the process of testing the model. Some of the answers might not look like answers you have gotten in other math classes, and, sometimes, more than one answer will be right. By the end of this course you may have a different definition for the "right answer."

One of the most important reasons for learning the modeling process is that this skill will help you in just about any career that you might choose. It will also help you make good life decisions and be more aware of local, national, and global issues. We know it may be hard to think about this now, but you will be learning long after you leave high school. Learning how to model is learning how to learn.

You may have been told that you are "not good at math." Well here is a fresh start for you to see mathematics in a new way—and for you to demonstrate your ability to use mathematics to explore the same types of problems that scientists, engineers, and bankers face in their jobs and lives. Give it a chance…you may learn that you are a good math student after all.

Solomon Garfunkel
Executive Director, COMAP

Jerry Lege
Lead Author, COMAP

Roland Cheyney
Project Director, COMAP

Joyce Q. Collett
Clear Creek ISD
League City, TX

Ronald E. Bell, II
University of Texas
at Austin and MOVES, Inc.

Sandra Nite
Sweeny High School

Marsha Davis
Eastern Connecticut
State University

CONTENTS Mathematical Models with Applications

Mathematics IS Modeling

Can mathematical modeling help plan a better event?

MATHEMATICAL TOPICS

- input output variables
- slope
- linear functions
- solving linear equations
- working with expressions
- distributive property

APPLICATIONS/RELATED DISCIPLINES

- Event Planning
- Business
- Promoter
- Agent

Bones

How does statistics help solve the mystery of soldiers who are missing in action?

Mathematical Topics

- data analysis
- box plots
- histograms
- scatter plots
- residuals
- linear equations
- best-fit criteria
- equivalence
- independent/dependent

Applications/Related Disciplines

- Forensic Science
- Medical Technician
- Anthropology
- Nurse
- Marine Biology
- Radiologist

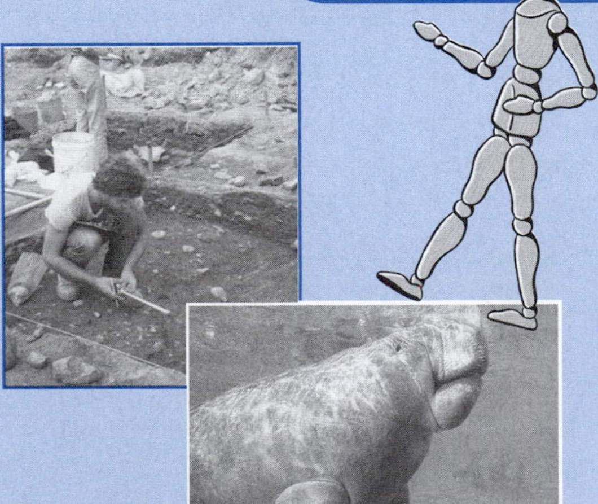

It Looks Like Rain

Why do ranchers and farmers receive a lot of help from geometry?

Mathematical Topics

- area problems
- polygons
- weighted averages
- Voronoi diagrams
- perpendicular bisectors
- Pythagorean formula
- estimation/significance
- volume
- Heron's formula
- Pick's formula
- coordinate geometry
- systems of equations
- reflection
- circumscribed circles
- distance formula

Applications/Related Disciplines

- Agriculture
- Ranching
- Resource Planning
- City Planning
- Civil Engineering
- Map Making

Testing 1, 2, 3

How is algebra used to make the Olympics fair to all athletes?

Mathematical Topics

- quadratic functions
- probability
- expected value
- vertex form of quadratics
- completing the square
- quadratic formula
- quadratic regression
- residual analysis
- polynomial expressions
- factoring

Applications/Related Disciplines

- Lab Technician
- Nurse
- Physician Assistant
- Pharmacist

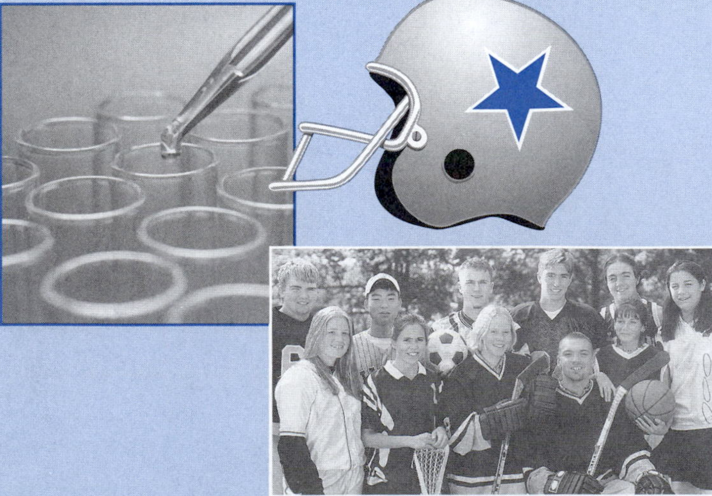

Art and Perspective

Is there a connection between mathematics and beauty?

Mathematical Topics

- scale
- proportion
- parallel lines
- 45-45-90 right triangles
- 30-60-90 right triangles
- sine, cosine, tangent
- similar triangles
- construction
- right triangle trigonometry
- parallelograms

Applications/Related Disciplines

- Graphic Artist
- Packaging Designer
- CAD/CAM Operator
- Technical Illustrator

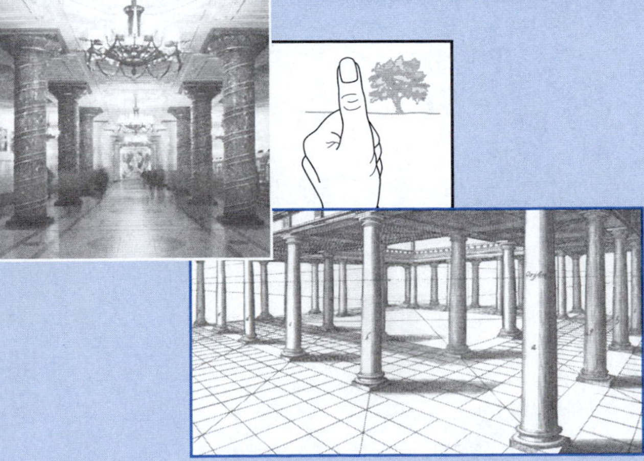

The Financial Ride of Your Life

Can mathematics make you a better consumer?

Mathematical Topics

- rate and ratio problems
- linear equations
- break even analysis
- amortization
- slope
- percentages

Applications/Related Disciplines

- Consumer Finance
- Buyer/Purchasing Agent
- Salesperson
- Marketing
- Business Manager
- Bookeeeping

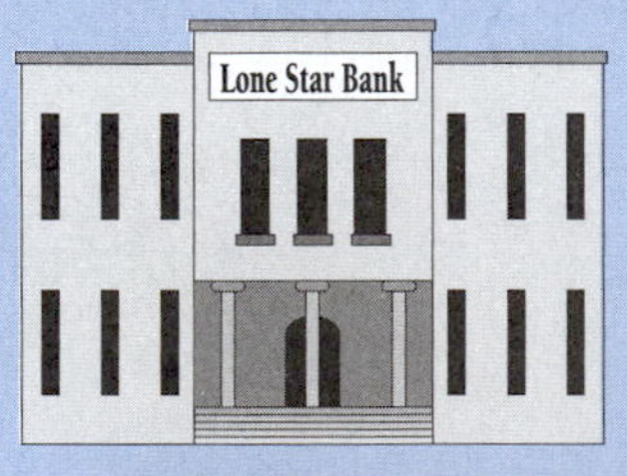

Growth: From Money to Moose

Why do some companies and people financially succeed when others fail?

MATHEMATICAL TOPICS

- exponential functions
- recursion relations/notation
- arithmetic/geometric sequences
- arithmetic/geometric series
- mixed growth sequences
- amortization
- half life/doubling time
- growth and decay
- laws of exponents

APPLICATIONS/RELATED DISCIPLINES

- Investment
- Banking
- Ecology
- Finance
- Accounting
- Insurance

Modeling Motion

How do functions help make action-packed movies and television?

MATHEMATICAL TOPICS

- rate of change
- periodic functions
- amplitude
- radian measure
- unit circle
- trigonometric graphs
- sinusoidal functions
- parabola

APPLICATIONS/RELATED DISCIPLINES

- Mechanical Engineer
- Special Effects Engineer
- Stunt Person
- Equipment Testing

Mathematics in the Music

Can you hear the mathematics in music?

Mathematical Topics

- transformations
- symmetry and reflection
- sinosoidal graphs
- sequence and series
- golden ratio

Applications/Related Disciplines

- Musician
- Sound Technician/Engineer
- Instrument Tuner
- Audiologist

CHAPTER 1

Modeling IS Mathematics

Can mathematical modeling help plan a better event?

In a world that is getting smaller and more complex, mathematics is more important than ever. Mathematics plays a powerful role in many areas: banking, business, engineering, technological research and manufacturing, the sciences, and practically every other career. Mathematics is used every day to make such major decisions as

- whether Delta Airlines should take over another airline
- how to finance and build Houston's new football stadium
- how to build the International Space Station and keep it in orbit.

These decisions are based on analyzing enormous amounts of data. For example, so much data is needed to run the international space station that it uses 52 computers. These computers look for patterns in the data that can be used to make predictions. To analyze a problem such as operating the space station, scientists use many of the same mathematics skills that you learn in school.

In this course you will use mathematics to solve nine real problems from your world. As part of your investigations, you will discuss and write about the problems, and use your imagination to analyze problems and find solutions. You may create pencil-and-paper solutions, or you may use a graphing calculator or a computer as tools to help do the computations. Although you use different tools, your primary goal is to understand a problem by investigating it mathematically. As you progress through the year, read this introduction periodically. Your understanding of mathematical modeling will change as you build models.

PREPARATION READING

The Modeling Process

The process of studying and gaining an understanding of a problem is called **modeling**. If the understanding is the result of using mathematics, the process is known as **mathematical modeling**.

MATHEMATICAL MODEL

A mathematical model is a way of describing a real problem using mathematics. A mathematical model may come in many forms:

- a description in words
- a graph, such as the graph of a straight line
- an equation, such as the equation for the area of a square
- a table of values, such as a table of rectangle dimensions with area 24"

Length (l)	12	6	4	3	2
Width (w)	2	4	6	8	12

- a drawing, such as a scale drawing of a room in a house
- a diagram, such as an arrow diagram or flow chart
- a computer simulation that imitates the real situation, such as a flight simulator

Once a model is built, you can use mathematics to find a solution to the problem. This will be *much* clearer after having done it once. For mathematical modeling, you learn by doing. Following is a general summary of the main steps in mathematical modeling.

STEPS IN MATHEMATICAL MODELING

To create a mathematical model, use the following steps.

Step 1. Identify the situation. Read and ask questions about the problem. Identify issues you wish to understand so that your questions are focused on exactly what you want to know.

Step 2. Simplify the situation. Make assumptions and note the features that you will ignore at first. List the key features of the problem. These are your assumptions that you will use to build the model.

Step 3. Build the model and solve the problem. Describe in mathematical terms the relationships among the parts of the problem, and find an answer to the problem. Some ways to describe the features mathematically include:

- define variables
- write equations
- draw shapes
- measure objects
- calculate probabilities
- gather data and organize into tables
- make graphs.

Step 4. Evaluate and revise the model. Check whether the answers make sense, and test your model. Go back to the original situation and see if the results of the mathematical work make sense. If so, use the model until new information becomes available or assumptions change. If not, reconsider the assumptions you made in step 2 and revise them to be more realistic.

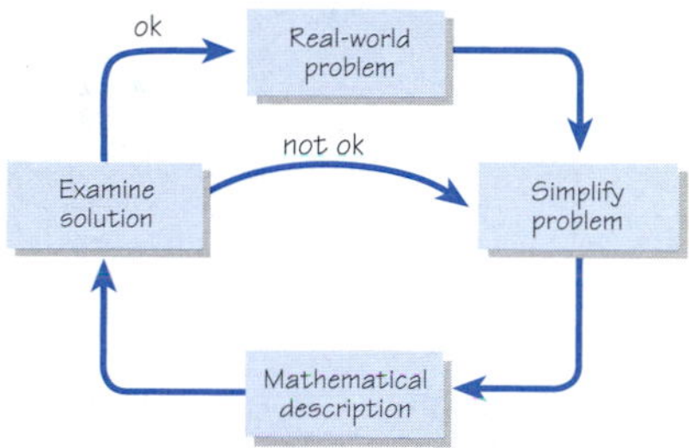

FIGURE 1.1.

Another way of visualizing the process of mathematical modeling is shown in **Figure 1.1**.

Mathematical modeling is called a process because it's like dance steps: You might go through a revision process several times before you are done making the model. One important principle that guides all modelers is *keep it simple*. In general all models ignore something, and first models usually ignore several factors. As long as the assumptions are clearly stated as part of the model, the only criticism can be that it's too simple!

In Chapter 1 you will build a model to find a solution to the following question:

Sam Arteste spends so much time thinking up "get-rich-quick" schemes that his nickname is 'The Scam.' Recently he decided that his future lies in concert promoting. He has three acts he is thinking of booking:

- "Who's That?" (aging rock band)
- "Ms. Teak" (rap performer)
- "Dixie Chickens" (country/folk duo).

He also has narrowed the place for the concert down to two choices, both in Dallas: The Cotton Bowl or the Starplex Amphitheater. Finally he has a report from a consumer group that suggests ticket prices affect how many people will come to a concert. The "bottom-line" problem for Sam is: how much should he charge for admission to the concert in order to make the most money?

The first task of a modeler is to identify your problem. Sam must make decisions about which band to book, which stadium to rent, and how much to charge for tickets. Decisions about the band and the stadium will depend on how much Sam is willing to spend. These costs will have an important effect on whether Sam can afford to put on the show. In order to pay for these costs, Sam must earn enough money from ticket sales. The money earned in ticket sales and the money spent in costs will play important roles in the model. If Sam is able to earn more than he spends, then he will make some "big bucks!" In Activity 1.1, you will start building a model by looking only at the money earned.

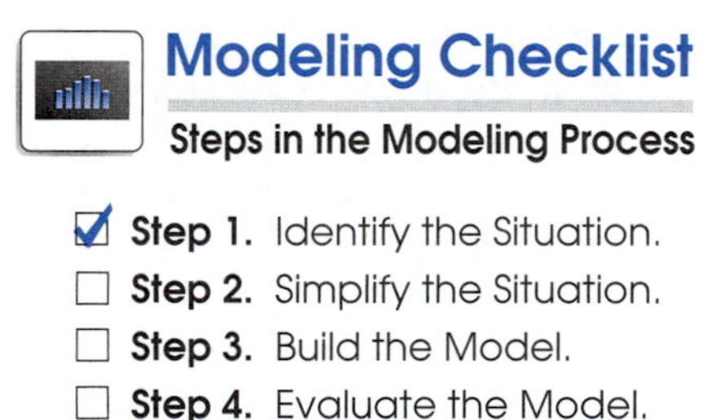

ACTIVITY 1.1

Keep It Simple

Figuring out which band to hire, where to schedule the concert, and how much to charge for tickets can be a complicated process for a concert promoter. At the same time, making decisions about these factors is fairly complex. Here's an opportunity for mathematical modeling. In this activity you will make some assumptions to simplify the problem so that you can start to build a model.

WHAT COMES IN

Modeling Checklist

Steps in the Modeling Process

- ☐ **Step 1.** Identify the Situation.
- ☑ **Step 2.** Simplify the Situation.
- ☐ **Step 3.** Build the Model.
- ☐ **Step 4.** Evaluate the Model.

To plan the concert, you must make decisions about the following three elements: The band, the arena, and how much to charge for tickets. Let's begin by concentrating on the basic issue for Sam:

"How much should he charge for admission to the concert in order to make the most money?"

You can simplify Sam's problem by considering for now how much money he can earn; that is, how much his *revenue* will be. You will consider Sam's costs later. Also for the moment you don't need to worry about whether the ticket prices are so high that few people will come.

Your first task is to make some assumptions so that you can build a model for Sam's revenue. This revenue model can take the form of a table, a graph, or an equation. Suppose the ticket price is \$20.

1. a) Make a table like the one in **Figure 1.2** and fill in the missing values.

FIGURE 1.2.

No. of Tickets Sold (n)	1	2	5	10	48	260
Money Collected (R)	20	40	100	200	\$960	\$5200

b) Explain how you determined the amount of money collected when ten tickets are sold. 20n

c) Explain how you determined how many tickets were sold when \$5200 in revenue was collected. $\frac{5200}{n}$

In question 1 you calculated the revenue for each quantity of tickets sold. Since the number of tickets in the table changed, you can represent that quantity by a variable. A letter that represents some quantity is called a **variable.** In general, the letters x and y are often used as variables. For real-world problems, you can choose another name that reminds you of the quantities being used, such as n for number of tickets.

In this chapter the variable n always represents the number of tickets sold. Because the ticket sales determine how much money Sam earns (and not the other way around), the number of tickets is the **input variable**.

There is a limit on what numbers n can represent. This variable can use only integer values between zero and the capacity of the stadium. This description of conditions on the values of an input variable is called the **domain of the variable.** For example, since the Cotton Bowl has a seating capacity of 25,704, then the domain of n for this situation is the set of integers from 0 to 25,704.

CHECK THIS!

For each value of a variable, be sure to include the units, which describe what kind of information the variable uses. For n, the unit is tickets sold, or just tickets.

2. a) What is the other quantity in Figure 1.2 that changes in value?

   b) What letter is being used to name that variable? Why use *that* particular letter?

   c) What are the units for that variable?

   d) Assuming the seating capacity of the Cotton Bowl is 25,704, find the maximum revenue that can be earned if each ticket costs $20.

The quantity analyzed in question 2, the revenue, also can be represented by a variable. The variable R will be used to represent revenue. Since this variable depends on the number of tickets sold, this variable is called the **output variable.** For example, when the input variable $n = 1$, then the output variable is $R = \$20$ because the revenue for selling one ticket is $20.

The **range** of the output variable is all the possible values of the variable. If R represents the revenue at the Cotton Bowl and each ticket costs $20, the range of the output variable R is multiples of 20 from 0 to $514,080.

GRAPHS AND EQUATIONS AS MODELS

Another way to build a model is with a graph in a **coordinate plane**. A coordinate plane consists of a horizontal number line, called the ***x*-axis,** and a vertical number line, called the ***y*-axis** (see **Figure 1.3**). We locate a point in the coordinate plane using an **ordered pair** (x, y). The first number in the ordered pair is a value of the input variable. The second number in the ordered pair is a value of the output variable.

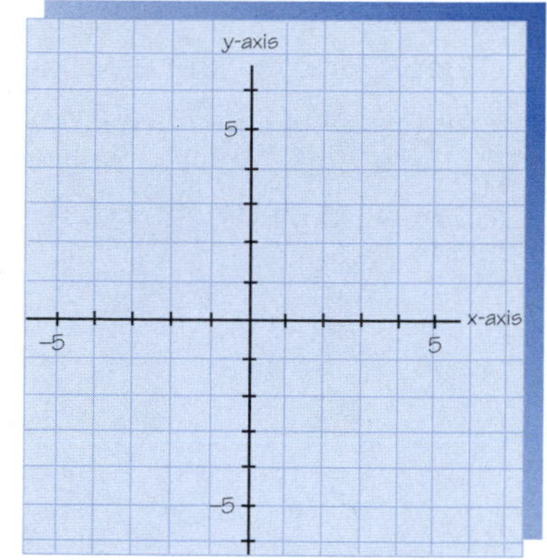

FIGURE 1.3. A coordinate plane.

3. a) Write ordered pairs of the form (n, R), using the first four data pairs from Figure 1.2.

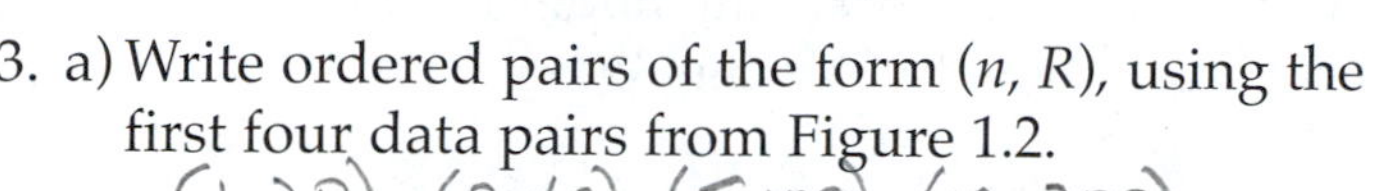

   b) Plot the point that represents each of these ordered pairs on a coordinate grid like the one in **Figure 1.4**. Since seven tickets cost $140, the point that represents (7, 140) is shown in Figure 1.4.

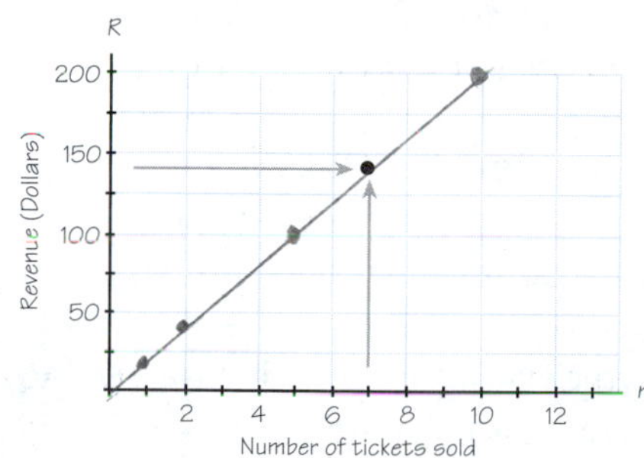

FIGURE 1.4.

To graph the point for the ordered pair (7, 140), move 7 units across the horizontal axis for n and 140 units up (the direction for R).

   c) Describe the pattern formed by the points that were graphed. If all possible values from the domain were graphed, what would the pattern look like?

4. a) Draw the line through the points you graphed in question 3. Use this graph to estimate how many tickets must be sold in order to earn $50.

   b) Does that answer make sense in the context of revenue from ticket sales?

Variables such as n and R can be used to solve an entire *class* of problems. Using variables to represent known and unknown quantities, a modeler can write an equation to describe a relationship between variables. You can write an equation to build a revenue model for ticket sales.

Modeling Checklist

Steps in the Modeling Process

- ☐ **Step 1.** Identify the Situation.
- ☐ **Step 2.** Simplify the Situation.
- ☑ **Step 3.** Build the Model.
- ☐ **Step 4.** Evaluate the Model.

EQUATION

When two expressions that contain variables are set equal to each other, an **equation** is formed. Equations are convenient ways to communicate the mathematical steps involved in a model. A **solution** to an equation is a set of values for the variables in the equation that makes a true statement.

5. a) We want to describe the calculations in question 1 using our variables for the revenue model. What is the equation that describes how to find the revenue from the number of tickets sold? $R = \$20n$

   b) Use your equation to predict the revenue obtained from selling 2000 tickets. Explain how to use the equation to determine that answer. $R = \$20(2000) = \$40{,}000$

   c) Use your equation to determine how many tickets must be sold so that the revenue is $13,000. What equation did you solve? $\$13000 = \$20n$ $n = 650$

The revenue model that you developed in question 5 uses the variables n and R. However, calculators only understand equations that use the traditional variables x and y. Thus you must be comfortable working with equations in both formats. You can use technology to explore properties of the model or to generate graphs *once* you have an equation.

6. a) Now put the power of technology to work. What equation must you enter into your calculator in order to see the graph of the relationship between the number of tickets and revenue?

   b) Use the following calculator steps to explore the revenue model.

   - Make the WINDOW settings on the calculator match the axes in the Figure 1.4 graph.
   - Display the continuous graph of the revenue model on your calculator.
   - Use the TABLE feature to check your calculations in the Figure 1.2 table.
   - Use the TRACE feature to check the points you graphed in the Figure 1.4 grid.

THE MORE THINGS CHANGE

In this activity you have assumed that the revenue from ticket sales is a multiple of the number of tickets sold. This type of model is an example of a **direct variation**, which is one kind of linear model. A relationship between variables x and y is a direct variation if it can be represented by an equation of the form $y = kx$ for some nonzero number k. (We also say that x and y **vary directly**.) The number k is called the **constant of proportionality.**

7. a) According to the model, if Sam sells 5000 tickets he can collect \$100,000. Calculate the revenue made by selling twice as many tickets. How did you get your answer?

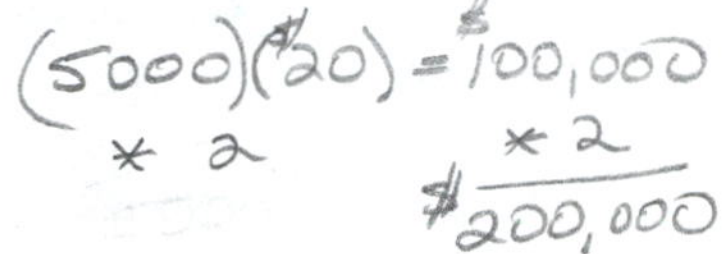

   b) For each ordered pair of numbers in Figure 1.2, calculate the ratio R/n. What do you notice about the answers obtained?

For a direct variation, the ratio of the output variable to the input variable is constant. For the ticket sales, this ratio is $R/n = \$20/\text{ticket}$. You can think of this constant as a **unit rate** (in this case, the revenue generated by selling a single ticket).

The rate of change is another common ratio. It is based on how much each variable value *changes*. The change in the value of a variable n is written Δn and read '**delta n**'. (The triangle is the capital letter 'delta' from the Greek alphabet, and means "the change in.")

8. a) According to the revenue model, selling 20 tickets makes \$400 in revenue, whereas selling 25 tickets makes \$500. In these two situations, the number of tickets sold *and* the revenue both changed their values. In this case what is Δn?

   b) In the situation described in 8(a), the change in the output variable, revenue, is written ΔR. What is ΔR for the values in (a)?

   c) What is the value of $\Delta R/\Delta n$ for the values you found in parts (a) and (b)? What units should the answer have?

9. a) Make a table like the one in **Figure 1.5**. Enter the values of Δn, ΔR and $\frac{\Delta R}{\Delta n}$ from question 8 in the second row of the table. Then find the values of R, Δn, ΔR, and $\frac{\Delta R}{\Delta n}$ for each of the next three rows when compared to the first row.

FIGURE 1.5.

n	R	Δn	ΔR	$\frac{\Delta R}{\Delta n}$
20	\$400	—	—	—
25	\$500	5	100	\$20
30	\$600	5	100	\$20
45	\$900	15	300	\$20
70	\$1400	25	500	\$20

b) What is $\Delta R/\Delta n$? How does this value compare with k, the constant of proportionality for this direct variation?

Question 9 shows that the rate of change of the input and output variables of a direct variation is constant, and equal to the constant of proportionality. A constant rate of change is an indicator of a linear relationship.

TEST FOR A LINEAR RELATIONSHIP

The relationship between x and y is linear if and only if the ratio of $\Delta y/\Delta x$ is constant for any pair of (x, y) values that satisfy the relationship.

Since a direct variation is one type of linear model, the graph of a direct variation is a line. The other identifying feature of a direct variation is that the line goes through the origin (0, 0).

10. a) Does the graph you made in question 3 go through the origin?

b) What does the ordered pair (0, 0) mean in the context of ticket sales?

You have found that ticket price is an important factor in the revenue model. So far you have considered a model in which the ticket price is \$20. It makes sense to ask, "How does changing the price affect the revenue?"

11. a) Now change the ticket price to \$25. Is the *process* used to calculate the revenue from selling 100 tickets different? $R = \$25n$

b) Will the actual revenue generated by selling 100 tickets change? If so, what will be the new revenue? $R = \$2500$

c) Will the model still be a direct variation? yes

d) Will the equation describing the revenue model change? If so, what will be the new equation? $R = \$25n$

e) Will the graph of the revenue model be the same as before? If not, in what way will the graph change?

f) Will the WINDOW settings needed to see the graph of the revenue generated for up to 10 tickets sold change?

g) Will the maximum revenue change? If so, what will be the new maximum revenue? $R = (25)(25704) = \$642,600$

Modeling Checklist

Steps in the Modeling Process

- ☐ **Step 1.** Identify the Situation.
- ☐ **Step 2.** Simplify the Situation.
- ☐ **Step 3.** Build the Model.
- ☑ **Step 4.** Evaluate the Model.

SUMMARY

In this activity you reduced Sam's problem to considering only the revenue from selling tickets so that you could build a revenue model.

- You developed an equation and a graph to represent a revenue model.
- You learned about direct variation, which is represented by an equation of the form $y = kx$. The value k is called the constant of proportionality.
- You discovered that the graph of a direct variation is a line that goes through (0, 0).

So far you have gathered some important tools to build a revenue model. However, this model only describes part of Sam's problem. In the next activity you will consider another issue: The costs that Sam must pay to put on a concert.

HOMEWORK 1.1 Rolling in the Money

x	y
1	
2	
4	
8	
10	

FIGURE 1.6.

1. a) Make a table like the one in **Figure 1.6** and use the equation $y = 1.5x$ to find the values of y.

   b) Use the values in the table to make ordered pairs, and to plot the corresponding points. Then draw the graph of the equation.

2. For each of the following, decide whether the relationship between the variables is a direct variation. Explain how you got your answer.

   a) The table of values shown in **Figure 1.7**.

x	0.2	0.6	2.5	6.1
y	1.2	3.6	15.0	36.6

FIGURE 1.7.

   b) The table of values in **Figure 1.8**.

t	0	1.6	2.4	3.0
d	0	3.2	4.2	5.1

FIGURE 1.8.

   c) The equation $xy = 12$.

   d) The equation $y = 0.3875x$.

   e) The graph in **Figure 1.9**.

   f) The graph in **Figure 1.10**.

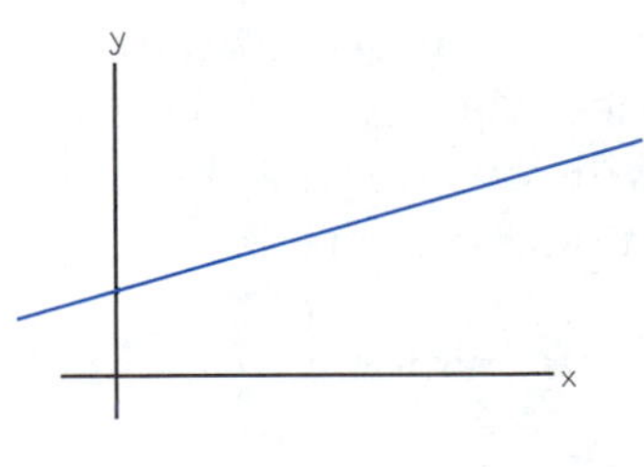

FIGURE 1.9.

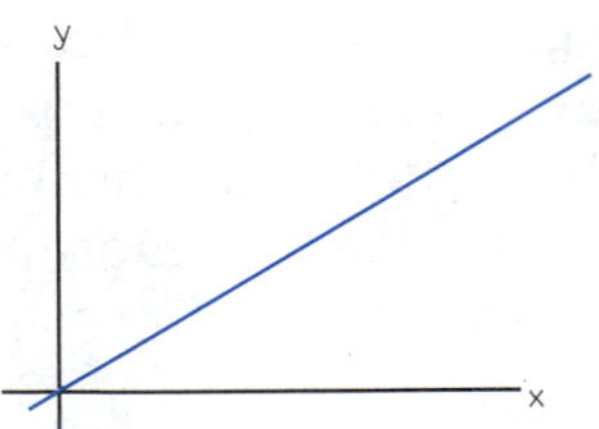

FIGURE 1.10.

3. a) Suppose that the relationship between two quantities x and y is a direct variation. Write an equation for this relationship.

   b) What is the value of the ratio that must remain constant if $y = 12$ when $x = 5$?

   c) If $x = 11$, what is the value for y?

   d) Predicting values of this direct variation can be done using **proportions** (two fractions that are equal to each other). Using the values $x = 5$, $y = 12$, and $x = 11$, we write the following expression: $\frac{12}{5} = \frac{y}{11}$. Why must the two fractions be equal?

   e) One algebraic method to solve a proportion involves "cross-multiplying" in the manner shown in **Figure 1.11**, and setting the two products equal to each other.

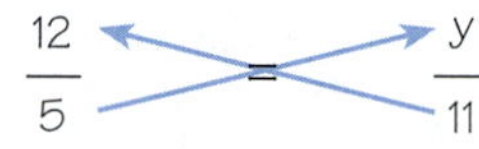

FIGURE 1.11.

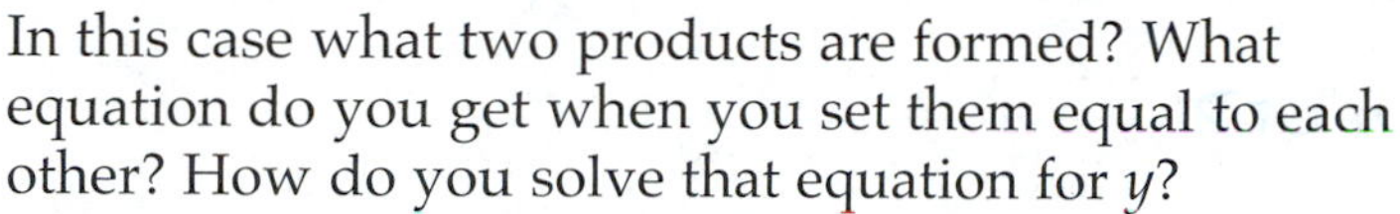

   In this case what two products are formed? What equation do you get when you set them equal to each other? How do you solve that equation for y?

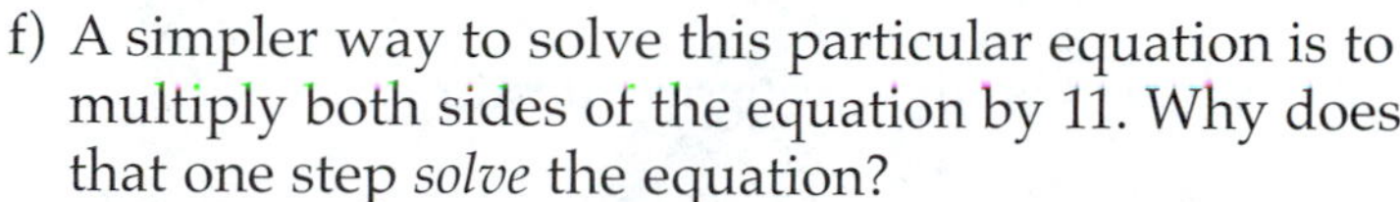

   f) A simpler way to solve this particular equation is to multiply both sides of the equation by 11. Why does that one step *solve* the equation?

   g) In order to find the value of the answer in (f), must you use the same arithmetic steps as the ones in (e)?

4. The problem in question 3 can be set up in a different way. This time compare the two y-values and the two x-values using a proportion again: $\frac{11}{5} = \frac{y}{12}$. Solve this proportion. Is the answer the same as the previous two times?

5. Solve each of the following proportions:

   a) $\frac{4}{x} = \frac{15}{23}$

   b) $\frac{7}{6} = \frac{y}{25}$

   c) x and y vary directly. When $x = 13$, $y = 5$. Find x, when $y = 13$.

There are a variety of other situations in which there is a constant **rate** (ratio of two different quantities). These may be described by a direct variation.

6. a) On a map scale, 1 cm represents a distance of 25 miles. Two cities on the map are 3.2 cm apart. What is the actual distance between the two cities?

   b) On a typing test Jane typed 98 words correctly in two minutes. Assume she can type for longer periods of time at the same rate without losing her concentration. How many minutes would it take her to type an essay that contains 343 words?

   c) Hanging three washers on a spring will stretch the spring 4.5 cm. If 13 washers are placed on it instead, how far will the spring stretch? (Assume that the spring can be stretched several feet without damaging the spring.)

7. Each of the following situations describes a relationship that is a direct variation. Write an equation that describes each situation. Use variables that make sense within the problem. Be sure to include units with the constant of proportionality.

   a) The map scale in question 6(a).

   b) The typing test in question 6(b).

   c) The washers on the spring in question 6(c).

8. a) Use the revenue equation $R = 15n$ to find the missing information in **Figure 1.12**.

n	2	20		1000	
R			3000		30000

FIGURE 1.12.

   b) For each of the five pairs of entries in the table, calculate the ratio R/n. What do you notice?

   c) Find $\Delta R/\Delta n$ for the situation when n changes from $n = 2$ to $n = 20$. What units does your answer have?

   d) Would the graph of this revenue model be steeper or flatter than the graph of the revenue model in question 1 of Activity 1.1? Explain.

9. The Staplex Amphitheater has a total capacity of 20,111, but there are two kinds of admission tickets available. There are 7533 reserved seats at $40 each and 12,578 lawn seats at $20 each. What is the maximum revenue at these ticket prices?

10. a) **Figure 1.13** shows a modeling diagram that describes the work done in Activity 1.1. Write a description of what the diagram means.

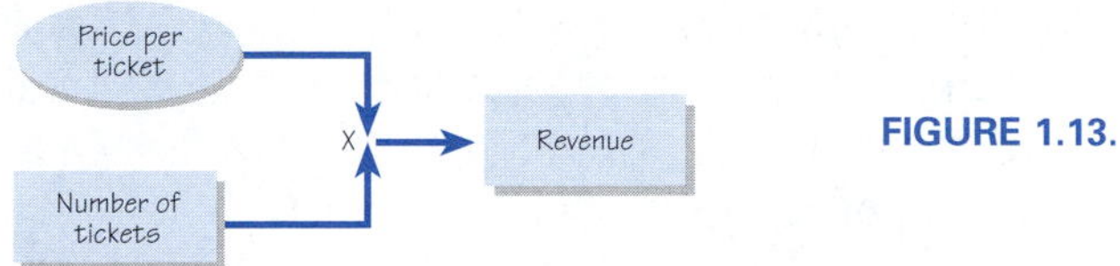

FIGURE 1.13.

b) What information does the model use in determining an answer? What quantity does the model predict?

c) Why do you think a different shape was used in Figure 1.13 for the price per ticket?

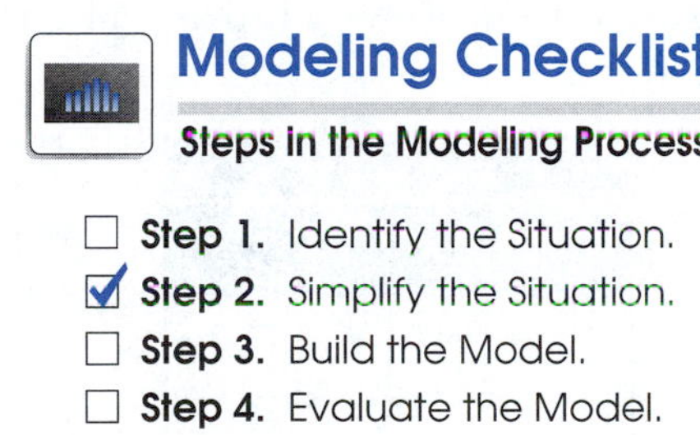

ACTIVITY 1.2

Building a Model for Cost

One powerful feature of modeling is that a complex problem can be changed into a simpler one by making assumptions. In the model in Activity 1.1, we only considered the money that is coming in. In becoming a concert promoter, Sam has discovered that money definitely goes out, and rather quickly! In this activity you will build a model of Sam's costs.

WHERE DOES THE MONEY GO?

In Activity 1.1 you found that the revenue from a concert varies directly with the number of tickets sold. If the cost of a ticket is \$20, then the revenue R may be modeled by the equation $R = 20n$. A complete model of Sam's concert promotion also must consider the costs of putting on the show. **Figure 1.14** contains some of the costs that he must pay.

Arena Cost

Location	Rental Fee
Cotton Bowl	\$75,000
Starplex Amphitheater	\$60,000

Band Cost

Performer	Booking Cost
Who's That?	\$48,000
Ms. Teak	\$75,000
Dixie Chickens	\$88,000

Other Costs

Other Expenses	Cost
Advertising	\$30,000
Ticket Agency	\$25,000

FIGURE 1.14. Costs of arena, band, advertising, and ticket agency.

1. a) Use the costs in Figure 1.14 to find the total cost for Sam to book "Ms. Teak" at the Cotton Bowl.

   \$75,000 + 75,000 + 30,000 + 25,000 = \$205,000

   b) What is the total cost of booking "Dixie Chickens" at the Starplex Amphitheater?

   \$60,000 + 88000 + 30,000 + 25000 = \$203,000

   c) What is the total cost of booking "Dixie Chickens" at the Cotton Bowl?

   \$75,000 + 88,000 + 30,000 + 25,000 = \$218,000

Sam Arteste
34 Star Drive
Dallas, TX 75370

982

Date March 14, 2001

Pay to the Order of Dixie Chickens \$ 88,000.00

Eighty eight thousand dollars only Dollars

First Union Trust

For Sam Arteste

2344 560000 00 982 9999

All the costs in Figure 1.14 do not change, no matter how many people attend the concert. If 100 or 10,000 people go to the concert, these costs remain the same. Costs that do not change are called **fixed costs.**

2. What's the **break-even** point—the number of tickets that must be sold so the promoter does not lose money?

VARIABLE COSTS AND DIRECT VARIATION

Sam got another lesson in concert promoting when he found out about the union contract that covers security guards, parking lot attendants, and concession workers. The contract requires that the number of workers should be based on concert attendance. The contract language is written in terms of a daily rate—the cost for an entire day's work—as shown in **Figure 1.15**:

Labor Group	Number of Attendees per Worker	Daily Cost per Worker
Security Guards	500	$800
Parking Lot Attendants	1000	$400
Concession-Stand Workers	200	$500

FIGURE 1.15. Costs that depend on the number of people who attend.

The important thing to recognize is that these costs are not a fixed amount like the stadium rental fee or advertising. The labor costs depend on the number of people who attend the concert. Since these costs vary, they are called **variable costs**.

3. a) If 15,000 people attend the concert, how much will it cost to hire security guards? Explain how you got your answer.

   b) If 12,800 people attend the concert, how much will it cost to hire security guards?

   c) What problem do you encounter in finding the answer to (b)?

4. a) Another way to look at the cost for security guards is to determine the cost per concertgoer. This is called a **unit rate**. If *each* person attending the concert has to pay her/his fair share of this cost, how much would that be?

   b) Use the unit rate to determine how much the cost of hiring the security guards would be if 12,800 people attend the concert. Explain how you got your answer.

   c) Compare your answer to (b) with the result from question 3(b). What would the cost actually be if the attendance is 12,800?

CHECK THIS!

Recall that a unit rate acts as the constant rate for a direct variation. For example, if y varies directly with x and $y = mx$, then m is a unit rate.

You probably noticed a problem in using a unit rate for the cost calculations. Every time Sam hires another guard, his costs go up by $800. (It would be difficult to hire part of a person!) An estimate of the true cost using the unit rate will be off by at most $800—small change for a big-time promoter.

As a mathematical modeler, you know that using a unit rate allows you to treat the relationship between cost and number of concertgoers as a direct variation. You can write an equation of the form $C = mn$, with C the cost, n the number of tickets, and m the unit rate for that cost. In order to use such an equation, make another simplifying assumption for your model:

The cost is an estimate determined on a per-person unit rate, rather than what is stated in the union contract.

CHECK THIS!

Since each unit rate is a cost per concert-goer, we will use the term *unit cost* to refer to each unit rate. For example, the unit rate for security guards will be called the unit cost for security guards.

5. a) Using this assumption, what is the unit cost for parking lot attendants?

   b) What is the unit cost for concession-stand workers?

Now you can build a model that determines the total cost from the number of people who attend the concert (that is, the number of tickets sold). Assume the band "Who's That?" is being booked at the Cotton Bowl. Use the fixed costs listed in Figure 1.14 and the unit costs for security guards, parking lot attendants, and concession workers.

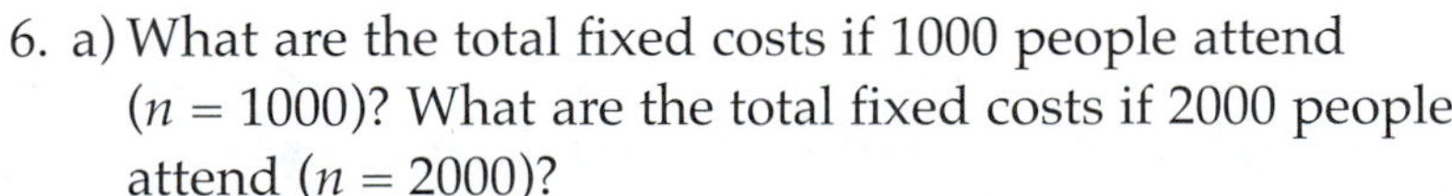

6. a) What are the total fixed costs if 1000 people attend ($n = 1000$)? What are the total fixed costs if 2000 people attend ($n = 2000$)?

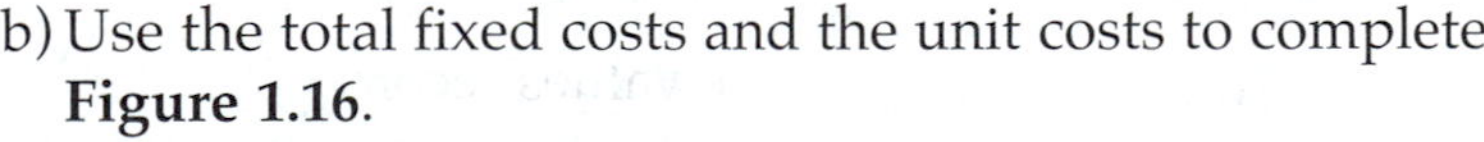

   b) Use the total fixed costs and the unit costs to complete **Figure 1.16**.

Number of Tickets	Variable Costs			Fixed Costs	Total Costs
	Security Guards	Parking Attendants	Concession Workers		
1000					
2000					

FIGURE 1.16.

In your model for total cost, you will need two variables, an input variable and an output variable. Think of the calculations you did in filling Figure 1.16. One quantity was used to determine the other quantity's value.

7. a) What is the input variable in this situation?

   b) The output variable should represent the amount that the model predicts. What is the output variable?

   c) Use the patterns in the Figure 1.16 table values to predict the various costs when $n = 3000$. Explain how to get each answer.

8. a) Use the names *SG*, *PA*, and *CW* to represent each of the variable costs, and write three separate equations that determine each cost from the number in attendance, n.

   b) Is the relationship between n and each variable cost a linear variation? How can you tell?

   c) If you combine the three variable costs together, what unit rate would you need? Use the name *VC* to represent variable cost, and write an equation to determine *VC* from the number of tickets n.

   d) Use your equation to calculate *VC* for $n = 1000$ and $n = 2000$.

   e) Now find the total of the variable costs in each row of Figure 1.16. How do these results compare with your answers to (d)?

9. Let C represent the total cost (both fixed and variable costs). Find $\Delta C/\Delta n$, the rate of change between the total cost and the number of tickets, for the values recorded in Figure 1.16. (Don't forget to include units!)

You can write an equation to describe total costs, just like you did for describing the total of the variable costs. Combining the variable costs into a single calculation, and doing the same thing for the fixed costs, can accomplish this.

10. Continue working with the scenario of booking "Who's That?" at the Cotton Bowl.

a) Fill in the **Figure 1.17** table. Check to make sure that the results agree with the previous calculations in Figure 1.16.

Number of Tickets	Variable Costs	Fixed Costs	Total Cost
0			
1000			
2000			
3000			
4000			
5000			

FIGURE 1.17.

b) Use the values in Figure 1.17 for number of tickets and total costs to make six ordered pairs. Use a grid like the one in **Figure 1.18** to graph the ordered pairs. How can you tell which quantity is the input variable and which quantity is the output variable from the graph?

c) If the points form a definite linear pattern, draw the line that goes through the points.

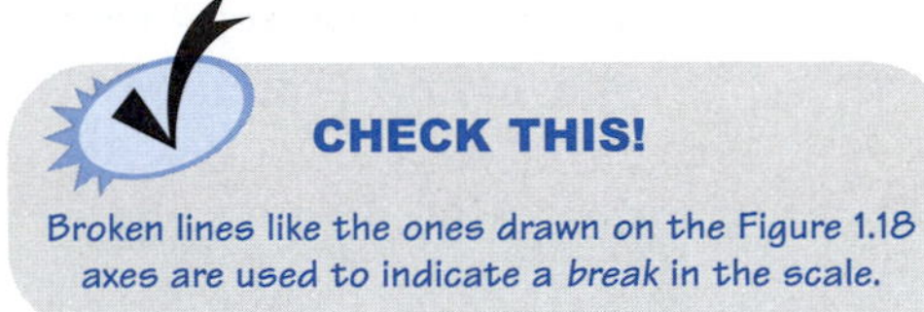

CHECK THIS!

Broken lines like the ones drawn on the Figure 1.18 axes are used to indicate a *break* in the scale.

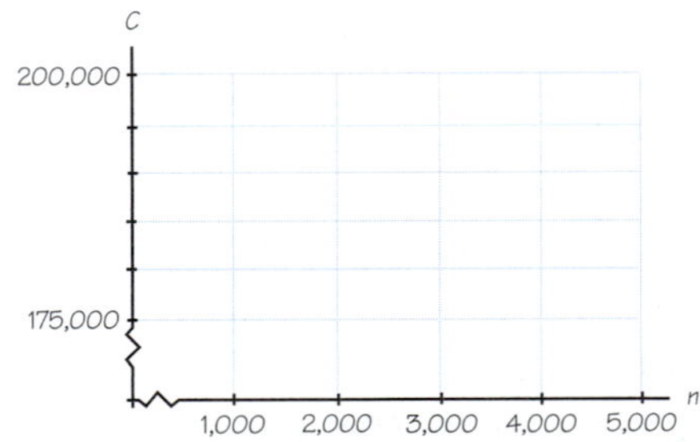

FIGURE 1.18.

d) Explain how you can calculate the variable cost (VC) in a single step. What is the equation that determines VC from the number of tickets n?

Modeling Checklist

Steps in the Modeling Process

- ☐ **Step 1.** Identify the Situation.
- ☐ **Step 2.** Simplify the Situation.
- ☑ **Step 3.** Build the Model.
- ☐ **Step 4.** Evaluate the Model.

e) What quantity do you add to variable costs to obtain total costs? What value should you use for this quantity?

f) Write an expression that adds the quantity you found in (e) to the expression for VC. This new expression is equal to the total cost C. Write the equation for total cost.

The result in question 10 is typical of a different type of linear model. The output variable C for that situation is determined by the following operations:

> Multiply n, the number of tickets sold, by the unit rate for the variable costs, then add the fixed costs.

Figure 1.19 shows an **arrow diagram** and an equation form of the operations in this type of equation:

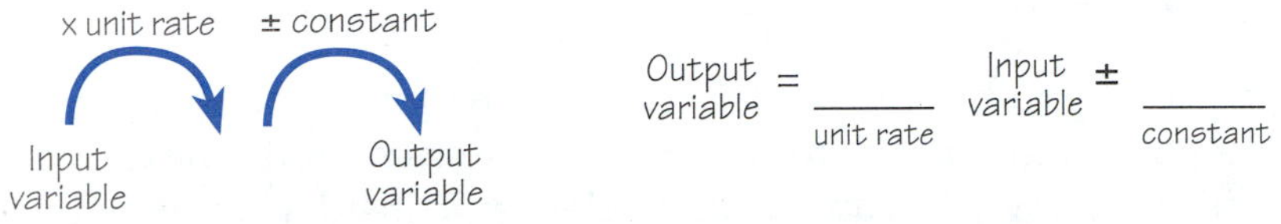

x unit rate ± constant

Input variable → Output variable

Output variable = ______ (unit rate) Input variable ± ______ (constant)

FIGURE 1.19.

In general, the letter for the unit rate is m (think of it as the *multiplier* value), and the constant is represented by b (think of it as the *beginning* value). A linear equation of this form is called a **general linear equation.**

GENERAL LINEAR EQUATION

A general linear equation has the form $y = mx + b$ where m is the unit rate and b is the value that corresponds to $x = 0$. The graph of a general linear equation is a line.

11. You can use the equation developed in question 10 to predict the number of tickets sold that would require a total cost of \$186,325.

    a) The value provided in this problem belongs to which variable?

    b) Substitute that value in place of the variable that represents it in the total cost equation. What new equation is made?

An arrow diagram can help in solving equations. If you know the value of the input variable, you simply follow the steps indicated by each arrow, in order. If you know the value of the output variable, you must work backwards—always doing the opposite operation, as shown in **Figure 1.20**.

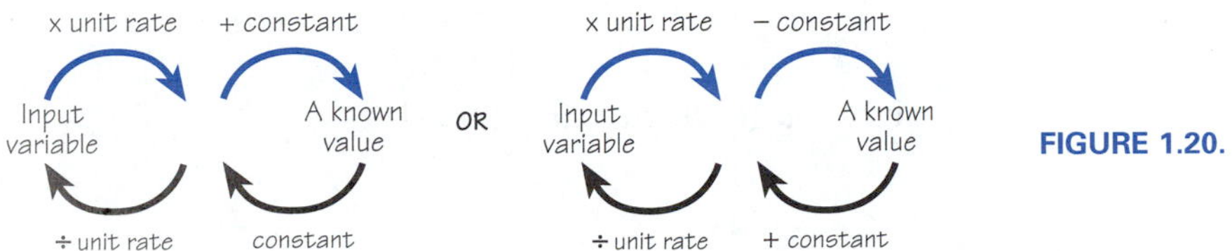

FIGURE 1.20.

    c) Solve the equation identified in (b) using the steps suggested in Figure 1.20.

SUMMARY

In this activity you considered another important factor for a concert promoter: the costs.

- You learned about fixed costs and variable costs.
- You put these costs together to find a model for the total cost.
- You learned about general linear equations that have the form $y = mx + b$.

Now that you have created models for revenue and for cost, you have some means for comparing the two. In the next activity you will expand your models by considering revenue and cost together.

HOMEWORK 1.2

The Cost of Doing Business

1. Each of the following situations can be described by a linear model. Identify whether the situation involves a direct variation or a general linear equation $y = mx + b$ where $b \neq 0$. Calculate the unit rate.

   a) Traveling on interstate highway I-10, with data shown in the **Figure 1.21** table.

   b) Fare in a taxicab, in which it costs \$2.00 to get in the cab, and an additional \$0.20 for each 1/4 mile traveled. (Hint: make a table!)

Time Elapsed (hr)	Distance Traveled (mi)
0	0
3	180
6	360
9	540

FIGURE 1.21.

   c) Temperature conversions shown in the **Figure 1.22** table.

   d) Money spent on many cans of coffee, when the sale price is \$1.89/can.

Celsius Scale (°C)	Fahrenheit Scale (°F)
5	41
10	50
15	59
20	68

FIGURE 1.22.

2. For each of the following, an equation of the form $y = mx + b$ is provided. Use the equation to find the values of y. Then calculate Δx, Δy, and the rate of change ratio $\Delta y/\Delta x$. (In these examples, $\Delta y/\Delta x$ has no units.)

   a) $y = 3x - 8$ (use **Figure 1.23**).

   b) $y = 1.2x + 15$ (use **Figure 1.24).**

   c) $y = 4.1x - 12$ (use **Figure 1.25**).

x	y
4	
8	
12	
16	

FIGURE 1.23.

x	y
5	
7	
9	
11	

FIGURE 1.24.

x	y
13	
19	

FIGURE 1.25.

x	y
3	16
5	13
7	10
9	7

FIGURE 1.26.

3. Reflect back on the work done in question 2 and Activity 1.2.

   a) Do the values for the input variable (x) *have* to increase by a constant amount in order to find the ratio $\Delta y / \Delta x$?

   b) What is the smallest number of ordered pairs needed to be able to calculate the ratio $\Delta y / \Delta x$?

   c) What is the relationship between the rate of change ratio $\Delta y / \Delta x$ and each equation given in question 2?

   d) How would you calculate Δx, Δy, and the ratio $\Delta y / \Delta x$ for the values in **Figure 1.26**?

Number of Months	Account Balance
0	\$220
1	\$250
2	\$280
3	\$310
4	\$340

FIGURE 1.27.

4. A bank savings plan book keeps track of the amount of money in the account at the end of each month. Those figures are recorded in **Figure 1.27**.

   a) Explain how the balance can be something *other* than \$0 after 0 months.

   b) Write the data in Figure 1.27 as ordered pairs. Make a grid like the one in **Figure 1.28** to graph these ordered pairs. Include the line that goes through all the points.

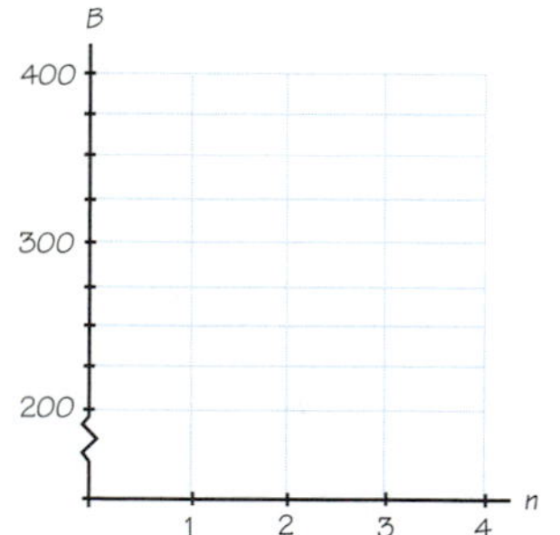

FIGURE 1.28.

   c) Suppose another row was added to the Figure 1.27 table, and the pattern in the numbers in the table was extended to this row. What would be the values for n and B? How did you determine those numbers?

   d) How would you use the pattern produced by the graph of the points already drawn to locate the graph of the new ordered pair for the new data?

   e) Write the equation for the line whose graph has been drawn.

5. a) The total of the variable costs you calculated in Activity 1.2 *could* have been written as the expression: $(1.60n + 0.40n + 2.50n)$. Simplify this expression.

   b) The total of the fixed costs *could* have been written as the expression: (75,000 + 48,000 + 30,000 + 25,000). When simplified, what did this expression become?

   c) The final equation model was: $C = 4.50n + 178{,}000$. Why can't you combine the two "parts" of the sum $4.50n + 178{,}000$ into a single term?

Expressions with variables can be simplified by combining **like terms**—terms that have the same variables raised to the exact same powers. The terms may have different **coefficients**, which are the numbers in front of the term. To combine like terms, add or subtract the coefficients.

6. Which of these pairs of expressions (that can show up in linear relationships) are like terms?

   a) $4x$ and 6

   b) $3n$ and $8n$

   c) $3x$ and y

7. Simplify each of the following expressions by combining like terms.

   a) $5y + y + 8$ (Hint: how many y's are in the second term?)

   b) $3x + 8 + 7x - 3 + 13$

   c) $12 - 2p + 7p - 5 - 4$

   d) $4y + 2 - 2x + 6 + 9x + y$

8. a) Find the fixed costs of booking each of the bands in **Figure 1.29** at each location.

	Who's That?	Ms. Teak	Dixie Chickens
Cotton Bowl			
Starplex Amphitheater			

FIGURE 1.29.

   b) Write the equation that determines the total cost of booking "Ms. Teak" at the Cotton Bowl, based on the number of tickets sold, n.

   c) What is the total cost model for booking "Ms. Teak" at the Starplex Amphitheater instead? Assume the variable costs are the same as those at the Cotton Bowl.

   d) If Sam wants to keep the costs as low as possible, which band should he book? And where should the concert be held?

9. If Sam wants to book "Ms. Teak" at the Cotton Bowl, and wants to limit the total costs to \$292,750, what would be the maximum number of people who could attend the concert? What equation do you solve?

10. **Figure 1.30** shows the graphs of the revenue model $R = 20n$ that was studied in Activity 1.1, and the cost model that was developed in Activity 1.2.

a) In terms of concert promotion, explain what the graphs suggest will happen for various attendance figures.

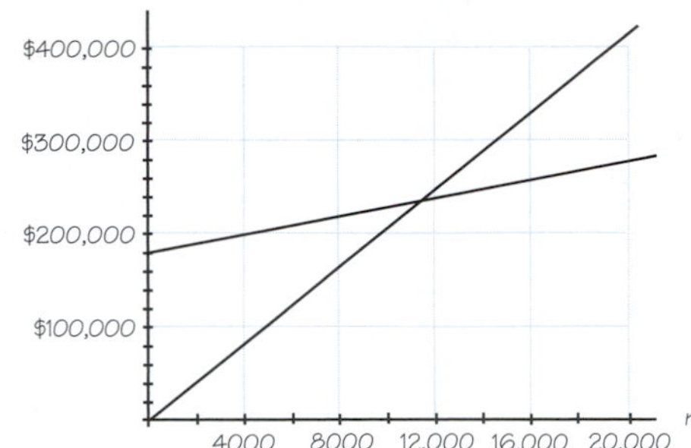

FIGURE 1.30. Graphs of cost and revenue models.

b) Estimate where the two lines intersect. In the context of concert promoting, what does the intersection point mean?

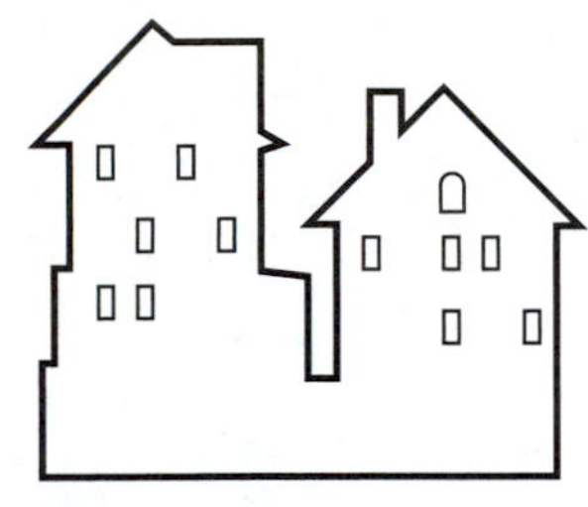

11. The variable costs of running a carpet-cleaning business are:

- $50 for cleaning supplies
- $250 for worker's wages
- $20 for gasoline for the truck

The yearly costs of running a carpet cleaning business are: $5000 for advertising in the yellow pages, $12,000 for rent, and $2500 for repairs on the van used.

Number of Houses (n)	Total Cost (C)
0	
1	
2	
3	

FIGURE 1.31.

a) Calculate the total costs for cleaning each number of houses shown in **Figure 1.31**.

b) Build a model that describes the yearly total cost C of running the business, in terms of the number of houses cleaned, n. Explain how you arrived at your answer.

c) What will happen if you charge $250 to clean each house?

d) If you charge $500 for each job, how many houses will you need to clean so that the business expenses (costs related to running the business) are paid off? Explain.

12. In Homework 1.1, a modeling diagram was introduced as another way to describe the relationships in a revenue model. Draw a similar modeling diagram for the various costs involved in Activity 1.2.

ACTIVITY 1.3

The Bottom Line: Revenue and Cost Together

In Activity 1.1, the assumption that there are no costs allowed you to build a revenue model. In Activity 1.2 you ignored revenue so that you could build a cost model. These pieces are not very realistic when treated separately. In this activity you will put the revenue and cost together to form a better model.

WHERE SAM LEARNS HOW TO GET RICH

Modeling Checklist

Steps in the Modeling Process

- ☐ **Step 1.** Identify the Situation.
- ☑ **Step 2.** Simplify the Situation.
- ☐ **Step 3.** Build the Model.
- ☐ **Step 4.** Evaluate the Model.

To reassemble Sam's problem, you need to look at the "bottom line," which is the profit to be made. The term **profit** refers to the amount of revenue left after all costs are paid. The mathematical relationship (model) that describes profit is $P = R - C$. In this activity you will explore a profit model in a variety of ways and, as a result, learn more about linear equations.

1. With so many choices it helps to consider a specific plan. Suppose Sam books "Who's That?" at the Cotton Bowl, and charges $20 per ticket for admission. (In this scenario, the unit cost was $4.50 per ticket and the fixed costs were $178,000.)

CHECK THIS!

Why bottom line? This term refers to the last line of an accountant's report where the profit is usually calculated. This calculation often appears at the bottom of this report, called an income statement. This report usually reflects the financial condition of a company.

Tickets Sold (*n*)	Revenue *R*	Variable Costs(*VC*)	Fixed Costs	Total Costs *C*	Profit $P = R - C$
0					
4000					
8000					
12,000					
16,000					
20,000					

FIGURE 1.32.

a) Calculate how much profit is made from various ticket sales amounts. Organize your work in a table like **Figure 1.32.** Assume all ticket holders attend the concert.

b) As a concert promoter, the "bottom line" is the break-even point. Explain how to find it from the calculations made in (a).

2. a) What was the input variable and the output variable for the revenue model in Activity 1.1?

   b) What was the input variable and the output variable for the cost model in Activity 1.2?

   c) Considering the table of values in Figure 1.32 and the input variables for the revenue and cost models, what should be the input and output variables for the profit model?

3. a) Do the values of n in Figure 1.32 increase by a constant amount? If so, what is the value of Δn?

   b) Does the profit column also go up by a constant amount? If so, what does ΔP equal?

   c) Calculate the ratio $\Delta P/\Delta n$ for various pairs of values in Figure 1.32. From that work, predict what the graph of the profit model should be.

4. Form ordered pairs (n, P) from the data in Figure 1.32 and plot points on a grid like the one in **Figure 1.33**. Draw the graph that describes the relationship between profit and the number of tickets.

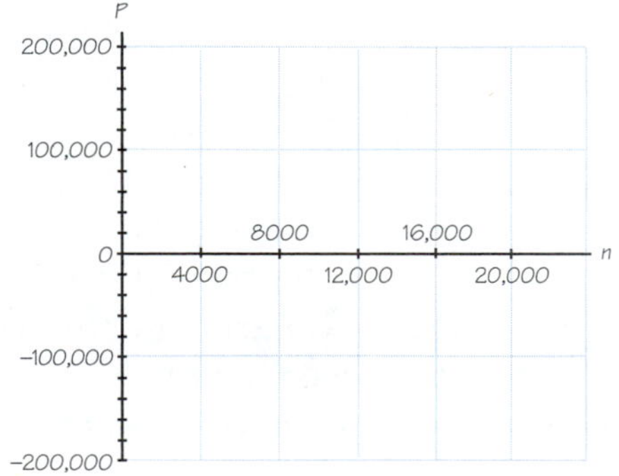

FIGURE 1.33.

5. a) How do the first values of n and P in Figure 1.32 affect the domain and range of the variables in the profit model?

   b) How do they affect where the graph was drawn in Figure 1.33.

In Activity 1.2, we showed that the general equation of a linear model has the form $y = mx + b$. Since the relationship between the variables in the profit model is linear, you can use a linear equation of the form $y = mx + b$ for the profit model. You will use the data in Figure 1.32 and the information you have gathered so far to determine the specific values of m and b.

6. a) In this situation, which variable is playing the role of x (the input variable)?

   b) Which variable is playing the role of y (the output variable)?

Modeling Checklist

Steps in the Modeling Process

- ☐ **Step 1.** Identify the Situation.
- ☐ **Step 2.** Simplify the Situation.
- ☑ **Step 3.** Build the Model.
- ☐ **Step 4.** Evaluate the Model.

Recall that the multiplier, m, in the linear equation is the same as the rate of change $\Delta y / \Delta x$, which is a comparison between how much the output variable changes and how much the input variable changes.

   c) For the two ordered pairs (12,000, 8000) and (16,000, 70,000) in Figure 1.33, what does Δy equal? What does Δx equal? What is the multiplier value, m?

   d) The beginning value, b, is the output value that corresponds to an input value of 0. In this case, what is the value of b?

   e) Put it all together! For this situation write an equation of the form $y = mx + b$ that describes the profit made in terms of the number of tickets sold, which is the bottom line for Sam.

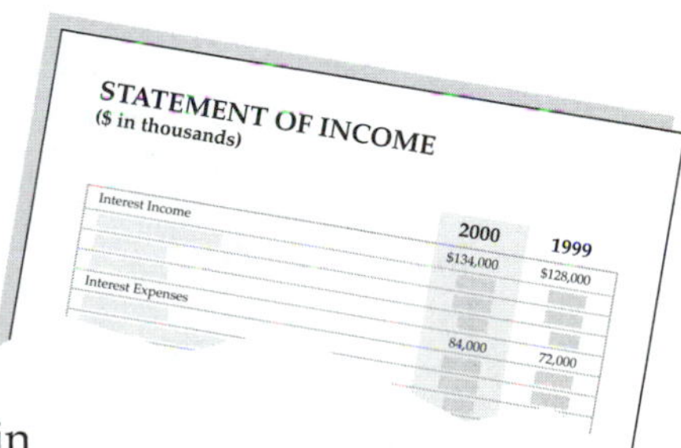

7. Also you can make a profit model by describing the situation in terms of the revenue and the cost, and then doing a little algebra.

   a) First write an equation that describes the revenue, R, in terms of the number of tickets sold, n.

   b) Write an equation that describes the total cost, C, in terms of the number of tickets sold. (Remember: All ticket holders are going to the show!)

   c) Substitute the expressions for R and C into the definition of profit ($P = R - C$), and write an equation for P.

   d) Simplify your answer by combining like terms; be careful to subtract all of the terms that belong to the total cost expression. Check to make sure your equation matches the one that was obtained in question 6.

SLOPE

We have seen that the multiplier m provides some essential information about a linear model. Since this multiplier is so important, it is given a special name. It is called the **slope** of the line. **Figure 1.34** illustrates two important methods for determining m for two ordered pairs lying on a line.

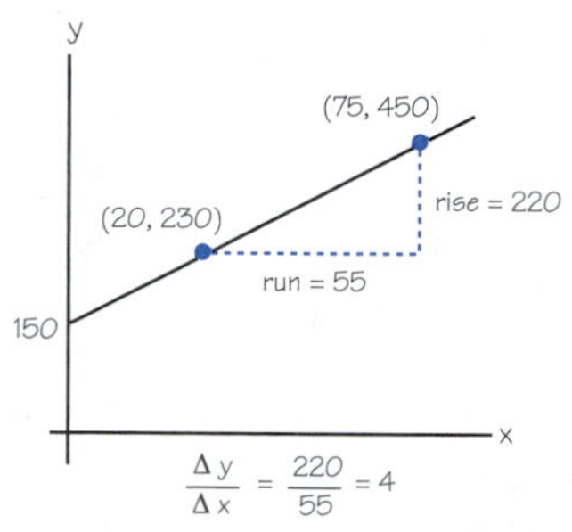

FIGURE 1.34
Ways to calculate slope.

CHECK THIS!

It's important to subtract coordinates in the same order! If you calculate Δx in Figure 1.34 using the difference 75 − 20, then you must calculate Δy in the same way: $\Delta y = 450 - 230$.

The slope can be determined from the "rise" and the "run" of the graph, or from the change in the x- and y-values of ordered pairs. With either choice, you only need two ordered pairs. You get the same answer using either approach.

SLOPE FORMULA

If $P(x_1, y_1)$ and $Q(x_2, y_2)$ are two points, then the slope of the line going through P and Q is calculated by the formula:

$$m = \frac{\Delta y}{\Delta x} = \frac{y_2 - y_1}{x_2 - x_1}$$

x	y
0	120
5	100
10	80
15	60

FIGURE 1.35.

8. Use the information provided in **Figure 1.35** to answer the following questions.

   a) How are the values for x changing? How do the values of y change?

   b) For the values of x and y in the third and fourth rows of the table, calculate the values for Δx, Δy, and the rate of change $\Delta y / \Delta x$.

   c) Write the values of x and y in the second and fourth rows as ordered pairs of the form (x, y). Use the formula to find the slope of the line that goes through the two points.

d) If you were to graph the information in Figure 1.35, how would the direction of the line be different than the line in Figure 1.34?

The value b in the linear equation also provides some important information. This value is called the y-intercept, because it marks the place where the graph crosses the y-axis. It is also the value that corresponds to $x = 0$ in a table. To find the value of b, substitute $x = 0$ into the equation, and solve for y.

9. a) Use the graphing calculator to check your work so far. What equation should be entered for Sam's profit model?

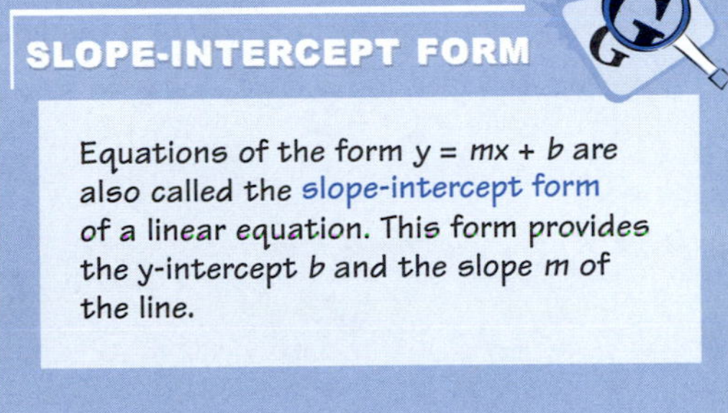

b) What WINDOW settings should be used, so that the graph matches the one made in Figure 1.33? Does the graph drawn in Figure 1.33 match your calculator screen?

c) Use TABLE or TRACE to check the coordinates of the points plotted in Figure 1.33.

10. The equation of the profit model is convenient for making calculations. Remember that the seating capacity at the Cotton Bowl is 25,704 people. What is the maximum profit when you take costs into consideration? How did you determine the answer?

One important goal is to make sure that the amount of money Sam earns (the revenue) is just enough that he doesn't lose any money after paying his costs. This situation is called the **break-even point.**

11. Recall that the break-even point is the situation in which the money that Sam earns (the revenue) is just enough to pay for the costs involved.

a) Use the equation for the profit model to find the break-even point.

b) Explain how you can find the break-even point from your graph of the profit model (question 4).

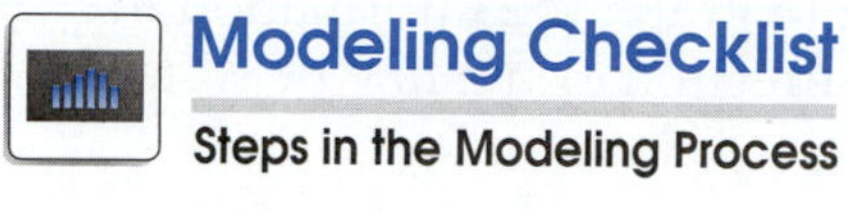

- ☐ **Step 1.** Identify the Situation.
- ☐ **Step 2.** Simplify the Situation.
- ☐ **Step 3.** Build the Model.
- ☑ **Step 4.** Evaluate the Model.

One powerful result of building models is that you can compare two plans no matter how much or how little they have in common.

12. Consider another possible concert plan: booking "Dixie Chickens" at the Cotton Bowl, and charging $23 per ticket.

    a) What model describes the profit P for this situation?

    b) What's the break-even point for this plan?

    c) What's the maximum profit for this plan?

    d) Assuming the concert is a sellout, which would be better for Sam: booking "Dixie Chickens" and charging $23 per ticket, or booking "Who's That" and charging $20 per ticket? Explain.

SUMMARY

In this activity you put the revenue and cost pieces together to build a better model.

- As part of building your model, you learned that the slope of a line is given by the equation $\frac{\Delta y}{\Delta x} = \frac{y_2 - y_1}{x_2 - x_1}$.
- You also found that the slope-intercept form of an equation of a line is $y = mx + b$.

Now that you have a profit model, you can use it to explore different plans for Sam's concert. In the next activity you'll discover some surprising things about this model.

HOMEWORK 1.3

A Profit in Your Own Land

1. a) Find the slope of the line whose equation is $y = \frac{4}{3}x + 12$.

   b) Find the slope of the line through the points $P(4, 9)$ and $Q(7, 21)$.

   c) Find the slope of the line through the points $P(12, 5)$ and $Q(2, 8)$.

2. For each of the following equations, use the information provided by the slope and y-intercept to make a sketch of the graph using a coordinate grid like the one in each of the following figures.

   a) $y = 3x - 2$ b) $y = \frac{3}{2}x + 5$ c) $y = -\frac{5}{2}x + 12$

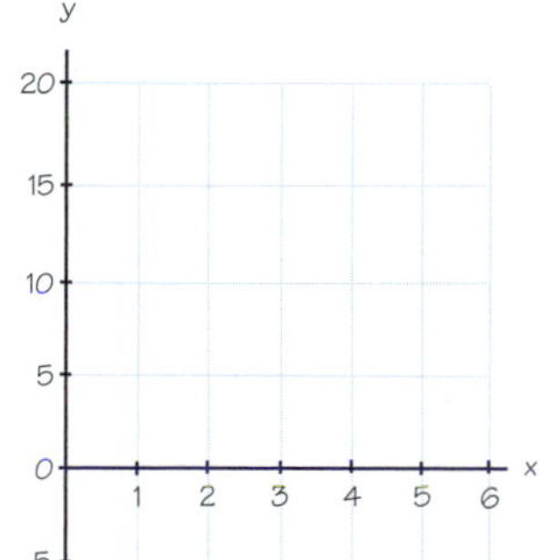

FIGURE 1.36.

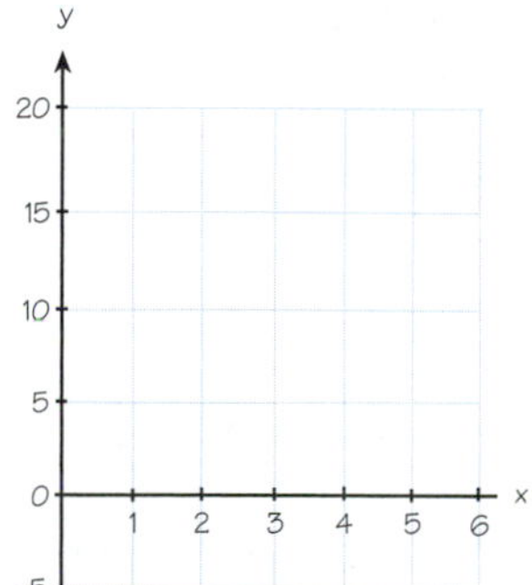

FIGURE 1.37.

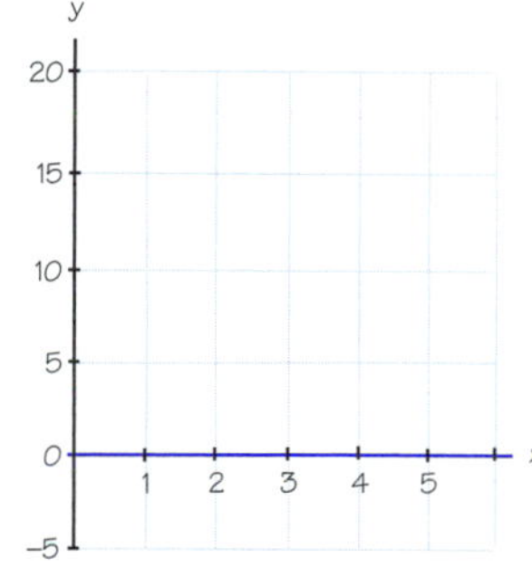

FIGURE 1.38.

3. The following equations were displayed using a computer drawing utility: $y = 1.2x + 5$, $y = 0.8x + 5$, $y = 2.3x - 5$, $y = -0.5x + 10$ and $y = -1.4x + 10$. The graphs are shown in **Figure 1.39**.

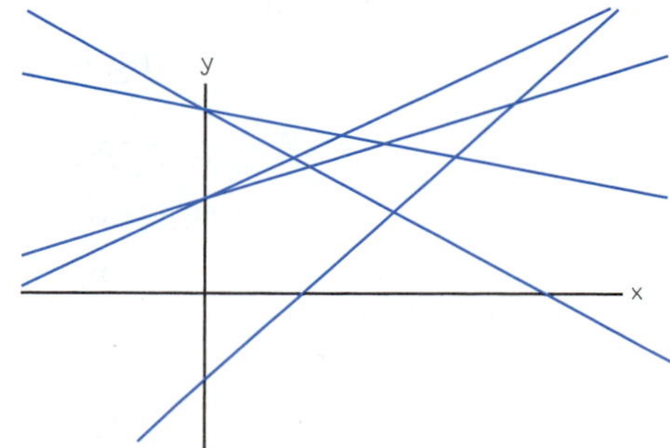

FIGURE 1.39.

Unfortunately the lines were not labeled and the axes don't show any scale. Explain how you can identify the graph of each equation.

When you obtained the profit equation in Activity 1.3, you used the **Distributive Property.**

4. Use the Distributive Property to eliminate the set of parentheses in each of the following expressions. Then simplify by adding like terms (if necessary).

   a) $2(3x - 4)$

   b) $-(x + 5)$

   c) $-3x + 5(2x - 4)$

   d) $5x - 2(x + 12)$

   e) $3(4x - 5) - 2(2x - 8)$

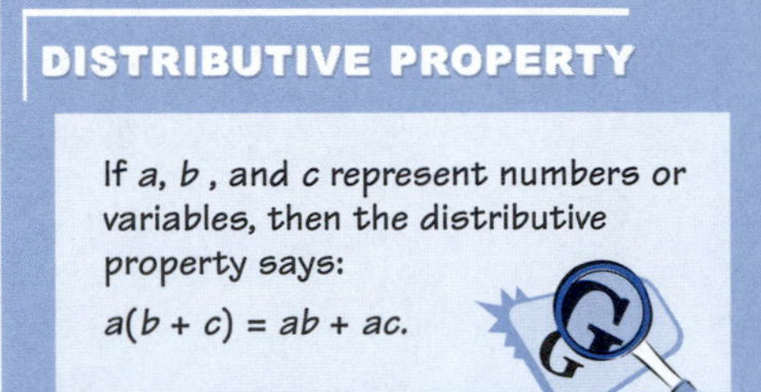

5. For each of the following equations, describe the general strategy that would be applied to solve the equation for the unknown variable. Then apply your strategy and solve it!

   a) $5.25n = 7749$

   b) $\frac{15}{8} = \frac{28}{x}$

   c) $8.20n + 15{,}600 = 26424$

   d) $1485 = 5(4x + 23)$

   e) $5n - 18 = 2n + 33$

6. Sometimes an equation that predicts one quantity can be rearranged to predict another quantity. To obtain the new equation you must solve an equation for a variable instead of a number. Arrow diagrams can help with this task.

   a) Using the same profit model from Activity 1.3, $P = 15.50n - 178{,}000$, draw an arrow diagram that starts with the number of tickets, n, and shows the numerical operations used to calculate the profit, P.

b) Now reverse the process; draw the arrow diagram that starts with the profit, and shows the numerical operations that calculates the number of tickets. Be sure to do the opposite operations in the opposite order.

c) Using the arrow diagram, write an equation that models the situation in which you are predicting the number of tickets from the profit.

7. a) Suppose a profit model $P = 15n - 2.45n - 8500$ describes the use of a different concert location and band. What is the price per ticket? What assumptions have you made?

b) What is the associated equation describing the total cost?

c) Explain how you can rewrite the profit model equation as a linear equation in slope-intercept form.

8. a) Suppose you book "Ms. Teak" at the Cotton Bowl, and charge $20 per ticket. Write an equation to describe the profit model for this situation. (For this scenario, the fixed costs will change a little.)

b) What is the break-even point for this case?

c) What is the maximum profit for this concert?

d) A non-profit organization will buy $50,000 worth of concert tickets to give away as a promotion. Write an equation that describes this situation, and then solve the equation. (Hint: Is money coming in, going out, or both?)

Recall that the Starplex Amphitheater has two kinds of seating. In order to build a profit model, assume that the ticket price is the same for both types of seats. The total capacity of the Starplex Amphitheater is 20,111. The fixed costs for booking the bands are shown in **Figure 1.40.**

	Who's That?	Ms. Teak	Dixie Chickens
Starplex Amphitheater	$163,000	$190,000	$203,000

FIGURE 1.40.
Fixed costs for booking three bands.

9. a) Build a profit model that accounts for revenue, fixed costs, and variable costs for booking "Who's That?" at the Starplex Amphitheater and charging $25 per ticket.

b) Build a profit model that accounts for revenue, fixed costs, and variable costs for booking "Dixie Chickens" at the Starplex Amphitheater and charging $35 per ticket.

ACTIVITY

1.4

Get Greedy and Go Broke

Now that you have learned how to determine the profit model for Sam's concert, you're ready to develop some ways to figure out what price to charge for admission. In this activity you will explore how revenue and profits change as the ticket price changes.

EVALUATING THE PROFIT MODEL

Suppose Sam books "Who's That?" at the Cotton Bowl, and the concert is a sell-out. The revenue and total cost were described by the equations

$R = pn$ (for any ticket price p), and

$C = 4.50n + 178{,}000.$

Ticket Price (p)	Revenue (R)	Total Costs (C)	Profit (P)
\$10			
\$20			
\$30			
\$40			
\$50			

FIGURE 1.41.

1. a) For each of the ticket price values in **Figure 1.41**, use your models to determine the profit.

   b) How much should Sam charge for admission?

Based on your work in question 1, you might think Sam's best strategy is to "raise the roof" on the price of tickets. It follows from our models so far that the profits continue to get bigger as the ticket price increases. But what happens if Sam sets the ticket price ridiculously high? Would you spend \$500 to see a concert? What about \$5000?

As the price of the ticket goes up it is likely that ticket sales will start to drop off. To model this situation, we must consider the relationship between ticket price and ticket sales—what is called **sales demand**.

2. Assume that "Dixie Chickens" are playing at the Cotton Bowl. The sales demand can be modeled by the equation: $n = -205p + 27740$, where p is the price of the ticket and n is the number of tickets sold.

a) Use the sales demand model to predict the number of tickets sold for the ticket prices in **Figure 1.42**.

Ticket Price	$10	$20	$30	$40	$50	$60	$70
No. of Tickets Sold							

FIGURE 1.42.

b) Does the sales demand equation have the general property that ticket sales drop off as the price of the ticket goes up? Is there a point at which there would be *no* interest in the concert?

c) According to the model, how many tickets would be "sold" at a free concert (where the tickets were given away)? How does that answer compare to the seating capacity of the stadium? Explain.

3. Suppose "Dixie Chickens" are going to perform at the Starplex Amphitheater instead. Is it reasonable to assume that the same sales demand model will apply at a different site? Identify any factors that might affect your decision to attend a concert at one location, but not at another one.

CHECK THIS!

To obtain a sales demand model, some real data often are gathered using surveys. A modeler would ask "Is the relationship between the price and the number of tickets linear? What equation models that relationship?"

Assume that at $20 per ticket for the "Dixie Chickens" concert, you can be assured of a sellout crowd of 20,111 people. Reliable sources estimate only 12,800 tickets would be sold if the price is $70 per ticket. We write these data as the ordered pairs (20, 20111) and (70, 12800). Also assume that the demand for tickets does drop off in a linear pattern. The graph of this situation is shown in **Figure 1.43**. You can use this information to find an equation for the sales demand.

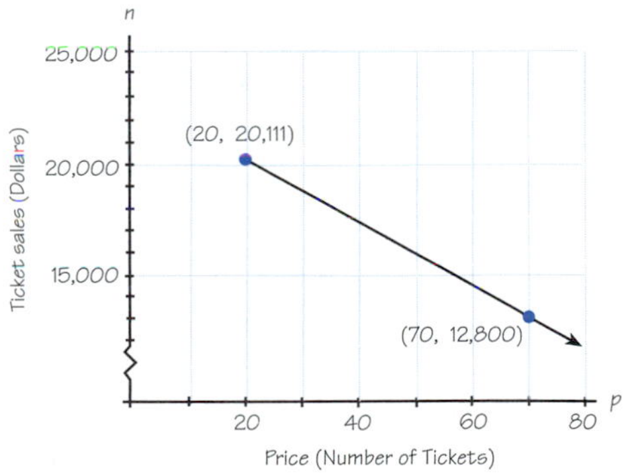

FIGURE 1.43.

4. a) What is the slope of the line? What unit does this number have?

b) Use the slope-intercept form $y = mx + b$ to find an equation for the sales demand. Replace m with the value of the slope of the line, and replace x and y with the variables needed for the model.

c) Now you need to find the value of b. Substitute the coordinates of one of the ordered pairs of numbers (either one will work). What new equation is produced?

d) Solve the equation in (c) for b.

e) Using your answers for (a) and (d), write the equation (in slope-intercept form) for the linear relation that models the sales demand described above.

THE POINT-SLOPE FORM

Another form of a linear equation is called **point-slope form**. Use this form when you know the coordinates of any point that is *on* the line and the slope of the line.

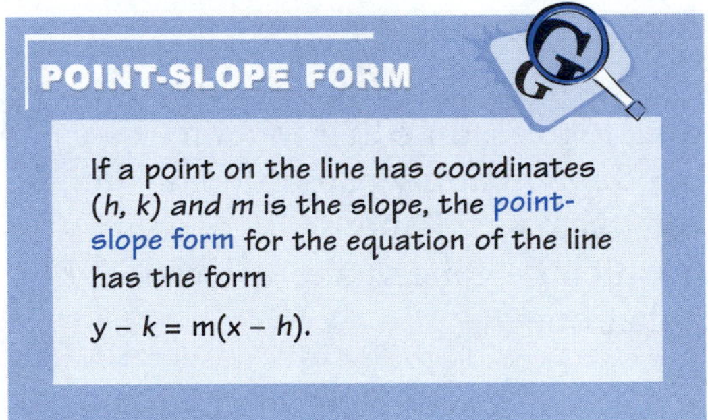

5. a) Use your answer from question 4(a) and either of the two ordered pairs in Figure 1.43 to write the equation for the line in point-slope form.

   b) Use algebra to write your answer from (a) in slope-intercept form. Does this answer match the answer from question 4(e)?

6. a) Most models are valid for a range of values that make sense for the situation. According to the assumptions, what is the lowest ticket price that should be used with this model?

   b) What is the highest possible price that makes sense in this model? How did you determine that answer?

7. a) Now see how ticket price affects Sam's profits. Consider again the plan of booking "Dixie Chickens" at the Starplex Amphitheater. Use the sales demand model from question 4 to predict the number of tickets that can be sold for the ticket prices in **Figure 1.44.**

   b) Now calculate the revenue generated for each ticket price.

   c) What equation describes the total cost in terms of the number of tickets sold? Use that equation to find the total cost for each ticket price.

   d) Calculate the profit to be made for each of the ticket prices.

Ticket Price (p)	Number of Tickets (n)	Revenue	Total Costs	Profit
$0				
$20				
$40				
$60				
$80				
$100				
$120				

FIGURE 1.44.

8. a) Based on your calculations in question 7, do you think Sam can get as rich as he wants simply by increasing the ticket price?

   b) What do your calculations suggest Sam should do to make the profits as big as possible?

9. Use the list capabilities of a graphing calculator to check the calculations you recorded in the Figure 1.44 table. Specifically:

   - Type the ticket prices into list L1.
   - Use the sales demand model to calculate the number of tickets sold using list L1.
   - Calculate the various revenues using the lists L1 and L2.
   - Use the cost model to calculate the total cost from knowing the attendance figures.
   - Calculate the profit from knowing the revenue and the total costs.

   Do the numbers in the calculator display agree with those recorded in the Figure 1.44 table?

SUMMARY

In this activity you used a sales demand model to explore the effect of ticket price on profit.

- You discovered that Sam cannot raise the price of a ticket as high as he wants. If the price is too high, fewer people will attend.
- As part of your investigation, you learned about the point-slope form of a linear equation.

In the final activity you will put all this information together to help Sam determine the ticket price that makes the most money from the concert.

HOMEWORK

1.4

Modeling "Tricks" of the Trade

1 a) Find the equation of the line that goes through the two points $P(-4, 12)$ and $Q(2, 3)$. Express your answer in slope-intercept form.

b) Find the equation of the line that goes through the two points $P(3, 8)$ and $Q(7,15)$. Express your answer in slope-intercept form.

2 a) Find the equation of the line that goes through the two points $P(5, 12)$ and $Q(9, 14)$. Express your answer in point-slope form.

b) Find the equation of the line that goes through the two points $P(4, 14)$ and $Q(8, 4)$. Express your answer in point-slope form.

3. Each of the following equations is written in point-slope form. Write the equation with y alone on one side of the equation (called calculator-ready form). Then use algebra to rewrite the equation in slope-intercept form.

a) $y - 8 = -3(x + 4)$

b) $y + 1 = \frac{2}{3}(x - 15)$

4. You can always tell whether a point is on a line. Just use the coordinates of the ordered pair as the values for x and y in the equation. If you get a "true" sentence, then the point *is* on the line. For each of the following equations, determine whether the point $P(2, 3)$ is on the graph:

a) $y - 3 = 2(x - 2)$

b) $y - 3 = 5(x - 2)$

c) $y - 3 = -4(x - 2)$

d) $y - 3 = m(x - 2)$

e) Using the previous results, explain why the point $P(h, k)$ would have to be on the graph of the line whose equation in point-slope form is $y - k = m(x - h)$.

5. Suppose that Garth Brooks could sell out the Cotton Bowl when the ticket price is \$50. A crowd of 16,000 is estimated when the admission is raised to \$80. Assume the interest in tickets drops off in a linear pattern as the prices increase.

   a) Find the equation that describes the ticket sales in terms of the ticket price for this situation.

   b) Do you think the most revenue will be made when the concert sells out? Explain. (If you're not sure, try calculating the revenue for a few situations, and see if the ticket price makes any difference!)

6. a) In Activity 1.4 you extended your work with profit models by including a new model called the sales demand. What was the input variable for the sales demand? What was the output variable?

   b) There were actually two input variables in the model that determined profit from the concert. What were they?

Think of the models you developed as conveyor belts transporting items that change as a result of an interaction with the model, as shown in **Figure 1.45.** For example, for the demand model $n = -205p + 27740$, a value of p is used by the model to calculate n as a result.

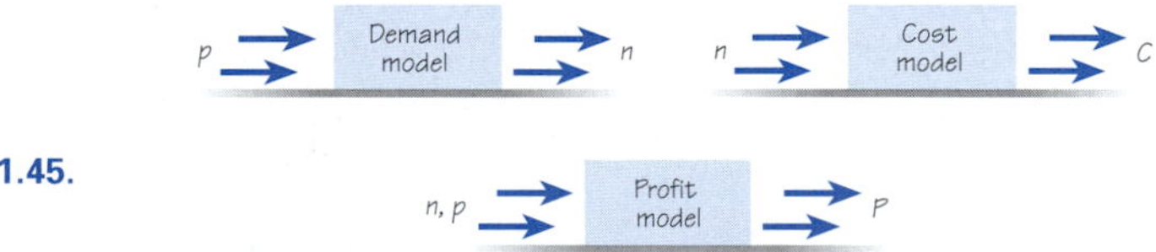

FIGURE 1.45.

It would be easiest if there were only one input variable, and all the models were described either in terms of that one variable, or expressions that use that one variable.

7. a) What was the input variable for the cost model? What was the output variable?

   b) Suppose we wanted the demand and cost models to each have n as the output variable. Does it make sense to change the cost model to make C the input variable of this model? In other words, does the amount of money spent in costs determine how many people attend?

c) Does it make sense to use the number of tickets sold n to determine the ticket price p? In other words, could the roles of the variables in the demand model be switched? Why might a promoter want to set the attendance at something less than a sell-out?

8. a) In Activity 1.4, the demand equation for booking "Dixie Chickens" at the Cotton Bowl was: $n = -205p + 27,740$. How can you tell from looking at the equation that the ticket price is the input variable and the number of tickets sold is the output variable?

b) In evaluating the equation for a particular ticket price such as \$15, what mathematical operations should you do, and in what order? What is the number of tickets sold at this price?

A nice way to visualize order of operations is through the use of an **arrow diagram**. The model used in question 8, and how to evaluate it for a ticket price of \$15, is shown in **Figure 1.46**.

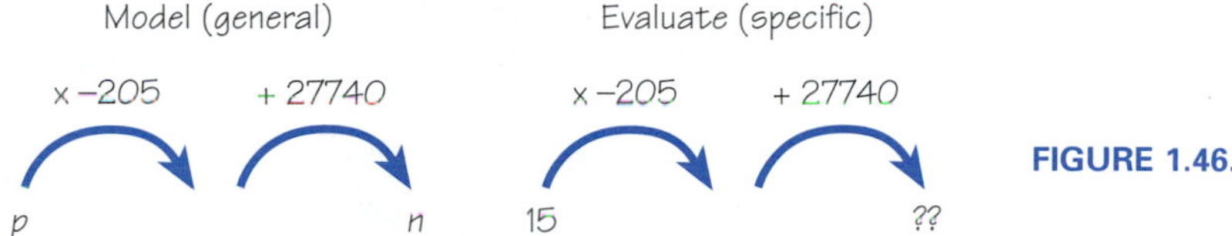

FIGURE 1.46.

9 a) In a number sense, switching the roles of the variables means doing the opposite operations in the opposite order. For the operations in the equation $n = -205p + 27740$, what are the opposite operations, done in the opposite order (last is now first, etc.)?

b) Start with the number of tickets you found in question 8(b) and go through the steps you've identified in 8(a). Do you get \$15 as an answer?

c) Switching the roles of the variables means *solving the equation for the other variable*. Use algebraic cancellation properties until the variable p is alone on one side of the equation. What expression do you get?

d) If necessary, rewrite your answer to (c) so that it is in slope-intercept form.

The arrow diagrams can help to visualize the algebra steps involved in solving a linear equation for the other variable. **Figure 1.47** shows how to work "backwards."

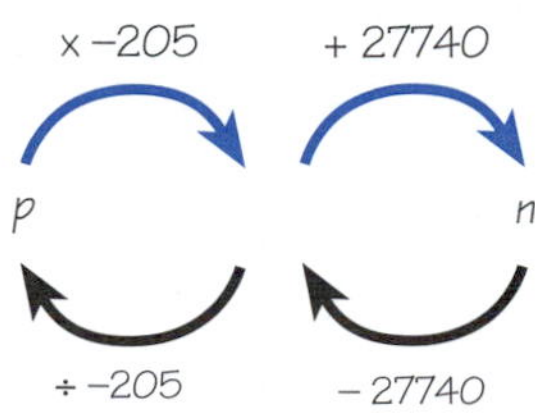

FIGURE 1.47.

In Activity 1.4, the profit model you built for booking "Dixie Chickens" at the Starplex Amphitheater was established by first assuming that a sold-out crowd would occur at a ticket price of $20. As a result there is a restriction placed on the use of the model.

10. a) What attendance would you expect if the admission ticket price was $15? What if it was only $10? Only $5? What about if the concert were free?

    b) What equation describes the pattern produced by your answers to (a)?

    c) How could the graph shown in Figure 1.43 of Activity 1.4 be modified to include the answers to (a)?

    d) Since there are two different patterns, identifying the model involves describing both the patterns formed and the conditions under which to use each one. (This is an example of a **piecewise-defined** equation.) Write the two equations that form the pattern in question 2. Give the domain for each equation (the possible ticket prices for which it is appropriate to use each equation).

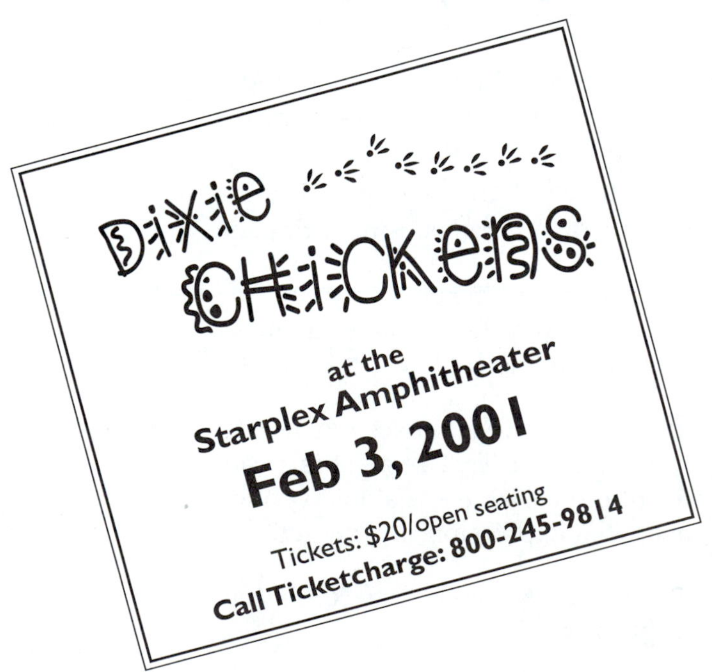

ACTIVITY 1.5 Building the Model Using Survey Data

In Activity 1.4 you learned how ticket sales affect profit. You used a sales demand model to help determine profit. In this activity, you will use data from a consumer survey to find a sales demand equation so that you can determine what price Sam should charge for his extravaganza.

FINDING THE DEMAND EQUATION

Modeling Checklist

Steps in the Modeling Process

- ☐ **Step 1.** Identify the Situation.
- ☑ **Step 2.** Simplify the Situation.
- ☐ **Step 3.** Build the Model.
- ☐ **Step 4.** Evaluate the Model.

How can you determine a sales demand equation for concert tickets? We'll start with some data. A consumer research group did surveys of local residents to learn how much they are willing to pay for concert tickets. **Figure 1.48** shows the potential ticket sales at each price based on these surveys.

Starplex Amphitheater		Cotton Bowl	
Price (p)	Ticket Sales (n)	Price (p)	Ticket Sales (n)
\$10	20,000	\$10	25,000
\$20	18,000	\$20	24,000
\$30	17,000	\$30	22,000
\$40	16,000	\$40	19,000
\$50	14,500	\$50	16,000
\$60	13,000	\$60	14,000
\$70	10,500	\$70	10,500

FIGURE 1.48. *Survey data of potential ticket sales*

The people were asked about their willingness to pay to see a typical band, not a "hot" performer. The sales demand model must take into account all the data in these tables. We'll consider the data for the Starplex Amphitheater first.

1. a) Write the Starplex Amphitheater data as ordered pairs of the form (p, n). Plot the ordered pairs on a grid like the one in **Figure 1.49**. A graph of the ordered pairs taken from data is called a **scatter plot.**

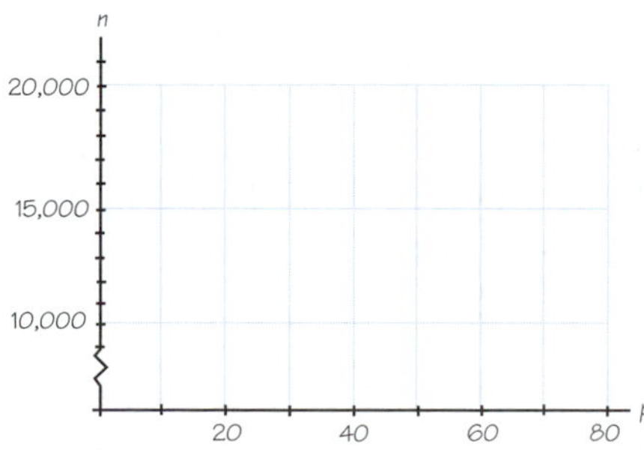

FIGURE 1.49.

CHECK THIS!

The line that approximates the data doesn't have to go through any of the points. Choose a line so that most of the points are reasonably close to the line.

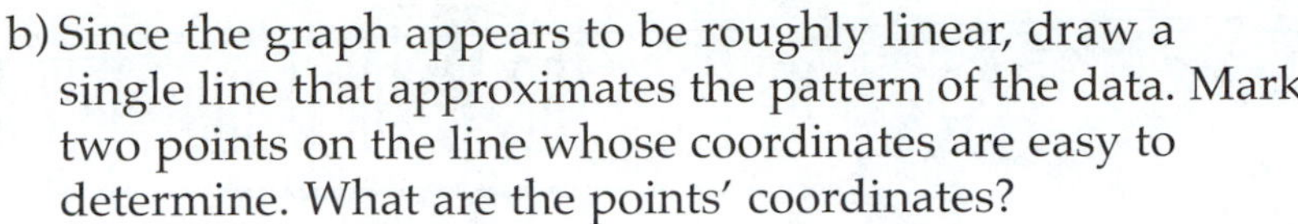

b) Since the graph appears to be roughly linear, draw a single line that approximates the pattern of the data. Mark two points on the line whose coordinates are easy to determine. What are the points' coordinates?

c) Find the equation of the line going through your two points. Write the equation for the sales demand in slope-intercept form.

Once again, it is time to put the power of calculator technology to work. Because we are starting with *data*, we need to work with lists.

2. Set up your calculator to work with statistical lists. (Clear old lists and equations, and check StatPlot settings.) Put the ticket price information in list L1 and the ticket sales information in list L2.

   a) Set your WINDOW settings to match the scatter plot graph you created in question 1(a). What WINDOW settings did you use?

   b) Display a scatter plot of the data on your calculator, and make a rough sketch of what appears. Does the calculator scatter plot look like the one you drew by hand? Explain.

Many calculators that work with statistical lists can determine a **regression equation**, which is the equation whose line is the best equation for a particular kind of equation. You can use your calculator's regression feature to build a model for ticket sales based on the data.

FITTING DATA

Choosing a graph to approximate some data points is called fitting the data. The line that best approximates the data points is called the line of best fit or the best-fitting line.

CHECK THIS!

Regression equations and their properties will be explored more fully in Chapter 2 "*Bones*."

3. a) Apply linear regression to the data in the calculator. What is the equation for the line of best fit? How does it compare to the model developed in question 1?

   b) Display the graph of the line of best fit with the scatter plot on your calculator. Make a rough sketch of the screen display. Does this graph roughly agree with the one you drew in question 1? Explain.

4. Over what range of values for the ticket prices should this model be used? Explain.

Now you have a sales demand model you can use to help Sam plan the concert. You need to find out how the new model affects Sam's profits. Assume that "Dixie Chickens" are booked at the Starplex Amphitheater.

5. a) Use your sales demand model to calculate the number of tickets that can be sold at the ticket prices in **Figure 1.50**. Enter your results into a table similar to the one in the figure.

Ticket Price	Number of Tickets Sold	Revenue Generated	Total Costs	Profit
$40				
$50				
$60				
$70				
$80				
$90				

FIGURE 1.50.

b) Calculate the revenue generated for each ticket price in Figure 1.50.

c) What equation models the cost involved in booking "Dixie Chickens" at the Starplex Amphitheater? (You may refer to the work you did in Homework 1.2.)

d) Use the cost model to calculate the total costs associated with each ticket price.

e) Finally, calculate the profit that can be made from each ticket price.

f) Based on the numbers in the Profit column, what price should Sam charge for a concert ticket? At that price, how much profit will he make?

Modeling Checklist
Steps in the Modeling Process

- ☐ **Step 1.** Identify the Situation.
- ☐ **Step 2.** Simplify the Situation.
- ☑ **Step 3.** Build the Model.
- ☐ **Step 4.** Evaluate the Model.

6. a) It is said that "A picture is worth a thousand words." Use a graphing calculator to display a graph of the model that predicts profits from the ticket prices. Which quantity (column) is going into the Xlist (L1)? Which is going into the Ylist (L2)?

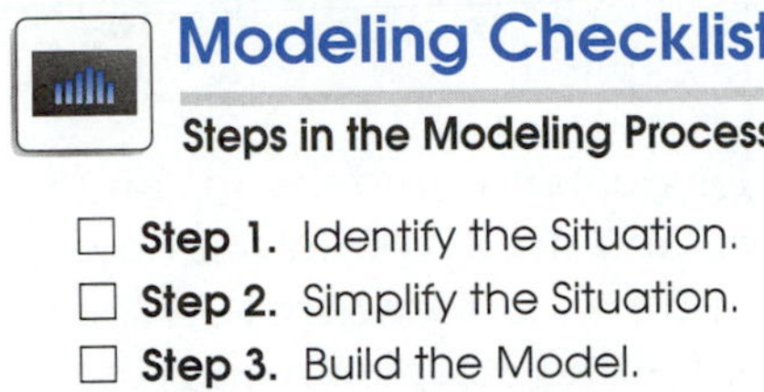

- ☐ **Step 1.** Identify the Situation.
- ☐ **Step 2.** Simplify the Situation.
- ☐ **Step 3.** Build the Model.
- ☑ **Step 4.** Evaluate the Model.

b) What WINDOW settings should you use for the scatter plot graph of this data?

c) Display the scatter plot graph, and make a rough sketch of the screen display. Give a general description of the graph that is produced.

d) Explain how you can use the graph to answer the question, "What price should Sam charge for a concert ticket?"

As you might suspect, the profit model based on the scatter plot in (c) is no longer as simple as the one explored in Activity 1.3. This graph is an example of another kind of mathematical model, called the quadratic, which will be studied in Chapter 4, "Testing 1, 2, 3."

SUMMARY

In this activity you learned how to obtain a model from some data.

- Using your graphing calculator, you found the line of best fit for some data.
- You used this model to determine how many tickets should be sold to earn the most profit.

In the modeling project you will explain the entire process you used to build this model.

HOMEWORK 1.5 From Data to Model Solution

1. A concert promoter always keeps track of advance ticket sales. These sales can be used to estimate the attendance at the performance, and to help determine how many workers are needed. **Figure 1.51** compares ticket sales four weeks in advance of several concerts with the actual attendance on the day of the performance.

Advanced Ticket Sales (*A*)	Actual Attendance (*n*)
5628	13,581
7043	15,902
8912	19,873
9117	21,683
9741	22,705

FIGURE 1.51.

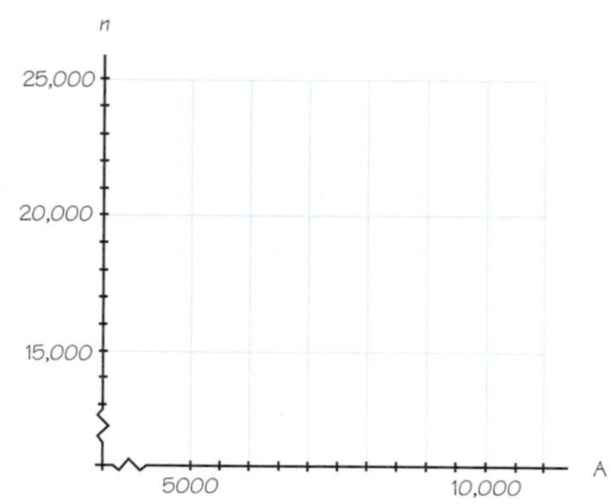

FIGURE 1.52.

a) Using a grid similar to the one in **Figure 1.52**, make a scatter plot that shows the relationship between the actual attendance and advanced ticket sales.

b) Draw a line that approximates the pattern in the data on your scatter plot. Find the equation that describes the line.

c) If one of the performances featured a popular act, would you expect the data point for this concert to be above or below the line? Explain.

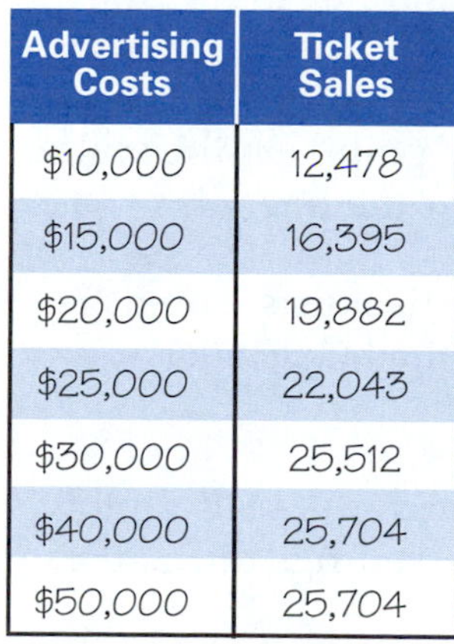

Advertising Costs	Ticket Sales
$10,000	12,478
$15,000	16,395
$20,000	19,882
$25,000	22,043
$30,000	25,512
$40,000	25,704
$50,000	25,704

FIGURE 1.53. Comparing advertising costs and ticket sales.

We have assumed that the amount spent on advertising is $30,000. What if this amount changes? How does the amount spent on advertising affect the number of tickets sold? To examine that relationship further, Sam gathered data relating the advertising costs and ticket sales from previous concerts at the Cotton Bowl. That information is displayed in **Figure 1.53**.

2. a) The pattern in the table shows that ticket sales increase as advertising costs go up. However, when "big bucks" are spent advertising a show there isn't any increase in ticket sales. What could explain this behavior?

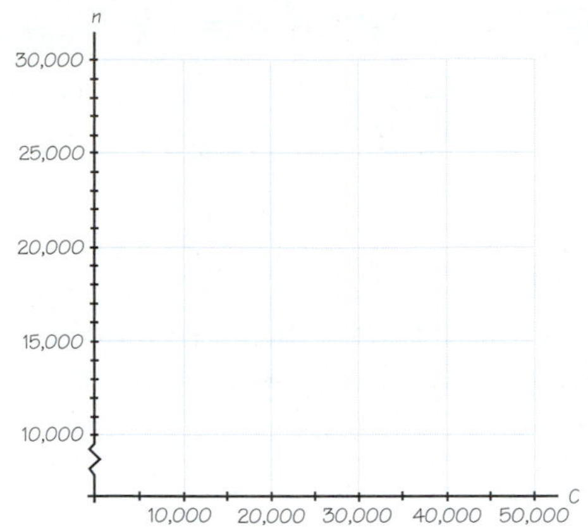

FIGURE 1.54.

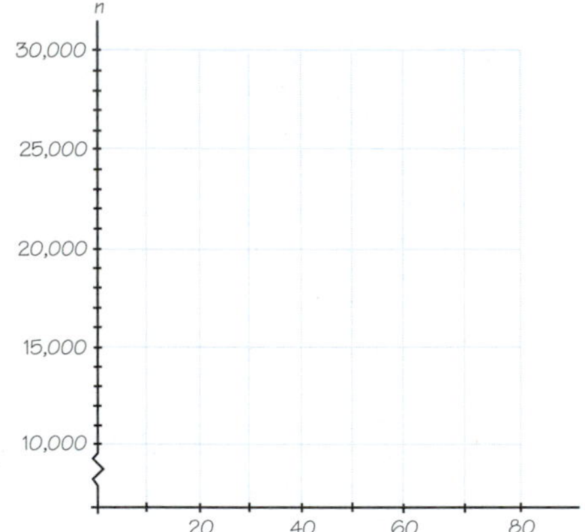

FIGURE 1.55.

b) Using a grid like the one in **Figure 1.54**, make a scatter plot graph to show how the ticket sales are related to the advertising costs.

c) The pattern for the first five data points is fairly linear. Draw a line that approximates these points on the scatter plot. Find the equation that describes the line.

3. Using the statistical features of a graphing calculator, find the regression equation for each of the following:

   a) The model for the relationship between attendance and advanced ticket sales in question 1.

   b) The model for the relationship between ticket sales and advertising costs in question 2.

In Activity 1.5, the data from the Starplex Amphitheater was analyzed, and a sales demand model was built using the data. Following that, the profits were calculated to see how that demand affected the problem of finding the right ticket price. In questions 4–7, you will perform a similar analysis using the data from the Cotton Bowl.

4. a) Work with the data from the Cotton Bowl in Figure 1.48 in Activity 1.5. Make a scatter plot graph of ordered pairs of the form (p, n), using a grid similar to the one in **Figure 1.55**.

   b) Draw a line that approximates these points on the scatter plot. Find the equation that describes the line.

5. Using a graphing calculator with statistical lists, enter the ticket price information in list L1 and the ticket sales information in list L2.

   a) Adjust the WINDOW settings to match the scatter plot grid in Figure 1.55. What WINDOW settings did you use?

   b) Apply linear regression to the data in the calculator. What is the equation of the line of best fit? How does it compare to the equation you developed in question 4?

c) Display a scatter plot of the data on your calculator, and make a rough sketch of what appears. Does the scatter plot and line look roughly like the line you drew to approximate this data?

Just like in Activity 1.5, you need to see if this sales demand model affects Sam's profits. Assume that "Who's That?" is booked at the Cotton Bowl.

6. a) Use the sales demand model from question 5 to calculate the number of tickets that can be sold at various ticket prices in **Figure 1.56**. Record all your calculations for this question in the table.

   b) From the information in the table, calculate the revenue generated for each ticket price.

Ticket Price	Number of Tickets Sold	Revenue Generated	Total Costs	Profit Made
$40				
$45				
$50				
$55				
$60				
$65				

FIGURE 1.56.

   c) What equation models the cost involved in booking "Who's That?" at the Cotton Bowl? (Refer to your work in Activity 1.2.)

   d) Use the cost model to calculate the total costs associated with each ticket price.

   e) Finally, calculate the profit that can be made from each ticket price.

   f) Based on the numbers in the Profit Made column, what price should Sam charge for a concert ticket? At that price how much profit will he make?

7. a) Use a graphing calculator to display a scatter plot graph of the model in question 6 that predicts profits from the ticket prices. Which quantity is going into the Xlist (L1)? Which is going into the Ylist (L2)?

   b) What WINDOW settings should you use for the scatter plot graph of this data?

   c) Display the graph and make a rough sketch of the screen display. How does this graph compare with the one from Activity 1.5 using the data from the Starplex Amphitheater?

Modeling Project How to Profit from Modeling

If a friend asked you what mathematical modeling is about, what would you say? You've *done* it, but the "it" will become clearer over time. As you experience more models you will have a better idea of what to look for and what the "rules of the game" are. Also writing about what you did will help build an understanding of mathematical modeling.

The modeling project for this first chapter is an opportunity to organize everything about your profit model for Sam Arteste's concert in a way that makes sense.

Here are the components of the model that are important to explain in your report.

- A description of the problem
- The information and data that was provided
- A modeling diagram
- The assumptions you made, especially simplifying assumptions
- The equations, tables, and graphs that relate pieces of the model
- The equations, tables, and graphs that describe the overall model
- Your calculations
- Your solution

There are many possible ways to organize your report, such as:

- **Time Sequence**—what was done each day in turn, like a diary, and where that day's work fits into the overall model development
- **Modeling Flow**—organizing your work around the various steps of the modeling process, especially how the various pieces of the model are actually the result of simplifying assumptions
- **Top/Down Design**—organizing your work by identifying what the problem is, then what you need to solve the problem, followed by what you need to solve that, etc.
- **Bottom/Up Design**—identifying all the various pieces of the model and how they were developed; then describing the links between the pieces to build the overall model
- **Dear Sam Letter**—writing a letter to Sam Arteste, telling him everything you've found out about his problem, and your recommendations on what he should do

The choice of the type of report is up to you. Choose the type that you think is the best way to organize your work. Good luck!

Practice Problems

Review Exercises

1. Several ways of representing the model that describes revenue from setting the ticket price at $30 are shown in **Figure 1.57**.

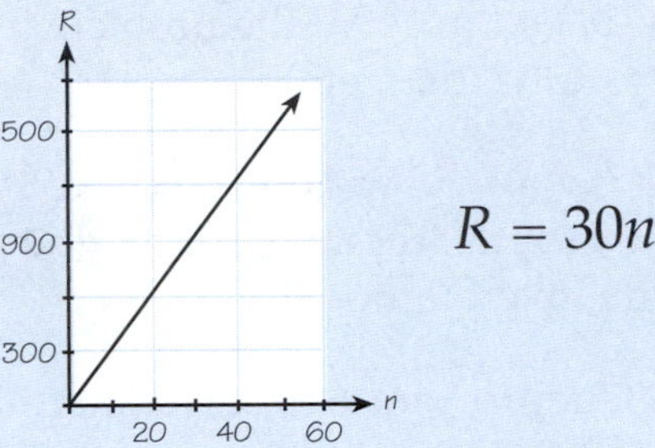

$$R = 30n$$

n	R
10	$300
20	$600
30	$900
40	$1200
50	$1500

FIGURE 1.57.

Explain how to tell the ticket price is a direct variation from:

a) The table of values

b) An equation

c) A graph

2. For each of the following situations, find the slope (m) or rate of change ($\Delta y / \Delta x$).

a) The table pattern shown in **Figure 1.58**.

x	y
5	23
9	29
13	35
17	41
21	47

FIGURE 1.58.

b) The slope of the line going through the points $P(5, 13)$ and $Q(10, 10)$.

c) The slope of the line shown in **Figure 1.59**.

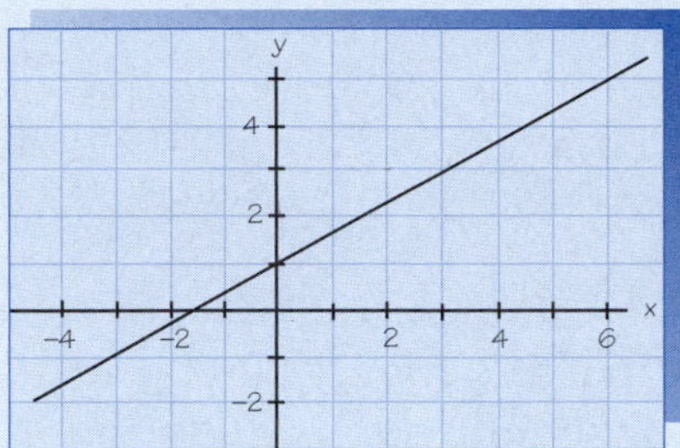

FIGURE 1.59.

3. Solve each of the following equations for the indicated variable.

   a) $\frac{5}{12} = \frac{x}{18}$, for x.

   b) $4.20n + 67000 = 76030$, for n.

   c) $C = 2.45n + 108000$, for n.

4. Two long-distance telephone carriers have quoted costs contained in the table in **Figure 1.60**.

Company	Monthly Service Charge	Rate (per minute)
NCO	\$5.00	\$0.20
Jog	\$3.00	\$0.25

FIGURE 1.60.

   a) Write equations describing the total cost associated with each company plan for one month. Use t as the variable to represent the amount of time spent on the phone.

   b) When is it cheaper to use the NCO plan? When is it cheaper to use Jog as the long-distance company?

5. Suppose the revenue model from question 1 was used in planning the concert with John Denver that set an attendance record of 17,829 at the Frank Erwin Center on the University of Texas campus in Austin. If that was a benefit performance, with the profits used to aid a worthy cause, the total costs associated with holding the concert would be "fixed" at \$48,600.

   a) Develop a profit model for this concert.

   b) Calculate the break-even point. Explain how you arrived at that answer.

c) Calculate how much money would be donated to charity. Explain how you arrived at that answer.

6. Prior to the same John Denver concert from question 5, an analysis had estimated that only 13,750 people would show up for the concert if the ticket price were $50.

   a) Use that information, and the actual attendance figures created by only charging $30 per ticket, to create a model that predicts the number of tickets sold from knowing the ticket price. Assume the drop-off in attendance follows a linear pattern.

   b) Rewrite the previous model so that the equation predicts the ticket price from knowing the number of tickets sold.

7. South Park Meadows in Austin holds concerts for large crowds every summer. **Figure 1.61** table shows a breakdown on possible costs involved in staging a concert there.

Labor	Wages/Day
Security (1 per 200 people)	$800
Ticket Booths (50 total)	$300
Medical (1 per 1000 people)	$1000
Sound Crew (20 total)	$2000
Cleanup (1 per 500 people)	$400
Parking Lot (40 total)	$300
Concessions (1 per 200 people)	$400

Incidental Fees	Cost
Band Contract	$75,000
Rent Facilities	$50,000
Ticket Agency	$20,000
Advertising	$30,000

FIGURE 1.61.

Develop a model that describes the total cost in terms of the number of concertgoers.

8. Suppose that a concert will have a sell-out attendance of 20,000 people when the admission ticket price is $25, but after that the attendance drops off by 2000 people for every extra $5 added to the ticket price.

   a) What is the model that describes the sales demand and predicts the number of people in attendance for any ticket price? What is the domain for that model?

   b) Expand the model to include all situations that produce a sell-out crowd by defining the model as two pieces. What is the equation and domain for the other piece?

c) To maximize profits, what should be the ticket price? What will the profits be?

d) If a total of $10,000 is spent as fixed costs, should the ticket price change (in order to maximize profits)? Will the revenue change? Will the profits change?

9. For a particular slope, explain the relationship between direct variations and lines in both slope-intercept form and point-slope form by comparing:

a) Order of operations (use arrow diagrams, lists, or spreadsheet layouts)

b) Equation forms

c) Graphs

10. Data on various attendance projections for a concert being promoted are provided in **Figure 1.62**.

Ticket Price	$10	$20	$30	$40	$50
Ticket Sales	25000	23500	22250	21100	19900

FIGURE 1.62.

a) Make a scatter plot of the attendance projections using a grid similar to the one in **Figure 1.63**.

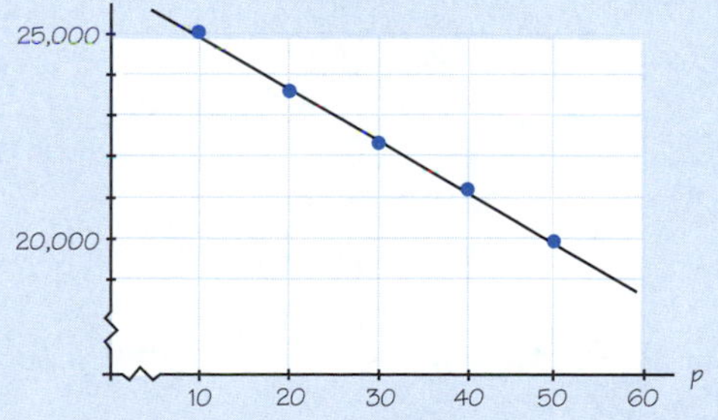

FIGURE 1.63.

b) Draw either a line on your graph that approximates the data and find the equation that describes the line, or use the calculator regression capability to find the best-fitting line for this data. What equation describes the ticket sales in terms of the ticket price?

11. The various relationships involved in maximizing profit that were developed specifically for use in this chapter can be summarized using labels. For each expression below, identify what is being calculated and what each label stands for.

a) $P = R - C$

b) $C = C_v + C_f$

c) $C_v = (r_1 + r_2 + r_3)n$

d) $C_f = C_1 + C_2 + C_3 + C_4$

e) $R = pn$

f) $p = k_1 n + k_2$

REVIEW

Modeling Profit

Model Development

In this chapter the first step of the modeling process—identify the problem—was provided in the Preparation Reading. In exploring the problem, you experienced what the modeling process is about. In the beginning that usually means "dabbling"—asking questions related to the situation, examining "what-if" scenarios, or working with simple number examples to see what calculations are needed. This is the second step of the modeling process, in which you try to understand the situation well enough to describe it mathematically. All modeling begins with a "Let's see what we can find" attitude, and any work done is probably on scratch paper.

On a different level that same exploration becomes more systematic with the types of assumptions you make. It was no accident that the first activity began by pretending no costs were involved. That was a simplifying assumption that led to an understanding of revenue. But it also allowed the calculations involved to be described formally, using diagrams and equations. Variables were found to represent what goes "in" and comes "out" of the model, and equations were used to solve problems. Technology was brought into use to explore how changes in variables (such as the ticket price) affected the revenue. You might have used statistical lists, spreadsheets, or graphs as tools for *understanding* during this phase, as well as for making the work easier. This analysis was done for the profit model being developed, but also for the *pieces* forming the model—revenue, cost, and sales.

This cycle (making an assumption, developing an equation, and solving a problem) was repeated many times, with the assumptions changed to produce a better model. Having the profits equal to revenue (because there were no costs) was not realistic. So a new profit model was developed to describe what happens when there *are* costs. Even that was not realistic enough, since the price of a ticket affects how many of them will be sold. A final model for profit was built that included the affect of "demand" on profits. This revision process comes from understanding the problem better and wanting it to reflect the situation more accurately. But it also reflects the style modelers use—starting with a situation simple enough to describe it in mathematical terms!

Mathematics Used

The revenue model involved a type of mathematical relationship called a direct variation. This is produced when calculations on the input variable are limited to only a single multiplication operation, and had the following form: $R = pn$, where p is the ticket price. On the calculator, that same form would be: $y = kx$, where k is the number by which you multiply. Many properties of direct variations were discovered. The two variables in the equation grow proportionally, and a ratio between the two quantities for any pair of associated values produces a constant answer equal to that number k. The rate of change between two specific conditions also produces a value equal to the same number k. The graph of a direct variation is a straight line going through the origin, with a slope also equal to k.

Cost models were found to have variable costs associated with hiring workers using a formula based on rates. They also had fixed costs that were affected by the band and location, and other factors. By combining like terms, the total cost could be found by just two arithmetic operations—multiplying the number of tickets sold by a unit rate r, and adding the total of the fixed costs. The unit rate multiplier was found to be the value of the rate of change between two specific conditions. Two-step equations that represent cost models could be described in the general form as $y = mx + b$, with graphs that were straight lines starting on the y-axis where the fixed costs were located.

A profit model was made by simply taking away costs from the revenue, and is another example of two-step equations that produce a linear graph. The equation multiplier combined the fact that ticket sales increase both the revenue and the variable costs. The other step in the equation was to reduce the profits by the amount paid in fixed costs. In general, a profit model equation would also look like this: $y = mx + b$, where x and y are the input and output variables, m is the number by which you are multiplying, and b is the number being added (or subtracted) afterwards. Equations like this are said to be in slope-intercept form, because their graphs produce lines that have a slope equal to the number m, and they cross the vertical axis at the point where $y = b$.

A refinement to the profit model took into consideration the fact that ticket prices will affect the number of tickets sold. At first, two pieces of information were used to create a "sales demand" model, which required being able to describe a line going through two points whose coordinates were known. If two points *P* and *Q* have coordinates (x_1, y_1) and (x_2, y_2), respectively, the slope of the line can be calculated using the formula

$$m = \frac{y_2 - y_1}{x_2 - x_1}$$

which is equal to the rate of change calculation $\Delta y / \Delta x$ between the conditions determined by the coordinates of the two points. The slope value and the coordinates of one of the points were used to describe the equation for the line, using the point-slope form; if the same point *P* were used, that equation would be $y - y_1 = m(x - x_1)$. Numerically this equation form involves going through three arithmetic steps to calculate an answer. It is possible to change the equation of a line from point-slope to slope-intercept form by distributing the multiplication over the parentheses, and then solving the equation for y. This algebraic process can also rewrite the model to calculate the ticket price from knowing the ticket sales.

A final refinement of the profit model used data to create the sales demand. Since the data was not exactly a straight-line pattern, it needed to be "fit" with a line that would best describe that relationship. Regression, which is a statistical calculation performed on the calculator, was used to find an equation to describe the best linear relationship between two sets of data. Through tables and graphs, it was found that introducing sales demand created a new type of model in which the profit would rise as the ticket price increased but only to a certain point. After that the profit would start going down; high ticket prices would discourage people from attending the concert. Since the chapter problem was to determine the ticket price at which the profit was the greatest, it was finally solved by using either tables or graphs. In Supplemental Activities, it was also done by using spreadsheets or algebraic equations.

Arrow diagram: A picture representation of the order of operations in evaluating or solving an equation.

Axes: Number lines drawn at right angles to locate points in a coordinate plane.

Break-even point: The term given to the number of ticket sales necessary to avoid losing money.

Capacity: The maximum number of people who can attend an event.

Coefficient: A number written in front of a variable term as a result of multiplying.

Constant of proportionality: The constant ratio formed by calculating y/x for (x, y) values that are solutions to a direct variation.

Coordinate plane: A system for identifying the location of points in the plane, using axes that have scales.

Coordinates: A pair of numbers that describe the location of a point. (The first number is always the x-coordinate.)

Data "fit": An equation that matches the pattern and general location of a scatter plot of data points.

Δ: A Greek capital letter, delta, used to symbolize "the change in."

Direct variation: A special kind of linear equation in which the two quantities grow proportionally. Graphs of direct variations go through the origin.

Distributive Property: An algebraic process for multiplying the sum of more than one term by something, all at once.

Domain: The description of conditions on the values an input variable can have.

Fixed cost: The costs that occur on a "one-time" basis.

Input variable: The variable used as the basis for a model or for describing the starting value of a calculation.

Like terms: Terms that have the same variables, and exponents for each of those variables.

Mathematical modeling: The process of taking a real-world problem, making assumptions and conditions on it, describing those using mathematical tools, and solving the problem. What distinguishes it from problem solving is that it is an ongoing process with new tools used, more realistic assumptions made, and the solution is described in more general terms.

Modeling diagram: A picture representation of the relationship between various elements of a model.

Output variable: A variable used as the basis for a model prediction or for describing the solution to a calculation.

Piece-wise defined equation: An equation that consists of a set of rules (equations) and a description (domain) of the values that each particular rule can work with.

Point-slope form: A particular form for the equation of a line that uses the slope of the line and coordinates for a point on the line. If $P(h, k)$ is the point, and m is the slope, the point-slope equation has form: $y - k = m(x - h)$.

Profit: How much money is made on an event; the difference between revenue and costs.

Range: The description of the possible values (or conditions) that the output variable has.

Rate: The ratio of two different quantities.

Rate of change: The ratio between the change taking place in the output variable and the corresponding change in the input variable, for two specified solutions to an equation.

Regression equation: The best-fitting equation for a set of data and a specified type of mathematical relationship.

Revenue: How much money comes in from an event such as a concert.

Sales demand: A description of how ticket sales are affected by ticket price.

Scatter plot graph: A graphical representation of the relationship between two quantities, in which a point contains one paired value from each of the quantities.

Slope: A measure of the steepness or direction of a line. It is determined by calculating rise over run, or $\Delta y / \Delta x$. Formally, if $P(x_1, y_1)$ and $Q(x_2, y_2)$ are two points on a line, the slope of the line is calculated by the expression:
$(y_2 - y_1)/(x_2 - x_1)$.

Slope-intercept form: A particular form for the equation of a line that uses the slope of the line and location where the line crosses the y-axis. If m is the slope and b is the y-axis intercept, the slope-intercept equation has the form: $y = mx + b$.

Unit rate: Expressing an amount, such as cost, on a per item or per person basis.

Units: These describe the kind of quantity or information that a variable uses.

Variable: A quantity whose value is allowed to change.

Variable cost: The costs that occur in a way that is determined by how large or small the input variable value is.

***y*-intercept:** Where a graph crosses the y-axis; also the term symbolized by b in the slope-intercept form for the equation of a line.

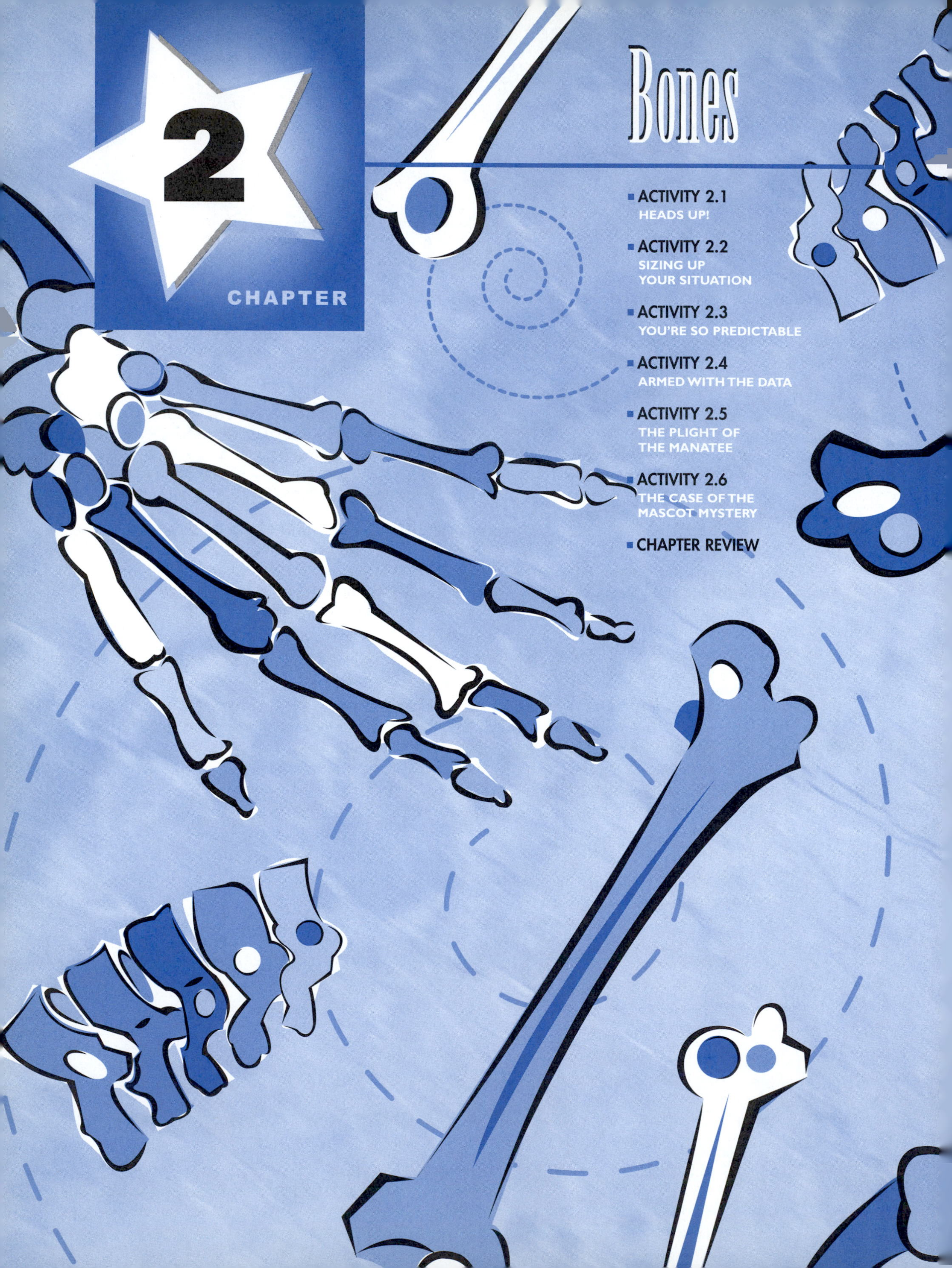

CHAPTER 2

Bones

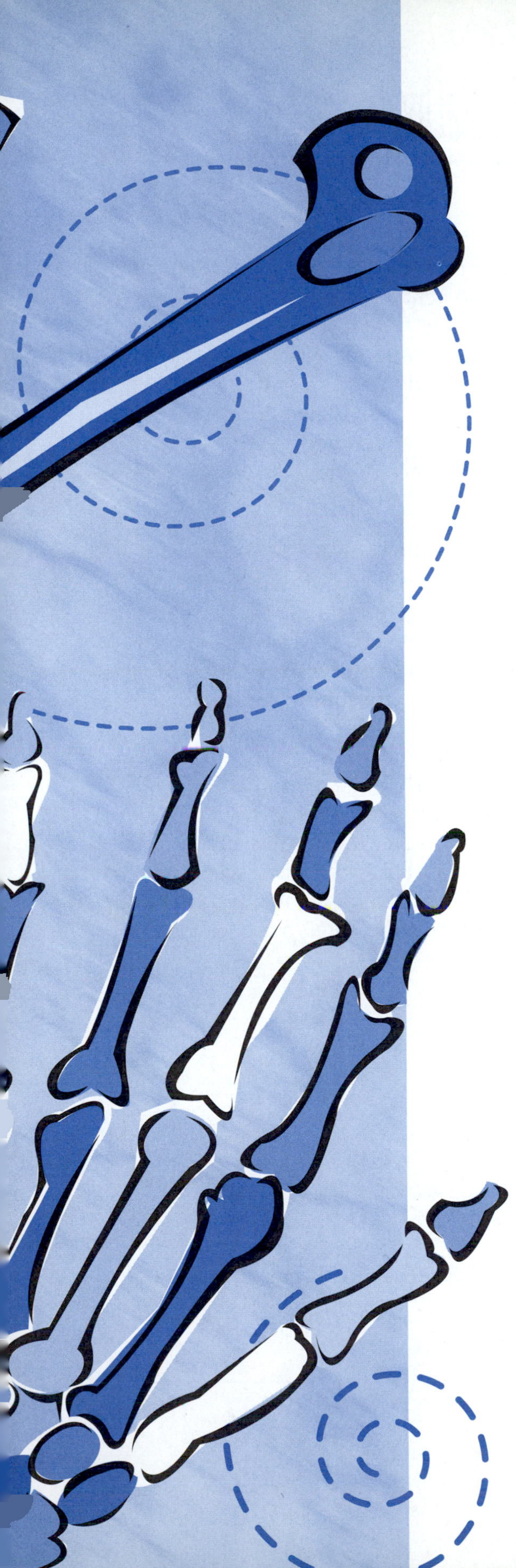

How does statistics help solve the mystery of soldiers who are missing in action?

Have you ever wondered how medical examiners, police laboratories, and anthropologists identify a dead person from some bones? Often, they use mathematical models that predict a person's height from the length of the long bones in the body. Here you will learn how these models are built.

You will gather data and build models that predict height. Model building often requires graphing data and looking for patterns. The graphs of the data you collect will have a linear shape, and you will learn ways to approximate the data with a line. Using the methods you learned in Chapter 1, you will find the equation of the line. Also you will learn to evaluate the precision of the predictions you make with your model.

PREPARATION READING

The Disappearance of Amelia Earhart

Amelia Earhart is probably the best-known woman aviator of all time. She was the first woman to fly solo across the Atlantic Ocean and later across the Pacific Ocean. On June 1, 1937, Amelia, accompanied by her navigator Frederick Noonan, set off on a flight around the world. On July 2, her plane disappeared while enroute from Lae, New Guinea, to Howland Island. The U.S. Navy searched for the lost flyers but failed to find any trace of them. To this day their fate remains a mystery.

The International Group for Historic Aircraft Recovery (TIGHAR, pronounced tiger) has investigated the Earhart mystery for more than 15 years. TIGHAR's theory is that Earhart crashed on the tiny island of Nikumaroro in the Republic of Kiribati, a new nation in the Pacific Ocean east of New Guinea. In the summer of 1997, a member of TIGHAR discovered a file of papers in Kiribati's national archives. The file reported a 1940 discovery of bones on Nikumaroro. The bones were sent to a medical school in Fiji where they were examined by Dr. D. W. Hoodless. Dr. Hoodless concluded that the bones were from a male approximately 5 feet 5 inches tall.

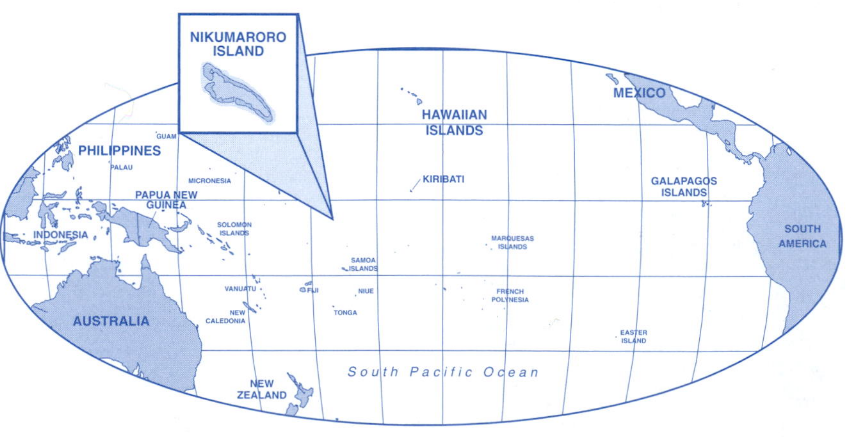

Certain statements in Dr. Hoodless' report raised doubt about his knowledge of the human skeleton. Sadly, the bones analyzed by Dr. Hoodless have disappeared. However, his report contained handwritten notes with measurements and observations on the bones. Some of his measurements are given in **Figure 2.1.**

Bones	Length (cm)
Humerus	32.4
Tibia	37.2
Radius	24.5

FIGURE 2.1.
Bones analyzed by Dr. Hoodless.

Forensic anthropologists Dr. Karen Burns and Dr. Richard Jantz analyzed these measurements. Using data from the Forensic Anthropology Data Bank at the University of Tennessee, they built models to predict height and determine gender and ethnic background. For this work, they assumed that Hoodless measured the bones in the same way they were recorded in the data bank. They concluded that the bones were most likely from a white female of European background who was about 5 feet 7 inches tall. This description fits Amelia Earhart. Could the bones examined by Dr. Hoodless have been Amelia's? This question cannot be answered until the bones are found or until TIGHAR discovers additional evidence on its next expedition to Nikumaroro.

In this chapter you will be asked to think like a forensic anthropologist. Given measurements of a set of bones, you will investigate the clues about a dead person. For the final project you will build models to predict height and determine gender. Your models will be based on some of the same data used by Burns and Jantz.

CHECK THIS!

A forensic anthropologist is a scientist who uses information about the human body and its bones to try to identify a dead person based on bones and teeth. This is one of many kinds of work that a forensic anthropologist does.

ACTIVITY 2.1

Scientists often use mathematical models to help investigate human remains. In this activity you will explore models that describe the relationship between the length of a person's head and the person's height. Then you will make some tentative predictions about bones that were found.

MYSTERIOUS FINDINGS

From time to time, bones are found in rugged areas. A hiker in Arizona's Superstition Mountains finds a skull, eight long bones, and many bone fragments. He notifies the local police who send a team of specialists to investigate. The team records information about the bones, such as their size and general condition. **Figure 2.3** shows some information that is probably similar to what the team recorded.

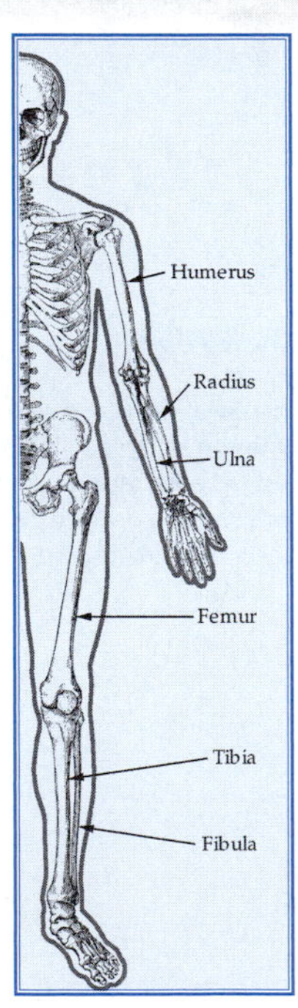

FIGURE 2.2.
A human skeleton.

Bone Type	Number Found	Length (cm)
Femur	3	41.5, 41.4, 50.8
Tibia	1	41.6
Ulna	2	22.9, 29.0
Radius	1	21.6
Humerus	1	35.6
Complete Skull (including jaw)	1	23.0
Fragments	More than 10	From 3.0 to 5.0 cm

FIGURE 2.3.
Sample record of bones found at site.

1. Use all the information you have so far to answer these questions.

   a) Study the data in Figure 2.3. The team reports that these bones belonged to at least two people. How do they know for sure?

   b) Which bones do you think belonged to the same person? What assumptions did you make to get your answer? Explain your answer. (To make it easier to classify the bones, refer to the dead people as Skeleton 1, Skeleton 2, and so on. You will need to refer to the answers to this question in Homework 2.1.)

c) Do you think the dead people were male or female? What evidence did you use to get your answer? (Remember, Dr. Hoodless concluded that the bones in Figure 2.1 belonged to a male, but Burns and Jantz disagreed with his findings.)

d) Do you think the dead people were young children or adults? Defend your answer.

e) Guess the heights of the dead people. How accurate do you think your guesses are?

A MODEL FOR ESTIMATING HEIGHT

One place to look for help in estimating heights is an artist's guidebook. Artists have found that the rule of thumb, "a 14-year-old is about 7 head-lengths tall" helps them draw teenagers with heads correctly proportioned to their bodies.

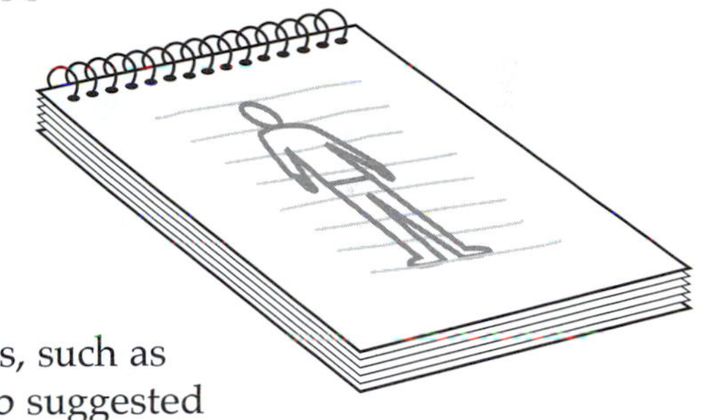

2. See how closely the dimensions of real students, such as those in your class, match the ideal relationship suggested by artists.

   a) Within your group, measure the length of each person's head (from chin to the top of the head). Record your data in the first two columns of a table like **Figure 2.4**. Be sure to give the units of measure you used for head length at the top of the second column.

Name	Head Length	Predicted Height	Actual Height

FIGURE 2.4. Group head-length and height data.

   b) Use the relationship "height = 7 head lengths" to predict the height of each person in your group. Record your results in the third column of your table.

c) Next measure each person's actual height and record your results.

Whenever you make predictions based on data, check the accuracy of your predictions when possible. One way to evaluate your predictions is to calculate what are called the residual errors.

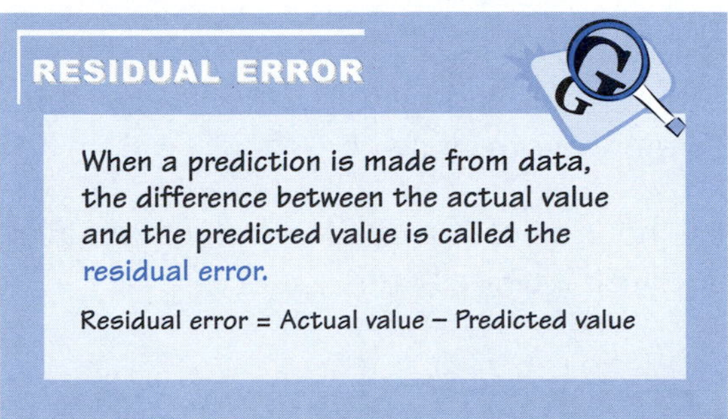

3. a) Calculate the residual errors for your data in question 2. Subtract the predicted height of each student in column 3 from the actual height in column 4. Add a fifth column to your table and use the heading "Residual Error" for this new column. Record your results in column 5. So that you have enough data to recognize patterns, collect data from at least one other group and add the data to your table.

   b) If a residual error is positive, does the prediction overestimate or underestimate the actual height?

   c) What if a residual error is negative? What if a residual error is zero?

   d) Are the residual errors fairly evenly divided between positive and negative values? How well did the relationship "height = 7 head lengths" do in predicting the actual heights of students?

4. a) Would a multiplier different from the multiplier 7 you used in question 2 do a better job? If so, what multiplier would you choose? Explain how you found this multiplier.

   b) Use your multiplier to complete a table like the one in question 2. Make sure to include a fifth column for the residual error.

   c) Why do you think your multiplier gives a better prediction than the multiplier 7?

MODELS OF THE FORM $y = mx$

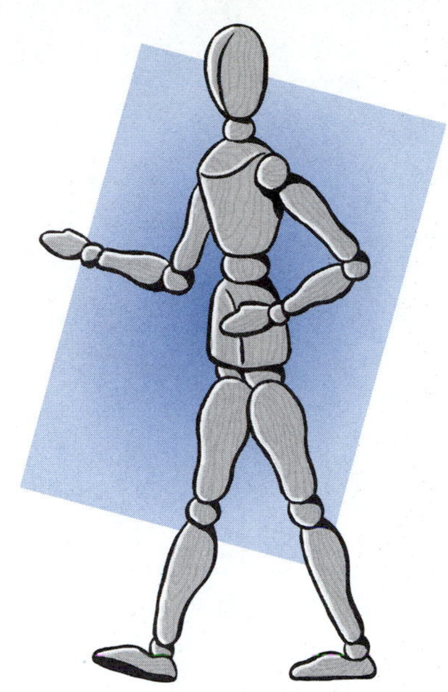

The relationship between height and head length changes with age. Therefore, artists adjust their guideline based on the age of the person they are drawing. When drawing sketches of adults (ages 18–50), artists follow this guideline: The height of an adult figure should be approximately 7.5 times the head length.

5. a) Write a formula that describes the relationship between height, H, and head length, L, according to the artists' guideline for drawing an adult.

   b) Write a formula representing guidelines for drawing sketches of 14-year-olds. Write another formula for drawing sketches of the students in your class. (As you did in (a), use the variables H and L.)

   c) In (a) and (b), you have written three relationships between H and L. Use your calculator to graph all three formulas in the same viewing window. (You will have to rename variables H and L to y and x, when you enter the formulas into your calculator.) Change the window settings to the following settings: Xmin = –5, Xmax = 30, Ymin = –50, and Ymax = 200. Then make a careful sketch of your three graphs. Be sure to label each graph with its equation and indicate the scale on each axis.

   d) How are the three graphs the same and how are they different?

   e) What effect does changing the value of the multiplier have on the graph?

6. a) You may have had trouble answering question 5(d) because the values for m in your three models were close together. Try graphing $y = 2x$, $y = 4x$, and $y = 6x$. How are the three graphs the same and how are they different?

   b) What effect does changing the value of the multiplier have on the graph?

   c) Predict what the graph of $y = 8x$ looks like.

MODELS OF THE FORM $y = mx$

Each of the models you graphed in questions 5 and 6 are in the $y = mx$ family.

- The graph of a model in the $y = mx$ family is a line that passes through the origin.
- The appearance of each of these lines is controlled by the number m, which is the slope of the line.

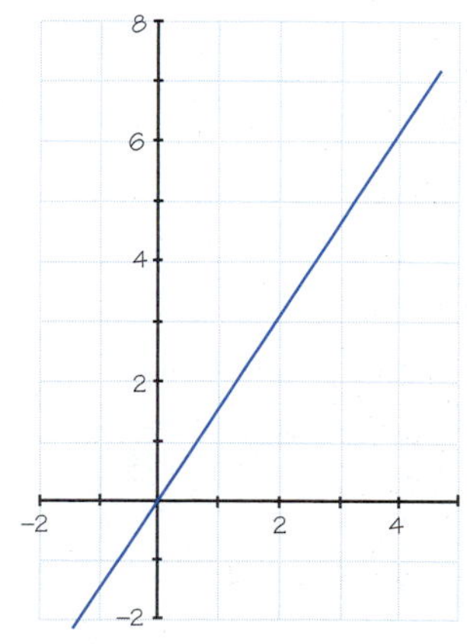

FIGURE 2.5.
A graph of a model in the $y = mx$ family.

7. a) Using the artists' guideline for adults, predict the height of a person whose head length measures 23.0 cm.

   b) Without doing further calculations, would your estimate be higher or lower if you knew the person was only 13 years old?

   c) Explain how you could use your graphs from question 5(c) to answer the preceding question.

   d) What is the meaning of the slope of the line that represents the equation H = 7.5L in the context of predicting height from head length?

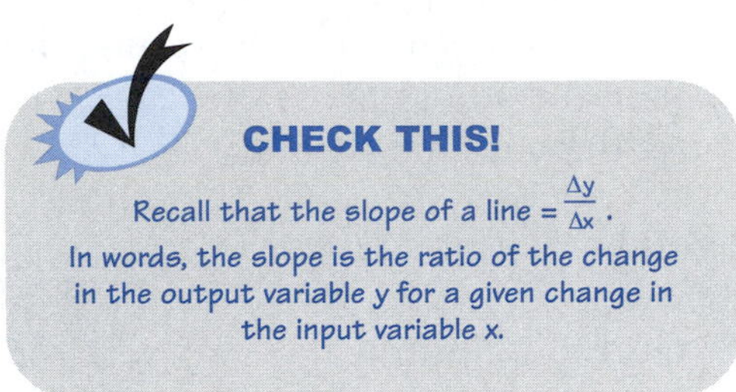

8. Juan is drawing a picture of his mother standing by a window. He follows the artists' guideline for drawing adults. He makes a rough sketch, but decides the figure is too small. In his final sketch, the head of his figure is 1 cm longer than the head in his rough sketch. How much taller than his preliminary sketch is Juan's final sketch? Justify your answer.

So far, you have explored a few models that describe the relationship between height and head length.

- models based on artists' guidelines for drawing figures
- a model of the relationship between height and head length that your group determined.

9. Think about how you might use one of these models to make a rough prediction of the height of the person whose skull length was recorded in Figure 2.3.

   a) What assumptions might you make in order to make your prediction?

b) Recall that the skull measured 23.0 cm in length. Predict the height of the person in centimeters (cm). Describe the process you used in making your prediction.

c) Does your prediction result in a reasonable height for a person? Explain.

d) Do you think your prediction is likely to be close to the actual height of the person? Why?

10. What information do you think might be helpful in determining better estimates of the heights of the dead people whose bone lengths are recorded in Figure 2.3? How or where might you obtain this information?

SUMMARY

In this activity you examined and interpreted equations that predict height.

- Using some artists' guidelines, you found some formulas that predicted height if you knew the length of the person's head.
- You also gathered data and used the data to find a formula to predict height based on the length of the person's head.
- You used the residual errors to compare the actual and predicted values.

Since these models give rough estimates of a person's height, you will need to do more investigating to find a better model that gives more precise predictions. One way to improve your models is to collect data and use the data to test your model. Collecting and analyzing data forms the basis of the next two activities.

HOMEWORK

2.1

I Think We Have Your Thigh Bones

1. Early models (from the late 1800s) that predicted height from the length of long bones were based on ratios. (See Figure 2.2 that shows some of the long bones.) For example, the ratio of the height H and femur length F is given by the formula $\frac{H}{F} = 3.72$.

   a) Based on this model, estimate the heights of the people whose femur lengths are given in Figure 2.3 in Activity 2.1.

   b) Suppose, for most adults, femurs range in size from 38 cm to 55 cm. According to this model, what is the range of heights of most adults? Are these reasonable heights for adults? (Recall that 2.54 cm = 1 in.)

   c) What formula would you enter into your graphing calculator in order to graph this relationship between height and femur length?

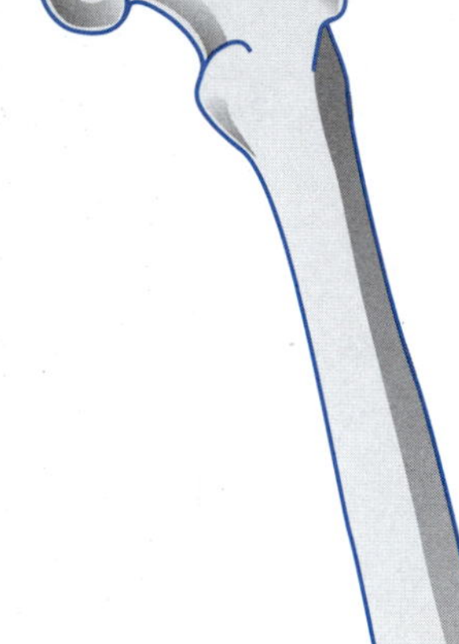

FIGURE 2.6. The femur (thighbone).

The formula that you wrote for 1(c) should be a member of the $y = mx$ family, as were the first models used to predict height from the lengths of long bones. Dr. Mildred Trotter (1899–1991), a physical anthropologist, was well known for her work in the area of height prediction based on the length of the long bones. She refined earlier models by adding a constant, thereby producing models of the form $y = mx + b$.

Here is one of the relationships proposed by Dr. Trotter that we call the first formula.

$$H = 2.38F + 61.41$$

where H is the person's height (cm) and F is the length of the femur (cm).

2. Again assume that most adults' femurs range in size from 36 cm to 55 cm. According to Dr. Trotter's formula, how tall is a person with a 36-cm femur? How tall is a person with a 55-cm femur?

H(36) = 147.09 cm
57.9 inch

H(55) = 192.3 cm
75.7 in

3. On graph paper, draw a set of axes similar to the axes shown in **Figure 2.7**.

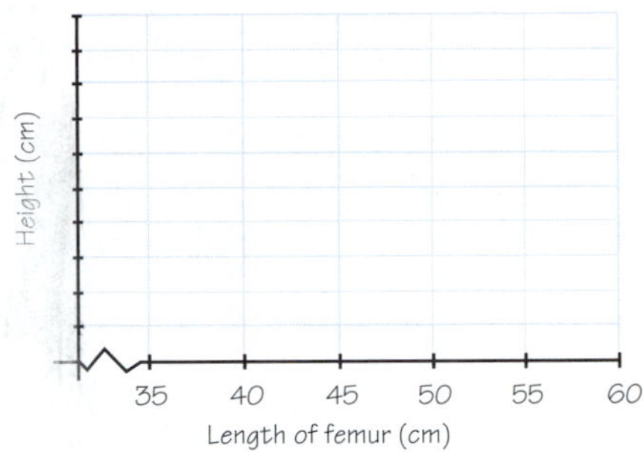

FIGURE 2.7. Axes for height and femur length.

Notice that the horizontal axis is scaled from 35 cm to 60 cm (a slightly wider range than the minimum and maximum femur lengths) with tick marks every 5 units. A zigzag has been added to indicate that there is a break in this scale between 0 and 35.

a) Draw a scale on the vertical axis that would be appropriate for data on adult heights (in cm).

b) Sketch a graph of Dr. Trotter's first formula on the set of axes you have drawn. (You may want to plot several points before drawing the graph.)

4. Jackson's femur measures 39 cm and his brother's measures 40 cm. Based on Dr. Trotter's first formula, predict the difference in the two brothers' heights.

5. The femurs of two women differ by one centimeter. Predict the difference in their heights. Explain how you were able to determine your answer even though the lengths of the two women's femurs were not given. In addition, tell how you could read off your answer from Dr. Trotter's first formula.

6. Suppose a man is 178 cm (about 5 ft 10 in) tall.

a) Explain how you could use your graph to estimate the length of his femur. What is your estimate?

b) Write a set of algebraic steps to solve Dr. Trotter's first formula, $H = 2.38F + 61.41$, for F. (A medical doctor might use such an equation to check that the length of a person's femur is normal for a person of that height.)

c) Use your equation in (b) to predict the length of a man's femur if the man is 178 cm tall. Compare your answer to the femur length you estimated in (a).

7. Another of Dr. Trotter's equations predicts height from the tibia length (this is called the second formula):

   H = 2.52T + 78.62, where H and T are measured in cm.

   $H = 2.52(41.6) + 78.62 =$

   a) The length of the tibia listed in Figure 2.3 is 41.6 cm. Using Dr. Trotter's second formula, predict the person's height. Is your answer a reasonable height for a person?

   b) Use your predicted height from 7(a) and your formula from 6(b) to predict the length of the femur of a person with the 41.6-cm tibia. Which femur length from Figure 2.3 is closest to your prediction?

   c) Write a set of algebraic steps to solve Dr. Trotter's second formula, H = 2.52T + 78.62, for T. (A medical doctor might use such an equation to check that the length of a person's tibia is normal for a person of that height.)

8. In a third formula, Dr. Trotter used both the tibia and the femur to predict height:

   H = 1.30(F + T) + 63.29. (All measurements are in cm.)

   a) Suppose students measure the femur and tibia of a skeleton. They determine that the femur is 42 cm long and the tibia is 43 cm long. Predict the height of the person using Dr. Trotter's formula, H = 1.30(F + T) + 63.29.

   b) Compare the prediction in (a) with the predicted height using Dr. Trotter's equation, H = 2.38F + 61.41.

   c) Compare the predictions in (a) and (b) with the predicted height using Dr. Trotter's formula, H = 2.52T + 78.62.

   d) You should have found a fairly large discrepancy between your predictions in (a)–(c). One possibility is that the students did not get precise measurements of the bone lengths. Suppose a man is 175 cm tall (about 5 ft 9 in). Based on Dr. Trotter's equations in (b) and (c), would you expect his tibia or his femur to be longer and by how much?

e) Repeat (d) for a person whose height is 160 cm.

f) Based on Dr. Trotter's equations, is there any evidence that indicates the students may have made faulty measurements? Explain.

9. a) Use one or more of Dr. Trotter's equations to estimate the heights of two of the people whose bones are described in Figure 2.3.

   b) Using her equations do you think these bones might have belonged to at least three people?

   c) Do your calculations give you cause to change any of the assumptions you made in 1(b) in Activity 2.1? If so which assumptions?

ACTIVITY

2.2

Sizing Up Your Situation

Your analysis in Activity 2.1 and Homework 2.1 were steps in a process known as mathematical modeling. The first step in this process is to identify a problem that you need to solve. Collecting and analyzing data often is the next step. In this activity you will design methods for collecting data. Then you will learn ways to analyze data.

WHAT TROTTER FOUND

During World War II, the armed services sometimes had problems identifying the remains of dead soldiers. Dr. Mildred Trotter was asked to help. She wondered if there was a relationship between the height of a person and the length of one of the person's long bones.

After asking this question, Dr. Trotter's next step was to collect data. She decided to measure each person's height and the length of each person's femur. Using these data, she created the model

$H = 2.38\ F + 61.41$ where height, H, and femur length, F, are in cm.

This equation expresses the relationship between the height measurements and femur measurements she observed from her data.

Since Dr. Trotter's models depended on data, her models were only as good as the accuracy of her data. Dr. Trotter took special care to check that her assistants followed detailed instructions for taking the measurements. In this way she was able to keep to a minimum the errors caused by the way the measurements were made.

DOING IT YOUR WAY

Dr. Trotter used the length of the femur bone in her model for predicting height. You can't directly measure your femur, but measuring your forearm gives a good estimate of the length of your ulna.

Now you will collect data on height and forearm length from each student in your class. You will need to figure out a method for taking the measurements. Remember, the accuracy of your model depends on the quality of the data that you collect. Everyone taking the measurements must use the same method and then record the data to the same degree of precision (for example, to the nearest eighth of an inch, or to the nearest millimeter).

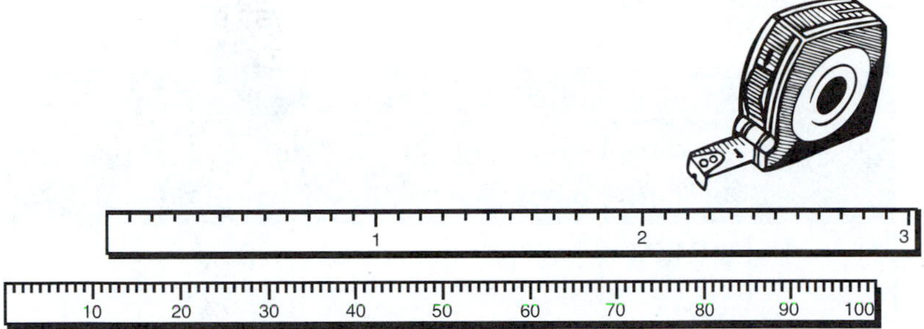

1. With members of your group, discuss methods for measuring (a) the heights of students and (b) the lengths of their forearms.

2. Test your methods as follows:

   a) Have two different students use your method to measure the height of the same student. Are both height measurements roughly the same? Are they recorded to the same degree of precision? If not, modify your method and test it again. Keep modifying your method until there is only a small amount of variation in the measurements taken.

   b) Repeat (a), but this time measure forearm length.

3. Discuss various groups' methods for measuring height and forearm. Then select one method. Write a brief description of the method that the class will use to collect the data.

4. Take measurements of classmates' forearms and heights. Record your results on Handout 2.2, *Class Data Recording Sheet*. Leave the last column blank. (You will collect more data from your class later.) Be sure to record the units you used for height and forearm length at the top of those columns.

STARTING YOUR ANALYSIS

One important characteristic of some data is the amount of variability in the data, that is, the amount that the data are spread out. Let's take a look at an example.

5. **Figure 2.8** shows the forearm length and height of each student in Ms. Jaczinko's class. What can you say about the precision of the recorded data?

Female			Male		
Name	Forearm Length (cm)	Height (cm)	Name	Forearm Length (cm)	Height (cm)
Alicia	24.0	157	Ahmed	26.5	173
Bia	24.5	166	Brian	27.0	177
Christi	27.0	164	Jesús	27.0	174
Chantalle	24.0	164	Davis	31.0	192
Coral-Mae	23.0	161	José	28.0	172
Juanita	27.5	164	Kelvin	29.0	180
Ji-Hyun	27.0	167	Lenny	27.0	174
Kim	26.0	162	Luis	28.0	175
Kristen	26.0	175	Mike	32.0	185
Maria	28.5	166	Pedro	30.0	185
Tianna	26.5	172	Sang	30.0	178
Teresa	25.5	176			

FIGURE 2.8. Height and forearm data from Ms. Jaczinko's class.

6. a) List the student heights from shortest to tallest.

   b) Look over the data from Ms. Jaczinko's class. By how much do the heights vary from the shortest student to the tallest?

   c) Suppose another tenth-grader joined Ms. Jaczinko's class. Is it reasonable to predict that the tenth-grader would be between 164 cm and 180 cm tall? Explain your answer.

7. a) List the girls' heights from the shortest to the tallest.

   b) By how much do the girls' heights vary from shortest to tallest?

c) If you know the new student in Ms. Jaczinko's class is a girl, what would you say about this person's height based on these data?

8. a) Another girl who is 196 cm tall joins Ms. Jaczinko's class. How does this new girl's height compare with the other heights in the class?

   b) What effect does this new height have on the variability of the girls' height data?

Sometimes one or more data are much larger or smaller than the rest of the data. Since these data can have a large effect on your analysis, statisticians have given these data a special name.

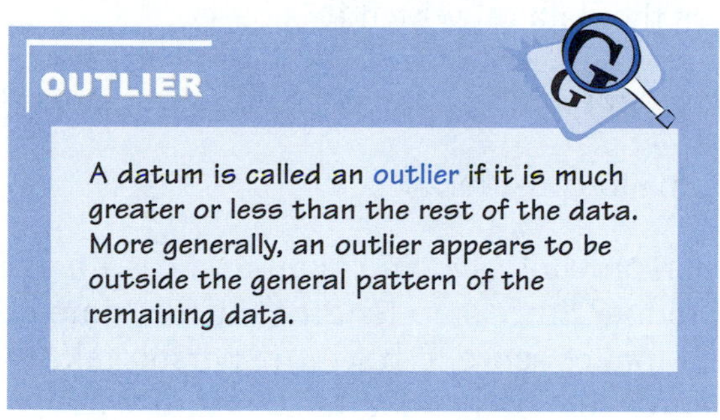

OUTLIER

A datum is called an outlier if it is much greater or less than the rest of the data. More generally, an outlier appears to be outside the general pattern of the remaining data.

9. a) List the boys' height from the shortest to the tallest.

   b) Give a boy's height that would be greater than the other heights of boys in Ms. Jaczinko's class so that it is an outlier. Explain why.

   c) Give a boy's height that is lesser than the other heights of boys in Ms. Jaczinko's class so that it is an outlier. Explain why.

SUMMARY

To build and test a model, you often need to collect data. In this activity you collected data, did preliminary analyses of the data, and learned about outliers. In the next activity you will learn more techniques to analyze data and make predictions.

HOMEWORK

2.2 Making Great Strides

Sometimes all that's left at a crime scene is a few footprints. However, the length of a person's stride is also related to the person's height. In this assignment you will develop a method for measuring a person's stride. Later you will gather these data and then use your measurements to form a model to predict height.

To collect reliable data, you need to plan carefully the method that you will use to collect the data. Remember, your model will only be as good as the data on which it is based.

Design a method for measuring the length of a person's stride.

Here are some items to consider.

- How will the person walk? Do you plan to measure from heel to heel or heel to toe? Since step lengths for the same person can vary, does it makes sense to have the person take more than one step and average the results? If so, how many steps should she take?
- Determine the measurement instrument (ruler, tape measure, meter stick, etc.) you will use to make the measurement.
- Specify the precision of the measurement.

After you have decided on your method, test it as you did the methods for measuring height and forearm length in Activity 2.2.

When you are satisfied with your method, describe it with a set of written instructions. Give your instructions to a friend to see if someone else understands what you mean. If necessary, revise your instructions. Save them until your class is ready to collect the stride-length data needed later in this chapter.

ACTIVITY 2.3

You're So Predictable

In Activity 2.2 you made a rough prediction of a new student's height based on the height and forearm data from Ms. Jaczinko's class. In this activity you will learn new ways to analyze these data so that you can make better predictions.

THE DOT PLOT

Let's go back to the data you studied in Activity 2.2. The table in **Figure 2.9** shows the heights and forearm lengths of students in Ms. Jaczinko's tenth-grade class.

Female			Male		
Name	Forearm Length (cm)	Height (cm)	Name	Forearm Length (cm)	Height (cm)
Alicia	24.0	157	Ahmed	26.5	173
Bia	24.5	166	Brian	27.0	177
Christi	27.0	164	Jesús	27.0	174
Chantalle	24.0	164	Davis	31.0	192
Coral-Mae	23.0	161	José	28.0	172
Juanita	27.5	164	Kelvin	29.0	180
Ji-Hyun	27.0	167	Lenny	27.0	174
Kim	26.0	162	Luis	28.0	175
Kristen	26.0	175	Mike	32.0	185
Maria	28.5	166	Pedro	30.0	185
Tianna	26.5	172	Sang	30.0	178
Teresa	25.5	176			

FIGURE 2.9. Height and forearm data from Ms. Jaczinko's class.

Often you can use graphs to examine data. Graphing the height data might help you analyze the data. One way to graph data is to make a **dot plot**.

Draw a number line and label the line with the heights shown in Figure 2.9. To make a dot plot, place a dot above each number that corresponds to a student's height. If two heights are the same, place one dot directly above the other. Dots for Alicia's, Bia's, Christi's, and Chantalle's heights have already been plotted in **Figure 2.10**.

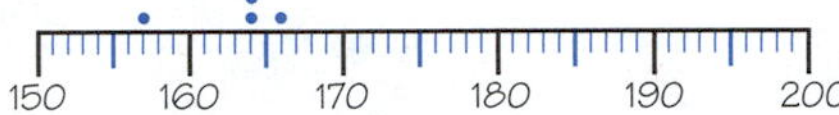

FIGURE 2.10.
Partial dot plot of height data from Ms. Jaczinko's class.

1. On a sheet of graph paper, draw a dot plot like the one in Figure 2.10. Complete the dot plot by graphing the data for the remaining students.

One way to predict a new student's height is to take the **average** of all the numbers.

To calculate the average height, add all the heights and divide by the number of students.

2. a) Find the average of the students' heights. Mark the average with an "X" on your dot plot. Use this average as a prediction of a new student's height.

   b) Suppose the new student is as short as the shortest student in Ms. Jaczinko's class. How far off was the prediction in (a)?

   c) Suppose the new student is as tall as the tallest student in Ms. Jaczinko's class. How far off was the prediction in (a)?

   d) Do you think that taking the average helped to make a good prediction? Explain. Can you suggest a better one?

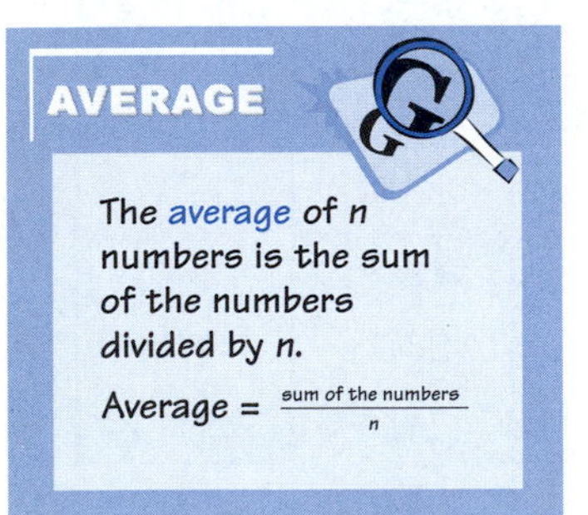

Notice that the height data graphed on your dot plot appear to separate into two groups, one to the left of 170 and the other to the right. Do you think this separation shows the split in height by gender? **Figure 2.11** shows number lines for two dot plots, one for only girls' heights and the other for only boys' heights. The two dot plots use the same scale so that you can compare the girls' heights to the boys'.

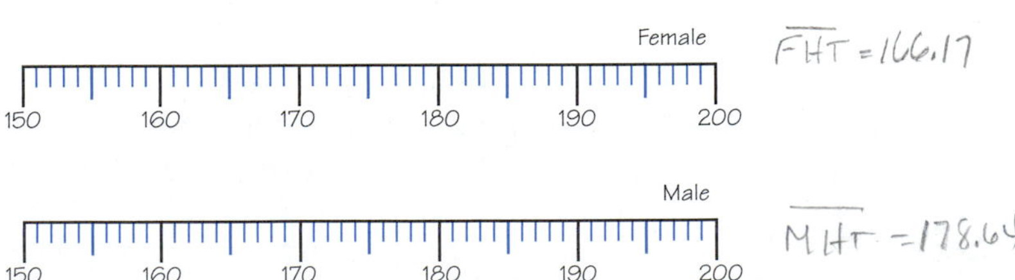

FIGURE 2.11.
Number lines for comparative dot plots.

3. a) On your own paper, draw these number lines, and graph the height data on the two dot plots.

   b) What do these dot plots tell you about the heights of Ms. Jaczinko's tenth-grade girls and boys? Do these girls or boys tend to be shorter?

4. a) Suppose the new student's name is Melissa. Realizing that the new student is a girl may change your prediction. Find the average for girls' heights in Ms. Jaczinko's class. Use this average to predict Melissa's height.

   b) If Melissa is as short as the shortest girl in Ms. Jaczinko's class, how far off is your prediction? What if Melissa is as tall as the tallest girl?

   c) Do you think this is a better prediction than the prediction made in question 2? Explain.

5. a) Suppose the new student turns out to be Martin (a boy), not Melissa. Choose a method for predicting Martin's height. Give your prediction and describe your method.

   b) If the new student's height is somewhere between that of the shortest boy and the tallest boy, what is the largest possible error that could have resulted from your prediction?

THE MEAN

Using the height data from Ms. Jaczinko's class, you have computed at least two and possibly three different averages:

- an average of all the data
- an average of the girls' heights
- an average of the boys' heights

The term **mean** is another name for average. For the remaining questions, when you are asked to calculate the mean (or average), just calculate the average using the formula.

Shorthand	Notation Meaning
Σx	The sum of the data.
n	The number of data.
$\bar{x}$	The mean, the sum of the data divided by the number of data.

FIGURE 2.12.
Table of mathematical symbols.

If you have entered the data into one of your calculator's lists, you can use a built-in calculator command to compute the mean. First, you will need to know some mathematical shorthand in order to understand what your calculator is telling you. (See **Figure 2.12.**)

6. Suppose you want the mean height of only a small group of students in Ms. Jaczinko's class. After entering the data into your calculator and pressing a few keys, the screen in **Figure 2.13** appears.

   a) What is the sum of these data?

   b) How many people are in this small group?

   c) What is the mean height for the people in this group?

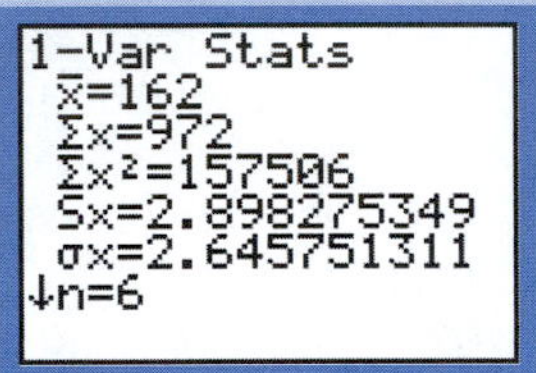

FIGURE 2.13.
One-variable statistics screen.

7. a) Now compare the heights of students in your class to the heights of students from Ms. Jaczinko's class. Make two dot plots for your class data similar to the ones that you made for question 3(a), one for the boys' heights and one for the girls' heights.

   b) Enter the boys' heights and girls' heights into separate lists in your calculator. What is the mean height for the boys? What is the mean height for the girls?

   c) Based on your dot plots and the means of boys' heights and girls' heights, do the boys in your class tend to be shorter or taller than the girls?

   d) Compare the data from your class to the data from Ms. Jaczinko's class. Describe the similarities and differences between the two data sets.

If you find it helpful, you may use your calculator's built-in statistical capabilities to calculate the means in the remaining problems.

WAY OUT THERE

In the next question, one data point is much greater than the other data points, that is, one data point is an outlier. Let's see how these data affect the analysis.

8. A researcher gathered data on the number of gray hairs on the heads of 25-year-olds. These are the data she found.

   0 23 45 6 8 9 33 15 0 2 4 10

   12 13 34 67 40 38 27 25 0 13 34 23

   56 34 7 780 44 6 4 0 31 22 5 16

   17 11 2 1

   a) Graph these data in a dot plot. (How do you plan to deal with the largest data point?) Then use your dot plot to help you list your data from the smallest to the largest.

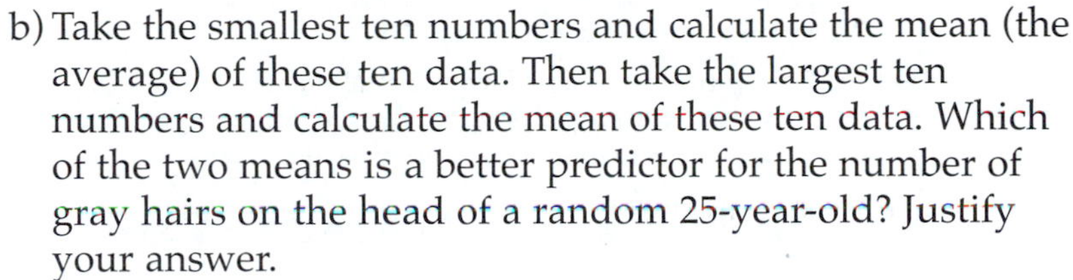

   b) Take the smallest ten numbers and calculate the mean (the average) of these ten data. Then take the largest ten numbers and calculate the mean of these ten data. Which of the two means is a better predictor for the number of gray hairs on the head of a random 25-year-old? Justify your answer.

   c) Next, calculate the mean using all the data.

9. a) How does the outlier in the data set in question 8, 780, affect the mean? To find out, calculate the mean again, this time leaving out the outlier.

   b) You have calculated four means that are marked on **Figure 2.14**:

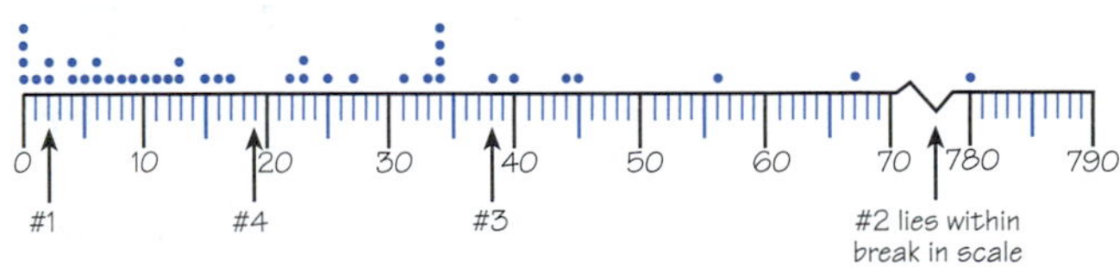

FIGURE 2.14. The four means you have calculated.

   Mean 1: the mean of the smallest ten numbers (question 8(b));

   Mean 2: the mean of the largest ten numbers (question 8(b));

   Mean 3: the mean of all the data (question 8(c));

   Mean 4: the mean of all the data leaving out the outlier (question 9(a)).

   Which of these means do you think is the best predictor of the number of gray hairs on a 25-year-old? Why?

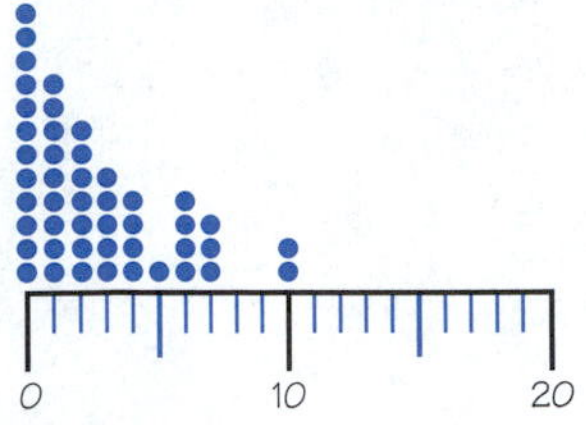

FIGURE 2.15.
Number of gray hairs on the heads of 20-year-olds.

10. When the researcher (from question 8) gathered data on the number of gray hairs on the heads of 20-year-olds, the data looked quite different from that for the 25-year-olds. Her data are displayed in the dot plot in **Figure 2.15**.

    a) Suppose a 20-year-old student teacher will visit your class tomorrow. Predict the number of gray hairs on the student teacher's head.

    b) If you had to make a prediction of the number of gray hairs on the head of a 25-year-old or on a 20-year-old, which prediction do you think would be closer to the actual count? Why?

VARIABILITY'S EFFECT ON PREDICTIONS

- If the data have a lot of variability (in other words, the data are very spread out), then it is difficult to make precise predictions.
- If the variability in the data is small (in other words, the data are very concentrated), it is easier to make more precise predictions.

SUMMARY

This activity showed that graphing data using dot plots can help you analyze the data. In addition, you learned some methods for analyzing data:

- Using the mean (or average), you made some predictions.
- You estimated how far off your predictions might be.
- Finally you discovered how an outlier can change your predictions considerably.

Now that you know some methods for analyzing data, you are ready to build models that predict height.

HOMEWORK 2.3

Group Comparisons

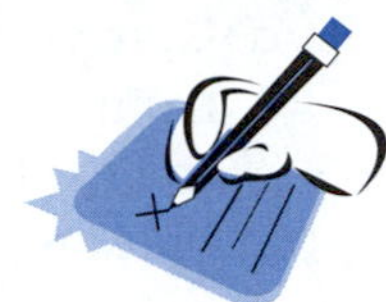

Each of the items in this assignment provides an opportunity to compare data from two groups. When you make comparisons to analyze data, use what you have learned from Activity 2.3 as well as common sense.

1. The table in **Figure 2.16** lists the weights of babies at birth for two groups of babies. The first group is babies whose mothers never smoked. The second group shows babies whose mothers smoked at least ten cigarettes per day. From these data, does it appear that smoking has an influence on a baby's birth weight? Explain your answer.

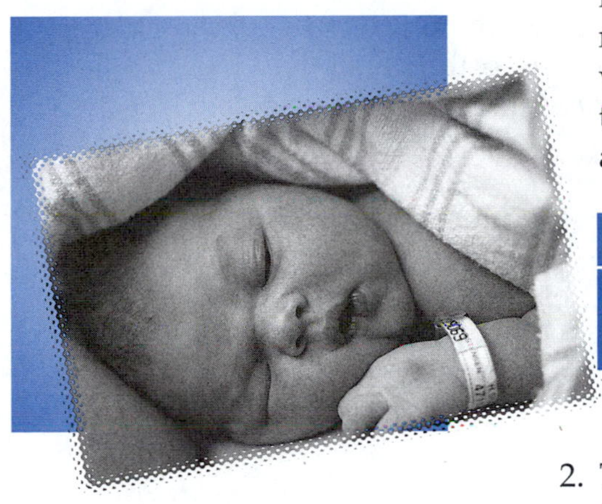

Never Smoked	6.3	7.3	8.2	7.1	7.8	9.7	6.1	9.6	7.4	7.8	9.4	7.6
Smoked Ten or More Cigarettes per Day	6.3	6.4	4.2	9.4	7.1	5.9	6.8	8.2	7.8	5.9	5.4	6.3

FIGURE 2.16.
Babies' birth weights.

2. Two groups of high school students were asked how much they typically spend on a date. The first group includes 12 students who did not exercise; students in the second group exercised at least twice a week. **Figure 2.17** shows the results.

FIGURE 2.17.
Cost of a date (in dollars).

Does Not Exercise	10	5	20	4	20	20	15	0	8	40	8	15
Does Exercise	15	15	15	5	10	5	5	6	30	25	30	60

a) From the data in Figure 2.17, make two dot plots using the same scaling on each. Place one dot plot directly above the other.

b) What can you learn from your dot plots?

c) Predict the amount spent on a date by a person who exercises. Explain why the average amount spent by the exercise group might not be a good choice for your prediction.

d) Complete the following sentences: "I predict that a person from the "does not exercise" group will spend _______ on the next date. However, given what this group has spent on dates in the past, this person might spend as little as _____ or as much as _____. So my prediction might be as far off as _______ ." (Add any additional comments that you think shed light on your prediction.)

e) Now compare your predictions. According to your predictions, which of the two groups of students spends more on a date? Does your dot plot support the same conclusion?

f) Make a scatter plot of these data. (In other words, plot the points (15, 10), (15, 5), and so forth.) Label the vertical axis "does not exercise" and the horizontal axis "does exercise."

g) Is it valid to claim that there is a direct connection between the exercise and non-exercise groups? Could you use the typical amount spent by someone in the "does exercise" category and predict how much a person in the "does not exercise" category would spend? Explain.

ACTIVITY

2.4

Armed With the Data

In Activity 2.1 you predicted the heights of the skeletons based on what you know about people's heights. Then you collected and analyzed data on student heights. In this activity you will build models based on the relationship between forearm length and height.

LOOKING FOR A PATTERN

In Activity 2.3 you predicted a student's height based on data from a single variable, students' heights. Later you used your knowledge of whether the student was male or female to improve your predictions. Now you will look for connections between the length of someone's forearm and height.

CHECK THIS!

In Chapter 1, we used the terms independent variable for the explanatory variable, and dependent variable for the response variable. You should become comfortable with these terms because they are used often in mathematics.

Since you want to estimate the height of a person based on the length of the forearm, forearm length is called the **explanatory variable**. Because height changes in response to changes in forearm lengths, height is called the **response variable**. On a scatter plot, the explanatory variable is represented by the horizontal axis, and the response variable by the vertical axis.

1. On graph paper, make a scatter plot of Ms. Jaczinko's class data (see Figure 2.9 in Activity 2.3). Let forearm length be represented by the horizontal axis and height be represented by the vertical axis. Remember to label each axis with its variable and use an appropriate scale for that variable. To distinguish the boys from the girls, use two colors (or different shapes), one to represent the boys' data and the other to show the girls'.

2. Use your scatter plot to make the following predictions.

   a) If a girl the same age as the students in Ms. Jaczinko's class has a forearm that measures between 25 cm and 27 cm, what would you predict for her height? How accurate do you think your prediction would be?

   b) Predict the height of a tenth-grade boy with a 28.5-cm forearm. Explain how you determined your answer.

   c) Predict the height for a tenth grader who has a forearm that measures 33 cm. How did you do this?

THE 16% MODEL

Often data can be used to test the validity of a model. Next you will use the data from Ms. Jaczinko's class to examine and refine a model that represents the relationship between height and forearm length.

3. Archaeologists study ancient human life. Like artists, archaeologists frequently use general rules of proportions. For example, an archaeologist might use the proportion that the forearm of a typical female teenager is 16% of her height.

   a) Translate this relationship into an equation that predicts height, y, from forearm length, x. We call this equation the "16% model." Test a few points to make sure this model makes sense.

   b) Sketch a graph of the 16% model on the same set of axis as your scatter plot.

   c) Compare the 16% model with the data from Ms. Jaczinko's class. Is the model true for all the people in Ms. Jaczinko's class? Justify your answer based on your graph.

One important part of the modeling process is testing a model. One way to test the model is to calculate the residual errors. Recall that you checked some predictions in Activity 2.1 by calculating the residual error, which is the difference between the actual value and the predicted value. You can test the 16% model using the data from Ms. Jaczinko's class.

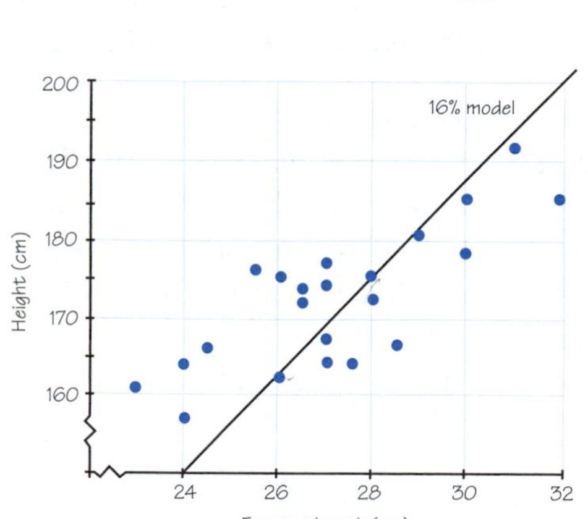

FIGURE 2.18.
Residual = actual value − predicted value.

CHECK THIS!

Remember that the residual error is

Residual error = Actual value − Predicted value

For the height data, the actual values are the actual heights of students in Ms. Jaczinko's class. The predicted values are the heights predicted by the 16% model.

4. a) Calculate the residual errors for the 16% model using the data from Ms. Jaczinko's class. Prepare two tables like the one you created in Activity 2.1, one for the girls' data and one for the boys' data.

   b) Do you think that the 16% model is a better predictor of the girls' heights or the boys' heights? Justify your answer based on the model's residual errors.

LINES THAT APPROXIMATE DATA

The plot of the height-forearm data from Ms. Jaczinko's class is fairly spread out, but it appears to have a linear form. This large amount of variability makes it difficult to pick a line that you can use to make good predictions for heights. The following two methods will help you choose a line that approximates the pattern of the height-forearm data from Ms. Jaczinko's class.

METHOD 1

1. Pick a point that appears to lie in the middle of the points displayed in your scatter plot. The point doesn't have to be a data point. What are the coordinates of this point?
2. Using this point as one point on your line, adjust the slope of your line until you find a line that you think best describes the pattern of the data.
3. Find the equation of the line you have selected using the point-slope formula. How did you decide which line is the best?

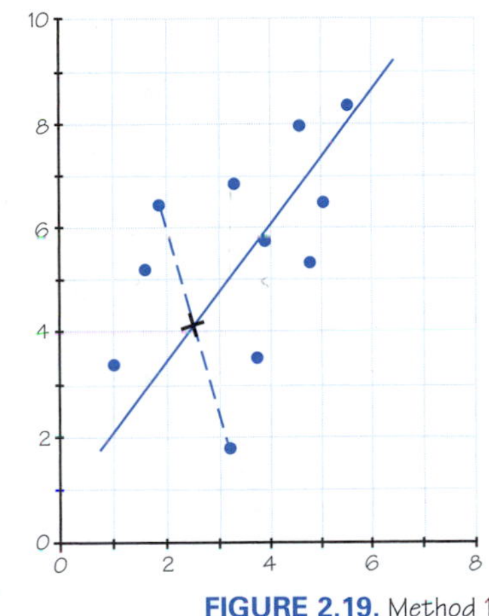

FIGURE 2.19. Method 1.

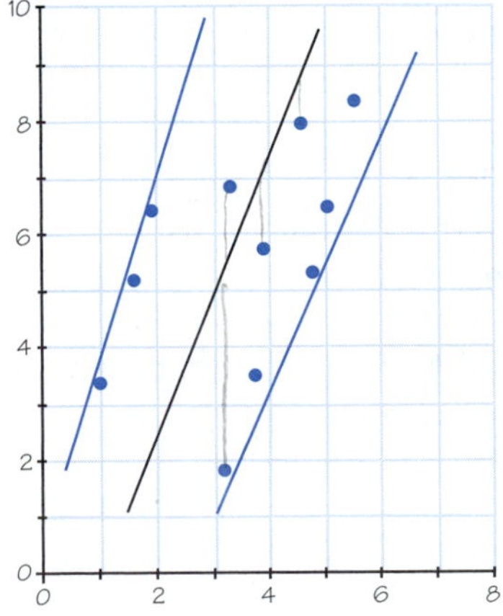

FIGURE 2.20. Method 2.

METHOD 2

1. Draw two lines in such a way that the points on your hand-drawn scatter plot are squeezed as tightly as possible between these lines. (The lines don't have to be parallel.)
2. Now draw one line halfway between the two lines that you have drawn.
3. Find the equation of this line by choosing a point on the line and using the point-slope formula. How did you decide which line is closest to the middle of the two outer lines?

5. Divide your group in half. Half of your group should use Method 1 to find a line that approximates the pattern of the height-forearm data from Ms. Jaczinko's class. The other half should use Method 2 to find a line that approximates the data.

6. a) In question 5, your group used two methods to determine an equation that is a model of the data in your scatter plot. Express your models in slope-intercept form ($y = mx + b$) if they are not already in that form. Compare the models that you determined by the two methods.

$\hat{HT} = 90.47 + 3.01\,FL$
$r = .784$

   b) Which model, the one from Method 1 or the one from Method 2, appears to describe the pattern of the data better? Explain why.

   c) Calculate the residual errors for each model using the data from Ms. Jaczinko's class. For each model, make a table like the one you created in question 4.

   d) What were your criteria for choosing the better model?

   e) Using your criteria, does your selected model from (b) appear to fit the data better than the 16% model? Explain.

7. Use your model from question 6(b) or the 16% model (whichever you think is better) to make the following predictions.

   a) Predict the height of a student whose forearm is 27 cm. Use the data from Ms. Jaczinko's class to check the precision of your prediction.

   b) Predict the height of a student whose forearm is 33 cm. Do you think the data provide any clues to suggest how precise this prediction might be? Explain.

   c) The forearm lengths of two students differ by 1 cm. Predict how much their heights differ. What if their forearm lengths differed by 2 cm? Justify your answers.

   d) Interpret the value of the slope of the line that represents your model in the context of predicting height from forearm length.

CHECK THIS!

Recall that the slope m of a line may be written as the fraction $\frac{m}{1}$. The slope $\frac{m}{1}$ means that when the explanatory variable increases 1 unit, the response variable increases m units.

SUMMARY

In this activity you compared some height data with the 16% model to test the validity of the model. Then you used the data to look for other models that approximate the data better.

- You calculated the residual error to test the 16% model.
- You used two methods to find lines that approximate the pattern in the data and calculated residual errors to test these models.

Your problem is how to find a model that has the least residual error. Activity 2.5 will show you how.

HOMEWORK

2.4

Looking for a Positive Relationship

In this assignment you will make scatter plots of some data sets, and in some cases fit a line to the scatter plot and make predictions. Pay particular attention to the characteristics of the relationship between the variables.

Do the points in the scatter plot appear to be scattered on either side of a straight line? Then the scatter plot has **linear form** and it makes sense to describe it with a linear equation (a member of the $y = mx + b$ family).

Does the pattern made by the points move upward as you look from left to right? If so, the two variables are **positively related** (as one variable increases the other tends to increase). If the pattern drifts downward, then the two variables are **negatively related** (as one variable increases the other tends to decrease).

1. Linda heats her house with natural gas. She wonders how her gas consumption is related to how cold the weather is. The table in **Figure 2.21** shows the average outside temperature (in degrees Fahrenheit) each winter month and the average amount of natural gas Linda used (in hundreds of cubic feet) each day that month.

FIGURE 2.21. Gas usage and temperature data.

Month	Sep	Oct	Nov	Dec	Jan	Feb	Mar	Apr	May
Outdoor temperature °F	48	46	38	29	26	28	49	57	65
Gas used per day x 100 cu ft	5.1	4.9	6.0	8.9	8.8	8.5	4.4	2.5	1.1

a) Make a scatter plot of these data. Which is the explanatory variable and which is the response variable? How did you decide?

b) Describe the characteristics of the relationship between outside temperature and natural gas consumption. Why does the relationship have this direction?

c) Draw a line that you think best describes the pattern of these data. What is the equation of your line?

d) Use your equation from (c) to predict the gas used during a month when the average temperature is 60° F.

2. The 11 members of a college women's golf team play a practice round, then the next day play a round in competition on the same course. Their scores appear in **Figure 2.22**. (A golf score is the number of strokes required to complete the course, so low scores are better.)

Player	1	2	3	4	5	6	7	8	9	10	11
Practice	89	90	87	95	86	81	105	83	88	91	79
Competition	94	85	89	89	81	76	89	87	91	88	80

FIGURE 2.22.
Golf scores.

a) Make a scatter plot that allows you to study how well the competition score can be predicted from the practice score.

b) Describe the characteristics between practice and competition scores. In particular, is there a positive or negative relationship? Explain why you would expect the scores to have a relationship like the one you observe.

c) One point falls clearly outside the overall pattern. Circle this point in your plot. A good golfer can have an unusually bad round, or a weaker golfer can have an unusually good round. Can you tell from the data given whether the unusual value is produced by a good player or a poor player? What other data would you need to distinguish between the two possibilities?

d) You might expect a player to have about the same score on two rounds played on the same course. Draw on your graph the line that represents the same score on both days. Does this line fit the data well when you ignore the outlier? If you don't like this line, draw a line that you would prefer to use to predict the competition score from the practice score.

e) Another golf team member shot a 95 in practice. Predict her score in competition.

When the relationship between two variables is **strong** and has linear form, then the points in a scatter plot will fall very close to a line. For weaker relationships, the data are more scattered.

3. **Figure 2.23** contains two displays (scatter plots) showing the relationship between human height and the ulna length (forearm-bone length). One of the scatter plots is based on real data and the other on fictitious data. Which do you think is which? In both displays, a line describing the pattern of the data has been added to the scatter plot.

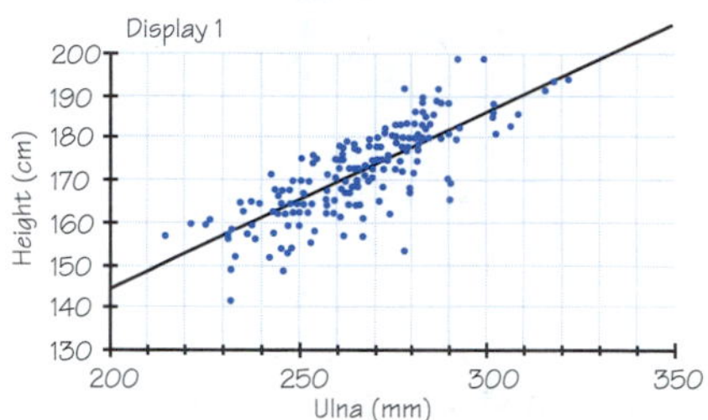

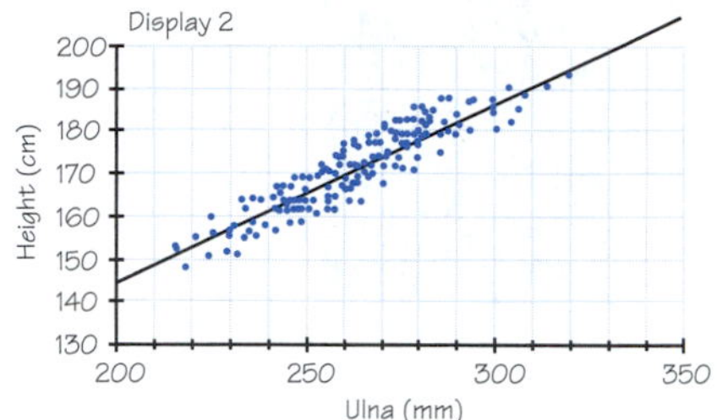

FIGURE 2.23. Two scatter plots showing height versus ulna length.

a) Both relationships have a linear form. Which of the two displays shows a stronger linear relationship? Explain.

$\hat{ht} = 60.2 + 4.2(29) = 182$cm

b) Suppose that a scientist used the model $y = 60.2 + 4.2x$ (where x is ulna length in cm and y is height in cm) to predict the height of a dead person whose ulna length of 29.0 cm was recorded in Figure 2.23. What did the scientist predict as his or her height?

c) How precise is the scientist's prediction if Display 1 shows the real data? How precise is the scientist's prediction if Display 2 shows the real data? Which display better supports the scientist's prediction?

d) Describe the connection between the strength of a linear relationship and the degree of precision with which you can make predictions using that scatter plot.

ACTIVITY 2.5

The Plight of the Manatee

In Activity 2.4 you learned two methods for finding a line that approximates the pattern of data. You used the equations of the lines you found as models of the data. In this activity you will learn about a method that statisticians have developed to find a line that is the best approximation of the pattern of data.

MANATEES AND POWER BOATS

On the coast of Florida lives the manatee, a large, appealingly ugly, and friendly marine mammal. However, the gentle Florida manatee does not live a carefree life. It is one of the most endangered marine mammals in the United States. One major threat to the manatee's survival is the large number of manatees killed each year by power boats.

Should the Florida Department of Environmental Protection limit the number of registered boats in order to protect the manatee population? Before they decide to limit the number of registrations, they will have to present a convincing argument to the public.

In this activity you must make a convincing case for whether or not to restrict power boats. These are the main steps in building a convincing argument to present to the authorities:

- Find a model that describes the relationship between manatee deaths and power boat registrations.
- Show that the model does a good job of describing the data.
- Use the model to make your prediction.

FINDING THE LEAST-SQUARES MODEL

Figure 2.24 contains data on the number of power boats registered in Florida (in thousands) and the number of manatees killed.

Year	Power Boat Registrations (in thousands)	Manatees Killed
1977	447	13
1978	460	21
1979	481	24
1980	498	16
1981	513	24
1982	512	20
1983	526	15
1984	559	34
1985	585	33
1986	614	33
1987	645	39
1988	675	43
1989	711	50
1990	719	47

FIGURE 2.24.
Power boat registration and manatee deaths in Florida, 1977–1990.

1. To make a recommendation about the number of power boat registrations, you must find a model that represents the relationship between manatee deaths and power boat registrations.

   a) Which measure, power boat registrations or manatee deaths, should be the explanatory variable and which should be the response variable? Explain why.

   b) Enter the data from the last two columns of Figure 2.24 into calculator lists (or columns A and B of a spreadsheet). Plot manatee deaths versus power boat registrations. Are the two variables positively or negatively related? What does this mean?

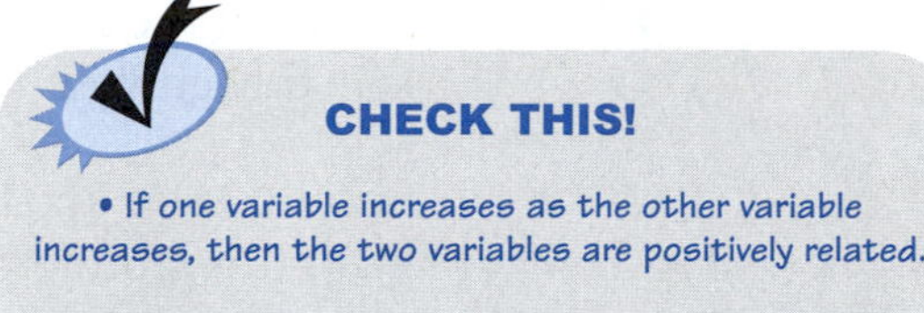

CHECK THIS!

- If one variable increases as the other variable increases, then the two variables are positively related.
- If one variable increases as the other variable decreases, then the two variables are negatively related.

Your graphing calculator (as well as most scientific calculators and spreadsheets) has built-in commands that calculate the equation of the **least-squares line**. Statisticians frequently select this line as the line that best approximates the pattern of the data. They call this line the **best-fitting line**.

How do statisticans find the least-squares line? According to the least squares criterion, each line that approximates the pattern of some data is given a score. To calculate a line's score, statisticians:

- calculate the residual errors
- square the residual errors
- sum the squared (residual) errors

Statisticians frequently refer to this number as the SSE (sum of squared errors). The line that has the smallest SSE is called the least-squares line.

Don't worry if you can't find the term "least-squares line" anywhere on your calculator. Most calculators do not use this term. The general name for determining an equation that fits the data is **regression**. Since you are looking for a linear relationship between the data, the technique used in fitting the "best" line is called **linear regression**.

2. a) Use your calculator (or spreadsheet) to calculate the values of the slope and *y*-intercept of the least-squares line. (If you have not done this before, your teacher will supply instruction.) Write its equation.

   b) What do your slope and *y*-intercept tell you about boats and manatees?

3. Plot both the data and the least-squares line so that you can see both graphs in the same window. Does the least squares line appear to fit the data well? Support your answer.

HOW GOOD IS MY FIT?

To determine whether a linear regression model adequately describes the relationship shown by a scatter plot, statisticians routinely ask if the line meets the following criteria.

- The sum of the errors, also called residuals or residual errors, should be close to 0.
- The pattern of data in the scatter plot should be randomly scattered above and below the line.

4. a) Use your calculator to compute a list of the residuals (errors) for the least squares line. Then calculate the sum of the residuals. Is the sum of the residuals close to 0? (How can you use the statistical features on your calculator to compute this sum most efficiently?)

   b) Does the least squares line appear to satisfy the second criterion above? What does this imply about the residuals?

   c) Using your calculator, make a scatter plot of the residuals versus the number of power boat registrations. Do the dots in your plot look randomly scattered above and below the horizontal line $y = 0$?

CHECK THIS!

A plot of the residuals versus the explanatory variable is called a **residual plot**. The residual plot gives you information about how well your line fits the data. A good scatter plot of the residuals looks like a bunch of dots thrown at a piece of paper in no particular order or pattern. The dots should appear randomly scattered above and below the x-axis.

d) Explain how the residual plot can help you decide if your line satisfied the second criterion.

If the dots in the residual plot do not appear randomly scattered, but instead form a relatively clear pattern, then your model does not describe the data adequately. In this case, you may need to search for another model. Perhaps you can try a line that has a new slope or new intercept or try a model that is not linear.

5. a) Why should you analyze whether your linear regression model adequately describes the data before you use your equation to make predictions?

   b) Does the least-squares line appear to do a good job of representing the relationship between the number of manatees killed and the number of power boat registrations? That is, can you use the least-squares line to make predictions on the number of manatee deaths?

MAKE YOUR RECOMMENDATION

Now that you have a model for the relationship between manatee deaths and power boat registrations, you are ready to make your recommendations to Florida officials.

6. Suppose you want to reduce the number of manatee deaths to about 30 per year. How many power boat registrations would cause about 30 power boat-related manatee deaths each year? Explain how you can use your model and algebra to answer this question.

7. a) Suppose, instead, that you set your recommendation for the number of Florida power boat registrations at 700,000 (slightly below the number of registrations in 1989). Predict the number of manatees that would be killed each year, on average, if this proposal were adopted.

   b) Use your scatter plot from question 3 to analyze the precision of your prediction.

8. Use your answer to question 2(b) to complete the following:

   Every time the number of boat registrations is raised by 50,000 registrations, one can predict that, on average, an additional ________ manatees would be killed by power boats each year. Justify your response.

SUMMARY

Statisticians often use the least-squares line as the best approximation of data.

- You used the residual errors to determine if this line adequately described the relationship between the variables.
- You used the least-squares line to make predictions.

But what if you have two models that use different explanatory variables? Activity 2.6 shows you how to select the one that gives more precise predictions.

HOMEWORK 2.5

Anscombe's Data

The tables in **Figures 2.25–2.28** present four sets of data prepared by statistician Frank Anscombe.

1. Enter each of the four sets of data into a calculator (or your spreadsheet). Notice that for three of the data sets the x-values are the same.

FIGURE 2.25. Data Set A.

x	10	8	13	9	11	14	6	4	12	7	5
y	8.04	6.95	7.58	8.81	8.33	9.96	7.24	4.26	10.84	4.82	5.68

FIGURE 2.26. Data Set B.

x	10	8	13	9	11	14	6	4	12	7	5
y	9.14	8.14	8.74	8.77	9.26	8.10	6.13	3.10	9.13	7.26	4.74

FIGURE 2.27. Data Set C.

x	8	8	8	8	8	8	8	8	8	8	19
y	6.58	5.76	7.71	8.84	8.47	7.04	5.25	5.56	7.91	6.89	12.50

FIGURE 2.28. Data Set D.

x	10	8	13	9	11	14	6	4	12	7	5
y	7.46	6.72	12.74	7.11	7.81	8.84	6.08	5.39	8.15	6.42	5.73

2. Determine the equation of the least-squares line for each of the data sets. Compare your equations for the various sets of data.

3. Make a scatter plot for each of the four data sets and draw the regression line on each of the plots.

4. In which of the four cases would you be willing to use your regression line to predict y given that $x = 14$? Explain.

5. What do you think Anscombe wanted students to learn from the four data sets he created?

6. Data Set D has an outlier, one data point that appears not to follow the general pattern of the data.

   a) Remove that outlier from the data and recalculate the least-squares equation.

   b) Does the least-squares line appear to fit the remaining data better than the equation that you calculated in question 2?

ACTIVITY 2.6

The Case of the Mascot Mystery

Sometimes you may have two or more models that predict some quantity. What if each model uses a different explanatory variable? For example, a model might predict someone's height based on forearm length, and another model might predict height based on the length of a person's stride. In this activity you will learn a method for selecting which model gives the more precise predictions.

WHICH MODEL IS BETTER?

Suppose you have the following two models that predict a person's height:

- A model of the relationship between height and stride length
- A model of the relationship between height and forearm length

Height (cm)	Stride Length (cm)	Forearm Length (cm)
166.0	58.2	28.5
164.5	55.9	27.2
175.0	59.1	28.6
184.0	68.9	30.5
161.0	72.5	26.5
164.0	*	28.2
171.0	*	29.0

How would you determine which model gives the more precise predictions? Let's explore some data to help you decide. Suppose you collected the data in **Figure 2.29** from a ninth grade class. The asterisks (*) indicate missing data.

FIGURE 2.29.
Data on height, stride length, and forearm length.

1. a) Make a scatter plot of height versus stride length. Make a second scatter plot of height versus forearm length. (For both plots, height should be on the vertical axis. Use the same scale on the vertical axis for each plot.) Which of the two scatter plots shows the stronger relationship? How can you tell?

CHECK THIS!

Recall that a linear relationship is a strong relationship if the dots on a scatter plot of the data fall close to a line that approximates the data.

b) Find the least-squares line that represents each of these relationships and record the residual errors in two calculator lists.

Recall that the least-squares line has the smallest SSE (the sum of the squared errors). One way to measure how well a line fits data is to calculate the average of the squared errors.

2. a) Calculate the average of the squared errors for each of your two least-squares lines.

 b) Explain why you would expect the line associated with the smaller average of squared errors to be the graph of the stronger relationship.

CHECK THIS!

Average of squared errors $= \frac{SSE}{n}$

n is the number of data points.

 c) Use the equations of your two least-squares lines to predict the height of a person whose stride is 73 cm and whose forearm length is 27 cm. Which of the two estimates is more reliable? Explain.

 d) State in your own words how you would choose, in general, between two different models that predict the same quantity.

Now that you have a method for selecting between two models that use different explanatory variables, together with the members of your group, solve the following problem.

THE MISSING MANATEE

A school's mascot is stolen (see **Figure 2.30**). The thief has left some clues: a plain blue sweater and a set of footprints under a window. The footprints appear to have been made by a man's sneaker.

The distance between the footprints, from the back of the heel on the first footprint to the back of the heel on the second, reveals that the thief's stride is approximately 58 cm. The length of the thief's forearm can be estimated by measuring the sweater from the center of a worn spot on the elbow to the turn where the cuff meets the sleeve. The thief's forearm is between 26 cm and 27 cm.

FIGURE 2.30.
The missing manatee.

School officials suspect that the thief is a student from a rival high school. You have two tasks: Gather data on stride lengths of students in your class to estimate the thief's height from his or her stride length. Then use the height and forearm data collected in Activity 2.2 to predict, as accurately as possible, the height of the thief from the length of his or her forearm.

3. With your group, discuss methods for getting reliable measurements for stride length. (If you have completed Homework 2.2, base your discussions on that work.) After your group discussion, the class must decide on the method that will be used to collect the footstep data. Write a brief description of the method that will be used to collect the stride-length data.

4. Collect the data. After you finish collecting the data, add your results to Handout 2.2, the *Class Data Recording Sheet* used in Activity 2.2.

5. What is a typical stride length for students in your class? Is there a difference between what is typical for a girl and a boy? Justify your answer based on the stride length data.

6. The footprint was made by a man's sneaker. However, sometimes girls wear men's sneakers, especially high-top sneakers. So you'll need to decide whether or not the thief was male. Use what you know about the thief's stride length and forearm length.

   a) Does the class stride-length data tend to confirm that the footprints belonged to a boy and not a girl? Support your answer using the data you gathered.

   b) Do the forearm data tend to confirm that the sweater belonged to a boy and not a girl? Support your answer using the data you gathered.

Now predict the height of the thief. You have two possible explanatory variables you can use for your prediction: stride length and forearm length. Which is better? Divide the work in questions 7 and 8 among the members of your group.

7. a) Determine a relationship between height and forearm length for your actual data. First, find a model based on the entire data set. Then use the girls' data to find a second model, and finally use the boys' data to find a third model.

   b) Compare the three models. Is there much difference between them? Explain.

   c) Analyze whether the relationship based on the class data is stronger than either relationship based on the single gender data. (What numeric measure will you use to assess the strength?)

   d) Which model gives more precise predictions if, in fact, the thief were male? Justify your answer using your data.

   e) Which model gives more precise predictions if, in fact, the thief were female? Justify your answer using your data.

   f) Make a residual plot for the class model. Do the dots in the plot appear to be randomly scattered, or is a clear pattern apparent?

8. Repeat question 6 for the relationship between height and stride length.

9. Select the best model for the job of predicting the height of the thief. Support your selection. Finally, use this model as the basis for completing the following clue.

   I predict that the thief is ______ cm tall. But the thief might be as short as _____ or as tall as _____.

SUMMARY

This activity showed you how to compare two models that predict a quantity but have different explanatory variables.

- You calculated the average of the squared (residual) errors to determine the strength of the relationship.
- Then you collected some data, built two models and used them to help identify several characteristics of the thief.

HOMEWORK 2.6

Not One of the Bunch

Selecting a line according to the least-squares criterion often produces a line with good properties. That's why selecting a line using the least-squares criterion is so popular. However, sometimes this line does a terrible job of describing the pattern of the data. In such cases you may have to adjust your model to improve how it fits the data.

1. What is the relationship between the number of calories a food actually has and how many calories people think it has? A food industry group surveyed 3368 people, asking them to guess the number of calories in several common foods.

Nutrition Facts

Serving Size 5 Crackers (15g)
Servings Per Container about 30

Amount Per Serving	
Calories 60	Calories from Fat 20
	% Daily Value*
Total Fat 2g	**3%**
Saturated Fat 0.5g	**3%**
Cholesterol 0mg	**0%**
Sodium 190mg	**8%**
Total Carbohydrate 10g	**3%**
Dietary Fiber Less than 1g	**2%**
Sugars 0g	
Protein 1g	
Vitamin A 0% •	Vitamin C 0%
Calcium 0% •	Iron 2%

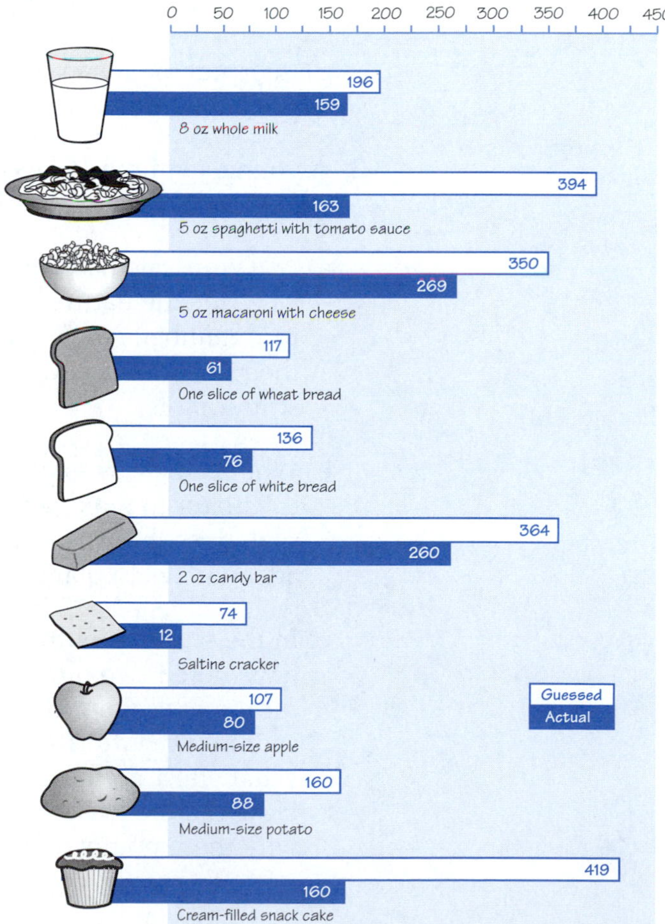

FIGURE 2.31.
Guessed calories and actual calories. [USA TODAY, October 12, 1983]

a) The goal is to predict the guessed calories based on the actual calories. Enter the data into your calculator and make a scatter plot with this in mind. (Which variable is the explanatory variable?)

b) Describe in words the most important features of the scatter plot.

c) Find the regression line for predicting guessed calories from actual calories. Then make a residual plot. Does the regression line adequately describe these data?

d) Would you classify any of the data as outliers? If so, identify them. What do they tell you?

e) If you found outliers, remove them and recalculate the regression line. Compare your new equation with the one in (c).

f) Do the calories in a food enable us to predict accurately what people will guess? Explain.

g) Interpret the meaning of the slope of your model for predicting guessed calories from actual calories.

2. A young swimmer's favorite stroke is the butterfly. Her times are listed in the table in **Figure 2.32**.

   a) Use your calculator to make a scatter plot of the data. Describe the nature of the relationship between time and race number. Are any outliers apparent? If so, describe their general location relative to the non-outliers in the scatter plot.

   b) What is the least-squares line for these data? Use your calculator to make a residual plot. Based on the residual plot, does this linear model appear to describe the data adequately? Explain.

   c) In the scatter plot that you observed in (a), you should have noted two outliers. These correspond to the times for the first two races. (This swimmer had just started swimming butterfly and her times were unusually slow.) What effect do these points have on the least-squares line? How can the least-squares criterion be used to explain why these points had this effect?

d) How do you think the least-squares line would change if the two outliers were removed from the data? Explain your reasoning.

e) Remove the outliers from the data. What is the equation for the least-squares line now? Again, use your calculator to make a residual plot. Based on the residual plot, does this linear model appear to describe the data adequately (with the exception of the outliers)? Explain.

f) Based on the model from (e), predict the swimmer's swim time for her 15th race and for her 16th race. On average, by how much are her times decreasing from race to race? Can this pattern continue indefinitely? Explain.

Race Number	50-Yard Butterfly Time (sec)
1	60.81
2	66.11
3	47.32
4	42.69
5	43.40
6	44.82
7	42.67
8	45.17
9	41.20
10	43.68
11	42.47
12	41.74
13	40.40
14	42.90

FIGURE 2.32.
Butterfly times.

Modeling Project Who Am I?

You are an anthropologist who finds bones of two or more individuals. The table in **Figure 2.33** contains the information about those bones.

Skeleton 1 (Possibly female)	Skeleton 2 (Taller of the two)	Uncertain
Femur: 413, 414	Femur: 508	Skull: 230
Ulna: 228	Ulna: 290	Humerus: 357
		Radius: 215
		Tibia: 416

FIGURE 2.33.
Classification of bones discussed in Preparation Reading. Measurements in millimeters (mm).

Figure 2.34 contains actual data from the Forensic Anthropology Data Bank (FDB) at the University of Tennessee. The FDB contains metric, nonmetric, demographic, and other kinds of data on skeletons from all over the United States. These individuals most likely came through the medico-legal channels as unidentified bodies, then went to forensic anthropologists for analysis and identification.

Use the data in Figure 2.34 to answer the following items. Present your findings in a formal report. All of your conclusions must be supported by statistical analysis using the data in Figure 2.34.

1. Determine several models to predict people's height from the lengths of various long bones in their arms and legs. Explain which of these models you would prefer to use and why.

2. Based on these data, do you agree that Skeleton 1 is female? Do the data provide any information that would help you determine whether Skeleton 2 is male or female?

3. Determine relationships between pairs of long bones that would help you decide whether the bones in the "uncertain" column belong to Skeleton 1 or Skeleton 2. (Or is there strong evidence that one of these bones belongs to a third person?)

4. Predict the heights of Skeleton 1 and Skeleton 2. Explain why you chose the model that you did to make your predictions.

1	168	307	240	258	448	384	368	1	165	307	230	248	452	363	355	2	178	337	272	272	475	393	390
1	178	336	247	261	463	404	390	1	163	297	240	260	435	356	356	2	172	344	255	281	470	400	393
1	161	294	213	227	413	335	322	1	143	282	216	233	398	334	318	2	188	360	269	283	510	422	416
1	155	324	262	279	465	395	375	1	154	297	228	248	423	344	334	2	189	347	272	283	547	432	445
1	165	314	243	258	432	364	364	1	171	342	272	290	485	418	407	2	177	330	246	262	462	386	370
1	168	303	223	244	441	355	342	1	162	303	237	262	433	367	364	2	166	322	242	258	442	373	374
1	165	311	231	254	436	362	360	1	150	308	220	247	383	352	341	2	186	332	267	283	478	391	388
1	173	312	248	266	483	405	401	1	157	288	201	215	429	363	350	2	177	322	245	265	457	397	395
1	165	322	229	246	448	368	352	1	158	314	239	263	432	371	358	2	176	332	259	274	458	382	378
1	163	298	221	245	443	355	361	1	162	306	250	268	444	355	352	2	180	323	251	275	448	390	387
1	153	280	218	234	410	345	344	1	159	310	238	255	449	362	352	2	173	335	253	273	497	404	389
1	165	294	220	235	448	354	353	2	169	337	254	273	460	396	385	2	175	330	253	274	470	384	382
1	170	311	235	253	440	360	347	2	153	296	223	243	407	337	338	2	169	313	252	265	472	391	385
1	160	316	214	226	437	356	348	2	175	339	256	271	470	390	381	2	175	336	256	274	464	388	377
1	159	292	223	233	419	346	336	2	179	343	242	263	464	378	371	2	181	390	284	303	521	440	435
1	163	315	228	251	438	356	347	2	179	352	253	269	484	407	397	2	193	356	297	318	522	451	433
1	165	303	237	249	451	356	348	2	198	354	263	292	508	417	412	2	182	362	275	293	499	424	405
1	165	308	234	248	439	348	344	2	173	327	256	276	463	383	387	2	169	322	249	266	426	366	356
1	165	315	227	240	448	363	353	2	180	357	268	278	494	401	390	2	180	337	265	281	482	412	399
1	175	316	244	260	473	390	374	2	178	344	254	269	464	371	366	2	185	363	286	302	520	429	420
1	180	333	256	278	475	391	381	2	175	339	245	272	456	374	366	2	180	355	274	292	490	422	424
1	168	321	230	248	450	365	362	2	177	343	250	266	483	361	365	2	170	378	272	291	512	404	390
1	163	299	219	236	435	357	339	2	180	353	260	281	490	420	415	2	180	370	278	292	523	429	420
1	165	304	246	264	467	392	383	2	170	303	235	249	435	366	361	2	175	333	260	273	484	398	386
1	160	309	236	248	432	364	358	2	191	364	263	278	511	430	417	2	168	342	262	280	484	404	385
1	158	319	246	268	442	371	364	2	188	349	269	288	498	427	423	2	170	347	269	291	476	396	393
1	165	325	242	250	448	378	365	2	179	323	256	276	486	398	400	2	166	315	240	260	456	377	362
1	170	335	248	263	474	400	382	2	180	350	263	280	480	419	418	2	185	363	295	309	524	446	427
1	182	334	254	273	514	420	407	2	181	350	263	282	488	391	381	2	191	382	299	316	537	479	466

FIGURE 2.34.
Data from Forensic Anthropology Data Bank (FDB).

Key to Data (in order from left to right): sex (1 = female, 2 = male), height (cm), humerus (mm), radius (mm), ulna (mm), femur (mm), tibia (mm), fibula (mm)

Practice Problems

Review Exercises

1. For each of the descriptions that follow, determine an equation of a line that satisfies it.

   a) The line through the point (2, 3) that has slope $\frac{1}{2}$.

   b) The line that has y-intercept 3 and slope $-\frac{1}{2}$.

   c) The line passes through points (3, 2) and (2, 5).

   d) The line has y-intercept 5 and is parallel to the line $y = 7x - 1$.

2. On December 18, 1994, three amateur spelunkers (cave explorers) stumbled across a cave in France, now known as Chauvet Cave, that contained ancient cave paintings. In later explorations, human footprints were found. According to prehistorian Michel-Alain Garcia, the footprints belonged to a boy who was about 4.5 feet tall and lived between 20,000 and 30,000 years ago.

   a) Scientists have used simple models to predict height from footprints since the mid-1800s. One model, still in use today, predicts height by dividing the maximum foot length by 0.15. Write a formula that describes this model.

   b) The model in (a) can be expressed as a member of the $y = mx$ family. What is the value of m?

   c) Using this model, predict the maximum length of the footprints found in Chauvet Cave. Give your answer in centimeters. (Remember 2.54 cm = 1 in.)

   d) Suppose that another footprint was found and it measured 1 cm more than the one discussed by Garcia. By how much would you increase your estimate for height? (Be careful of the units you use to report your answer.)

3. Anthropologists have refined early models for estimating a person's height from the length of a footprint. One revision suggests using different models depending on whether the

footprint was from the right foot or from the left. Here are two models designed for use when you assume the footprints are from adults.

For right foot: $H = 3.641L + 72.92$

For left foot: $H = 4.229L + 56.49$

where all measurements are in cm.

a) Suppose a footprint measured 22 cm. Predict the person's height, first by assuming the print was from the right foot and then by assuming it was from the left foot.

b) Use your calculator to graph each of these models. Based on your graphs, what length for a footprint leads to the same height prediction from both models? How could you determine this value using algebra? Try it and check to see if you get the same results.

c) Both the right-foot and left-foot models are from the $y = mx + b$ family. For each of the models, interpret the meaning of m in the height-footprint context. What does b mean in this context?

4. a) Solve the right-foot equation in question 3 for L in terms of H. Then do the same for the left-foot equation.

b) Suppose a person was 153 cm tall. Use your equations from (a) to predict the lengths of the person's right and left footprints. How much longer is the larger foot than the shorter foot?

5. Regular Chips Ahoy® chocolate chip cookies boast "1000 chips in every bag." You can also buy reduced fat Chips Ahoy cookies. Do you think both types of Chips Ahoy cookies contain roughly the same amounts of chips? To find out, a statistics class opened bags of regular Chips Ahoy and reduced fat Chips Ahoy cookies, randomly selected 15 cookies from each bag, and counted the number of chips in each cookie. Their data appear in **Figure 2.35**.

Reduced Fat	13	15	14	12	15	17	13	10	15	18	19	18	20	21	16
Regular	20	17	22	20	16	21	18	16	19	27	22	19	16	24	

FIGURE 2.35.
Chip counts in reduced fat and regular Chips Ahoy cookies.

a) From the data in Figure 2.35, make two dot plots using the same scaling on each. Place one dot plot directly above the other.

b) Suppose you selected a cookie from the regular Chips Ahoy bag. Predict the number of chips in the cookie. How far off might your prediction be?

c) Suppose instead, you selected a cookie from the reduced fat bag. Predict the number of chips in the cookie. How far off might your prediction be?

d) Does it appear that in order to produce a lower fat product, the number of chips is affected? Explain.

e) George wanted to make a scatter plot of the row 1 data in Figure 2.35 versus the row 2 data. Do you think his scatter plot would reveal useful information about these two types of cookies? Explain.

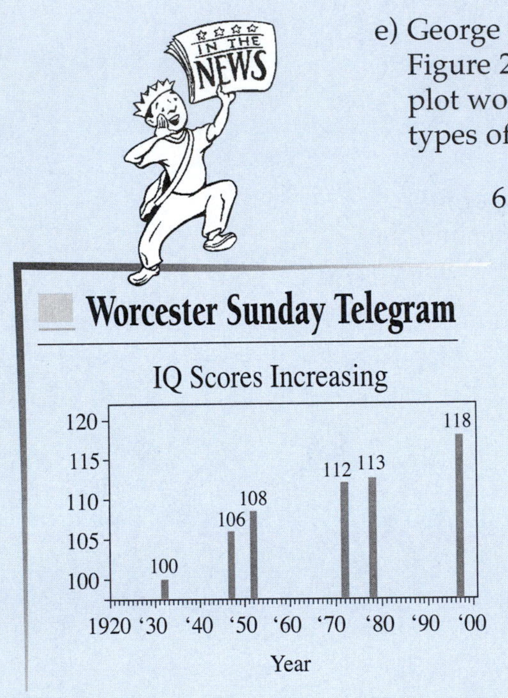

FIGURE 2.36. Average IQ scores from 1932 to 1997.

6. A newspaper article in the *Worcester Sunday Telegram* reported that scores on intelligence tests are going up at a rate of 3 IQ points every decade.

a) In 1932 the average IQ test score for Americans was 100 points. Use the information given in the article to write a model predicting the average IQ score for years after 1932. Let $x = 0$ represent 1932 to simplify your model.

b) **Figure 2.36** provides data on the average IQ scores taken by Americans. Make a scatter plot of these data. (Let $x = 0$ represent 1932.)

c) Based on these data, do IQ scores and years have a positive or negative relationship? Explain.

d) Fit a least-squares line to these data. What is the equation for this line? Add this line to your scatter plot in (b).

e) Based on the least-squares equation from (d), was the newspaper correct in reporting that IQ scores were rising at a rate of 3 points every decade? Explain.

f) You are considered to be a genius if you have an IQ over 140. In what year will the average IQ be at the level of genius? Do you believe this? Explain.

The Boston Marathon is the world's oldest and best-known marathon. The first Boston Marathon was held in 1897. Fifteen men participated. Over 17,000 people entered the 104th Boston Marathon held April 17, 2000. Questions 7–9 are based on the first-place times contained in **Figure 2.37**.

Women's	
Marathon Number	**Time (hr: min: sec)**
82	2:34:28
83	2:26:46
84	2:29:33
85	2:22:42
86	2:29:28
87	2:34:59
88	2:24:55
89	2:25:21
90	2:24:30
91	2:24:33
92	2:25:24
93	2:24:18
94	2:23:43
95	2:25:27
96	2:21:45
97	2:25:11
98	2:27:13
99	2:58:00
100	2:40:10

Men's	
Marathon Number	**Time (hr: min: sec)**
31	2:40:22
36	2:33:36
41	2:33:20
46	2:26:51
51	2:25:39
56	2:31:53
61	2:20:50
66	2:23:48
71	2:15:45
76	2:15:30
81	2:14:46
86	2:08:51
91	2:11:50
96	2:08:14

FIGURE 2.37. Men's and women's Boston Marathon times.

7. Focus on the men's marathon. The men's data in Figure 2.37 contain times of the first place finishes for selected marathons from 1927, marathon 31, to 1992, marathon 96.

   a) Make a scatter plot of the men's times versus the marathon number. (What unit did you use to measure time?)

   b) Fit a least-squares line to the data in your scatter plot. What is the equation? Add a graph of the least-squares line to your scatter plot.

c) Is the least-squares equation adequate to describe the pattern in your data? Base your answer on a residual plot.

d) Interpret the slope of your linear model in the context of the Boston Marathon.

e) Use your model to predict the time of the first-place finisher for the 104th Boston Marathon. The actual winning time was 2:09:47. How close was your prediction?

8. Also Figure 2.37 shows women's times for marathons 82–100, held in years 1980–1996. Enter the women's data into your calculator or spreadsheet.

   a) Fit a least-squares line to these data. Interpret the slope in the context of the marathon. Does this seem reasonable? Explain.

   b) Make a scatter plot of the women's data. Add the least-squares line to your plot. Does the least-squares equation do a good job of describing the women's data?

   c) Make a residual plot. Based on the plot, is the least-squares line adequate to describe these data? Do you think you could make good predictions using this model? Explain.

   d) Remove the times corresponding to the 99th and 100th marathons. Refit the least-squares line to the remaining data. What affect did the removal of these points have on the slope?

Men's	
Marathon Number	**Time (hr: min: sec)**
82	2:40:10
83	2:26:57
84	2:38:59
85	1:55:00
86	2:00:41
87	1:51:31
88	1:47:10
89	2:5:20
90	1:45:34
91	1:43:25
92	1:55:42
93	1:43:19
94	1:36:40
95	1:29:53
96	1:30:44
97	1:26:28
98	1:22:17
99	1:25:59
100	1:30:11

Women's	
Marathon Number	**Time (hr: min: sec)**
82	3:48:51
83	3:52:53
84	3:27:56
85	2:49:40
86	2:38:41
87	2:12:43
88	2:27:7
89	2:26:51
90	2:05:26
91	2:09:28
92	2:19:55
93	2:10:44
94	1:50:6
95	1:43:17
96	1:42:42
97	1:36:52
98	1:34:50
99	1:40:41
100	1:52:54

FIGURE 2.38. Men's and women's wheelchair times in the Boston Marathon.

9. a) Make scatter plots for the men's and women's wheelchair times in **Figure 2.38**. So that you can compare the men's and women's data, use the same scale on the axes of both plots.

   b) Is there a positive or negative relationship between marathon number and times for the men's data? What about for the women's data? Interpret what this means in the context of the marathon.

   c) Comment on the form of the data. Does the pattern of the data appear linear or nonlinear? Explain.

   d) Fit least-squares lines to both data sets and then make residual plots. Do the pattern of the dots in your residual plots confirm your answers to (b)? Explain.

The data in **Figure 2.39** provide information on the heights of people as children and again as adults. Use the data in this figure to answer questions 10–13.

Girl's Height at 1.5	Girl's Adult Height	Boy's Height at 2	Boy's Adult Height
78.0	157.0	89.0	178.0
79.4	158.4	89.9	177.1
80.4	161.4	90.3	179.6
81.3	164.7	90.8	181.8
81.3	160.4	90.9	184.0
82.1	163.7	91.0	180.5
83.2	164.4	91.1	182.0
83.2	170.2	91.2	183.1
83.9	170.5	91.4	180.1
84.9	166.5	91.5	185.1
86.2	171.3	92.9	182.0
87.9	170.7	93.3	186.3
88.2	179.7	94.7	187.4
89.4	176.9	95.4	187.9
90.1	176.9	96.1	189.4

FIGURE 2.39.
Height data on children and adults.

10. Suppose one of the people from this study planned to visit your school.

    a) If you find out the visitor is a woman, predict her height. How did you decide on your prediction?

    b) What if the visitor is a man?

11. Create a display that compares the men's heights with the women's heights. Write a description of the information your display conveys.

12. Suppose you wanted to predict how tall a 1 1/2-year-old girl would be when she reached adulthood.

    a) Which is the explanatory (independent) variable and which is the response (dependent) variable?

    b) Make a scatter plot of the relationship between women's adult heights and their heights when they were 1 1/2 year old.

    c) Would you describe the relationship between women's heights and girls' heights as linear or nonlinear? Positive or negative?

    d) Fit a least-squares line to the data on your scatter plot. Write its equation, and sketch its graph on your scatter plot.

    e) Make a residual plot. Based on your residual plot, does the least-squares line appear to describe the relationship between women's adult height and childhood height adequately? Explain.

    f) Use your equation to predict the adult height of a 1 1/2-year-old girl who is 82.5 cm tall.

13. a) Determine the least-squares line for predicting men's heights from their heights when they were 2 years old.

    b) If two 2-year-old boys differ in height by 1 cm, predict how much their heights will differ when they are adults.

    c) What if their heights as 2-year-olds differ by 2 cm?

    d) Does the y-intercept of the least-squares line have any meaning in this context? Explain.

    e) Does the slope of the least-squares line have any meaning in this context? Explain.

14. Suppose the grades in **Figure 2.40** are from students in a high school statistics class.

    a) What is the average grade on the midterm exam? On the final exam?

    b) Make a scatter plot of these data.

    c) Does the pattern in your scatter plot appear to have linear form? Are there any outliers? Does there appear to be a positive or negative relationship between students' midterm grades and students' final grades?

    d) Fit a least-squares line to the data. If you decide to remove any outliers before fitting the line, be sure to report them and explain why you removed them.

    e) Add a graph of your model from (d) to your scatter plot. Do you think your line does a good job in describing these data? Explain.

    f) One student missed the final exam and had to take it at a later date. She got 77 on the midterm. Predict her grade on the final exam.

Name	Midterm exam grade	Final exam grade
	50	54
	53	52
	57	59
	62	70
	62	68
	68	70
	69	68
	73	79
	73	82
	74	83
	74	80
	75	82
	78	88
	79	89
	79	90
	81	88
	81	89
	81	92
	82	91
	85	94
	87	98
	100	99
	100	100

FIGURE 2.40.
Midterm and final exam grades.

2 CHAPTER REVIEW

Predicting Height from Long-Bone Lengths

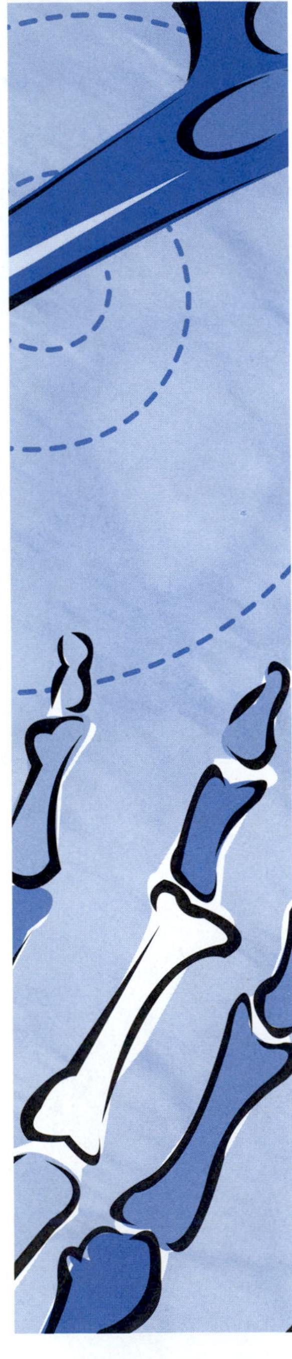

Scatter Plots

In many situations you might be asked questions such as "Are values of quantity 1 related to values of quantity 2?" For example, a forensic scientist might ask, "Is height related to femur length?" In general, such questions suggest the use of graphs called scatter plots.

Since the question implies that one quantity might help predict values of the other quantity, it is common to refer to the quantities as explanatory and response variables, respectively (or the independent and dependent variables, respectively). A scatter plot is a graph in which the values of the dependent variable are represented on the vertical axis and the values of the independent variable are represented on the horizontal axis. It is also referred to as a graph of the dependent variable versus the independent variable.

Choose the scatter plot when you want to look for patterns in a relationship between two quantities.

Linear Relationships

Linear relationships between two variables can be described by graphs, equations, and tables.

- Graphs of linear relationships are lines.
- The amount of "tilt" in the graph of a line is measured by the slope of the line. A line with slope of zero is horizontal; the farther from zero the slope, the steeper the graph of the line.
- Here are two common forms of linear equations:
  the slope-intercept form, $y = mx + b$,
  the point-slope form, $y - k = m(x - h)$.

Given any two points on a line, you can determine the value of m, the slope of the line, by computing the ratio $\Delta y / \Delta x$ between the two points. In each of these forms, the slope appears as the number multiplying the independent variable, x.

- A table of ordered pairs represents a linear relationship if a plot of the x, y-values lies on a straight line, or if the slope has the same value when calculated using any two ordered pairs from the data.

Equivalence

Two linear equations are equivalent if they have the same slope and both pass through the same point. For example, the graph of the equation

$$y - 5 = 3(x - 1)$$

is a line that passes through the point (1, 5) and has slope 3. The equation

$$y = 3x + 2$$

is equivalent because the slope is 3, and the point (1, 5) is a solution:

$$y = 3(1) + 2$$

$$y = 5.$$

Fitting and Evaluating Equations

The main question of this chapter is, "How can you identify and describe a relationship between two variables so that you can predict values of one variable from values of the other?"

First collect data on the two variables. As noted above, a scatter plot is a useful display for gaining insight into possible relationships. Next, from the scatter plot, check the direction (positive, negative, or neither) and the form (linear or nonlinear) of the relationship.

If a scatter plot has a linear form, you can fit a line to the data and use the equation of your line to make predictions. The principal tool in evaluating the fit of your line is the set of residual errors—the differences between the actual and predicted values of the dependent variable. Different criteria based on the residual errors can be used to determine the "best-fitting" line. Unfortunately the best-fitting line according to one criterion is not always the best according to another. However, a "good" fit should always have residuals that are randomly scattered around the horizontal axis.

One of the most commonly used criteria for determining the best-fitting line is called the least-squares criterion. The least-squares line has the smallest sum of the squared errors (residuals). Also called the regression line, it is popular because it generally does a good job of describing data that have a linear form. However, when outliers are present or when the scatter plot does not have a linear form, the least-squares line, or any other line, does a poor job of describing the pattern of a scatter plot.

A plot of the residuals versus the independent variable can be very helpful in spotting outliers or nonlinear data.This plot can display outliers more prominently than a scatter plot of the original data. Also if the data have a nonlinear form, a residual plot will show a strong pattern.

When outliers are present, removing the outliers and refitting a linear model to the remaining data may produce a better prediction model. However, when data have a nonlinear form, no line will adequately describe the pattern of the data. In this situation look for a different kind of model.

▪ The Precision of a Prediction

The precision of a prediction is linked to the variability inherent in the data. For example, suppose you had the following data on student heights (in cm): 150, 152, 154, 156, 158. If you were asked to predict the height of a student in this group, you might decide to chose the mean height of 154 cm for your prediction. In this case, the actual height could be as short as 150 cm or as tall as 158 cm; so you could be as far off as 4 cm. You can use a similar approach when dealing with relationships between two variables by examining the variability in the residuals.

▪ Choosing Between Two Linear Models

In some situations, you may have two independent variables that are linearly related to the same dependent variable. In this case, it is generally best to base your predictions on the independent variable that has the stronger linear relationship with the dependent variable. Strong relationships have low variability, so one way of determining the strength of the linear relationship is to use the sum of the squared errors. For example, you could select the least-squares line associated with the independent variable that has the smaller sum of square residuals. If the data on the two independent variables contain different numbers of observations, select the least-squares line associated with the independent variable that has the smaller average squared error.

Average: To find the average of a data set, sum the data and divide by the number of data in the set.

Dependent variable: The variable that is to be predicted; the variable that "responds" to changes in the independent variable. Mathematicians frequently use the letter y to represent this variable.

Dot plot: Display in which dots are placed above a number line to represent the values of data for a single variable.

Independent variable: The variable on which a prediction is based; the variable that explains the dependent variable. Mathematicians frequently use the letter x to represent this variable.

Least-squares criterion: Choose the line with the smallest sum of squared errors (SSE).

Least-squares line: The line that satisfies the least-squares criterion.

Linear equation: An equation relating two variables, x and y, that can be put in the form $y = mx + b$.

Linear form: The form of a scatter plot for which it is possible to draw a line that describes the general flow of the data.

Linear regression: Fitting a line to data using the least-squares criterion.

Mean: To compute the mean, sum the data and divide by the number of data. Sometimes "mean" and "average" are used interchangeably.

Negative relationship: A relationship between two variables in which one variable tends to decrease while the other increases.

Nonlinear form: The form of a scatter plot in which the general pattern of the data is not well described by a straight line.

Outlier: In a collection of data, an individual data point that falls outside the general pattern of the other data.

Point-slope form: $y - k = m(x - h)$; a form for a linear equation where (h, k) is a point on the line and m is the slope of the line.

Positive relationship: A relationship between two variables in which both variables tend to increase together.

Regression: Fitting lines or curves to data.

Residual errors: Actual value of the dependent variable minus the predicted value.

Residual plot: A scatter plot of the residuals versus the independent variable.

Scatter plot: A plot of ordered pairs of data.

Slope-intercept form: $y = mx + b$; a form for a linear equation where m is the slope and b is the y-intercept.

SSE: The sum of the squared residual errors.

Strong relationship: A scatter plot of the data lies in a narrow band.

Weak relationship: A scatter plot of the data does not lie in a narrow band; the data points are more scattered.

Versus: When used in the phrase y versus x, it describes a scatter plot of y and x in which y is the dependent variable and x is the independent variable.

3
CHAPTER
It Looks Like Rain
ACTIVITY 3.1
FIRST TIME THROUGH: MAKING A SIMPLE MODEL OF RAINFALL
ACTIVITY 3.2
GAUGING INFLUENCE
ACTIVITY 3.3
CAUTION: UNDER CONSTRUCTION
ACTIVITY 3.4
AREA OF CONCERN: CALCULATING AREA
ACTIVITY 3.5
GET THE POINT?
ACTIVITY 3.6
AT THE BOUNDARIES
CHAPTER REVIEW

Why do ranchers and farmers receive a lot of help from geometry?

Economists and government officials regularly use mathematical models to make predictions. For example, a prediction of the number of computer industry jobs created in the next few years is usually based on a model. Even weather forecasters build complicated mathematical models to predict the weather.

Weather has a large impact on many businesses and other organizations. For example, farmers need rain to grow their crops, and ranchers need rain for their herds. Even sports teams are directly affected by the local weather if they don't own domed stadiums. More importantly, public water systems rely on rain to replenish their water supplies. The Texas Natural Resources Conservation Commission (TNRCC) is the state agency that monitors drought conditions and helps public water systems manage their water supplies.

In this chapter you will investigate how to use the TNRCC rainfall data to calculate an estimate for rainfall. You will develop a simple model and test it to discover its weaknesses. Then you will use some geometry to develop an improved model. As part of the model building, you will learn ways to break the state into regions. Unfortunately, these regions usually are not simple triangles and rectangles. Thus you will need to learn how to calculate the areas of regions of various shapes. At the end of the chapter you will obtain an estimate for total rainfall in a state.

PREPARATION READING

Rain Falls Mainly on the Plain?

One of the most important duties of the Texas Natural Resources Conservation Commission (TNRCC) is estimating the rainfall for the entire state. This estimate is vital to many people in Texas, because the state relies on a series of reservoirs for its water. (A reservoir is a lake that is used to supply water to a region.) When there is not enough rainfall, many reservoirs may fall below safe levels. If this seems likely, Texas may have to purchase water from a neighboring state.

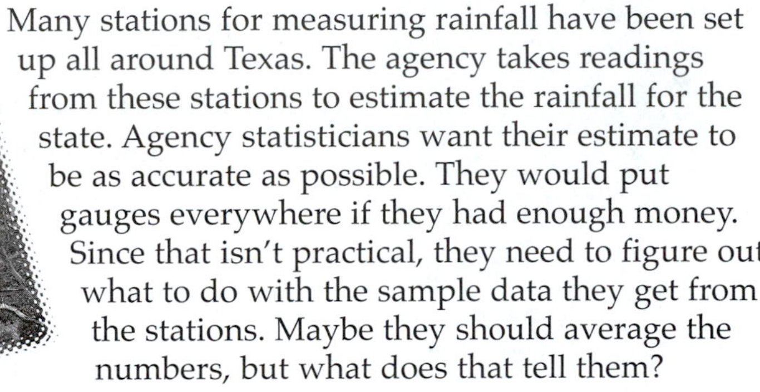

Many stations for measuring rainfall have been set up all around Texas. The agency takes readings from these stations to estimate the rainfall for the state. Agency statisticians want their estimate to be as accurate as possible. They would put gauges everywhere if they had enough money. Since that isn't practical, they need to figure out what to do with the sample data they get from the stations. Maybe they should average the numbers, but what does that tell them?

Before they make an estimate, the statisticians need to decide how to report rainfall. Here are two ways to interpret the term "rainfall."

- Rainfall is the *average depth* of the water that falls on Texas. In other words, if you think of the state as large as a wading pool, how deep would the pool be?
- Rainfall is the *total amount* of water that falls on the state.

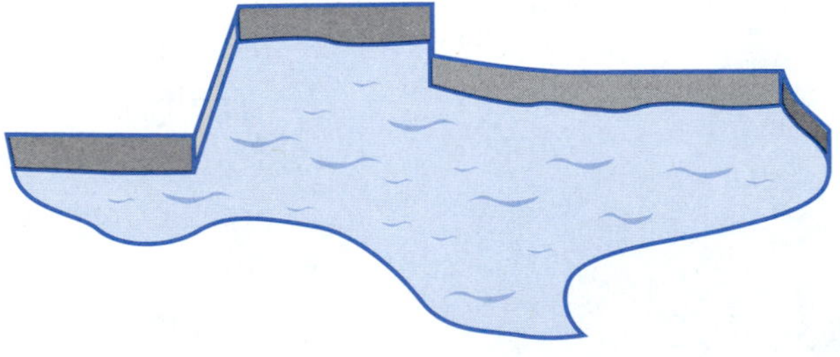

To solve a problem such as this, you need to create a mathematical model. As mathematical modelers you need to use the modeling process that was introduced in Chapter 1.

MATHEMATICAL MODELING

To create a mathematical model, use the following steps.

STEP 1. Identify the situation.
Read and ask questions about the problem.

STEP 2. Simplify the situation.
Make assumptions and note the features that you will ignore at first.

STEP 3. Build the model and solve the problem.
Describe in mathematical terms the relationships among the parts of the problem, and find an answer to the problem.

STEP 4. Evaluate and revise the model.
Check whether the answers make sense; i.e., test your model.

You should test your model to determine which assumptions control important parts of the model. You can change assumptions and go through the entire modeling process over and over. Ideally, you can improve your model to get "better" answers and more realistic mathematical descriptions of the problem. Good luck!

ACTIVITY

3.1

First Time Through: Making a Simple Model of Rainfall

We will begin by building a first model to estimate the amount of rainfall in a region. In this activity the problem faced by the TNRCC will be simplified so you can get a model started. This simplified problem will allow you to find some ways to tackle the Texas problem. After developing the model, you will test it and consider ways to improve it.

PRELIMINARIES

Sometimes it's a struggle to get started on modeling a complex situation. The steps of the modeling process are a useful guide to exploring a problem and describing it in mathematical terms. Because you need to know what you're looking for, the first step is to identify the problem.

1. Based on the Preparation Reading, what is the problem you are investigating in this chapter?

CHECK THIS!

You simplify a problem by changing it into a simpler one that you know how to solve. Solving the simpler problem may give you some ideas about the original one, or how to solve it.

To begin the modeling process, we will use a strategy of simplifying the problem. **Figure 3.1** contains data from *some* of the official rainfall gauges for Colorado. The readings are measurements of rainfall in inches.

FIGURE 3.1. Rainfall data for Colorado.

Gauge 1	Gauge 2	Gauge 3	Gauge 4	Gauge 5	Gauge 6	Gauge 7	Gauge 8
2.11	2.15	1.21	4.45	2.67	2.51	3.11	2.43

Water resource officials might use this data to estimate the total amount of rainfall that the state received over a period of time. In question 2 you will start the task of estimating the total amount of rainfall in Colorado using this data

2. a) Explain how the problem you identified in question 1 has been simplified using rainfall data from Colorado.

   b) Explain how the problem you identified in question 1 has been simplified using rain gauge data from eight stations.

ASK QUESTIONS AND MAKE ASSUMPTIONS

An important part of the modeling process is to look carefully at the information you are given. In this case rainfall in Colorado at eight different stations is provided in Figure 3.1. Let's take a closer look at these data.

3. For how long a period of time do you think the data were collected? Does that affect your solution to the problem?

Any assumption you make when building a model should be identified clearly from the beginning. Later on, you can come back and check how important each assumption was to the model. Another good practice is to list any critical information you may need to solve the problem. Ask yourself how (or where) you could obtain that information.

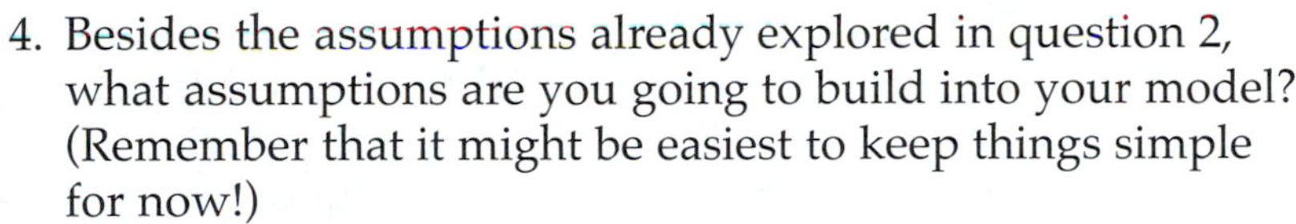

4. Besides the assumptions already explored in question 2, what assumptions are you going to build into your model? (Remember that it might be easiest to keep things simple for now!)

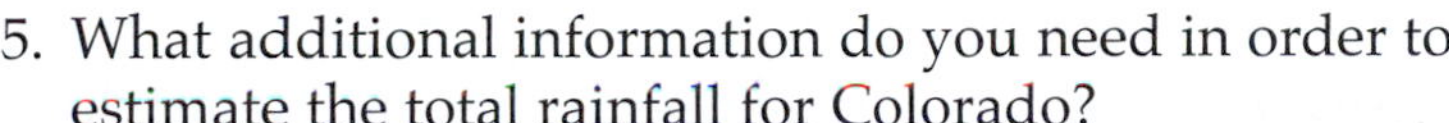

5. What additional information do you need in order to estimate the total rainfall for Colorado?

THE MAIN EVENT: FINDING A SOLUTION

Now that you have done the preliminary work, it's time to do the math. The next step of the modeling process is to develop a plan. Figure out *some* way to use the data and your assumptions to find a reasonable answer to the problem. Once you have a plan, you "just do it." Use your plan to find a solution to the problem.

CHECK THIS!

Each plan or model will use its own set of mathematical tools and calculations. In the modeling process common sense is as important as mathematical ability!

6. Describe what you are going to do to find an estimate for the total rainfall for Colorado. Average the rainfall from the 8 gauges

7. Carry out your plan and find a solution. What are the units for your answer? Convert your answer to a standard unit of cubic feet.

Average depth = 2.58" $V = \pi r^2 h$

Even though you have found an answer, the modeling process has only just begun. Remember that mathematical modeling is a continuing process. Since your model is supposed to solve a real problem, you must evaluate your model. That is, continue with the next steps in the modeling process.

- Check whether the model gives an answer that makes sense.
- Test the plan by looking at the assumptions to see which were important and how they affected the solution.
- Look for any new information that may help improve the plan and make it more realistic.

Your payoff will be a "better" solution that may lead to more and more ways to improve the model until it is "as good as it gets."

8. Your answer for the total rainfall has volume units, like cubic feet. Think of your model as a swimming pool that contains the total rainfall you found as your solution to question 7. (What would the pool look like?) How does the information used by the model determine the dimensions of the pool?

9. a) Take a few minutes to test your model. If there was an error in the rain gauge measurements of 0.01 inch (either high or low), how would that affect your answer?

   b) If there was an error of one mile in the measurement of the length of Colorado and an error of one mile in the measurement of the width, how would that affect your answer?

   c) Are the rain gauge measurements or the length and width measurements more critical to a good rainfall estimate? Explain.

WRAP UP

When you build a mathematical model, it is important to check whether the model gives you reasonable answers. Review your work in this activity as you answer the following questions.

10. a) Is your answer to question 7 reasonable?

    b) Does the method you used make sense for the real-world problem? Why?

    c) Is the limitation due to the process you used, or is it due to your assumptions?

11. a) Think about where you might want to go next in modeling this situation. What key feature should change so that the answer better describes the real world situation?

    b) How would you modify the activity to improve the model?

SUMMARY

In this activity you learned that one way to start modeling is to simplify the problem. You investigated a related situation that you know you can solve.

- You used the modeling process to ask questions and make assumptions about the simplified model.
- You created a plan, and calculated an estimate of the total rainfall in Colorado.
- You tested your model and discovered some of its limitations.

Unfortunately, your rainfall estimate would not be good enough for Colorado water officials. Nevertheless, the model did produce some useful results. Your results will help you look for ways to improve your model. Activity 3.2 will show you the way.

HOMEWORK 3.1

Rain in the Lone Star State

In this chapter the activities focus on estimating the total amount of rainfall in the state of Colorado, because that problem is a simplification of the chapter's central problem. In this assignment you will adjust the Colorado model to determine rainfall estimates for Texas.

1. Seventeen "first-order" stations are spread across the state of Texas. In addition, one station in Shreveport, Louisiana is used often. The total rainfall data for the 1997 calendar year are given in **Figure 3.2**; the measurements are in inches.

Gauge 1	Gauge 2	Gauge 3	Gauge 4	Gauge 5	Gauge 6	Gauge 7	Gauge 8	Gauge 9
27.08	43.27	24.95	46.79	36.21	44.10	36.16	23.01	45.00

Gauge 10	Gauge 11	Gauge 12	Gauge 13	Gauge 14	Gauge 15	Gauge 16	Gauge 17	Gauge 18
9.63	76.79	22.67	17.10	23.38	33.92	65.05	23.78	69.20

FIGURE 3.2.
Rainfall data for Texas.

a) Use your knowledge of the weather patterns in the state to predict where the gauges are located.

b) Do you think the measurements used in Activity 3.1 were yearly totals? If so, explain why. If not, estimate the time interval in which the data was collected.

c) Is the time interval an important factor in the modeling of this situation?

d) What is the average depth of the rainfall at the gauges listed in Figure 3.2?

e) Using the same method you used in Activity 3.1, estimate the 1997 total rainfall for Texas. You may need to look in an atlas or encyclopedia to get some information to answer this question.

2. A Texas company that specializes in digging wells has data on how deep each well must be to reach the water table. Could that information be used to estimate the amount of water available for use in the state? Explain.

3. Lake Austin is a reservoir near Austin that is formed by the Tom Miller Dam. The lake is 20.25 miles long and is 1300 feet across at its widest point. The dam is 100.5 feet high.

   a) If the lake was shaped like a rectangular swimming pool with those dimensions, what would be the volume of the lake in acre-feet? (1 acre = 43,560 sq. ft.)

   b) The actual capacity of the lake is *only* 21,000 acre-feet. What's wrong with thinking that the volume of the lake can be estimated by thinking of it as a rectangular pool?

   c) Was that same kind of error produced in your work on estimating the total rainfall for Colorado? Explain.

   d) While the dam front isn't triangular, it might be reasonable to assume that the rest of the lake is shaped like an inverted roof-top—gradually sloping from shallow to deep. **Figure 3.3** shows a cross section of the lake. Calculate the capacity of the lake again, assuming cross sections of the lake are in the shape of a triangle.

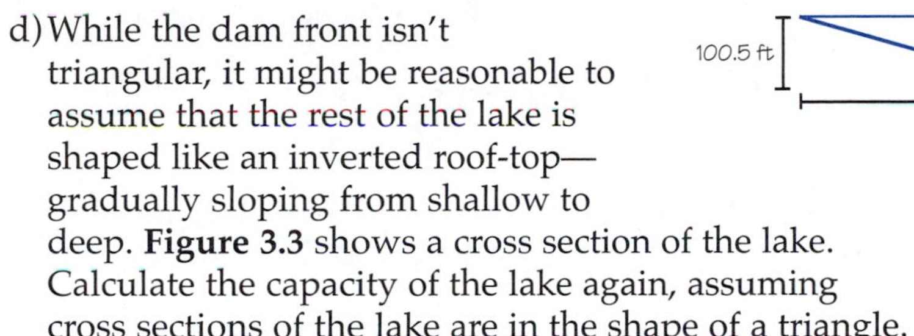

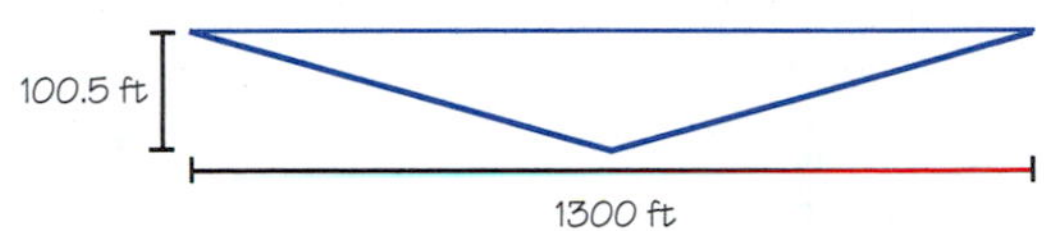

FIGURE 3.3. Cross section of Lake Austin.

   e) Suggest ways to improve the model describing the capacity of Lake Austin to make it more accurate.

Modelers try to verify their mathematical predictions by gathering evidence to support their model. As you work through this chapter, you can compare some of your results with weather patterns and water availability in your area. In the following question you gather some data to use in your comparison.

4. a) Contact the local weather service or broadcast media company to find out what a "typical" rainfall average is for your area. If the range of values doesn't fall within those shown in Figure 3.2 from question 1, the model might not be accurate.

   b) Delivering water to peoples' homes is the responsibility of either the city, county or state government, or a private company. Contact the water resources management or other government monitoring office in your area. Find out the holding capacity for local reservoirs. If your water supply comes from a river, try to get an estimate for the stream flow—broken down by season, preferably.

ACTIVITY 3.2

Gauging Influence

The model for estimating rainfall in Activity 3.1 was too simple, so we must continue. After all, you can't convince people that the average rainfall means the same amount of rain falls everywhere in the state. A possible next step is to let each gauge represent a region of the state. Instead of assuming that the same amount of rain falls everywhere in the state, let each gauge represent the rainfall in the region around it. In this activity you will explore how to build that suggestion into the model in a way that makes sense.

REPRESENTING EACH REGION IN COLORADO

Figure 3.4 shows the location of the eight Colorado rain gauges used in Activity 3.1. Your first task is to find which gauge measures the rainfall in each city.

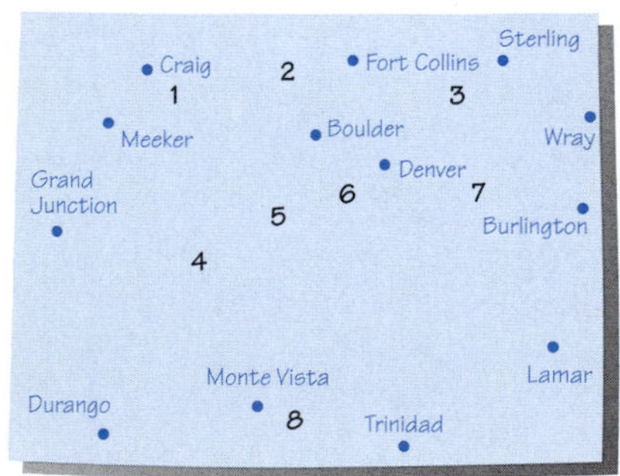

FIGURE 3.4.
Colorado map with rain gauge locations.

City	Gauge #
Denver	
Trinidad	
Sterling	
Wray	
Boulder	

FIGURE 3.5.

1. a) Identify the gauge(s) that best measures the rainfall amount of each city listed in **Figure 3.5.** Explain the method you're using, and why it's harder to find a gauge for some cities than for others.

   b) Use **Figure 3.6** to list the cities for which each gauge is the best measure of rainfall. Be sure to place all thirteen cities.

Gauge	Cities	Gauge	Cities
1		5	
2		6	
3		7	
4		8	

FIGURE 3.6.

c) Using Figure 3.4 and your answers to (a) and (b), rank the gauges by the amount of area they best represent. Record your rankings in **Figure 3.7**, putting the gauge that corresponds to the largest area at the top of the list, and the one that corresponds to the smallest area at the bottom.

Rank	Gauge
1st	
2nd	
3rd	
4th	
5th	
6th	
7th	
8th	

FIGURE 3.7.

REGIONS OF INFLUENCE

As you start to look at the regions around the gauges, some terminology will be useful. **Centers of influence** are the points representing an entire geographical area (the rainfall gauges). The geographical area surrounding the center is called a **region of influence** (see **Figure 3.8**). Each region of influence has a limit to how far it goes. We need a way to locate the boundary lines that separate regions of influence.

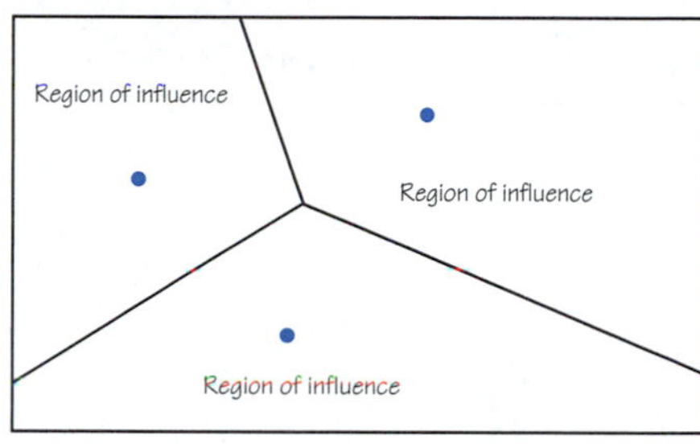

FIGURE 3.8.
Three regions of influence around three rainfall gauges.

2. If you knew the annual rainfall amounts for each rain gauge in Figure 3.4, how would you revise your method of determining the rainfall total in Activity 3.1?

How can the eight gauges in Figure 3.4 best represent Colorado? Divide the state into regions around centers of influence (the rain gauges). Draw each region so that everything inside the region is closer to its center of influence than to any other center of influence (see **Figure 3.9**). This kind of diagram is called a **Voronoi diagram** or a **Voronoi tiling**. Voronoi diagrams are named for the Russian mathematician M. G. Voronoi, who invented them in 1908.

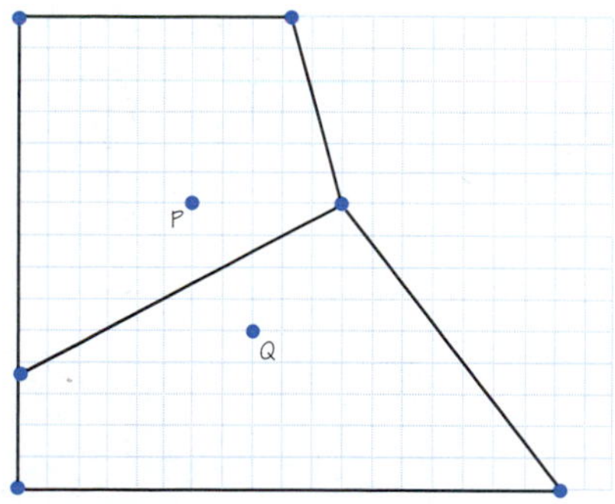

FIGURE 3.9.
All points in the region of influence around P are closer to P than they are to Q.

In the following exercises, each diagram shows two or more rainfall gauges in a part of the state. Each gauge measures the rain for the part of the state that is closest to it. Try to find a boundary that makes the diagram a Voronoi diagram.

- Draw one or more boundary lines so that everything in a region is closer to its center of influence than to any other center of influence. For example, all points inside the region around A should be closer to A than they are to B.
- Estimate the percentage (or fraction) of the diagram that is the area around each gauge.

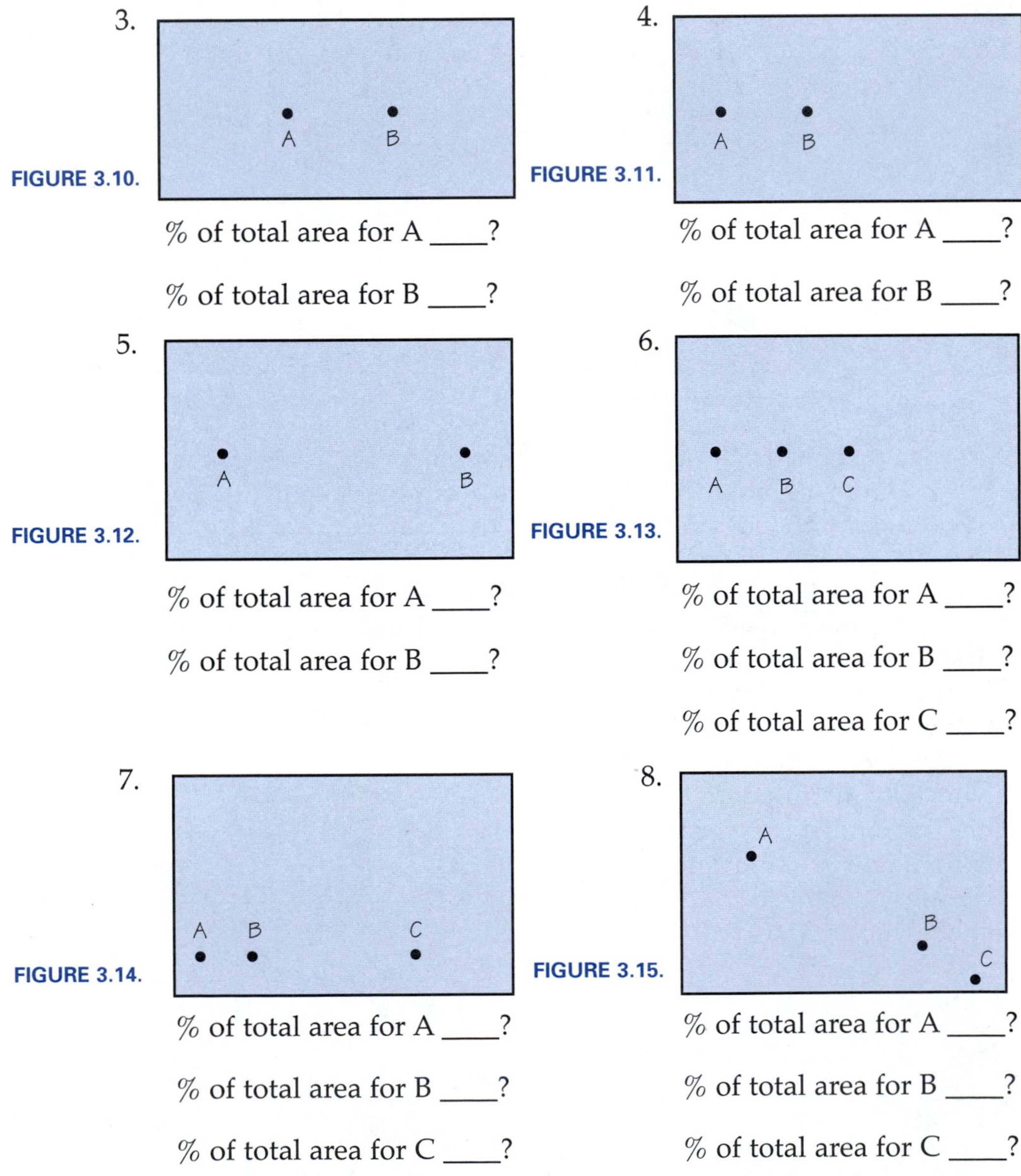

3. FIGURE 3.10.

% of total area for A ____?

% of total area for B ____?

4. FIGURE 3.11.

% of total area for A ____?

% of total area for B ____?

5. FIGURE 3.12.

% of total area for A ____?

% of total area for B ____?

6. FIGURE 3.13.

% of total area for A ____?

% of total area for B ____?

% of total area for C ____?

7. FIGURE 3.14.

% of total area for A ____?

% of total area for B ____?

% of total area for C ____?

8. FIGURE 3.15.

% of total area for A ____?

% of total area for B ____?

% of total area for C ____?

9. a) Look at the part of the state shown by the diagram in **Figure 3.10**. This part has two rain gauges. What percent of the total rainfall in this part is contributed by the rainfall on the region about A?

   b) For this same figure, what percent of the total rainfall in this part is contributed by the rainfall on the region about B?

   c) How would you use the percents you found in 9(a) and (b) to calculate an estimate of the total rainfall in the part of the state shown in Figure 3.10?

10. a) Look at the part of the state shown in **Figure 3.11**. What percent of the total rainfall in this part is contributed by the rainfall on the region about A?

    b) For this same figure, what percent of the total rainfall in this part is contributed by the rainfall on the region about B?

    c) How would you use the percents you found in 9(a) and (b) to calculate an estimate of the total rainfall in the part of the state shown in Figure 3.11?

    d) In general, how might the percentages calculated in questions 3–8 be used in determining the estimate for the total rainfall?

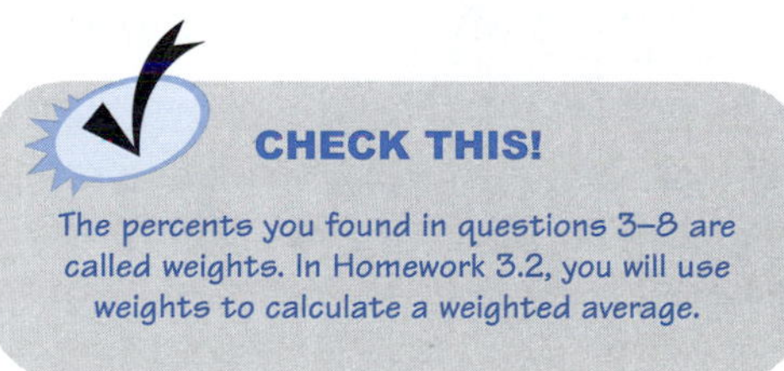

CHECK THIS!

The percents you found in questions 3–8 are called weights. In Homework 3.2, you will use weights to calculate a weighted average.

11. a) Look again at the Voronoi diagrams you created in questions 3–8, and imagine the line segment AB. In each diagram find the point where AB meets the boundary line, and call this point P. What can you say about the distance from A to P and the distance from B to P?

    b) Now check the angle formed by the line segment AB and the boundary line. What is the measure of this angle?

12. Using your answers to question 11, describe a general process for locating the boundary between two centers of influence in a Voronoi diagram.

    a) Test your procedure (from question 11) on **Figure 3.16**, a map of Colorado with only three rain gauge locations.

FIGURE 3.16. Three rain gauge locations in Colorado.

    b) How did you divide the state so that each part is represented by only one gauge? Explain your method.

SUMMARY

In this activity you found a way to divide up a state into regions.

- You learned about Voronoi diagrams.
- You discovered a way to divide a state into regions around rain gauges.

Now that you know how to divide a state into regions, you need to find ways to draw the boundary lines precisely. Remember, your rainfall estimate depends upon the accuracy of all your work. Activity 3.3 continues your search for a better model for estimating rainfall.

HOMEWORK 3.2

Weighted Average

1. Mathematicians are always looking for shortcuts for calculations. Suppose you wanted to find the average of the following numbers: 6, 8, 10, 5, 6, 15, 6, 8, 15, 6.

   a) One way to calculate the average is to add up all the numbers and divide by how many numbers there are. Write an expression for this calculation and then calculate the average.

   b) Another, faster method is to group the numbers by their value. For example, group the two 8s in the list together. Then you use multiplication as a quick way to add the numbers in each group. Write an expression for this calculation, and then use it to find the average again.

The most efficient way to calculate the average in question 1 is to use a **weighted average**. To find a weighted average, do the following:

- For each value in the list, find the percent of the total number of numbers that the particular value occurs. For example, the 8 appears 2 times out of a list of 10 numbers, so its **weight** would be 2/10, or 20%.
- Multiply each number by its weight and then add these products.

Value	Weight
5	
6	
8	$\frac{2}{10}$ = 0.2 or 20%
10	
15	

FIGURE 3.17.

2. a) Use **Figure 3.17** to find the weights for each number in the following list (from question 1): 6, 8, 10, 5, 6, 15, 6, 8, 15, 6. The weight for the number 8 is given.

   b) Use the weights to write a new expression to calculate the answer as a weighted average. Check to make sure you get the same answer as before.

3. a) Keesha wants to know her math grade. Her grades were: 89 for tests, 85 for quizzes, 99 for homework, and 75 for class participation. Compute the average of these scores.

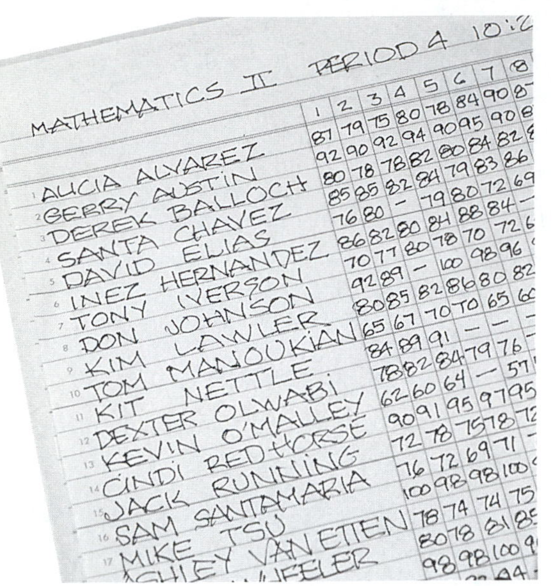

b) Keesha knows that tests count 50%, quizzes count 20%, homework counts 20%, and class participation counts 10%. Now compute the weighted average (using the percentages as weights).

c) Try changing each score (one at a time) by ten points and calculating the weighted average to find the impact each has. Which particular score affects the weighted average the most? Explain your answer.

d) Which grade had the least impact on the weighted average?

e) How is this problem related to estimating the rainfall for Colorado? In what way can weighted averages make a better estimate?

In the model for estimating rainfall, it seems reasonable that a larger area should contribute more to the calculation of average rainfall. In other words, maybe a weighted average should be used. Let's see how this idea can be used for some regions of Colorado using the rain gauge depths (see **Figure 3.18**) from Activity 3.1 again. (The measurements are still in inches.)

FIGURE 3.18. Rain gauge data for Colorado.

Gauge 1	Gauge 2	Gauge 3	Gauge 4	Gauge 5	Gauge 6	Gauge 7	Gauge 8
2.11	2.15	1.21	4.45	2.67	2.51	3.11	2.43

To estimate rainfall in the following regions of Colorado:

- Make the Voronoi diagram by dividing each region into two parts making sure that every point is closest to the gauge in that part.
- Estimate what percent of the region each part is.
- Estimate the average rainfall in the region using the percent of the total area as weights to calculate a weighted average.
- Finally, compare the weighted average to an estimate that uses the average of the two gauge measurements.

4. **Northwest region**: Gauges 1 & 2

   % of area for gauge 1: ____?

   % of area for gauge 2: ____?

   Rainfall using average: _____?

   Rainfall using weighed average: ________?

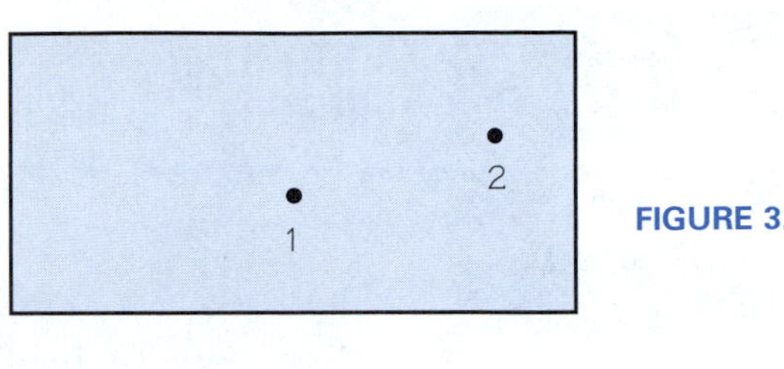

FIGURE 3.19.

5. **Northeast region**: Gauges 3 & 7

   % of area for gauge 3: _____?

   % of area for gauge 7: _____?

   Rainfall using average: _____?

   Rainfall using weighted average: ________?

FIGURE 3.20.

6. **Central region**: Gauges 5 & 6

   % of area for gauge 5: _____?

   % of area for gauge 6: _____?

   Rainfall using average: _____?

   Rainfall using weighted average: ________?

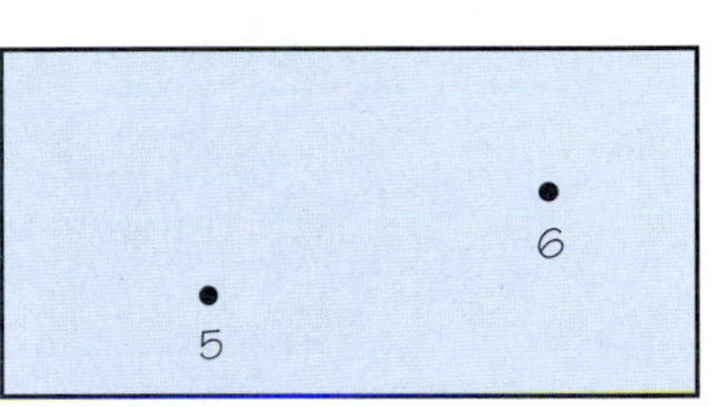

FIGURE 3.21.

7. a) Look over your work in questions 4–6. Under what conditions was there a large difference between the simple average and the weighted average?

   b) For the extra work involved, how can you be sure that the process has been improved?

ACTIVITY 3.3

Caution: Under Construction

Now that you have a method to divide the state into regions, you can calculate rainfall estimates region by region. Instead of using an average rainfall for the entire state, your new model will give a rainfall estimate for each region around a rainfall gauge. To do so, you need a way to draw the Voronoi diagram precisely. In this activity you will use geometry to locate the boundaries of Voronoi regions exactly.

AT THE BOUNDARY

From the work you have done so far, it's clear that the boundary lines of Voronoi regions have the following properties.

- The boundary line is always between two rain gauges.
- The boundary line forms a right angle with the line segment that connects the two gauges.
- If the point M is the point where the line segment meets the boundary line, then the distance from M to one gauge is equal to the distance from M to the other gauge.

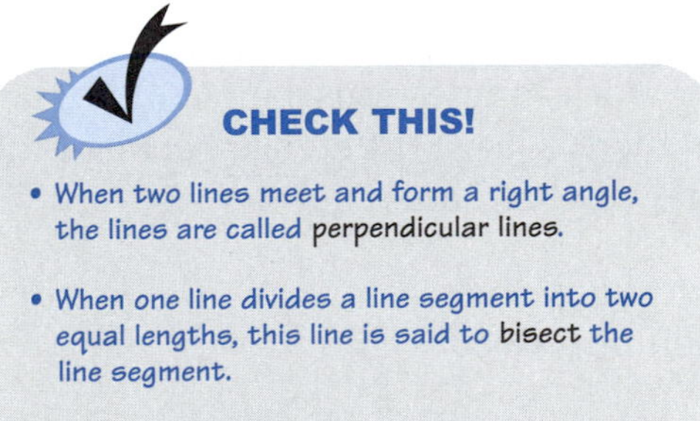

CHECK THIS!

- When two lines meet and form a right angle, the lines are called **perpendicular lines**.
- When one line divides a line segment into two equal lengths, this line is said to **bisect** the line segment.

Figure 3.22 shows these properties for the boundary line between the two gauges x_1 and x_2.

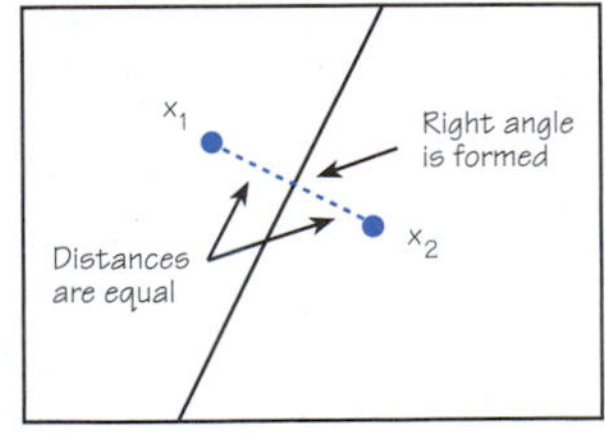

FIGURE 3.22. Properties of a Voronoi boundary line.

The boundary line and the line segment between the two gauges are **perpendicular**. Since the boundary line **bisects** the segment, the boundary line is a **perpendicular bisector** of the line segment.

PERPENDICULAR BISECTOR

A line is a perpendicular bisector of another line if the lines are perpendicular and the line divides the other line into two equal line segments.

The line divides the segment into two equal lengths. The point of intersection is the **midpoint** of the segment.

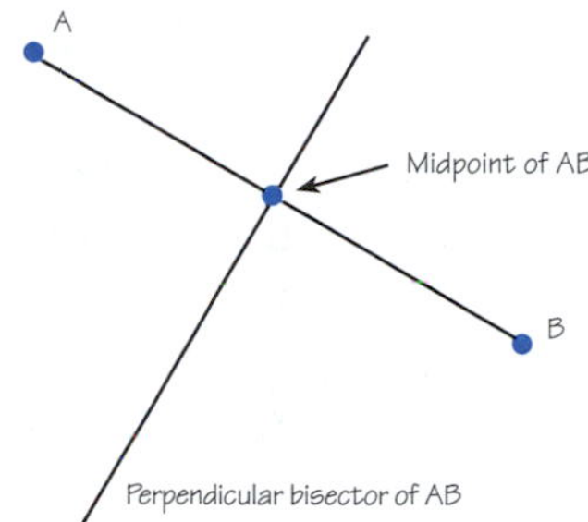

FIGURE 3.23.
The perpendicular bisector of the line segment AB.

EXPLORING GEOMETRIC PROPERTIES

Recall that in Voronoi diagrams everything inside a region must be closer to its center of influence than to any other center of influence. In the following exercise you will check that this property is true when the boundary line is a perpendicular bisector of the line segment between two centers of influence.

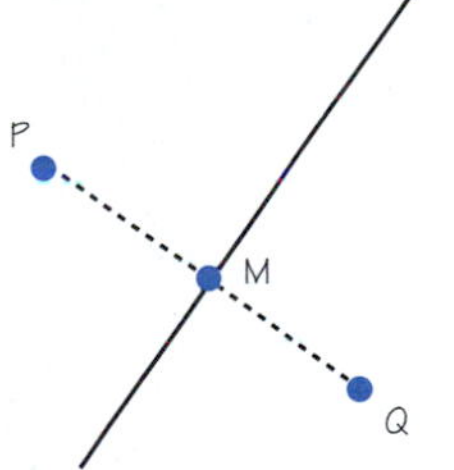

FIGURE 3.24.

1. Using **Figure 3.24**, add four points on the perpendicular bisector of line segment PQ, and label them A, B, C, and D.

   a) Using a ruler, measure the distances PA and QA, PB and QB, PC and QC, and PD and QD. Record them in **Figure 3.25**. What do you notice about the pairs of measurements?

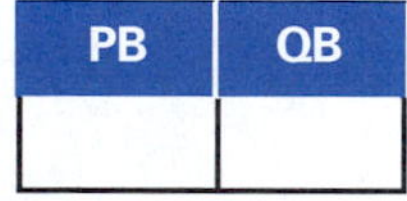
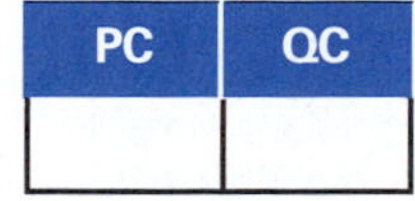
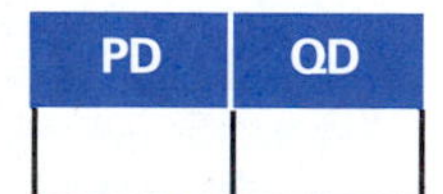

PA	QA

PB	QB

PC	QC

PD	QD

FIGURE 3.25.

   b) Use your result from (a) to write a general statement about the distance between any point on the perpendicular bisector and the endpoints of the line segment that is bisected.

   c) How can this result help locate the center of influence for any point in the region?

2. a) Let's try the problem in reverse. Suppose you knew that a point Z was the same distance from rain gauges located at points P and Q. (Refer back to Figure 3.9 in Activity 3.2, if necessary.) What might you conclude?

CHECK THIS!

Recall that an isosceles triangle is a triangle with two sides of equal length. The vertex angle is the angle formed by the two equal sides of the triangle.

b) Sketch an isosceles triangle and then draw an **altitude** from the **vertex angle** to the **base**. What else can you say about this altitude?

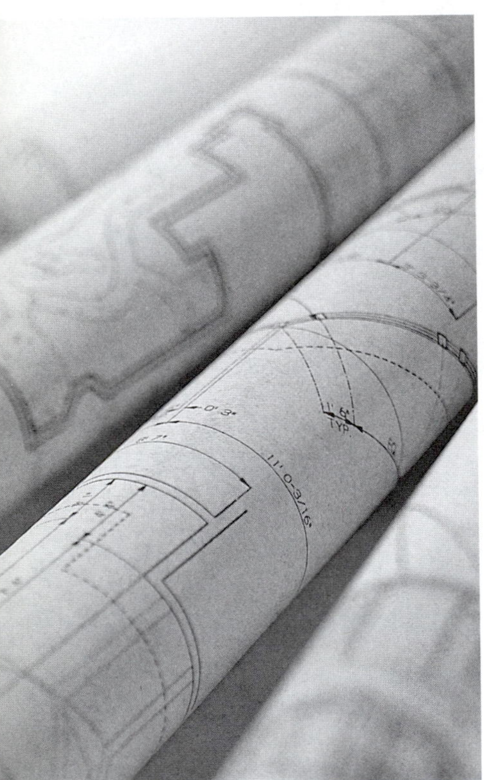

TOOLS FOR CONSTRUCTION

Architects and draftsmen often need to sketch geometric figures in their blueprints. Creating accurate diagrams of perpendicular bisectors takes work. Geometers, people who study and use geometry, use **construction techniques** to create precise sketches of geometric figures.

Construction techniques produce drawings that yield better data. These drawings become "recipes" that can be communicated to others and improved over time.

CONSTRUCTION TECHNIQUES

Construction techniques use mathematics and geometry facts together with a straightedge, compass, and other drawing instruments to draw accurate figures.

CHECK THIS!

You can use the following construction tools to help create accurate figures:

- a straightedge
- a compass
- a Mira®, a plastic tool for making reflections of an image
- a piece of wax paper.

The following exercises explore some construction techniques. Once you have completed the two constructions, you will be ready to move on to Voronoi diagrams.

3. a) Using the appropriate tools, try discovering a technique for constructing an angle bisector for a typical angle. Start with an angle like the one in **Figure 3.26**. Draw an isosceles triangle so ∠A is the vertex angle and one of the two equal sides is the line segment $\overline{AB}$.

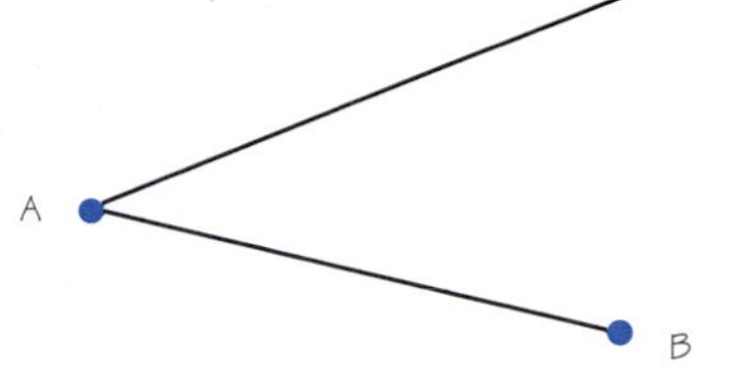

FIGURE 3.26.
Angle that should be bisected.

   b) Now use the results from question 2 to finish your construction and construct the angle bisector. Warning: You may need to make use of another figure to finish the construction.

4. Using the same tools, try discovering a way to construct a perpendicular bisector to a segment. Think of the line segment as the base of an isosceles triangle and use the information you found in question 2.

CONSTRUCTING THE BOUNDARY LINES BETWEEN RAIN GAUGES

Now you have all the information you need to construct the boundaries of a Voronoi diagram accurately. **Figure 3.27** shows a region that contains three rain gauges. It is 15 miles wide and 20 miles long.

5. a) Determine the regions that each rain gauge in Figure 3.27 represents by making the figure into a Voronoi diagram. Use a straightedge to indicate boundaries clearly. Use one or more of the construction tools you just explored to locate the boundaries.

   b) Explain how you constructed the boundary of the region around gauge B.

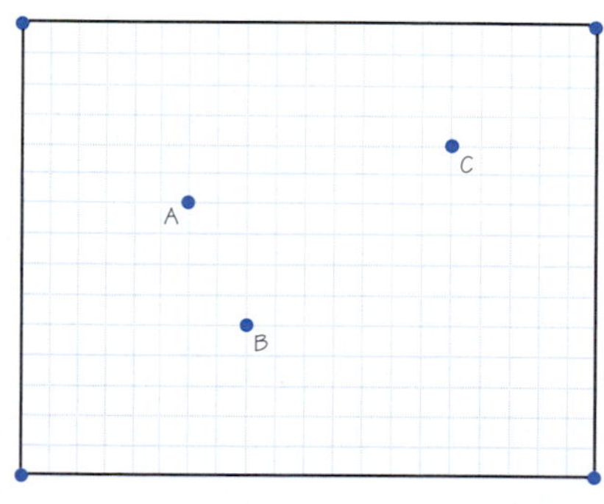

FIGURE 3.27.
A three-point Voronoi Problem.

SUMMARY

In this activity you used some principles from geometry to discover the properties of boundary lines between rain gauges.

- You explored some construction techniques to draw accurate boundaries.
- You learned how to draw accurate boundaries for a simple diagram having three rain gauges.

Now your new model is taking shape. You can draw the Voronoi regions accurately, but most of these regions are not simple figures, such as rectangles or triangles. You need a method to find the areas of these unusually shaped regions. Activity 3.4 will introduce methods to find the area of this kind of region.

HOMEWORK

3.3

Boundaries of All Shapes and Sizes

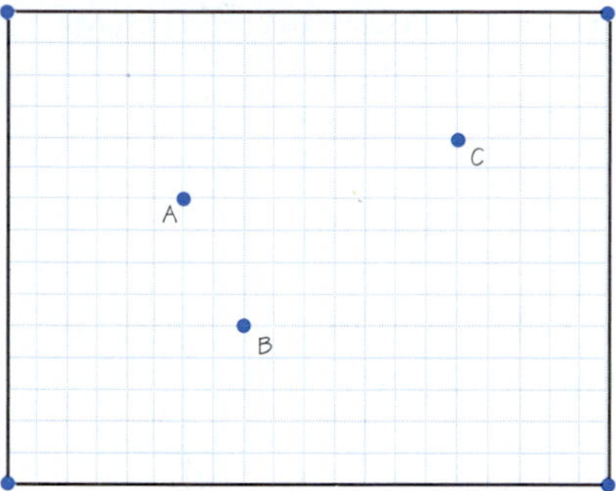

FIGURE 3.28.

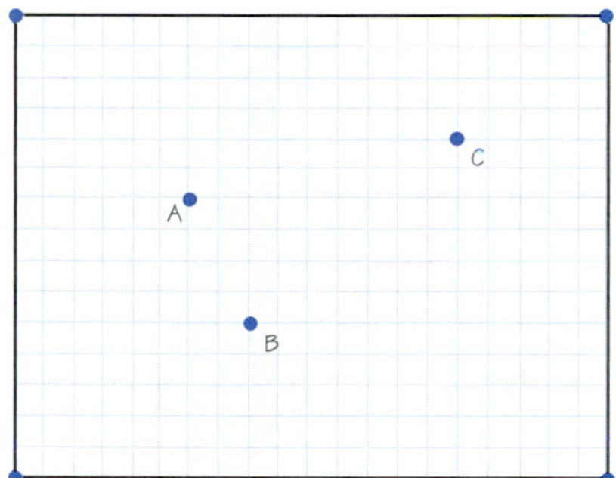

FIGURE 3.29.

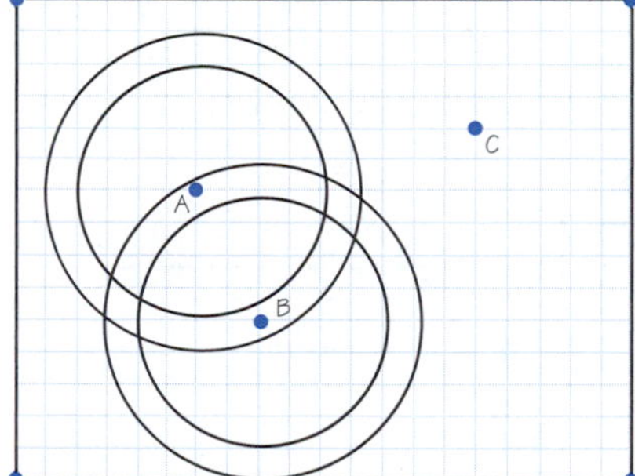

FIGURE 3.30.

1. **Figure 3.28** shows the locations of three competing helicopter services and their helipads. Draw a boundary of the region that contains all points within two miles of helipad A. Draw a second boundary of the region that contains all points within two miles of helipad B. (Assume the distance between grid marks is one mile.)

2. a) In **Figure 3.29**, draw a boundary of the region that contains all points within three miles of Helipad A. Draw a second boundary for all points that are within three miles of Helipad B.

   b) Is it clear which people living inside the region serviced by Helipad A are closer to A than they are to B? Explain.

3. **Figure 3.30** has two circles around the point A, one with a radius of four miles and another with a radius of five miles. Another two circles are drawn around B, one with a radius of four miles and another with a radius of five miles. Mark and label the points of intersection. How far is each of these points from the helicopters at A? How far is each point from B?

4. a) Draw a line segment through the points of intersection you found in question 3. Draw the line segment AB. What is the relationship between these two line segments?

   b) How would you describe each point of intersection in terms of how far people live from the helicopters located at A and B?

CHECK THIS!

Notice that questions 3 and 4 give you another way to draw a perpendicular bisector of the line segment between two points. You can use this method to construct the boundaries of a region of a Voronoi diagram.

5. a) A circle around A in Figure 3.30 has a radius the same length as the radius of a circle around B. If the circle around A touches—doesn't crossover—a circle around B, what is the length of the radius of A?

   b) Where is the point at which the circles meet?

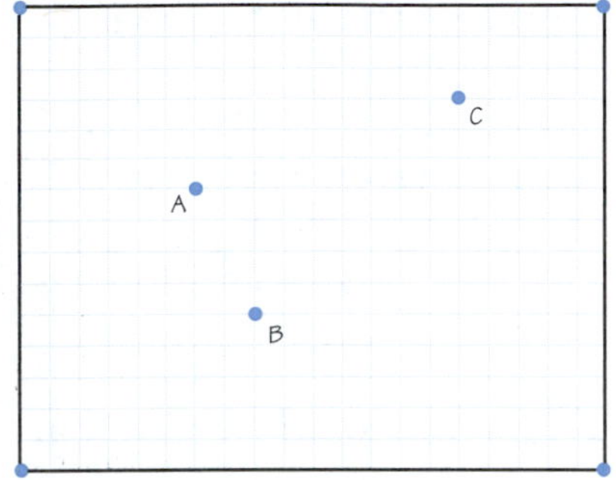

FIGURE 3.31.

6. In **Figure 3.31**, draw a pair of circles to find the perpendicular bisector of $\overline{AB}$, $\overline{AC}$, and $\overline{BC}$.

7. The three perpendicular bisectors you drew in question 6 intersect at a point. Mark that point V, and describe the location of that point.

8. a) Draw a circle centered at V that goes through one of the centers of influence (A, B, or C).

   b) What do you notice about the position of the other two centers of influence relative to the circle?

   c) Draw the segments $\overline{AB}$, $\overline{AC}$, and $\overline{BC}$. A circle is said to **circumscribe** a polygon if the vertices of the polygon lie on the circle. What do you notice about the circle you've drawn?

9. In **Figure 3.32** and **Figure 3.33**, construct perpendicular bisectors using any construction technique that you want. Label the center of the circle that contains the points A, B, and C, and draw the triangle ABC.

10. Use your results from question 9 to explain how the type of triangle formed by the three points affects the location of the center of the circumscribed circle for that triangle.

11. Are you ready to tackle the problem of finding rain gauge locations in Colorado (or Texas)? Let's see! Make **Figure 3.34** a Voronoi diagram so that each point is the center of influence of a region.

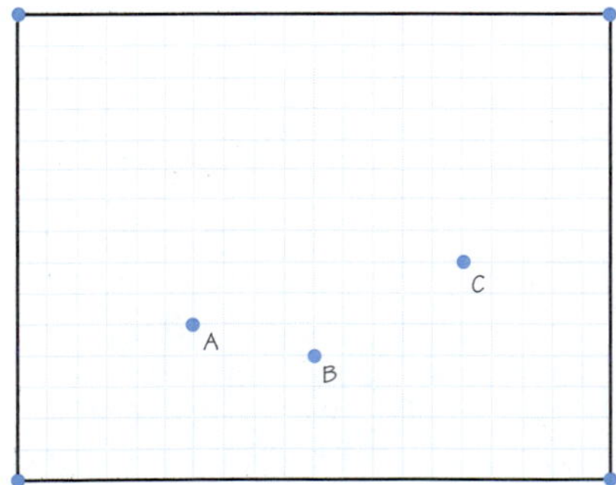

FIGURE 3.32.

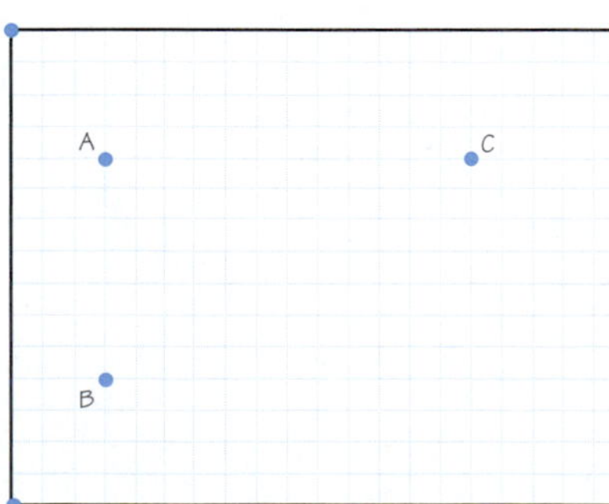

FIGURE 3.33.

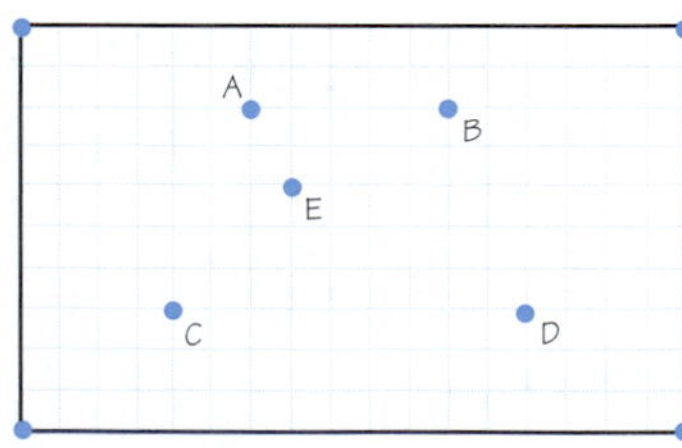

FIGURE 3.34.

ACTIVITY 3.4

Area of Concern: Calculating Area

Now that you know how to break up the state into regions, you are well on your way to developing an improved model. If you know the positions of some rain gauges in the state, you can subdivide the total area into smaller regions, called regions of influence or regions of proximity. Each region contains the gauge that measures rainfall for the entire region. How do you calculate the areas of these smaller regions? In this activity you will learn three new ways to calculate the area of a region.

DRIVE FRIENDLY - THE TEXAS WAY

CHECK THIS!

Why do you need to find the area of each region? Each region contributes to the total rainfall, and the amount of ranfall it receives is calculated by using its area as a weight.

CALCULATING AREAS BY SUBDIVIDING

Let's start with a simplified problem. The map in **Figure 3.35** shows a county that has three gauges. The dimensions of the county are 15 miles by 10 miles. The Voronoi boundaries for the map are already provided.

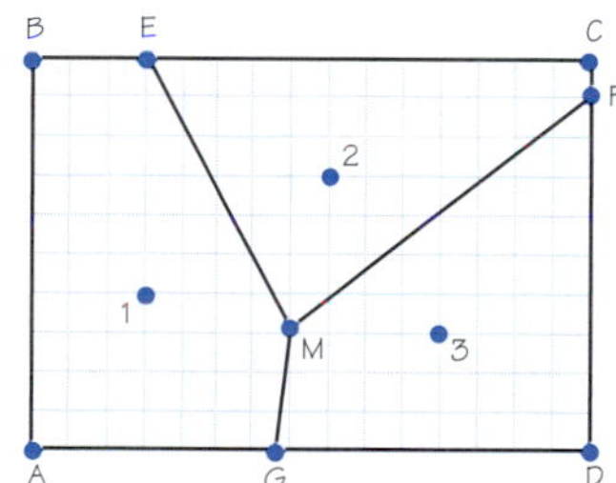

FIGURE 3.35.
Voronoi diagram of a county with three rain gauges.

1. What is the area of region 1, the shape defined by the points A, B, E, M, and G? Explain the method you used to find your answer.

2. If you compare your answer with other students in your class, do you expect there to be one right answer? Explain.

3. Do you think you have found the correct area of the region around gauge 1? If not, explain how you could do better.

4. Using your method from question 1 again, find the areas of regions around gauges 2 and 3.

5. The rainfall readings, in inches, for the three gauges in Figure 3.35 are given in **Figure 3.36**. Estimate the total rainfall for the county. Explain your method.

1	2	3
2.11 in.	2.15 in.	1.21 in.

FIGURE 3.36.

OTHER METHODS FOR CALCULATING AREA

We know that the total rainfall estimate depends on finding the areas of regions of the state. Since the shapes of regions often are different, one method of finding area might not be right for every region. You need more than one way to calculate area. Also if you know more than one way to calculate area, you can check to see how accurate each calculation is.

The following are three methods for determining area:

- Pick's formula
- Heron's formula
- Monte Carlo method

You will use these methods to estimate the areas of some regions in the county shown in **Figure 3.37**. In the figure the distance between two grid marks is one mile. The area of the first region you consider can be calculated using methods you already know.

6. a) Start with the region around gauge 3. Divide it into a rectangle and two triangles by drawing a line straight down from point P to the base of the rectangle. Draw another line going straight across from point E to your first line. Find the area of each part of the region around gauge 3, and then find the total area of the region.

   b) Using this method, does the way in which you divide the original region affect your answer?

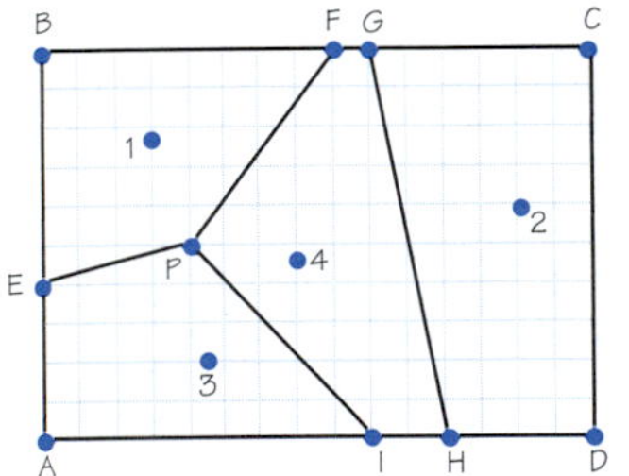

FIGURE 3.37.

Now consider the region around gauge 4 in Figure 3.37. Since the shape of this region is more complicated, you can try a new method called **Pick's formula.**

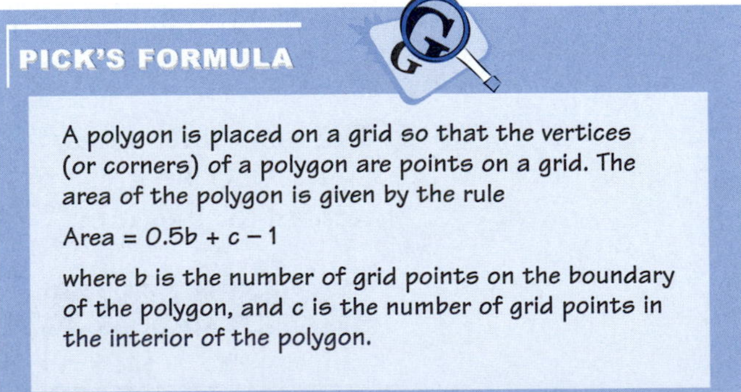

PICK'S FORMULA

A polygon is placed on a grid so that the vertices (or corners) of a polygon are points on a grid. The area of the polygon is given by the rule

Area = $0.5b + c - 1$

where b is the number of grid points on the boundary of the polygon, and c is the number of grid points in the interior of the polygon.

Example 

Use Pick's formula to determine the area of the region around gauge 4 in Figure 3.37.

Solution

To use Pick's formula, mark all the grid points that lie on the boundary of the region around gauge 4. **Figure 3.38** shows that step done for you. Start at any point on the boundary, go around the boundary in either direction, and count the number of points that you have marked. In this case, there are 11 points; that is your value for b.

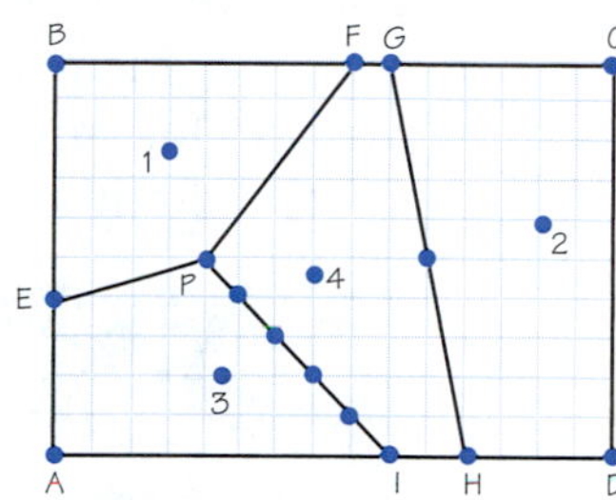

FIGURE 3.38.

Now count the number of grid points totally inside the region around gauge 4. There are 33 grid points inside this region; that is your value for c. Now, replace b with 11 and c with 33 in Pick's formula:

$$\text{Area} = 0.5b + c - 1$$

$$\text{Area} = 0.5(11) + 33 - 1$$

$$= 5.5 + 33 - 1 = 37.5 \text{ mi}^2$$

Using Pick's formula, the area of region around gauge 4 is 37.5 mi^2.

7. a) Use Pick's formula to estimate the area of region around gauge 3 in **Figure 3.38**.

   b) How does this result compare with the estimate you found in question 6?

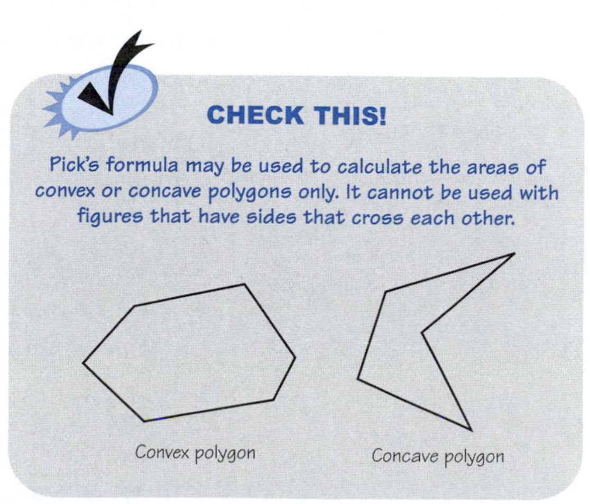

HERON'S FORMULA

You now have two methods for finding area of regions around rain gauges.

- One method is to divide a region into simpler shapes whose areas you can calculate.
- A second method is Pick's formula.

The next method is based on finding the area of a triangle. If a region of the county can be broken up into triangles, the easiest method for calculating area is to use the formula $A = \frac{1}{2}bh$, but another option is to use **Heron's formula**.

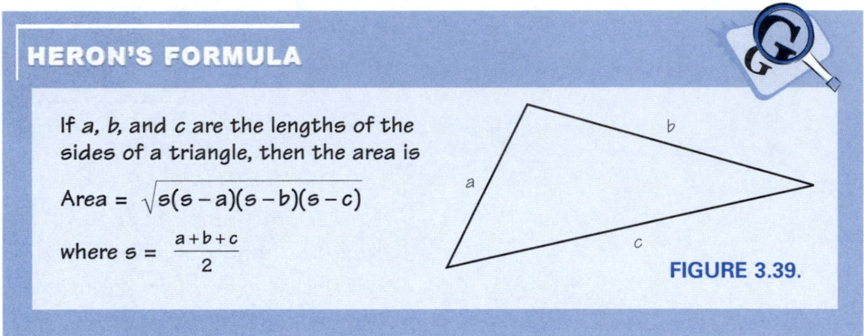

HERON'S FORMULA

If *a*, *b*, and *c* are the lengths of the sides of a triangle, then the area is

$$\text{Area} = \sqrt{s(s-a)(s-b)(s-c)}$$

where $s = \frac{a+b+c}{2}$

FIGURE 3.39.

The next question is a step-by-step guide on how to use Heron's formula to find the area of the region around gauge 1 in the county. (**Figure 3.40** shows the same county with four rain gauges whose areas you have been estimating.)

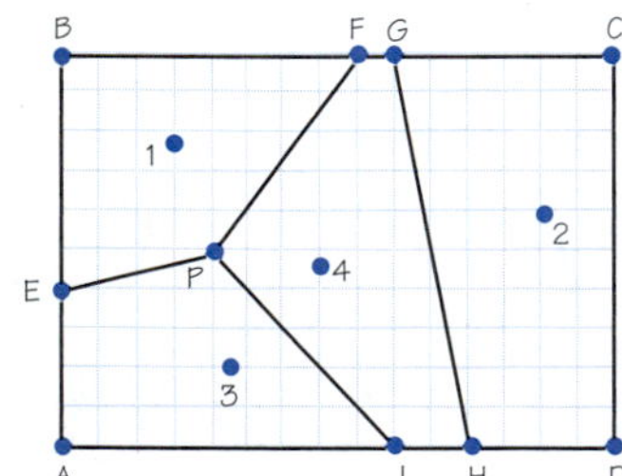

FIGURE 3.40.
The regions of influence around rain gauges.

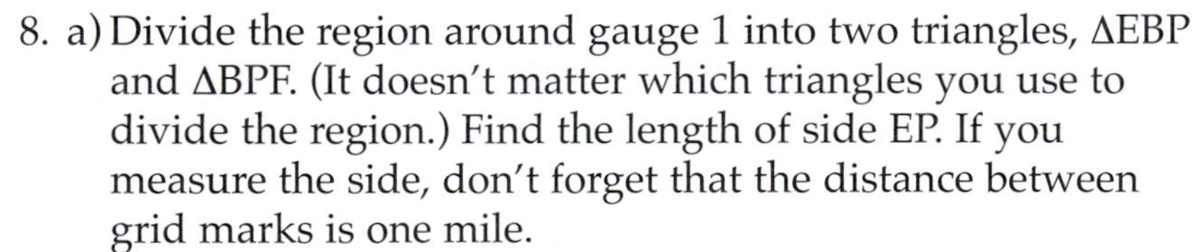

8. a) Divide the region around gauge 1 into two triangles, ΔEBP and ΔBPF. (It doesn't matter which triangles you use to divide the region.) Find the length of side EP. If you measure the side, don't forget that the distance between grid marks is one mile.

   b) How could you find the length of side EP without measuring?

   c) Find the lengths of the other sides that form ΔEBP.

   d) Add all the lengths and divide by two. Your result is the value of s. In this example, what is the value of *s*?

   e) Subtract each length from the value of *s*; that is, find the values of the terms $(s - a)$, $(s - b)$, and $(s - c)$. Multiply all those answers together, and then multiply that answer by s. Finally take the square root of that answer. What does Heron's formula predict for the area of the first triangle?

f) Repeat the previous steps for the other triangle. Then add the two areas together. What is the total area of Region 1?

g) Use the formula $A = \frac{1}{2}bh$ to calculate the area of those same two triangles. How do those answers compare to the ones found using Heron's Formula? Which method was easier for this problem?

9. a) Use Heron's formula to find the area of the region around gauge 3 in Figure 3.40.

   b) How does this result compare with the estimates you found using Pick's formula and by subdividing (questions 6–7)?

THE MONTE CARLO METHOD

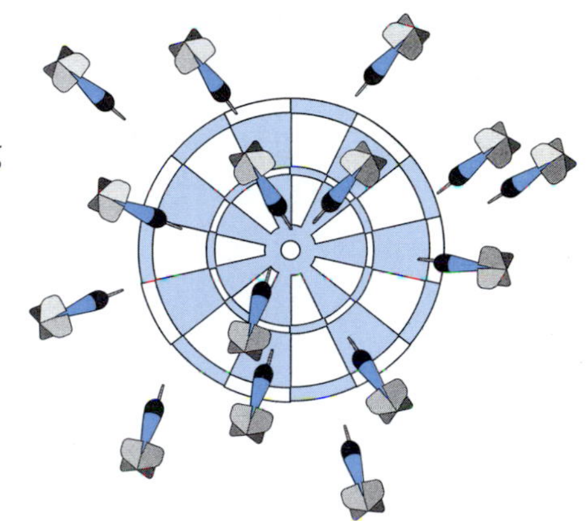

Another method for estimating the areas of regions of a county relies on the role of chance. Chance is connected to gambling and since Monte Carlo is famous for its gambling casinos, this method is known as the **Monte Carlo method**.

We want to find the area of region 2. Imagine throwing darts randomly at Figure 3.40. If you throw enough darts then the ratio

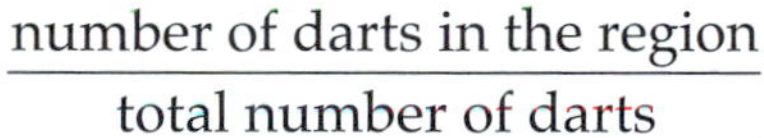

$$\frac{\text{number of darts in the region}}{\text{total number of darts}}$$

gives you an estimate of the percent of the entire area that lies within the region.

Consider the origin (0, 0) to be at point A in Figure 3.40. You can use your calculator to randomly generate coordinates of the ordered pair (x, y). These ordered pairs represent the randomly thrown darts on the grid. Since the grid is 15 miles by 10 miles, generate a random number in [0, 15) on your calculator for the x-coordinate of the ordered pair (x, y). Then generate a random number in [0, 10) for the y-coordinate. Keep track of the number of trials (how many ordered pairs you generate), and the number of times the point lies in the region. The ratio

$$\frac{\text{number of random points in a region}}{\text{number of trials}}$$

gives you an estimate of the percent of the entire area that lies within a region. The more trials you do the better the estimate becomes.

CHECK THIS!

On a TI-83 calculator, you can get the answer as an ordered pair with the command: {15*rand, 10*rand}. Keep pressing ENTER to get more ordered pairs. (It's a good idea to set the calculator to display one decimal place first!)

10. Try your luck at estimating the area of the region around gauge 2 in Figure 3.40 using the Monte Carlo method. Generate at least 50 ordered pairs. What area estimate do you get?

WRAP UP

Now it's time to compare and contrast the methods. Which method is the best? Well that depends on the situation. The following question asks you to make some judgments about the models.

11. a) Compare the three new methods. Which one might be the most accurate?

    b) Which method might have the most limited use?

    c) What method might take the most time?

    d) Which method might be the best to use with technology?

    e) Which method should you use when the number of regions gets "as big as Texas?"

SUMMARY

Now that you have found some ways to measure area, you can find the areas of the regions around rain gauges. But first, step back and review what you have achieved so far.

- In Activity 3.2, you learned how to divide the state into a Voronoi diagram.
- In Activity 3.3, you used construction techniques to locate the boundaries of the regions of influence precisely.
- In this activity you determined the areas of regions in a state using three different methods.

You have all the information you need to revise your model for estimating rainfall. Later in the the chapter you will be asked to estimate the rainfall in Colorado using these methods. In the next activity the problem will be tackled from an entirely new direction.

CHECK THIS!

Use the area and the rain gauge measurement of a region to find the annual rainfall in the region. The sum of the rainfall in the regions gives you the total state rainfall.

HOMEWORK 3.4 Area Still There?

1. You have found there are a lot of possible ways to subdivide a polygon into a collection of rectangles and triangles. Each collection involves calculations to determine the area. Some plans might take less work or involve simpler calculations. The following exercises explore ways to subdivide the polygon in **Figure 3.41**.

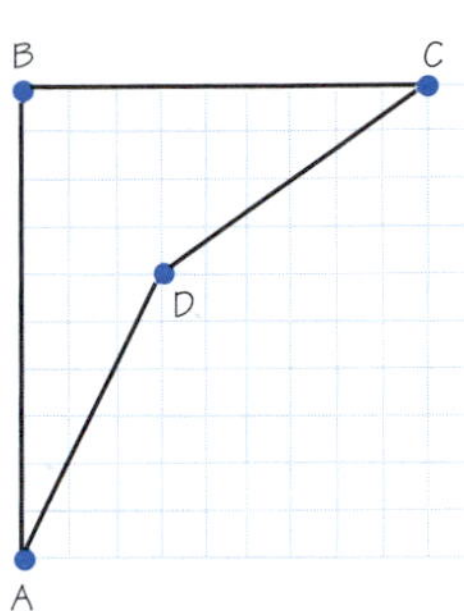

FIGURE 3.41.

a) The polygon is subdivided into the shapes shown in **Figure 3.41a**. Describe the strategy to find the area of the region.

b) The polygon is subdivided into the shapes shown in **Figure 3.41b**. Describe the strategy to find the area of the region.

c) The polygon is subdivided into the difference of the shapes shown in **Figure 3.41c**. Describe the strategy to find the area of the region.

FIGURE 3.41a.

FIGURE 3.41b.

FIGURE 3.41c.

d) Which of the three collections of shapes is the best strategy for finding the area of the region? Why?

2. Of all the methods introduced in Activity 3.4, which one would you use to determine the area of the region shown in **Figure 3.42**? Explain.

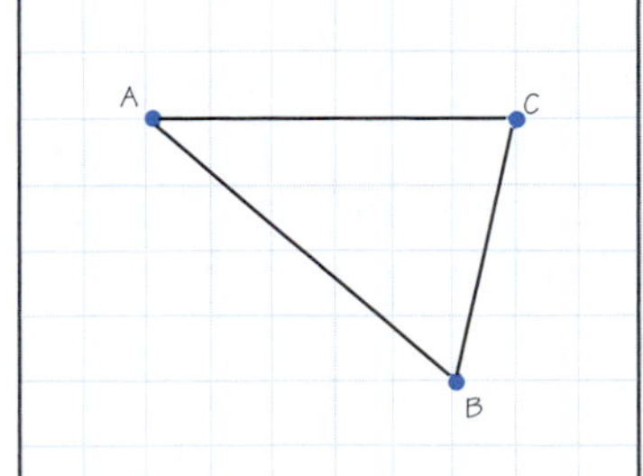

FIGURE 3.42.

3. The drawings in **Figure 3.43 a–d** are the result of four simulations that placed 100 dots at random within a rectangular domain with dimensions 15′ by 10′. Estimate the area of the region in the upper-right corner. Explain how you got your estimate.

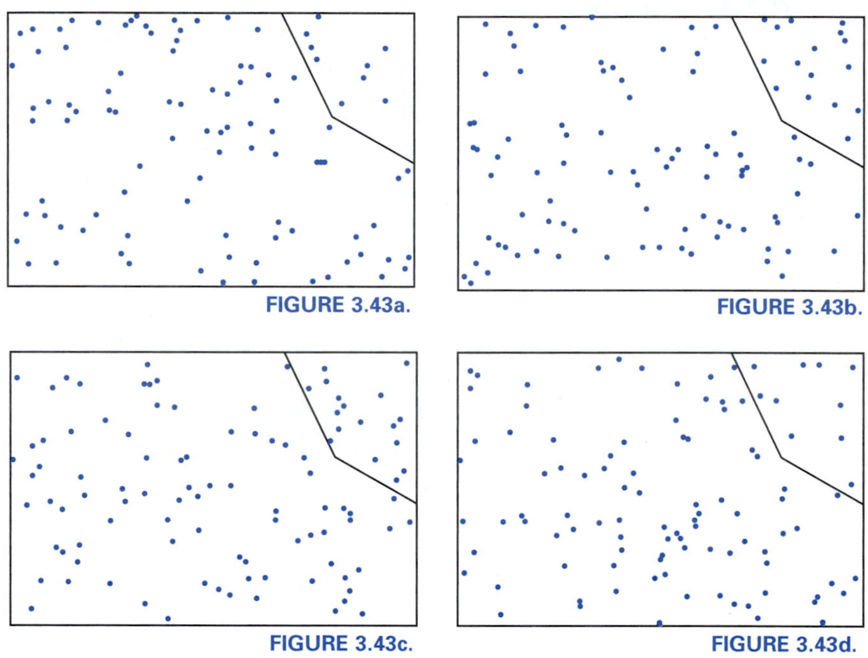

FIGURE 3.43a.

FIGURE 3.43b.

FIGURE 3.43c.

FIGURE 3.43d.

4. Now try to estimate the area of a region that does not have the shape of a polygon. Suppose you wanted to find the area of Lee County, shown as a rough sketch in **Figure 3.44**. (The scale of the drawing is about 22 miles = 1 inch.)

   Write a plan to estimate the area of Lee County. Use your plan to estimate the area. Be sure to record proper units.

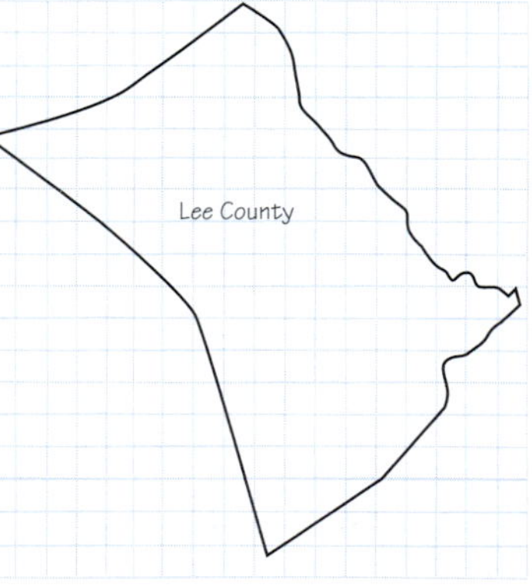

FIGURE 3.44.

Take a moment to think about how the new methods you learned will change your first model of rainfall. Look back over the mathematical process used in Activity 3.1 for solving the problem. Because of the availability of new information, that method needed to be revised. The work that you have done recently suggests that the newer model requires a different approach.

5. a) What assumption determines the region represented by each rain gauge?

   b) Describe the mathematical process used to determine the exact region that is represented by each gauge.

   c) What methods were developed to locate each region?

   d) In order to use these regions to estimate rainfall, what other information did we explore? What methods were developed?

   e) What kind of calculation will allow these regions to contribute to the total rainfall estimate in a way that is proportional to their importance?

6. Before we do the whole state of Colorado, let's try one of our approaches to find the rainfall for just one region. We will use rain gauge 3 from **Figure 3.45**.

1	2	3	4	5	6	7	8
2.11	2.15	1.21	4.45	2.67	2.51	3.11	2.43

FIGURE 3.45.
Rain gauge data from Colorado.

   a) Use Handout 3.1 to determine the part of the state represented by rain gauge number 3. Then use one of the methods (including the Monte Carlo simulation calculator program COLORADO) to determine what percent the region of rain gauge 3 is of the entire state of Colorado.

   b) Estimate the rainfall for the area in ft-acres for the region of rain gauge 3.

   c) Be ready to discuss and compare your answer in class. Are we ready to accurately find the rainfall for the entire state using these methods?

ACTIVITY

3.5

Get the Point?

So far in this chapter you have been given drawings that provide the locations of the rain gauges. You have used geometric principles and construction techniques to locate the boundary lines for the regions around those gauges. How would the process change if coordinates were used to describe the locations instead? Using coordinates, you can calculate the lengths of line segments and the area of regions with greater precision. In this activity you will learn how to find the equation for a perpendicular bisector so that you can locate the boundaries exactly.

PERPENDICULAR BISECTOR EQUATION

The task of this section is still to estimate rainfall, and the problem of finding the Texas totals is still simplified to work with Colorado instead. Before beginning let us review an algebra skill that will be needed later.

Example 

A boundary line between two rain gauges on a coordinate grid has slope of $m = \frac{1}{2}$ and a point on the line has coordinates (4, 3). Find the equation of the boundary line, and write the equation in the slope-intercept form $y = mx + b$

Solution

Use the point-slope form of the equation of the line $y - y_1 = m(x - x_1)$. Replace m with $\frac{1}{2}$, x_1 with 4 and y_1 with 3 in the point-slope equation.

$$y - y_1 = m(x - x_1)$$

$$y - 3 = \frac{1}{2}(x - 4)$$

Use the distributive law to simplify the right side of the equation. Multiply $\frac{1}{2}$ by both x and by 4, and then subtract the two expressions you get.

$$y - 3 = \frac{1}{2} \cdot x - \frac{1}{2} \cdot 4$$

$$y - 3 = \frac{1}{2}x - 2$$

Now add 3 to each side to get y alone on one side of the equatic

$$y - 3 + 3 = \frac{1}{2}x - 2 + 3$$

$$y = \frac{1}{2}x + 1$$

The slope-intercept form of the equation for the boundary line is $y = \frac{1}{2}x + 1$.

CHECK THIS!

Using coordinates such as (1, 3) to locate the position of a rain gauge does change the process you developed to estimate rainfall from the gauge data. You can use many of the methods you have used in this chapter. However, you will set up your calculations in a completely different way.

CHECK THIS!

Recall that the distributive law says if *a*, *b*, and *c* are numbers, then

$$a(b + c) = ab + ac$$

$$a(b - c) = ab - ac$$

This means you can rewrite $a(b + c)$ or $a(b - c)$ without parentheses by multiplying *a* by every term inside the parentheses.

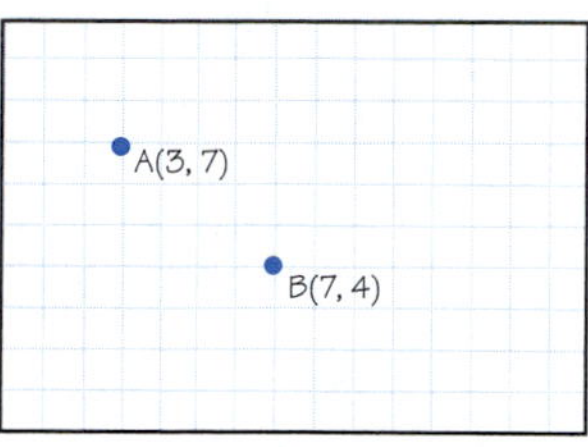

FIGURE 3.46.
Simplified map of Colorado.

Figure 3.46 shows two rain gauges on a grid with coordinates A(3, 7) and B(7, 4). To make the figure a Voronoi diagram, you need to find the boundary line between the region around A and the region around B.

We want to find the equation of the boundary line that is the perpendicular bisector of line segment AB. To write this equation, we need to find the following:

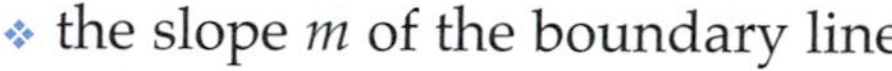

- the slope m of the boundary line
- one point (h, k) that is clearly on the line.

Then use the point-slope form of the equation of the line $y - k = m(x - h)$.

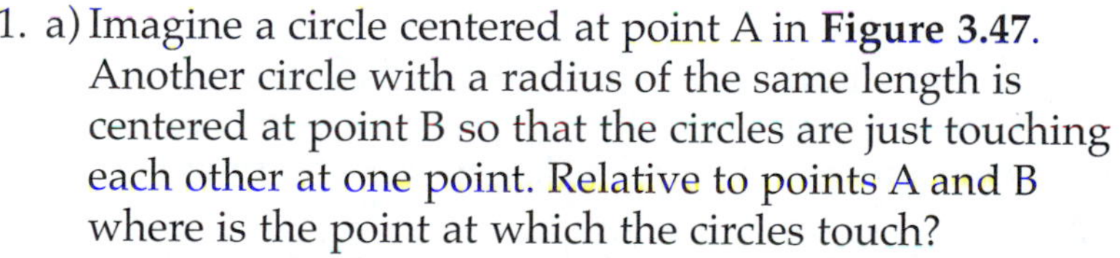

1. a) Imagine a circle centered at point A in **Figure 3.47**. Another circle with a radius of the same length is centered at point B so that the circles are just touching each other at one point. Relative to points A and B where is the point at which the circles touch?

   b) What is the point's position with respect to the perpendicular bisector of the line segment $\overline{AB}$?

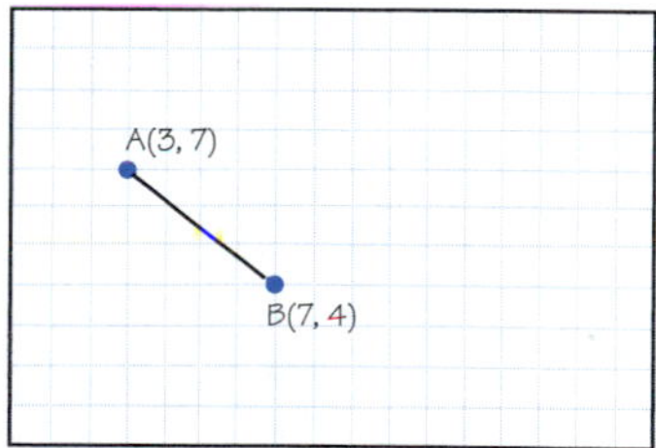

FIGURE 3.47.

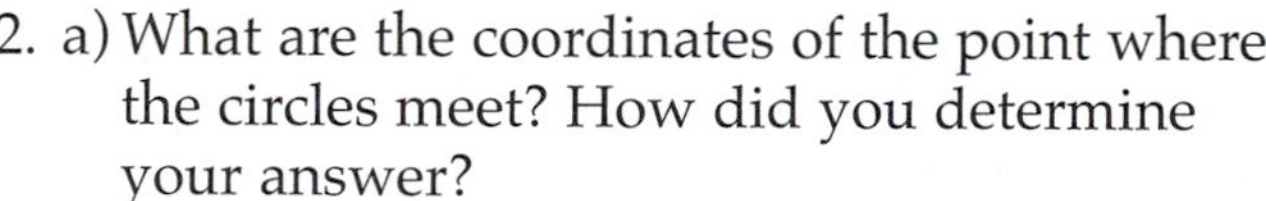

2. a) What are the coordinates of the point where the circles meet? How did you determine your answer?

   b) What is the connection between that point and line segment AB?

3. a) What is the slope of line segment $\overline{AB}$?

   b) How did you determine the slope?

CHECK THIS!

Remember that the boundary line between two rain gauges at A and B is the perpendicular bisector of the line segment $\overline{AB}$.

4. a) What is the slope of the perpendicular bisector to segment $\overline{AB}$?

   b) How did you determine that?

5. Use the point-slope formula to find the equation of the boundary line of the Voronoi region (perpendicular bisector to segment $\overline{AB}$).

6. Each of the points A and B is a center of influence of a region. Use algebra and your results from questions 2–5 to rewrite the equation of the boundary line between A and B in slope-intercept form: $y = mx + b$.

THE PERP CALCULATOR PROGRAM

The properties you discovered in questions 1–6 form the basis of the PERP calculator program. If you have a TI graphing calculator, either RUN the program or type in the program listing from Handout 3.4.

7. Use the PERP calculator program to find the perpendicular bisector of the line segment $\overline{AB}$ for the points A and B in Figure 3.47.

   - Enter the four coordinates for the two points.
   - Make sure you press ENTER after each number.
   - After the coordinates of the midpoint are displayed press ENTER again to get the slope and y-intercept of the perpendicular bisector.

8. Find the equations of the three perpendicular bisectors of the three line segments $\overline{AB}$, $\overline{AC}$, and $\overline{BC}$ defined by the three centers of influence shown in **Figure 3.48**.

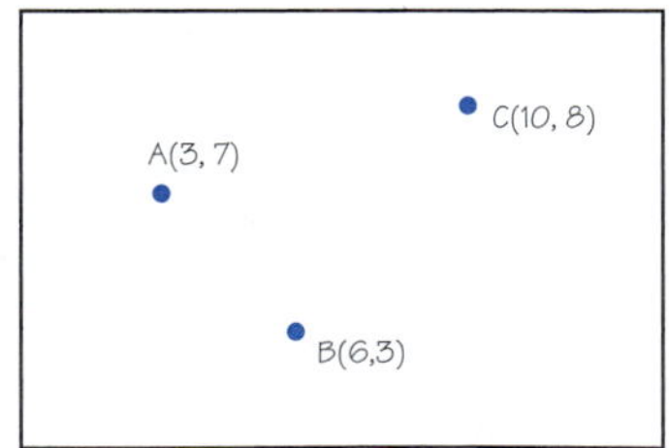

FIGURE 3.48.

SUMMARY

Your rainfall estimate depends on locating the boundary lines of proximity regions. Therefore, it is important to locate these boundaries precisely. A coordinate system may be used to find a boundary line precisely. In this activity you learned the following techniques.

- You discovered how a coordinate grid is used to plot the rain gauge locations.
- You used the point-slope equation to find the equation of the boundary line.

It's not enough to find just one boundary line. The boundary of a proximity region is usually composed of several line segments. To find the area of the region, you need to find all the boundary segments of the region. In addition, it is necessary to find the points where boundary segments meet, since these points create a corner of the region. Activity 3.6 shows you how to find these corners so that you can calculate the area of a region using the methods you studied in Activity 3.4.

HOMEWORK

3.5 Surveying Boundaries

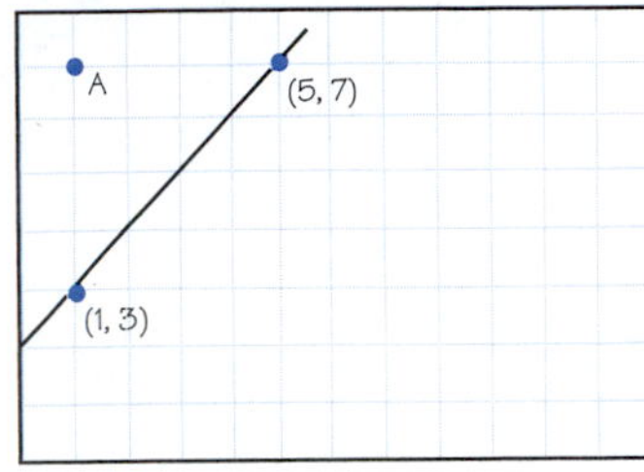

FIGURE 3.49.

1. a) The points (1, 3) and (5, 7) lie on the boundary line of a region of proximity. Explain how to find the equation of the boundary line.

   b) Find the equation of the boundary line in slope-intercept form $y = mx + b$.

   c) What are the coordinates for the center of influence that is labeled A in **Figure 3.49**?

   d) Suppose the point B is another center of influence. If the line segment $\overline{AB}$ is bisected by, and perpendicular to, the boundary line, what is the slope of the line segment?

   e) What is the equation of the line that contains the line segment $\overline{AB}$? Write the equation in slope-intercept form $y = mx + b$.

   f) Find the coordinates of B. Explain how you got your answer.

2. a) The equation of a boundary line for a region of proximity is $y = 2x + 3$. Find the equation of the line that is perpendicular to the boundary line and intersects the boundary at (2, 7). Write the equation in slope-intercept form.

   b) Suppose the boundary line is the perpendicular bisector of the line segment $\overline{AB}$ whose end points are centers of influence of proximity regions. If A has coordinates (–2, 9), what are the coordinates of B?

3. Find the equations of the lines that are the boundaries of the region surrounding the center of influence at B in **Figure 3.50**.

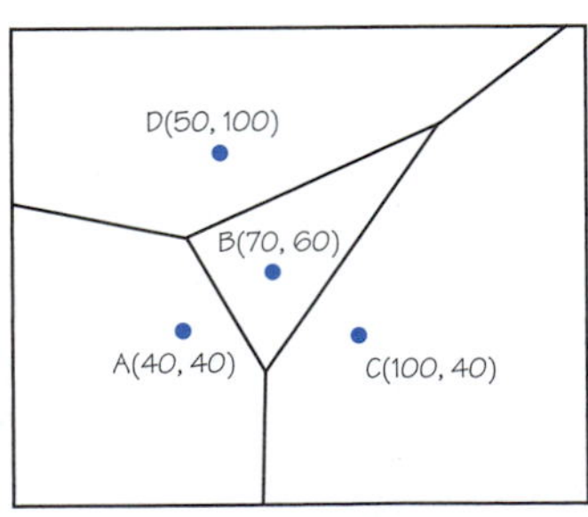

FIGURE 3.50.

ACTIVITY

3.6

At the Boundaries

In the last activity you found the equations of the boundary lines of a region around a rainfall gauge. Since two boundary lines of a region meet to form a corner of the region, it is important to locate precisely the point where these lines meet. Knowing the locations of the corner of a region will help you estimate the area of the region around a rain gauge. You can use the equations of two lines to find the point of intersection of the lines (see Figure 3.51).

FINDING THE INTERSECTION POINT

Figure 3.51 shows a Voronoi diagram with three centers of influence (A, B, and C) and two boundary lines. The equation of the boundary line between A and B is $y = 0.4x + 3.5$. The equation of the boundary line between B and C is $y = -2x + 22.5$.

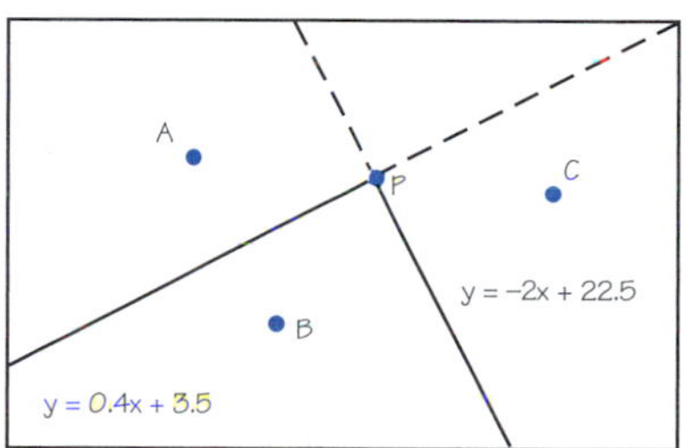

FIGURE 3.51.

To find the coordinates of the point of intersection P, you need to solve the pair of equations taken together. Two equations taken together are called a **system of equations**. In the case of the two lines in Figure 3.51, use the two equations $y = 0.4x + 3.5$ and $y = -2x + 22.5$ to form the system of equations

$$\begin{cases} y = 0.4x + 3.5 \\ y = -2x + 22.5 \end{cases}.$$

You may have studied a few methods for solving this type of problem. Since both equations have the variable y alone on one side, there is a simple method to solve this system of equations. The following example shows you how.

Example 

Solve the following system of equations.

$$\begin{cases} y = 4x - 5 \\ y = 2x - 3 \end{cases}$$

Solution

Notice that both equations have the variable y alone on the left side. The first equation means that the value of the variable y is the same as the expression $4x - 5$. At the same time, the second equation means the value of the variable y is the same as the expression $2x - 3$. At the point of intersection, the y-values are equal, so the two expressions must also be equal.

$$4x - 5 = 2x - 3$$

We can use algebra to solve this resulting equation for x. Let's try to get the variable x alone on the left side of the equation. First, collect all the terms in x on the left side by subtracting 2x from both sides.

$$4x - 2x - 5 = 2x - 2x - 3$$
$$2x - 5 = -3$$

We want to get all the terms with just numbers on the right side. Add 5 to both sides of the equation.

$$2x - 5 + 5 = -3 + 5$$
$$2x = 2$$

All that's left to do is divide both sides by 2 to get x alone on the left side.

$$\frac{2x}{2} = \frac{2}{2}$$
$$x = 1$$

Now that we know the value of x, find the value of y. Replace x with 1 in the first equation of the system of equations.

$$y = 4x - 5$$
$$y = 4(1) - 5$$
$$y = 4 - 5$$
$$y = -1$$

Since $y = -1$, the point of intersection has coordinates $(1, -1)$.

Businesses often need to solve systems of equations to analyze their operations. For example, these analyses were used to help an oil refinery run its operations more efficiently and to help an airline improve its service. You will need to use a system of equations to improve your rainfall estimate.

1. Solve a system of equations to find the coordinates of point P in Figure 3.51.

2. Run the calculator program INTRSECT to find the coordinate of P in Figure 3.51. Do you get the same answer you got to question 1?

3. Find the point of intersection of the two lines $y = x + 2$ and $y = -3x - 1$. Use algebra techniques, and then verify the answer using the calculator program.

CHECK THIS!

You can use the same method to find formulas for the coordinates of the point of intersection of any two lines. If the general equations of two lines are $y = m_1x + b_1$ and $y = m_2x + b_2$ and $m_1 \neq m_2$, then the formulas for x and y are

$$x = \frac{b_2 - b_1}{m_1 - m_2}$$

$$y = \frac{m_1b_2 - m_2b_1}{m_1 - m_2}$$

This formula is used in the INTRSECT calculator program to find the equation of the perpendicular bisector.

CALCULATING DISTANCES AND AREAS

Knowing coordinates helps you find the lengths of line segments with much greater accuracy. In fact, the real limit to an exact measurement is accurately determining the coordinates. Global positioning satellites (GPS) have become increasingly important because they can show your exact position on the Earth. A GPS satellite orbits around the Earth and signals to ground stations to determine your exact position.

We will take the approach again that a simple example can help reveal the general steps to a process. **Figure 3.52** shows two points. We would like to find the length of segment $\overline{AB}$ using the coordinates of A and the coordinates of B.

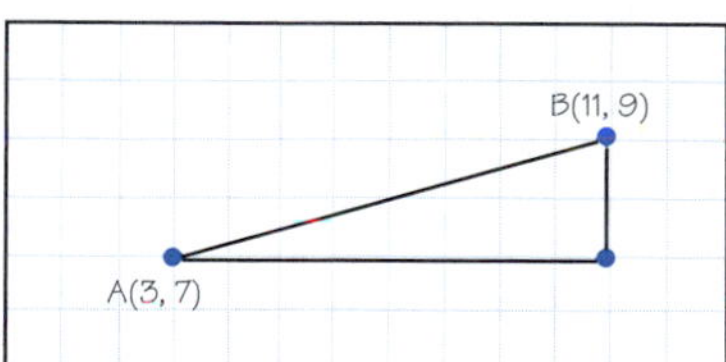

FIGURE 3.52.

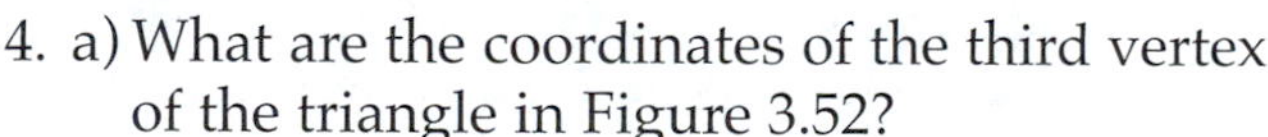

4. a) What are the coordinates of the third vertex of the triangle in Figure 3.52?

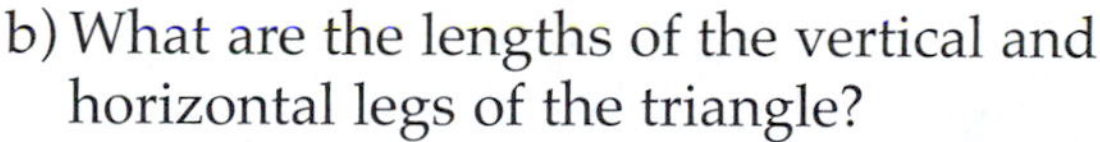

b) What are the lengths of the vertical and horizontal legs of the triangle?

CHECK THIS!

Recall that the Pythagorean Theorem says that in a right triangle $c^2 = a^2 + b^2$, where c is the measure of the hypotenuse and a and b are the measures of the two legs.

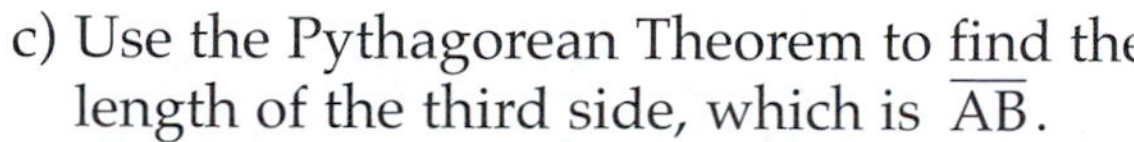

c) Use the Pythagorean Theorem to find the length of the third side, which is $\overline{AB}$.

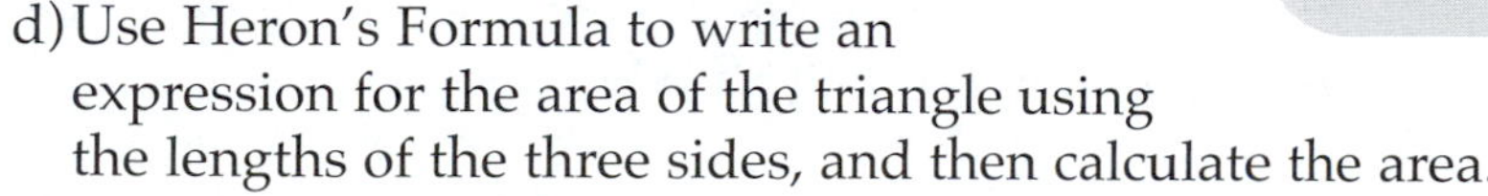

d) Use Heron's Formula to write an expression for the area of the triangle using the lengths of the three sides, and then calculate the area.

e) Find the area of the triangle using the formula $A = \frac{1}{2}bh$. Do the two formulas give the same answer? Is one easier to work with than the other?

Let's take another look at the coordinates of the points because those coordinates may be used to find the distance between the two points.

5. a) What is the difference between the x-coordinate of B and the x-coordinate of A?

   b) How does your result in (a) compare with the length of the horizontal leg of the triangle?

   c) What is the difference between the y-coordinate of B and the y-coordinate of A?

   d) How does your result in (c) compare with the length of the vertical leg of the triangle?

Your answers to question 5 show that, in general, for two points A (x_1, y_1) and B (x_2, y_2)

- the length of the base of the right triangle whose hypotenuse is $\overline{AB} = |x_2 - x_1|$
- the length of the height of the right triangle $= |y_2 - y_1|$

Using the Pythagorean Theorem and the expressions for these two legs of the triangle, we get the length of the line segment $\overline{AB} = \sqrt{(x_2 - x_1)^2 + (y_2 - y_1)^2}$. This expression is known as the **distance formula.**

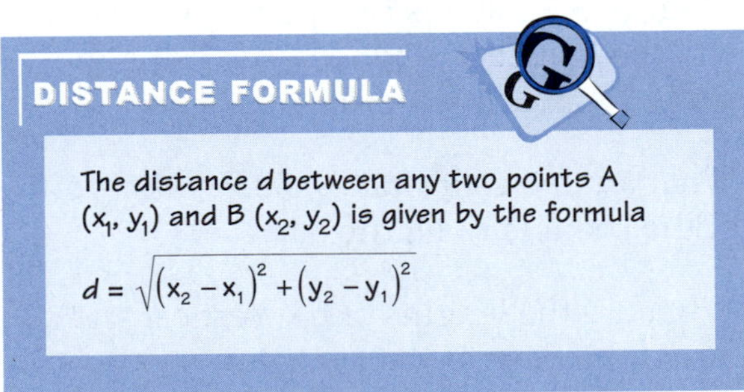

DISTANCE FORMULA

The distance d between any two points A (x_1, y_1) and B (x_2, y_2) is given by the formula

$$d = \sqrt{(x_2 - x_1)^2 + (y_2 - y_1)^2}$$

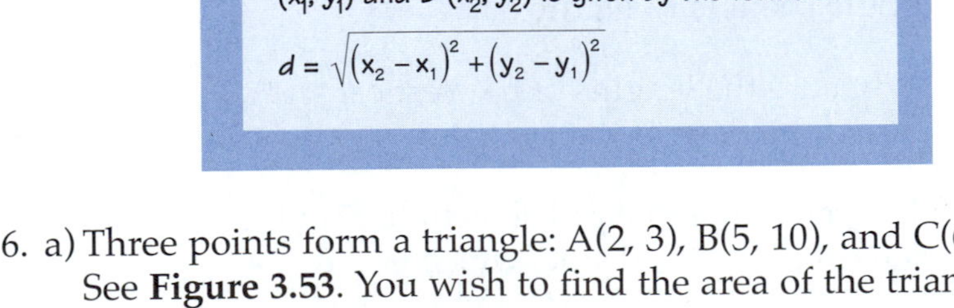

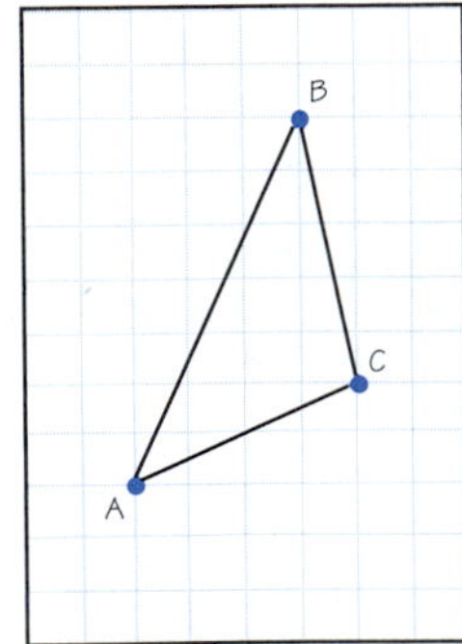

FIGURE 3.53.

6. a) Three points form a triangle: A(2, 3), B(5, 10), and C(6, 5). See **Figure 3.53**. You wish to find the area of the triangle using just the coordinates given. Can the formula $A = \frac{1}{2}bh$ be used in this situation? Why?

   b) Use the distance formula to find the lengths $\overline{AB}$, $\overline{AC}$, and $\overline{BC}$. Express your answers to two decimal places.

   c) Use Heron's Formula to find the area of ΔABC.

7. a) Run the calculator program HERON to calculate the area of the triangle in question 6. When prompted, choose 'COORDS' by pressing 0, and then ENTER. Does the calculator verify your answer from question 6(c)?

   b) Examine the program listing using EDIT mode (be careful not to alter the program), or look at Handout 3.6. Which program line(s) use the distance formula?

   c) Which lines use Heron's Formula?

PUTTING IT ALL TOGETHER

Now you have all the tools you need to find the areas of regions around rain gauges. **Figure 3.54** is a completed Voronoi diagram showing the location of four rain gauges, whose coordinates are A(10, 30), B(20, 20), C(30, 30), and D(17, 10). The space between grid marks represents a distance of five miles.

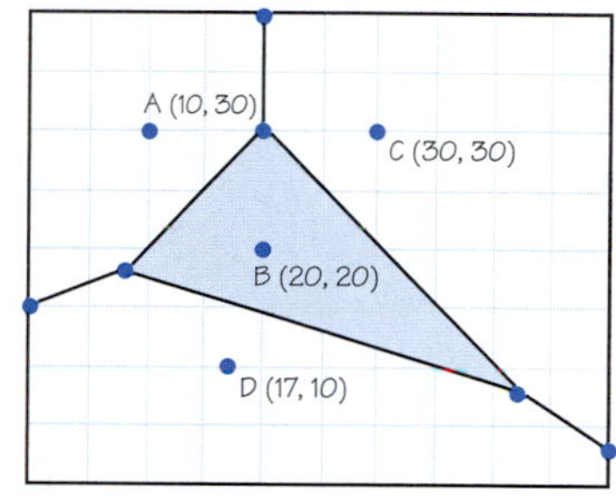

FIGURE 3.54. Completed Voronoi diagram.

8. a) We would like to know the area of the proximity region surrounding the rain gauge centered at B (shaded region). What are the equations of the boundary lines forming the proximity region?

   b) What are the coordinates of the three vertex points of the triangle?

   c) What are the lengths of the three sides of the triangle?

   d) What is the area of the triangle?

9. a) Which line segment lies on the boundary line between regions C and D?

   b) How might you find the coordinate of the point on the border of the domain and on the boundary line between region C and region D?

CHECK THIS!

Recall that the domain for this kind of problem is the area that is being divided—the total area of the state, the county, and other regions.

SUMMARY

In this activity you explored the final tools you needed to find the areas of regions around rain gauges plotted on a grid.

- You learned how to solve a system of equations.
- You solved systems of equations to locate precisely the intersection of boundary lines to determine the coordinates of the corners of a region.

This activity shows how you may find the areas of regions plotted on coordinate grids algebraically. This method provides another way to calculate the rainfall estimate for Texas. You will be asked to find the rainfall estimate for Texas in the Modeling Project at the end of the chapter.

HOMEWORK

3.6 Power in Those Rules

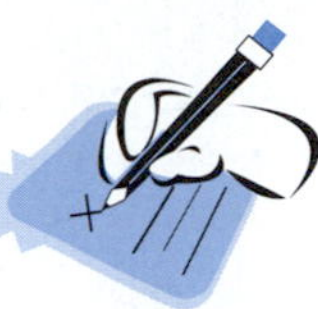

1. a) Find the distance between the point C in Figure 3.54 and the point where $\overline{BC}$ meets the boundary of the proximity region around B.

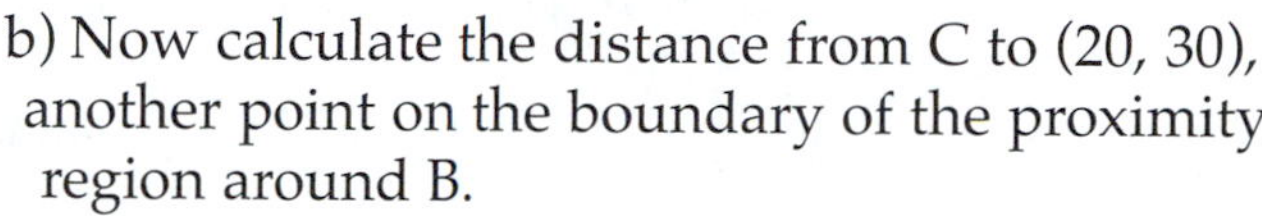

   b) Now calculate the distance from C to (20, 30), another point on the boundary of the proximity region around B.

   c) Now calculate the distance from C to (30, 20), another point on the boundary of the proximity region around B.

   d) Based on your answers to (a)–(c), what do you think is the shortest distance from C to the boundary of the proximity region around B?

   e) The distance from a line to a point P not on the line is defined as the shortest distance between point P and *any* point that is on the line. Explain how to find that distance for any line and point P.

2. Refer back to Figure 3.54. Find the distance that gauge D is away from the boundary line that divides the region it represents from the region represented by gauge B.

3. Figure 3.54 shows two lines that intersect; you were asked to find the intersection point knowing the specific equations. Now generalize what you did in solving that system of equations, only working with labels instead of numbers. Let the two equations be: $y = m_1x + b_1$ and $y = m_2x + b_2$.

   a) Set the two expressions for y equal to themselves.

   b) Subtract the variable term m_2x and the constant term b_1 from both sides of the equation and simplify. What does the equation look like now?

   c) Use the Distributive Property to rewrite the variable side in "factored" form.

   d) Solve for x by dividing both sides of the equation by the expression in front of the variable.

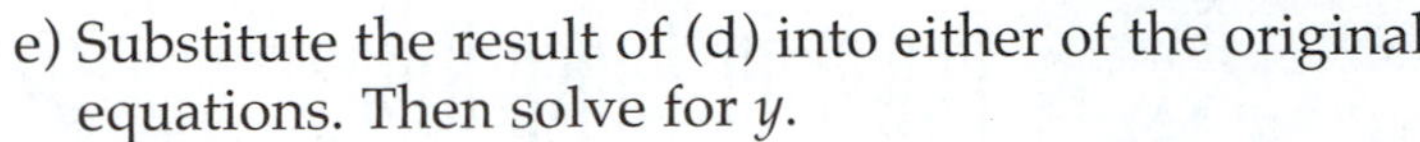

e) Substitute the result of (d) into either of the original equations. Then solve for y.

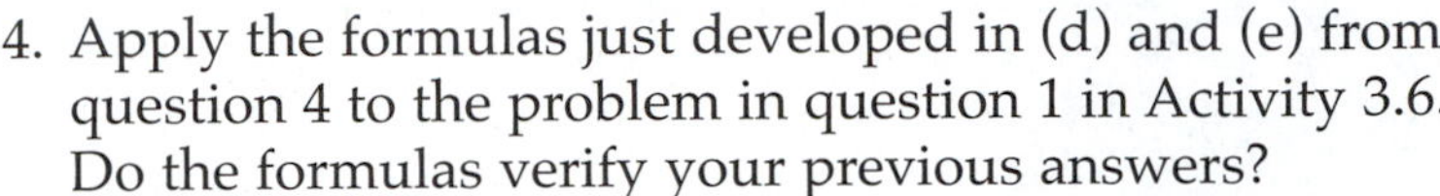

4. Apply the formulas just developed in (d) and (e) from question 4 to the problem in question 1 in Activity 3.6. Do the formulas verify your previous answers?

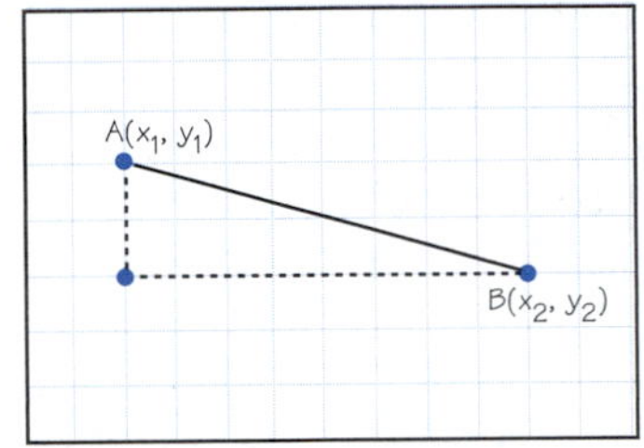

FIGURE 3.55.

In Activity 3.6, you found the lengths of two sides of the right triangle whose hypotenuse is AB (see **Figure 3.55**).

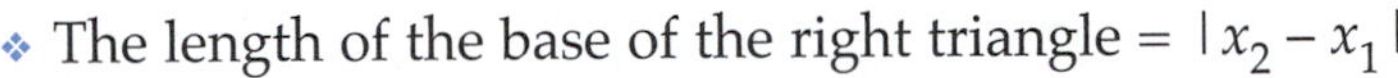

- The length of the base of the right triangle = $|x_2 - x_1|$
- The length of the height of the right triangle = $|y_2 - y_1|$

In the next question, use this information to prove the distance formula.

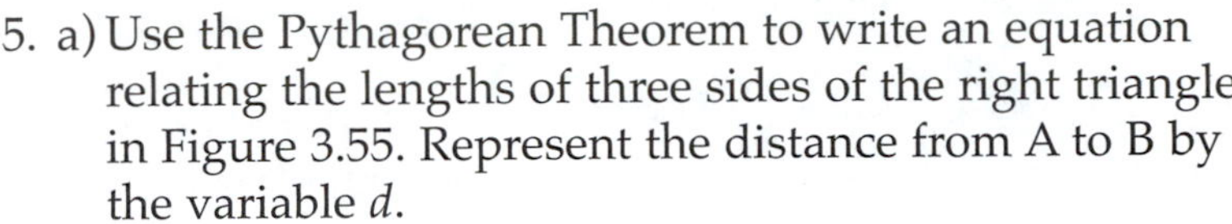

5. a) Use the Pythagorean Theorem to write an equation relating the lengths of three sides of the right triangle in Figure 3.55. Represent the distance from A to B by the variable d.

   b) Solve for d to get the distance formula.

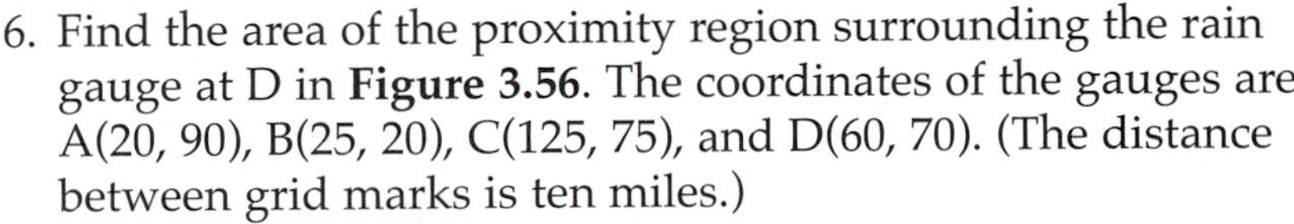

6. Find the area of the proximity region surrounding the rain gauge at D in **Figure 3.56**. The coordinates of the gauges are A(20, 90), B(25, 20), C(125, 75), and D(60, 70). (The distance between grid marks is ten miles.)

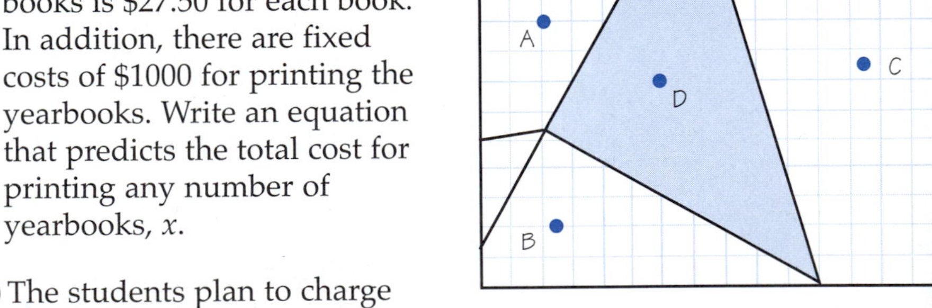

FIGURE 3.56.

7. a) Bright Springs High School yearbook editors know that the cost, c, of printing year books is $27.50 for each book. In addition, there are fixed costs of $1000 for printing the yearbooks. Write an equation that predicts the total cost for printing any number of yearbooks, x.

   b) The students plan to charge $30 for a copy of the yearbook. From past experience, the editors know they will receive another $600 from selling advertisements in the yearbook. What expression represents the total revenue (the amount they receive) for selling x yearbooks and for selling advertisements? Use the variable r for revenue.

c) The editors want to know how many yearbooks they must sell so the total cost equals the total revenue. Describe how to determine how many books they must sell.

d) How many yearbooks must they sell so that revenue equals costs?

e) If the students sell exactly enough books so the amount they receive is equal to their cost, how much did they receive?

f) If the editors wish to earn a profit of $2550, how many books must they sell?

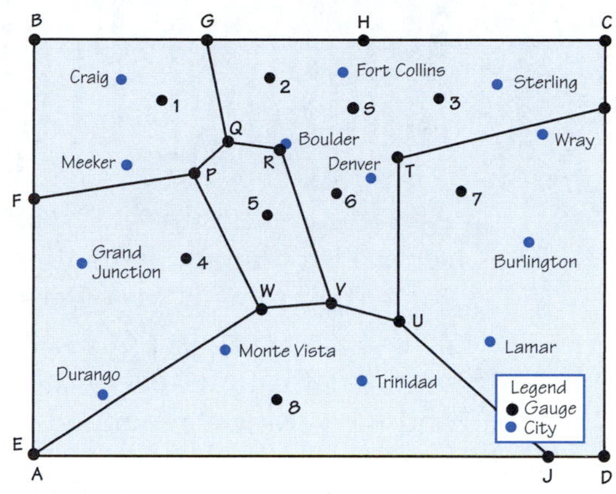

FIGURE 3.57.
Incomplete Voronoi Regions for Colorado Rain Gauges

8. We are now ready to answer the question: How much is the estimated rainfall in Colorado. We have done some of the calculations for you, but left some blank. It is your job to:

   a) Use the map of Colorado (**Figure 3.57**) to determine the Voronoi regions for rain gauges 2, 3, and 6.

   b) Complete the proximity regions area totals in **Figure 3.58**.

   c) Calculate the total area and rainfall for the state of Colorado.

Gauge Location	Location (miles)	Rain Gauge Reading (in inches)	Proximity Region	Area of Region
1	(83,234)	2.11	FGBQP	11,560 mi^2
2	(155,249)	2.15	?	?
3	(265,235)	1.21	?	?
4	(98,132)	4.45	EFPW	17,400 mi^2
5	(150,158)	2.67	PQRVW	5710 mi^2
6	(197,173)	2.51	?	?
7	(280,173)	3.11	TIDJU	24,650 mi^2
8	(159,36)	2.43	AEWVUJ	20,420 mi^2

FIGURE 3.58.
Incomplete Rain Gauge Data for Colorado.

Modeling Project Rainy Days in Texas

At various times in this chapter, you have been confronted with the central problem of estimating the total amount of rainfall in a state. To solve the problem, you were provided data on the amount of rain collected at certain points around the state, and you developed mathematical models for describing both the conditions of the problem and the solution generated by making specific assumptions.

Now we are ready to tackle the problem of finding the rainfall for the entire state of Texas.

The locations of the seventeen "first-order" measuring stations across Texas are provided in **Figure 3.59**. An 18th gauge, located in Shreveport, LA, has been added to the list; scientists often include it since there isn't a first-order station located in East Texas.

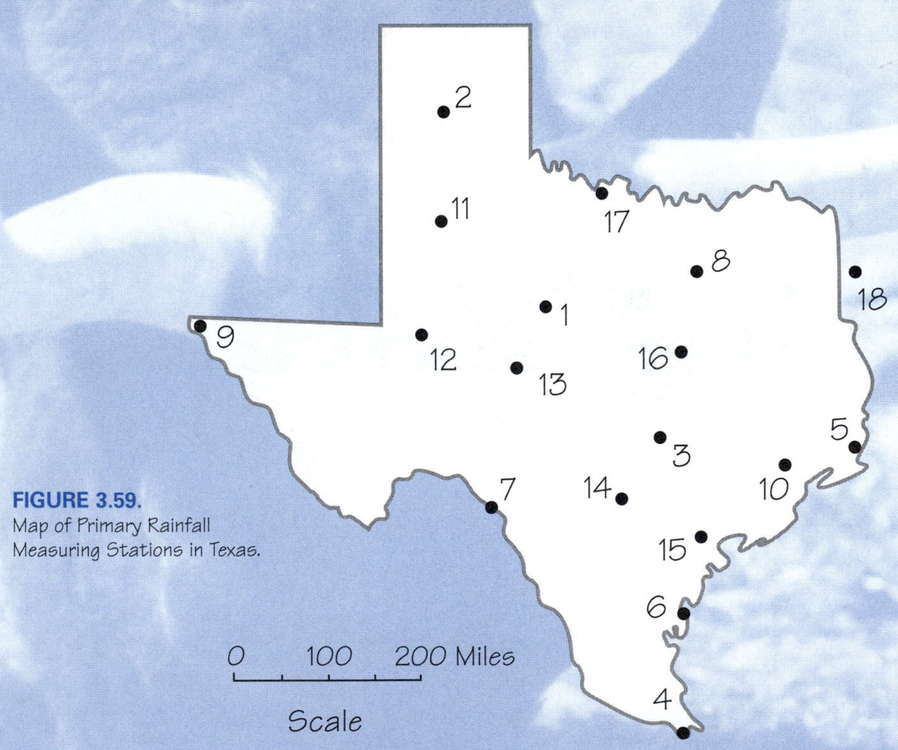

FIGURE 3.59.
Map of Primary Rainfall Measuring Stations in Texas.

Gauge Number	Location	Rainfall Reading	Gauge Number	Location	Rainfall Reading
1	Abilene	27.08	10	Houston	9.63
2	Amarillo	43.27	11	Lubbock	76.79
3	Austin	24.95	12	Midland/Odessa	22.67
4	Brownsville	46.79	13	San Angelo	17.10
5	Beaumont/Port Arthur	36.21	14	San Antonio	23.38
6	Corpus Christi	44.10	15	Victoria	33.92
7	Del Dio	36.16	16	Waco	65.05
8	Dallas/Fort Worth	23.01	17	Wichita Falls	23.78
9	El Paso	45.00	18	Shreveport, LA	69.20

FIGURE 3.60.
Legend and Rainfall Data from 1997.

1. Use any two methods you learned in this chapter to determine the estimated rainfall in Texas in 1997.
2. Compare the two methods you used. Do you believe one is more reliable than the other?

Practice Problems

Review Exercises

1. On a piece of graph paper create a rectangular domain with (0, 0) and (10, 8) as corner points, then plot the following four points: A(4, 6), B(2, 3), C(9, 3), and D(5, 1). Draw a rough sketch of the Voronoi diagram for this domain.

Use **Figure 3.61**, the Voronoi diagram, for a square domain with (0, 0) and (10, 10) as corner points, to answer questions 2, 3, and 4. The four points shown have coordinates: A(1, 3), B(2, 6), C(7, 4), and D(9, 5).

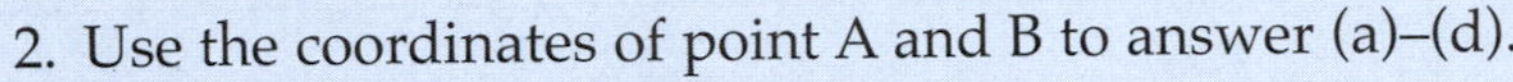

FIGURE 3.61. Use with questions 2, 3, and 4.

2. Use the coordinates of point A and B to answer (a)–(d).
   a) Find the coordinates for the midpoint of line segment AB.
   b) Find the slope of line segment AB.
   c) Find the distance between points A and B.
   d) Write the equation for the perpendicular bisector to AB in point-slope form, then use algebra to re-write the answer into slope-intercept form.
3. Line segment JF is described by the equation $y = 7x - 33$; line segment JG has equation $y = -2x + 20.5$.
   a) Write an equation that can be used to find the coordinates of point F and then solve it.
   b) Write an equation that can be used to find the coordinates of point G and then solve it.
   c) Write an equation that can be used to find the coordinates of point J and then solve it.
4. **Figure 3.62** contains calculated areas for the four proximity regions shown in Figure 3.61, and **Figure 3.63** has the rain gauge data corresponding to them.

Region	Area
A	18.35 mi²
B	30.23 mi²
C	29.48 mi²
D	21.97 mi²

FIGURE 3.62.

Region	Rainfall Depth
A	3.26 in.
B	2.81 in.
C	1.95 in.
D	2.24 in.

FIGURE 3.63.

a) Calculate a simple average for the rainfall depths, and use that to determine the total rainfall for the domain.

b) Calculate a weighted average for the rainfall depths, and use that to determine the total rainfall for the domain.

5. Suppose a one-acre farm receives 15″ of rainfall during the rainy season.

a) How many cubic feet of water is that? (One acre has an area of 43,560 square feet.)

b) How many gallons of water is Mother Nature providing the farm? (One gallon of water has a volume of 7.46 cubic feet.)

6. Consider the five-sided polygon as shown in **Figure 3.64**, with the origin O(0, 0) in the lower-left corner.

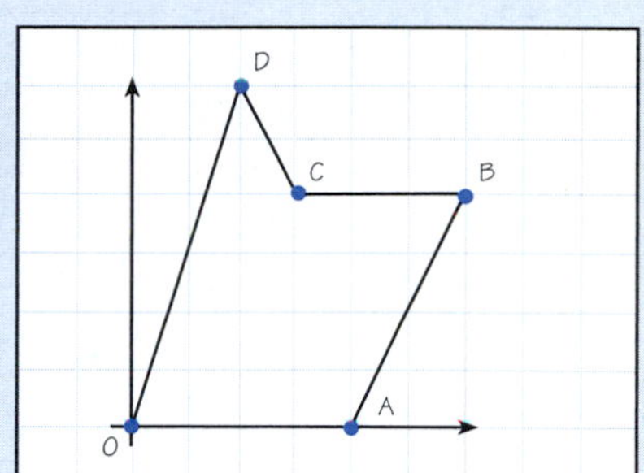

FIGURE 3.64.
Use with question 6.

a) If you want to calculate the area of the region by using the formulas $A_{rectangle} = bh$ and $A_{triangle} = (1/2)bh$, how would you sub-divide the region into pieces? Why? What is the area of each piece?

b) If you wanted to estimate the area by a Monte Carlo simulation, what would be the easiest shape to use for the region representing the total area? What kind of numbers would you want to generate on the calculator for the random x- and y-coordinates?

c) Use Pick's Formula to calculate the area of this region.

d) You're thinking of using Heron's Formula to find the area. How many triangles would you work with? How could you tell that from knowing the original shape is a five-sided polygon?

e) Which triangles would you use? What are the lengths of their sides?

7. The **center of gravity** of an object is the balance point for weights hanging on a support rod. **Figure 3.65** shows a yardstick with a 500-gram weight located at the 8″ mark, a 200-gram weight at the 14″ mark, and a 1000-gram weight at the 30″ mark.

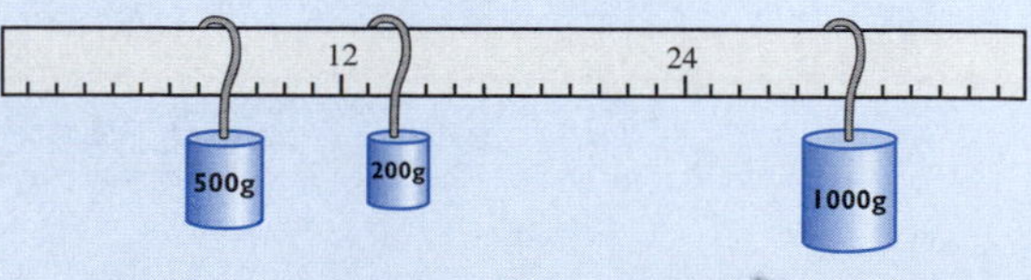

FIGURE 3.65.
Balancing weights on a yardstick.

Determining where the center of gravity is located is a weighted-average calculation. Think of the problem this way: *"A certain percent of the weight is located here, and another percent is located there…."*

a) What is the center of gravity for the situation shown in Figure 3.65?

b) Tommy and Timmy love to play on the teeter-totter, shown in **Figure 3.66**. Tommy weighs 180 lbs, and Timmy weighs 150 lbs. They have found that the 12-foot board will balance when their 90-lb. sister Tammy helps out.

To balance the weight, where should Tammy sit? (Hint: The center of weight should be at the balance point!)

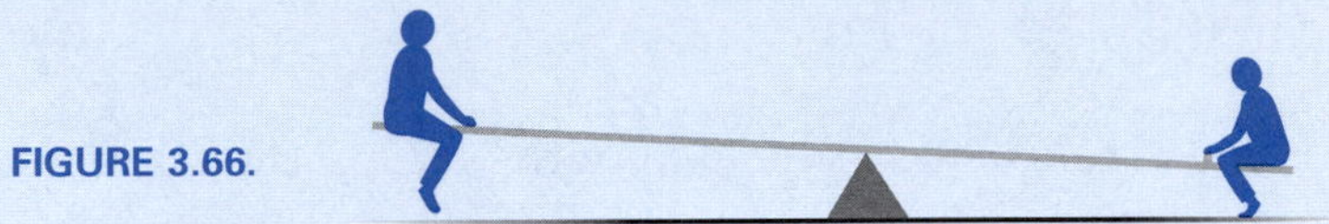

FIGURE 3.66.

8. A **kite** is a four-sided polygon with the property that two pairs of adjacent sides have the same length, as shown in **Figure 3.67**. Prove that diagonals of a kite are always perpendicular to each other.

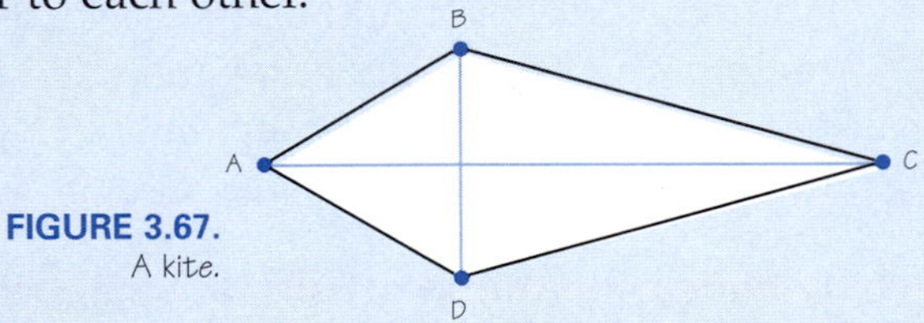

FIGURE 3.67.
A kite.

9. Estimate the area of Big Bend National Park, using **Figure 3.68**. (The scale used to draw the map is 4.25 miles/grid line.)

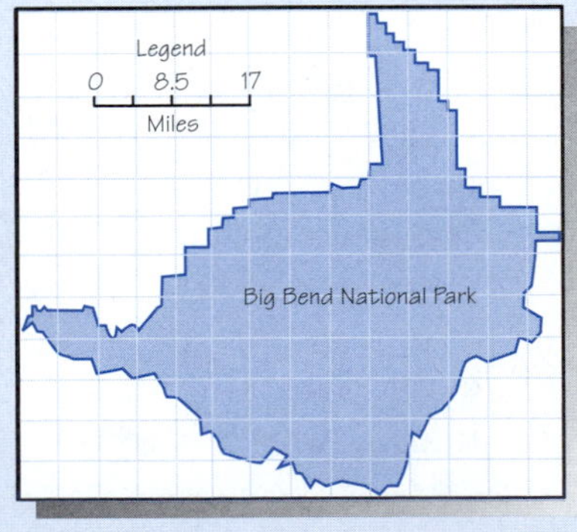

FIGURE 3.68.
Map of Big Bend National Park.

10. **Figure 3.69** shows a triangle with the given side lengths provided.

    a) Knowing that the triangle is *isosceles*, what relationship exists between the altitude of the triangle and its base? Use that relationship (and another famous one) to calculate the height of the triangle.

    b) Generalize the steps taken in (a) to develop a formula for calculating the height of an isosceles triangle with base b units long and side lengths c units long.

    Figure 3.70 shows a different triangle, also with given side lengths.

    c) Calculate the area using Heron's Formula, and use that result to find the triangle's height.

    d) Generalize the steps you took in (a), and write a formula for calculating the height of a triangle from knowing side lengths a, b (base), and c.

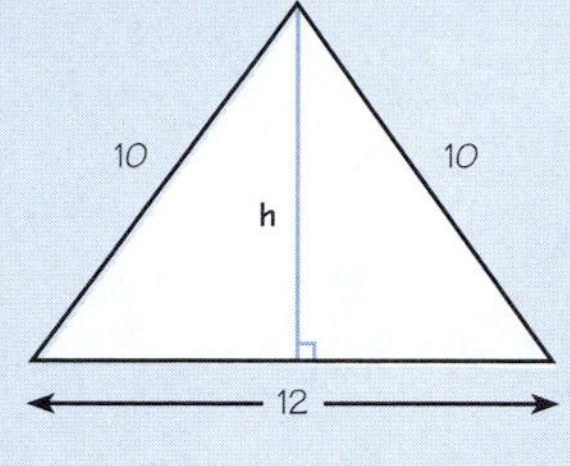

FIGURE 3.69.

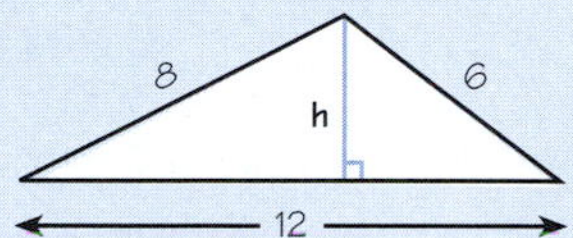

FIGURE 3.70.

11. A Voronoi region was broken up into triangles, with measurements (in miles) recorded as in **Figure 3.71**. Find the total area of the region.

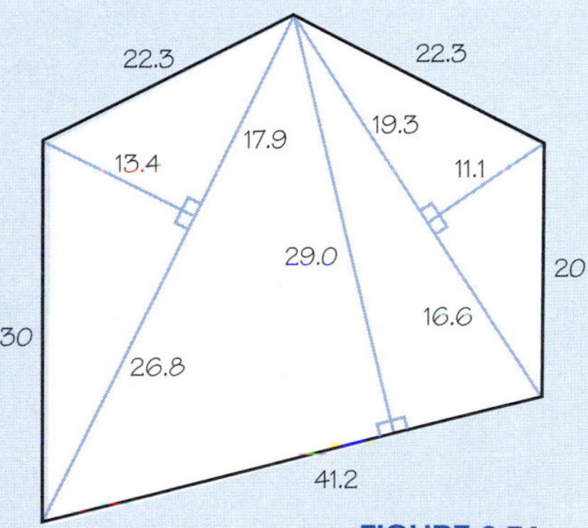

FIGURE 3.71.

12. A **trapezoid** is a quadrilateral with two sides parallel. It is customary to refer to the two parallel sides as **bases**, and call them b_1 and b_2, as shown in **Figure 3.72**.

    a) You can take two identical trapezoid shapes, fit them together (with one upside-down), and rearrange it to form a rectangle. **Figure 3.73** shows the steps.

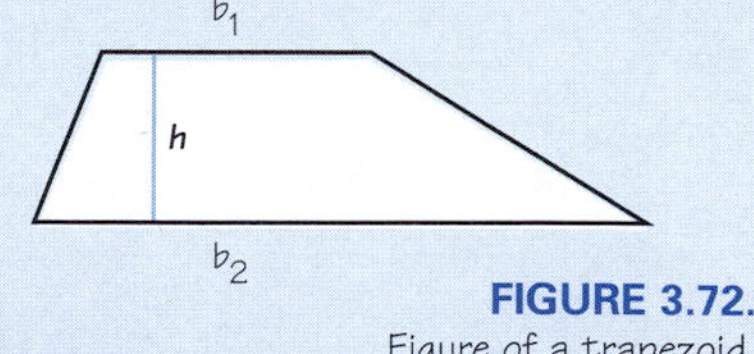

FIGURE 3.72.
Figure of a trapezoid.

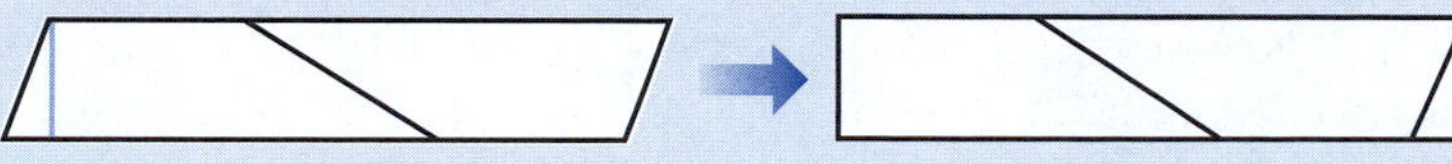
FIGURE 3.73.

Use this process knowing the lengths of the two bases and the height to develop a formula for finding the area of a trapezoid.

b) One method used to determine area estimates is to divide a region into trapezoids, find the area of each trapezoid, and then add all the areas together. A survey team, wanting to use that approach to find the area of the pond, measured the distance directly across the pond every five feet; their measurements are shown in **Figure 3.74**. Find the area of the pond.

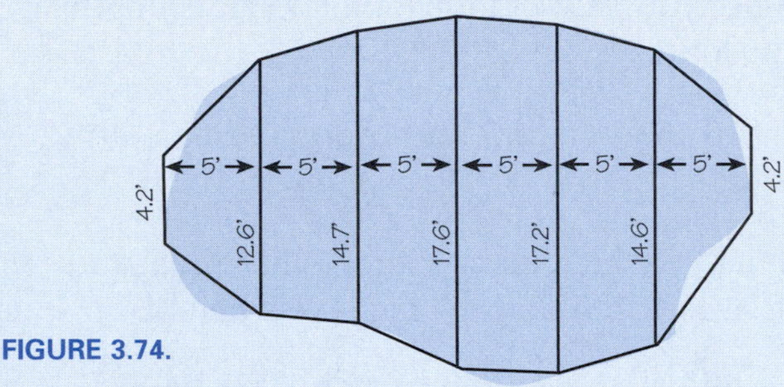

FIGURE 3.74.

13. A ground-based doppler radar unit was installed at Randolph A.F.B., outside of San Antonio, to provide early detection of thunderstorms. It has an effective range of 100 miles.

a) Victoria is approximately 74 miles east and 46 miles south of Randolph A.F.B. Will its location show up on the radar screen?

b) Several cities in the area are shown in **Figure 3.75**, a map having San Antonio as center. The grid marks represent a distance of 25 miles. Which cities can have the weather conditions monitored by the radar unit at Randolph A.F.B.? Explain.

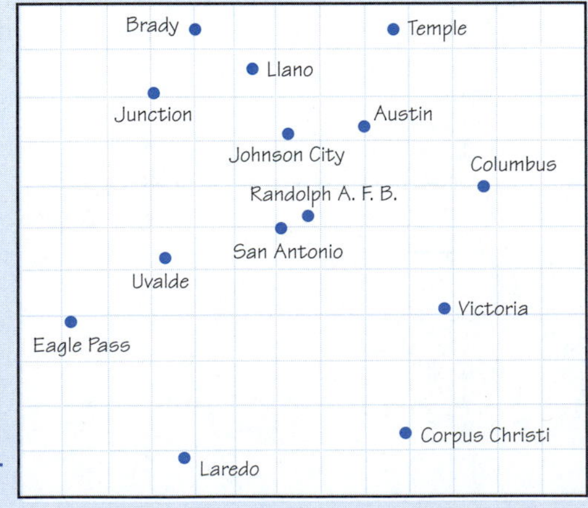

FIGURE 3.75.

Voronoi diagrams can be used in a variety of other situations besides estimating rainfall. In the next several problems, some of the many application areas are explored.

14. In **Figure 3.76**, a rectangular region with dimensions 10 x 8 has eight centers of proximity with the following coordinates: (2, 7), (7, 7), (3, 5), (5, 4), (8, 5), (1, 2), (6, 2), and (9, 2). The domain has been resolved into a completed Voronoi diagram.

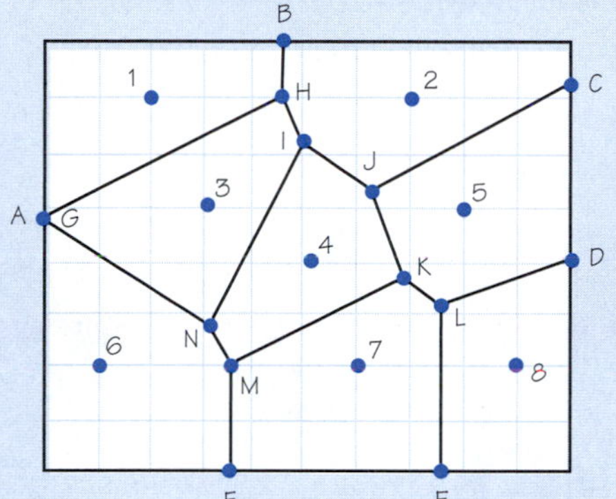

FIGURE 3.76.

a) Working with the drawing, if two regions share a boundary (edge), draw a line segment connecting their centers of influence. When done, describe the pattern made, and compare it to the original Voronoi diagram.

b) This kind of drawing is called the **Delaunay triangulation**. Do this again for other Voronoi diagrams including some of the ones you used in the chapter. You should find that the basic pattern is the same. Explain why it *has* to be that way.

c) Pick any one of the regions formed by the Delaunay triangulation, and imagine a circle going through the three vertex points. Is it possible for that circle to contain a center of proximity? Explain.

d) If you wanted to know which two centers of influence were closest to each other, just given the location of the eight centers, you'd have to compare distances from *each* center to *every* other one. How many calculations would that require?

e) If you used the Delaunay triangulation instead, how many distance calculations would you need? Can you use that information to determine which two centers of influence were closest to each other? Explain.

15. The pictures shown in **Figure 3.77** were taken from the same point. As you can see, the two images overlap a little. It would be nice to reconstruct the "view" that the pictures together represent. Other than cutting the pictures out and laying them on top of each other, can you think of a mathematical way to make a panoramic picture out of the two pieces?

FIGURE 3.77.

16. A proposal to locate a toxic waste site just west of San Angelo is being considered by county governments in the region. **Figure 3.78** shows the area affected by the proposal, and the towns in that region.

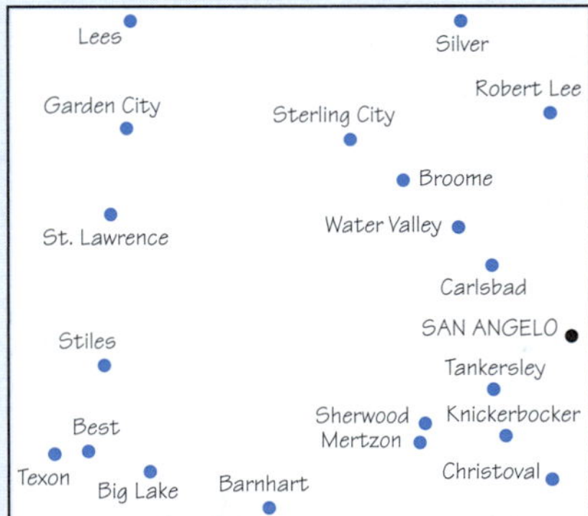

FIGURE 3.78.

Everyone agrees that it should be in the undeveloped region but no one can pinpoint an *exact* location for the disposal site. Where would you put it (assuming that it *has* to go somewhere on the map!)?

17. A remote-controlled land "rover" to be used on Mars is being tested in rugged territory near Big Bend National Park. Nicknamed SOL, the landscape map that was used to program its path is a 15 mi. x 10 mi. rectangle, with the origin located in the lower-left corner. Mountains are marked with a letter and coordinates: A(2, 8), B(6, 9), C(12, 8), D(8, 5), E(4, 3), and F(10, 2), as shown in **Figure 3.79**.

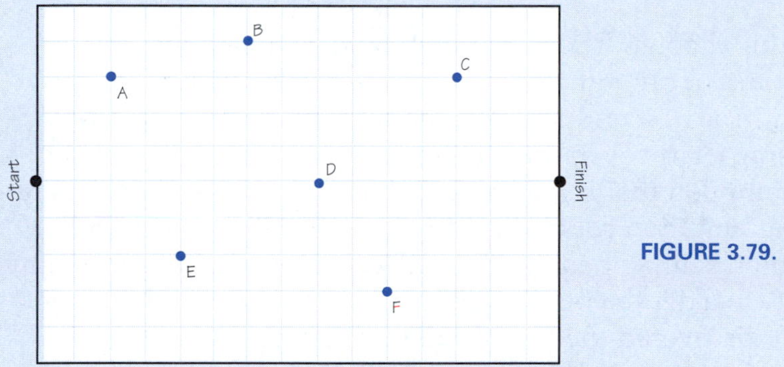

FIGURE 3.79.

The design engineers have designed a test for the navigational system, moving from the "Start" point to the "Finish" point, staying as far away from the mountains as possible. Your job is to provide the instructions to give the vehicle, including all directions (expressed as the equation of a line) and distances to be traveled.

3 CHAPTER REVIEW

Modeling Estimation of Rainfall Totals

Model Development

In this chapter you examined how to estimate the total amount of rainfall that falls on a region by using rain gauge data. The first and simplest model is made by taking an average. However, this process is based on the unrealistic assumption that the weather behaves the same everywhere—an assumption contradicted by the data. At first, the total rainfall is imagined as a "sheet" of water covering the entire state to a depth equal to the average. You tested the model and discovered the factor that most affected the answer was the average depth. As a result, it was necessary to revise the model.

Relative gauge locations force a change in assumptions and the need for an improved model. Assume that a gauge only represents a portion of the region, and that the gauge representing a *particular* location is the one closest to it. The process of calculating an estimate of the average rainfall changes to take advantage of these new assumptions. If the data points are far away from each other, or if they represent different amounts of the region, then it makes sense to calculate a weighted average to give some values more importance than others. Each gauge has its measurement weighted by the percent of the state that is closer to it than to any other gauge, or by the total amount of area assigned to that gauge.

Additional information spurs a final approach. The exact locations for the rain gauges are specified by the introduction of coordinates. This gain in precision doesn't change the plan of calculating a weighted average. However, it lets you write equations for the boundary lines, determine the exact locations for intersection points, and calculate both distances between those points and the area of proximity regions with increased accuracy. This dramatically improves the estimation of total rainfall, and the work on the model is completed. After this, to get a "better" answer, you'd need to process a lot more rain gauge data but the method used to calculate the answer won't change.

Mathematics Used

The "geometric" approach used with the improved model focused on making a Voronoi diagram, which is used in proximity problems like the rainfall estimation. This means locating boundaries that sub-divide the area into various regions of influence, with the gauges being the center of influence. The boundary between two Voronoi regions is the perpendicular bisector of the segment joining the centers of the two regions. These perpendicular bisectors can be constructed by four methods.

1. Fold a piece of paper and crease it so the two centers coincide.
2. Place a Mira so the reflection of one center coincides with the other center.
3. Draw intersecting compass arcs from the centers and join the two points of intersection.
4. Use the segment, midpoint, and perpendicular construction features of a drawing utility.

It was found that a Voronoi vertex, the place where three boundary lines come together, was also the center for a circle that circumscribes the triangle formed by the three gauges surrounding that vertex. Only a portion of the perpendicular bisector was used as a region boundary, and some pairs of centers weren't even considered. Therefore, care must be used when resolving a region into a Voronoi diagram. One strategy for working with complicated regions was to break them down into smaller, three-point problems.

Voronoi regions are usually polygons (unless the boundary of the domain is curved). The area of those regions needs to be determined in order to calculate a weighted average of rainfall. Four methods are explored for finding the area of these polygons.

1. Divide into triangles and apply Heron's formula,

   $A_\Delta = \sqrt{s(s-a)(s-b)(s-c)}$, where $s = \left(\frac{1}{2}\right)(a + b + c)$.

   This finds the area of a triangle from the length of its sides.

2. Apply Pick's formula that finds the area of any polygon (whose vertices are points on a grid) from the number of grid points that are on the polygon's border or interior,

   $A_{poly} = 0.5b + i - 1,$

   where b = number of grid points *on* the boundary and i = number of grid points in the interior.

3. Use a drawing utility's measuring features.

4. Estimate by designing and running a Monte Carlo simulation.

Once the area for the various regions is determined, the weighted average estimation for the total volume of rain needs to be calculated. This is found by multiplying the area for each region by the amount of rainfall in the gauge that represents that region, and summing those answers.

The analytic approach uses coordinates to describe the location of the various gauges. This leads to writing equations for the boundary lines by going through five steps.

1. Find the slope of the line segment connecting the two centers of influence.

2. Find the midpoint $P(h, k)$ of that line segment.

3. Write the slope of the perpendicular bisector line by taking the reciprocal of the first slope and changing its sign (negative reciprocal).

4. Write the equation of the perpendicular bisector line in point-slope form,

   $y - k = m(x - h),$

   with m the slope of the line, and the midpoint $P(h, k)$.

5. Change that resulting equation into slope-intercept form by using algebra.

After the equations of two intersecting perpendicular bisectors are found, the coordinates of the Voronoi vertex at which they intersect can be found by solving the system of their two equations. Assuming the equations are both in slope-intercept form, the following steps apply.

1. Set the two expressions for y equal to each other.

2. Solve this new equation for x.

3. Substitute the value for x back into either equation and evaluate y.

The intersection points of the boundary lines are the vertex points for the polygon whose area needs to be determined. One way to do this is to divide each region into triangles, use the distance formula to find the lengths of any unknown sides, and then use Heron's formula to find each triangle's area. The coordinates are important to locate because the distance formula requires the coordinates of the two points.

Coordinate geometry develops algebraic formulas for geometric objects from known properties of those objects. The results are used to develop computer or calculator programs that find equations of boundaries and areas of regions. The formulas of coordinate geometry give exact results. Therefore, the precision of answers obtained from coordinate methods is limited only by the precision of the measurements used in the formulas. Only the number of rain gauges used in the determination limits the accuracy of the answer. A decent computer program, similar to the calculator routines used in the chapter, could handle the 600+ weather observing stations in Texas with relative ease. Who knows, maybe you'll be writing that computer program!

Acute triangle: A triangle in which all of the angle measures are less than 90° (but more than zero degrees).

Altitude: The distance from a vertex of a triangle to the opposite side along a perpendicular. Its length is the height of a triangle.

Base: The side of a triangle that is perpendicular to the altitude. Also the name of the two parallel sides in a trapezoid.

Bisector: A segment or line that divides an angle or segment into two equal parts.

Center of gravity: The balance point for an object's weight.

Center of influence: A point used to establish boundaries of regions of influence. All points in a region are closer to that region's center than to any other region's center.

Circumscribed circle: A circle that goes through all vertices of a triangle.

Concave polygon: A polygon in which some of the sides, when extended, intersect other sides. At least one diagonal is outside the polygon.

Construction technique: A technique of mathematical argument that uses straightedge, compass, and basic mathematical facts.

Convex polygon: A polygon in which none of the sides, when extended, intersect other sides. All diagonals are inside the polygon.

Delaunay triangulation: A tiling pattern formed by connecting the centers of regions of influence that share a boundary.

Diagonal: A line segment connecting two non-adjacent vertex points of a polygon.

Distance formula: If you want to know the distance between A(x_1, y_1) and B(x_2, y_2), or the length of the segment AB, the formula is

$$d = \sqrt{(x_2 - x_1)^2 + (y_2 - y_1)^2}.$$

Domain: A region in which centers of influence are located. The domain is the area that is being divided into regions of influence.

Heron's formula: If a, b, and c are the lengths of the sides of a triangle, then the area of a triangle is calculated by using the formula

$$A = \sqrt{s(s-a)(s-b)(s-c)}, \text{ where } s = \left(\frac{1}{2}\right)(a+b+c).$$

Isosceles triangle: A triangle with two equal side lengths.

Kite: A convex quadrilateral with two pairs of adjacent sides having the same length.

Midpoint: A point that is halfway along a segment (equidistant from the segment's two endpoints). In coordinate geometry, the coordinates of a midpoint are found by averaging the coordinates of the two endpoints.

Obtuse triangle: A triangle with one angle that measures more than 90°, but less than 180°.

Perpendicular: Two lines (or segments) that form a right angle (90°).

Perpendicular Bisector: A line segment and a line (or another segment) that are perpendicular and the intersection point divides the line segment into equal pieces. The intersection point is the midpoint of the line segment.

Pick's formula: If the vertices of a polygon are lattice points on a grid, then the area of the polygon is found by using the formula

$A = 0.5b + i - 1$,

where b is the number of grid points on the polygon's border, and i is the number of points on its interior.

Point-slope form: Equation of a line that has the form: $y - k = m(x - h)$, where the slope of the line is m, and the coordinates of a point P (known to be on the line) are (h, k).

Region of influence: A region in which each point is closer to the region's center of influence than to any other center of influence. It's also called a **proximity region**.

Simple average: The average found by adding up all the numbers and dividing by how many numbers there are. (Also called the 'mean' or *just* the average.)

Slope-intercept form: The equation of a line that has the form $y = mx + b$, where the slope of the line is m, and the location where the line crosses the y-axis is b.

Trapezoid: A four-sided polygon (quadrilateral) with two sides parallel.

Vertex angle: An angle formed by the two equal sides of an isosceles triangle.

Voronoi boundary: A boundary between two centers of influence.

Voronoi center: A center of influence.

Voronoi diagram: A diagram composed of several centers of influence and their regions of influence. It's also called a **Voronoi tiling**.

Voronoi region: A region of influence.

Voronoi vertex: A point at which Voronoi boundaries intersect.

Weight: Relative portion of something that is used to make a **weighted average**.

Weighted average: The average found by multiplying each category by the decimal weight attached to that category and totaling the answers.

CHAPTER 4

Testing 1, 2, 3

- **ACTIVITY 4.1** IT'S ONLY A TEST
- **ACTIVITY 4.2** THE TESTING POOL
- **ACTIVITY 4.3** ATTEST TO THAT!
- **ACTIVITY 4.4** AUTOMATIC QUADRATIC
- **ACTIVITY 4.5** QUADRATICS ON THE MOVE
- **ACTIVITY 4.6** SOUNDS LIKE QUADRAPHONIC
- **CHAPTER REVIEW**

How is algebra used to make the Olympics fair to all athletes?

Imagine practicing gymnastics for several hours per day, six or seven days a week. Think what it must take to win a place on an Olympic team. Andreea Raducan dedicated her life to gymnastics and competed in gymnastics for her country, Romania, during the Sydney 2000 Olympic Games. She won a gold medal in the all-around competition. However, Olympic officials took away her medal because she tested positive for a banned substance.

Many athletes use performance-enhancing substances, such as anabolic steroids, to get that extra edge they feel they need. Since steroid usage is so widespread among athletes, Olympic officials regularly test all athletes before the Olympic games. Officials also test athletes during training.

Testing thousands of athletes requires an army of medical personnel. How do officials manage all these tests? In this chapter you will learn how drug testing is administered, and you will use mathematical modeling to help determine the best testing plan.

PREPARATION READING

Anabolic Steroids: Use & Effect

Anabolic steroids are synthetic compounds that act like male hormones. Many athletes such as bodybuilders, weight lifters, runners, swimmers, and football players believe that using steroids give them strength advantages.

The chemical structure of anabolic steroids is different from any vitamin or amino acid. Their chemical structure stimulates the DNA within cells. The DNA reacts to this stimulus by directing the production of specific new proteins. The biological response that occurs is dependent on the location and number of receptors. If the receptor is muscle, there will be a tissue-building effect. If the receptor is in the brain, there may be noticeable psychological and even behavioral effects.

The development of bigger muscles is a result of the anabolic component of the steroid. Unfortunately there are side effects such as acne, the growth of facial hair, and the premature closure of the space between parts of the body's long bones. Anabolic steroids can have harmful effects on the liver and the cardiovascular system (causing a greater risk for developing high blood pressure and blood-clotting). In addition, anabolic steroids can cause serious illnesses, such as cancer. There is growing concern that flooding a young adolescent with synthetic hormones may not only disrupt physical changes, but also the normal psychological and emotional growth that occurs during the teenage years.

Several studies have placed steroid use among high-school students at six to seven percent. Figures as high as 11% of 11th-grade boys using steroids have been reported. These studies suggest that between 250,000 and 500,000 high school students have used steroids. Even more disturbing are reports that approximately 40% of these students have used steroids for five or more cycles. Approximately 40% began using steroids before the age of 16. As expected, the majority of teenage users are participants in school athletics. However, as many as a third of the male users are not involved in sports and take steroids in an attempt to enhance their appearance.

There have been only a few surveys among the male college population, all of which show an estimated two percent usage. However, among male college athletes, the estimated usage rises to between five and 17%. The problem is not restricted to males. As many as one to two percent of senior high-school girls use steroids. Between three and five percent of Olympic and professional female athletes have admitted to using steroids at some time during their careers.

Source: Wright, J. E. and Cowart, V. Anabolic Steroids: Altered States. Carmel, IN: Benchmark Press, 1990.

ACTIVITY

4.1 It's Only a Test

The Preparation Reading gave you a lot of information about anabolic steroids, including where they come from, how they affect people, and how widely they are used. It is a mixed blessing to have so much information; one must look closely to find interesting questions that might be used to develop mathematical models.

WHAT'S THE PROBLEM?

The first step of the modeling process is to identify a problem you want to solve. When examining the use of anabolic steroids in society, there are a lot of problems that a mathematical modeler *might* consider.

- A modeler might build a model that predicts the increased demand in health services (hospitals, pharmacies, etc.), because steroids could cause illness.
- A modeler might build a model that determines whether the cost of an intervention program to educate people is worth the number of people it could help.

1. The people who request the tests for banned substances, such as the Olympic Committee, will have their own problem to consider. They may want to determine if someone has been using performance-enhancing substances like anabolic steroids and how many people are taking them. What other question would be important?

The Olympic Committee has two different ways to test for the use of performance-enhancing substances. During the four years between Olympic events, athletes are selected at random and tested for banned substances. During the Olympics all athletes are tested immediately before the Games begin. In addition all medal winners are tested immediately after their event is completed.

2. a) Why are athletes tested for several years before the Olympics even begin?

b) Why select athletes at random? Why not test all the athletes during that time?

c) Why do you think the random strategy isn't applied to the medal winners?

CHECK THIS!

Recall that selecting "at random" means choosing in such a way that each person is equally likely to be chosen.

Tests for anabolic steroids are expensive, but necessary. Suppose that the state legislature is considering a decision to require people to take a particular steroid test. Out of fairness it will test *everyone* in a targeted group of people, so a strategy such as randomized testing isn't an option. The state legislature has a fixed amount of money for administering the tests, and wishes to detect as many people as possible.

3. Assume that each of the following groups contains 100 people, so the number of people tested isn't an issue. In each case identify whether you think it's worth the money spent to test the group for steroid use. Provide reasons to support your answer.

   a) Professional wrestlers

   b) Golfers

   c) Stock car drivers

   d) High school football players

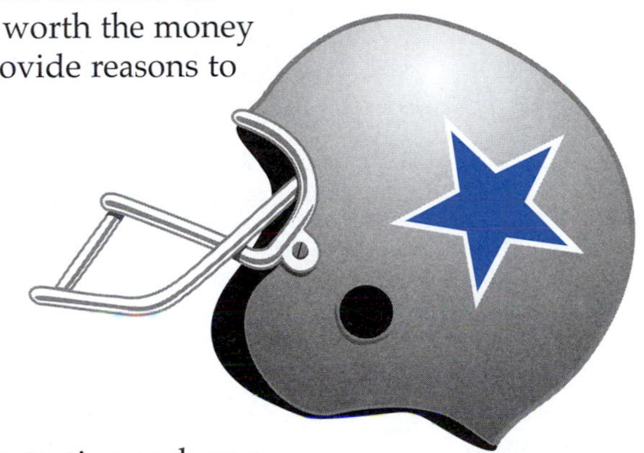

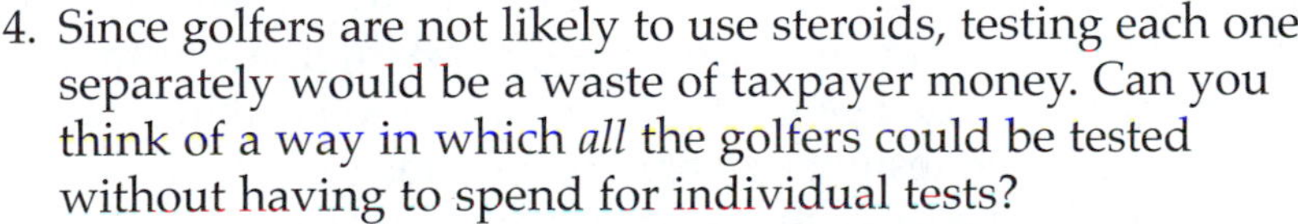

4. Since golfers are not likely to use steroids, testing each one separately would be a waste of taxpayer money. Can you think of a way in which *all* the golfers could be tested without having to spend for individual tests?

One strategy that might be applied is to use **pooled samples**. A pool, or mix, is created from a portion of the sample provided by several individuals, and then tested. This is especially common for Olympic event organizers. They must test hundreds of athletes to determine eligibility, which costs a lot of money. Over time they have adopted a pooled sample strategy, using a smaller number of samples—usually five to ten— from athletes.

5. a) Portions of samples from five athletes are mixed to form a pooled sample. The lab results come back negative, indicating that there is no steroid use among these athletes. How would the strategy of testing a pooled sample save money in this case?

b) Why use only a portion of the sample provided by each individual to create the pool?

c) A different pooled sample comes back with a positive result, indicating the presence of some steroids. What might the test administrator do next?

d) Whether you save money by using pooled samples depends on the group you are testing. Refer back to the groups identified in question 3. For which groups might it make sense to apply a strategy of pooling samples?

e) What feature (or characteristic) distinguishes the various groups in question 3, so that in some situations you would consider pooling the samples?

CHECK THIS!

If we expect that a person is not likely to be a steroid user, then we say that the *probability* of being a steroid user is low. The probability is a measure of the likelihood that something will occur. You will learn more about probability in Homework 4.1.

HOW MANY TESTS?

There are many ways to set up the tests for banned substances. Olympic testers usually mix samples of either five or ten athletes together in a pooled sample. If an entire football team is tested together at one time, testers would use a pool of about 50 samples.

A positive test result from a pooled sample tells testers that *somebody* is taking steroids. Unfortunately the test cannot identify which people, or how many people, are taking the substance. Additional tests are needed to determine who is taking the substance. When testing a pool that contains samples from many people, the follow-up testing and analysis become incredibly complicated. The following are some possibilities testers would consider.

- Test each athlete individually.
- Break the pool into smaller pools and test those pools.

When you start exploring this problem to build a model, it's best to keep things simple and make a simplifying assumption.

6. a) Any assumption simplifying the problem of testing for anabolic steroids should make sense. Can you use a pooling strategy if the sample contains only one athlete? Explain.

   b) For the simplest pooling strategy, how many athletes should be included in a sample?

c) This simplest pooling strategy will be the focus of your investigation. Explain why it's called a **paired sample** strategy.

CHECK THIS!

We will assume that the test is perfect . A positive result is assumed to be 100% accurate in detecting steroid use. A negative result is 100% accurate that there is no steroid use.

7. a) Two athletes are chosen at random to be tested. A paired sample is created by mixing together some of their samples. If the paired sample test comes back negative, what can you conclude? Do you save money in this situation?

b) If the paired sample test comes back positive, do you *have* to test each individual separately? (Remember, we're trying to save as much money as possible!)

c) Suppose the paired sample test is positive. If you test one athlete and that test comes back negative, what can you conclude? Does the paired sample strategy save you money in this case? How many tests were needed?

d) If the paired sample is positive, and the first individual test is *also* positive, do you need to test the other athlete? Does the paired sample strategy save you money in this case? How many tests were needed?

8. a) Now consider a larger group, such as the 100 stock car drivers in question 3. If you test all the people individually, you *know* you will have to do 100 tests. If you apply a paired sample strategy, what is the least number of tests you can give?

b) In addition to testing 50 paired samples, you might have to give more follow-up tests if some of the paired samples are positive. If you kept track of how many tests were given, how would you know if the paired sample strategy ended up being cheaper than testing each person separately?

9. When using a paired sample testing strategy, sometimes you must compare two groups that have different numbers of people being tested. For example, suppose that 40 pairs of football players require 65 tests, and 100 pairs of golfers require 161 tests. In both situations it is cheaper to pair the samples together than it is to test each person individually. How can you verify that statement?

10. Another way of comparing the costs of testing two groups with different numbers of people is to calculate how many tests on average are needed for each pair.

    a) The test that costs less to administer is the one with the smallest average number of tests per pair. Which group is cheaper to test—the football players or golfers mentioned in question 9? Explain.

    b) What is the average number of tests per pair of golfers if you test each one individually?

    c) What is the average number of tests per pair of football players if you test each one individually?

    d) How can the average number of tests per pair indicate whether it is cheaper to pair samples or to test individually?

SUMMARY

In this activity you explored some ways of testing people for banned substances.

- You learned how pooled samples can reduce the cost of testing.
- You compared testing paired samples with testing each individual separately.
- You discovered that the average number of tests is a useful measure when comparing testing groups.

These testing strategies may be applied to a wide variety of groups. However, the type of group tested affects which strategy works best. Activity 4.2 will begin the process of using mathematics to determine when a paired sample strategy should be applied. Mathematics can help you predict which strategy to use for a specific situation—no matter how many people are being tested.

HOMEWORK

4.1

Test Your Logic

In Activity 4.1 you carried out the first step of modeling steroid testing. You examined and studied the problem of steroid testing. Before answering the following questions, take a few minutes to look over the questions that were asked, your responses to those questions, and any notes you might have taken from the closing discussion.

1. a) What is the central problem that your model will solve?

   b) What assumptions did you make? Why were they introduced into the model?

2. A government agency requires that its new scientists must be tested for steroid use.

   a) Do you think the likelihood that these scientists use steroids would be low or high?

   b) Which testing method would be less expensive, the paired sample strategy or individual tests?

   c) Estimate the average number of tests per pair for these scientists.

3. a) You want to build a model in the form of an equation that has a output variable and an input variable. What should be the output variable (that is, the value you want the model to measure)?

   b) Explain what causes the value of the output variable you identified in (a) to change?

   c) Based on your two previous answers, what should be the input variable?

4. a) In Activity 4.1 you explored two strategies for testing people for anabolic steroid use. If you test two athletes individually, how many tests must be given?

b) If you test two athletes using a paired sample strategy, what possible outcomes can take place?

c) Using a paired sample strategy, how can you tell when you are saving money by not testing individually?

Since the result of a steroid test must be either positive or negative, the testing can be represented by a tree diagram like the one in **Figure 4.1**. In the figure, steroid test represents the *situation* and the test results are the *outcomes*.

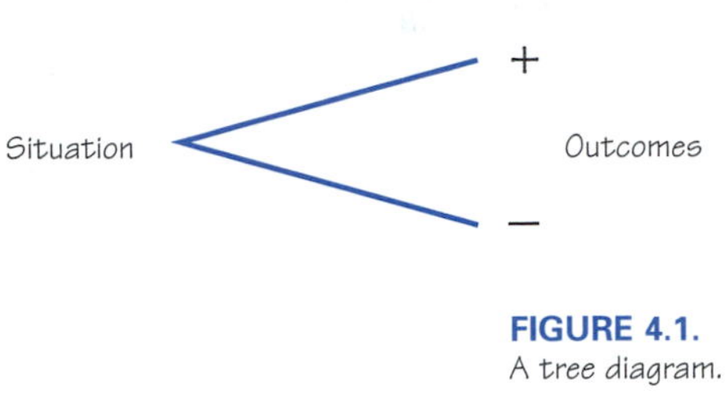

FIGURE 4.1.
A tree diagram.

Of course some outcomes might require more testing. For example, a positive result from the test of a pooled sample will require testing some or all of the samples that made up the pool. In that case, the tree would have one or more branches built on the tip of the + branch.

5. a) Draw a tree diagram that represents testing a pair of athletes using a paired sample strategy starting with the test of the pooled sample.

b) Draw a tree diagram that represents testing two athletes individually from the athletes' point of view.

The dice game described in the next question relies on the idea of a **probability**. Think of some event where you don't know the outcome. For example, the event could be one coin toss at the beginning of a football game to decide who gets the ball. There are two possible outcomes: heads or tails. The captains of the two teams try to guess the outcome; neither knows which way the coin will land. Probability measures the likelihood that each outcome will occur. Since there are two possible outcomes (heads or tails), and each can happen only in one way, the captains know the probability that the coin will land heads is $\frac{1}{2}$, and the probability that the coin will be tails is $\frac{1}{2}$.

PROBABILITY

For an event, the probability that a particular outcome will occur is the ratio

$$\frac{\text{number of ways that particular outcome may occur}}{\text{total number of ways that all outcomes may occur}}$$

A probability is a number from 0 through 1 and is usually written as a fraction or decimal.

6. A dice game has the following rules.

    - Before rolling two dice, one player predicts the sum of the numbers on the two dice.
    - If the sum of the dice matches the prediction, then points are awarded equal to the sum of the dice. If the prediction is not matched, no points are earned and the other person takes a turn.

    Mathematics can help determine a sound strategy for playing the game.

	Outcomes for Die #1					
Outcomes for Die #2 +	1	2	3	4	5	6
1						
2						
3						
4						
5						
6						

FIGURE 4.2.

    a) Copy the table in **Figure 4.2** that shows all the possible results of rolling two dice. In each cell of your table, write the payoff (sum of the two dice) that would result from each roll.

    b) The **probability** of an outcome (such as getting a payoff of five) is the number of times that particular outcome can happen, divided by the total number of possible outcomes. For each payoff listed in **Figure 4.3**, find the probability for the outcome.

Outcome Payoff	Outcome Probability	Outcome Worth
2		
3		
4		
5		
6		
7		
8		
9		
10		
11		
12		

FIGURE 4.3.

    c) The **probabilistic worth** of an outcome is defined as the value (or payoff) of an outcome multiplied by the probability of the outcome happening. Complete the table in Figure 4.3 by calculating the worth associated with each of the outcomes.

    d) Use a grid like **Figure 4.4** to draw a graph of the probabilistic worth (W) versus the number that represents the guess (and the payoff) in the game (n).

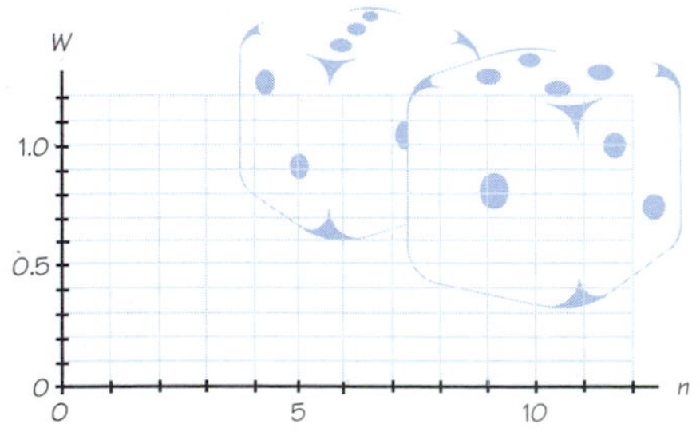

FIGURE 4.4.

    e) Based on the work you've done, what strategy should you use in playing this dice game?

7. Repeat the analysis in question 4 for a dice game with identical rules except that the points are awarded for guessing the correct *product*, not the sum. What strategy should you use in playing this game?

ACTIVITY

4.2

The Testing Pool

In Activity 4.1 you explored the problem of how to test people for banned substances. The idea of a pooled sample was introduced, and you studied some testing situations that used paired samples. You want to build a model that determines under what conditions a paired sample strategy is less expensive than individual testing. In this activity you build and test a first model that analyzes the costs of paired samples.

BUILDING A MODEL

The incidence of steroid use in the population was introduced as the explanatory variable (independent variable) for the model. Use the variable p to represent the probability that a person is a steroid user (or the percentage of the overall population that are steroid users). It is assumed that each test is an **independent** event—that one result does not affect another. You determined that the response variable (dependent variable) is the number of tests per testing pair that you would expect to need on average. Call that variable E, for the expected number of tests. Since you want a model that describes the relationship between p and E, let's explore how these quantities are related.

CHECK THIS!

In Activity 4.1, you completed the first two steps of model building. You identified the problem and made some assumptions to simplify the problem. Now you're ready for step 3—building the model.

Answer the following questions assuming that two samples are mixed together to form a paired sample. These conditions will give you values for p and E that you can use to form your model.

1. a) If *nobody* in the population is taking steroids, how many tests per pair do you expect to have to make?

   b) If p represents the probability that a person is using steroids and E represents the number of tests you expect, then what are the values of p and E for the situation in question 1(a)? Write the values as an ordered pair of the form (p, E). Enter these results into the table in **Figure 4.5**.

   c) If *everybody* is taking steroids (and what a crazy world that would be!), how many tests per pair do you expect? For this case, what are the values of p and E? Write these values as the ordered pair (value of p, value of E). Enter these results into a table like Figure 4.5.

Situation	Value of p	Value of E	Ordered Pair (p, E)
No one takes steroids			
Everyone takes steroids			

FIGURE 4.5.

Next explore what happens between the extremes.

2. a) If very few people are taking steroids (the value for p is near 0), how many tests per pair do you expect to make? What would you expect as a long-term average for the number of tests per pair of people tested?

   b) When p is near 0, as in question 2(a), what value should E be near?

   c) If almost all the people are taking steroids (p is near 1), how many tests per pair do you expect to have to make? Estimate the long-term average for this case.

   d) When p is near 1, as in (c), what value should E be near?

3. a) If exactly half the population were taking steroids, how many tests per pair do you think would be needed if the samples were paired? Explain your reasoning.

   b) Use your estimate for E from part 3(a) to write the ordered pair (p, E) for $p = 0.5$.

   c) Graph on a coordinate grid like **Figure 4.6** the points corresponding to the two ordered pairs you entered into the table in Figure 4.5. Also graph the point corresponding to the ordered pair you found in (b) on the same coordinate grid.

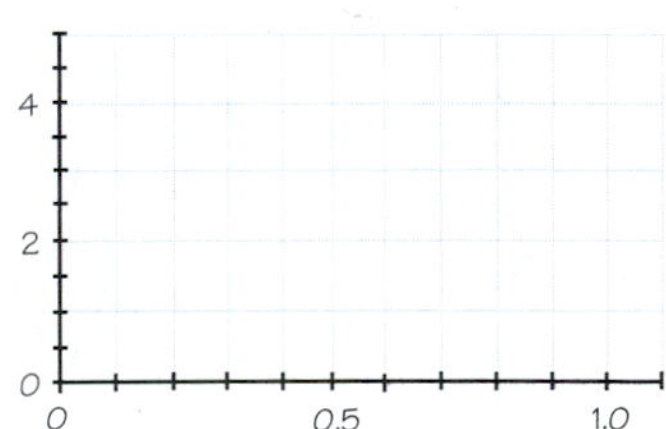

FIGURE 4.6.

   d) What pattern do you see in the three points you plotted?

4. Using your answers to questions 2 and 3, identify a mathematical relationship between the variable p and the variable E. What form would the equation for your model have?

5. a) Explain how you can use the ordered pairs you recorded in Figure 4.5 to find the equation of this model.

   b) Find an equation expressing the relationship between p and E.

6. What are the domain and range for your model?

CHECK THIS!

The **domain** for a model describes the numbers that are reasonable for the input variable. The **range** for a model describes the possible outputs produced when applying the model to a situation.

EVALUATING YOUR MODEL

Now that you have built a model, it's time to test it. In groups, you will simulate what happens when a paired testing strategy is applied to a situation in which half of the people have a certain trait.

7. a) Your model predicts E, the average number of tests per pair, based on the value of p, the probability that a person is taking steroids. The situation when half of the people are taking steroids is represented by what value of p?

   b) According to your model, what is the value of E for the value of p you found in (a)?

The simulation you will perform uses a mixture of 50% red objects and 50% blue objects. The paired testing strategy will be used to check for the presence of red. If your model predicts that an average of two tests per pair is needed when half of a population is taking steroids, this simulation should get a similar numeric result. Take a minute to read the directions.

SIMULATION OF A PAIRED SAMPLE STRATEGY

Two people will play the roles of Olympic athletes who are tested for being red. The third person will be the lab worker who gives the test.

1. Select some identical red objects (such as beads, tiles, or chips) and place them into a bag. Now place an equal number of blue ones (same identical objects) into the bag and mix the objects well.
2. Each athlete takes a turn drawing one object from the bag and showing it to the other athlete. Don't let the lab worker know the test results for now.
3. Put the object back into the bag before the other athlete takes a turn.
4. After each athlete has drawn and put back the object, the athletes get together to decide if the paired sample is positive or negative.
   - The paired sample is positive if at least one of the objects drawn is red.
   - The paired sample is negative if both the objects drawn are blue.
5. The athletes announce to the lab worker whether their paired sample is positive or negative.
6. If the paired test result is negative, the testing for that trial is done. If the test result is positive, then the lab worker must choose one of the athletes to be tested individually (it doesn't matter which one is selected). Don't draw a new object; just pick an athlete and let that person announce either positive (red) or negative (blue).
7. Test the other athlete individually in the same manner if necessary.

The simulation you will perform represents a paired sample testing strategy when half the people are taking steroids. Since your model predicts that on average two tests per pair are needed when half the people are taking steroids, the simulation should get a similar result.

8. Why does the bag have to have equal numbers of two different colors?

9. Why is it important to put the object drawn *back* into the bag before the other athlete draws the next one?

10. a) Assume that red means the person is a steroid user, and blue means the person is not taking steroids. Under what circumstances would only one test be needed?

    b) If the paired sample is positive, and the first person tested had a blue object, do you have to test the other athlete? Explain.

    c) If the paired sample is positive, and the first person tested is red, do you have to test the other athlete? Explain.

Repeat the simulation ten times. Record the actual status of each athlete, the result of the pooled test, and the result of any individual tests in Handout 4.2, the data table for this activity. Use '+' for a positive result or '–' for a negative result. In the last column, record the number of tests needed for each trial.

11. a) Look over the data recorded in Handout 4.2. How many trials needed only one test? Two tests? Three tests?

    b) Calculate the average number of tests needed for the ten trials.

ANALYZING CLASS RESULTS

In a table like **Figure 4.7**, record each group's average number of tests needed for ten trials.

Group	1	2	3	4	5	6	7	8	9	10
Average No. of Tests										

FIGURE 4.7.

12. As the tester, when would you be saving money using paired samples?

13. Based on the results of the simulation, is it *possible* to save money using a paired sample strategy when half the population is red?

14. a) How many groups have an average number of tests greater than two per pair?

    b) Based on these trials, is it *likely* that you would save money by pooling samples?

    c) How do the results of this simulation compare with the average number of tests predicted by the mathematical model that describes testing for steroid use?

15. Think back to the simulation. Are there problems or limitations in the way you conducted the simulation that might affect the results you got?

16. How could you modify your simulation to better understand the situation when half the population is taking steroids?

A MONTE CARLO SIMULATION

How can we be sure that drawing objects from a bag will yield the same result when a lot more experiments are performed? Technology allows modelers to design simulations by programming computers and calculators. Those programs go through the same steps as the hand-drawn simulation, only faster and without arithmetic errors.

In your group, run the calculator program DRACULA.

- When you are asked, "What is P? P = ?", enter 0.5 (remember, we're studying what happens when *half* the population is taking steroids).
- When asked the second question, "How many trials? N = ?" enter 100.

After a few seconds, the calculator will provide the average number of tests needed per pair, as well as the number of times each outcome (one, two, or three tests needed) occurred.

CHECK THIS!

Remember: the calculator may use the capital letter 'P' in instructing you to enter a probability number. The tradition in mathematics is to use the variable name *p* (lower case).

In a table like the one shown in **Figure 4.8**, record your group's results for the average number of tests needed for 100 trials, as well as the result from other groups in your class.

Group	1	2	3	4	5	6	7	8	9	10
Average No. of Tests										

FIGURE 4.8.

17. a) Compare the results from the DRACULA simulation with the hand-drawn experiment. How is the average number of tests affected by the number of times the experiment is performed?

    b) Do you think it's cheaper to use paired samples to test for steroid use when half the population is taking steroids? Explain.

    c) How do the results of the DRACULA simulation compare with the result predicted by your model?

18. The results of these simulations provide information about when you should apply a paired sample strategy. Also, the results question an assumption about your model. Review the problem and the original assumptions to see if you can determine what the results imply about the problem.

SUMMARY

In this activity you built a model and tested it.

- You gathered data about paired samples and concluded the model should be a linear equation.
- You performed simulations that you used to test your model.

The simulations you performed left you with questions about your model. In Activity 4.3 you will explore your model further and make adjustments.

HOMEWORK 4.2

Test on Friday? (Not!)

1. a) Dante Takum is trying to arrange tag-team matches for his wrestlers, half of whom are taking steroids. The event organizers insist on having them tested for steroid use. Each test costs $160. If you test each wrestler individually, what is the total cost?

   b) Based on the simulations done in Activity 4.2, how much would it cost (on average) to test the pooled sample from the pair of wrestlers on the tag-team first, and then test individuals if needed?

2. A high school testing its athletes for steroids found that one test was needed 28% of the time, and two tests were needed 31% of the time. Did the school save money by pooling blood samples?

3. The table in **Figure 4.9** shows results from repeating the hand-drawn testing simulation in Activity 4.2. Was the game fairly played? How do you know?

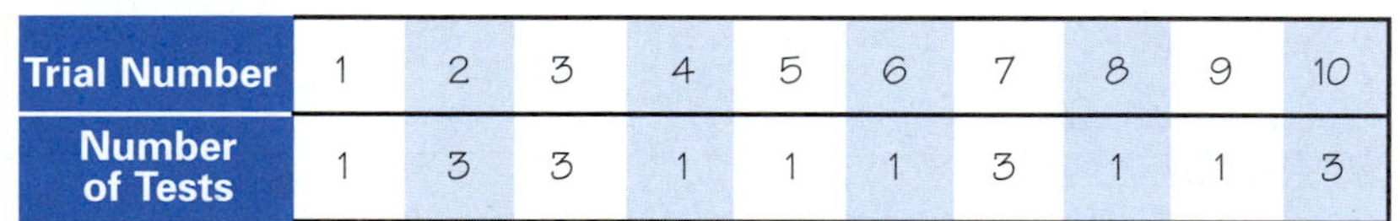

Trial Number	1	2	3	4	5	6	7	8	9	10
Number of Tests	1	3	3	1	1	1	3	1	1	3

FIGURE 4.9.

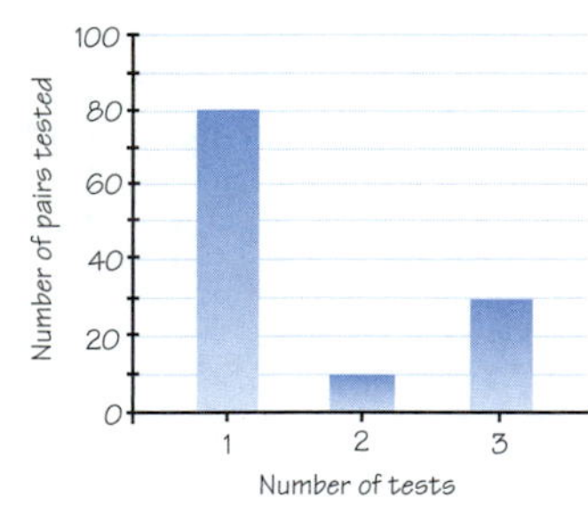

FIGURE 4.10.

4. a) The bar graph in **Figure 4.10** shows the results of steroid tests given to athletes at a large high school. The strategy of pairing samples was used. How many students were tested?

   b) Did the school save money by using the paired-sample strategy? Explain your answer.

   c) What was the average number of tests per pair needed to test these athletes?

5. **Figure 4.11** is a bar graph that describes steroid testing at another school. Did the school save money by using the paired-sample strategy?

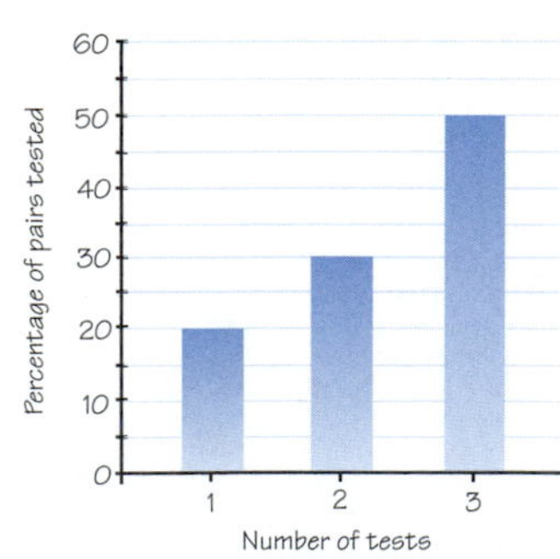

FIGURE 4.11.

6. In Activity 4.2 you modeled a situation in which half the population was taking steroids by using two colors of objects such as plastic chips. You used an equal number of chips of each color. Suppose red chips represent a steroid user.

   a) What is the smallest number of blue and red chips needed to model a population in which 75% use steroids?

   b) What is the smallest number of blue and red chips needed to model a population in which 15% use steroids?

   c) What is the smallest number of blue and red chips needed to model a population in which 23% use steroids?

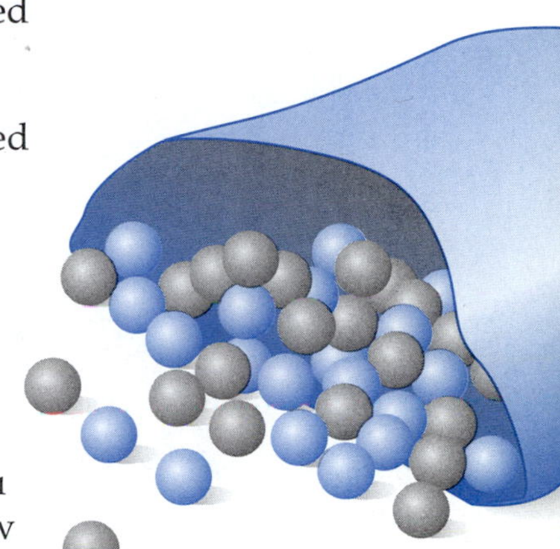

7. a) Suppose you wish to model a situation in which the probability of an individual taking steroids is 25%. Use the same color scheme as the previous question. What is the smallest number of chips needed?

   b) Assume you have exactly the collection of chips that you identified in (a). If you drew a blue chip on the first draw and did *not* replace it, what would be the probability of getting a blue chip on the second draw?

   c) Start with the same arrangement you identified in (a). If you drew a red chip on the first draw and did *not* replace it, what is the probability of drawing a red chip on the second draw?

8. a) Suppose you use a coin to simulate a situation in which the probability is 0.5 that a person is taking steroids. If you flip the coin twice, will you always get one head and one tail?

   b) If you flipped the coin a million times, would you always get 500,000 heads and 500,000 tails?

   c) Which do you think is more likely: The number of heads in ten tosses would be between four and six, or the number of heads in 100 tosses would be between 40 and 60?

   d) Suppose a fair coin was flipped ten times and happened to land heads eight of those times. If you flipped the coin another 100 times, how many heads would you expect to get out of the 100 tosses?

ACTIVITY

4.3

Attest to That!

Your results in Activity 4.2 suggested that your model for predicting the average number of tests when half the people are taking steroids ($p = 1/2$) is not very accurate. In Activity 4.3, you will test your model more using a variety of situations to measure the accuracy of the model.

WHAT YOU KNOW SO FAR

Figure 4.12 shows the information discovered so far about the problem of testing for steroid use by pairing samples.

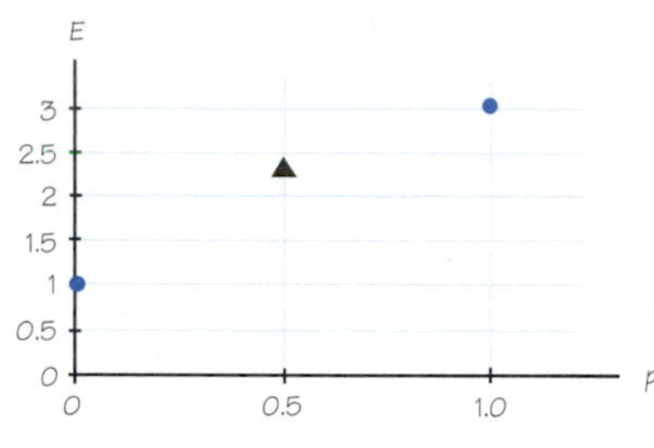

FIGURE 4.12.

- The left dot represents the known fact that one test per pair would be required every time if no one was taking steroids ($p = 0$; $E = 1$).
- The right dot represents the known fact that three tests per pair would be required every time if everyone was taking steroids ($p = 1$; $E = 3$).
- The triangle reflects the work done in Activity 4.2. There the probability was assumed to be 0.50 that a person selected at random is taking steroids. You probably found that the average number of tests per pair was slightly greater than two.

The model you developed in Activity 4.2 is $E = 2p + 1$. This model predicts that $E = 2$ when $p = 0.5$. Since the value of p obtained in the simulation in Activity 4.2 was greater than the predicted value, the model should be changed to account for the new findings.

Let's try continuing to use a linear equation as the model. To describe the problem more completely, you will need to collect more data. Then you will position a line to fit all data points, in a manner similar to what you did in Chapter 2.

DATA COLLECTION—DRACULA SIMULATION

As in Activity 4.2 the DRACULA calculator program can be used to show what happens when the paired sample strategy is applied to a variety of groups. Each group will have a different probability (or likelihood) that people use steroids.

Here's how to collect your data using the simulation. Make a data table similar to **Figure 4.13**.

1. When asked the first question, "What is P? P = ?" enter the first probability listed in the table.

2. When asked the second question, "How many trials? N = ?" enter 100. (Better results can be obtained with a larger number, but it will take much longer to run the simulation.)

3. The calculator will report the average number of tests per pair and the number of times each outcome occurred. Record the average number of tests *E* in the table.

4. Repeat the simulation with another probability. Continue until you find the average number of tests per pair for every probability in the table.

Probability (p)	Avg. No. of Tests (E)	Predicted No. of Tests	Residuals (Errors)
0.0			
0.1			
0.2			
0.3			
0.4			
0.5			
0.6			
0.7			
0.8			
0.9			
1.0			

FIGURE 4.13. Comparing actual and predicted average number of tests.

CHECK THIS!

Recall that $p = 0.5$ means exactly half the population is using steroids. So if a sample of 100 people are tested, we estimate that 50 of them are steroid users. In that case:

$$p = \frac{\text{number of steroid users}}{\text{total number of people}} = \frac{50}{100} = 0.5.$$

5. Using the data collected, make a scatter plot graph of Average Number of Tests (E) versus Probability (p) with a grid like the one shown in **Figure 4.14**.

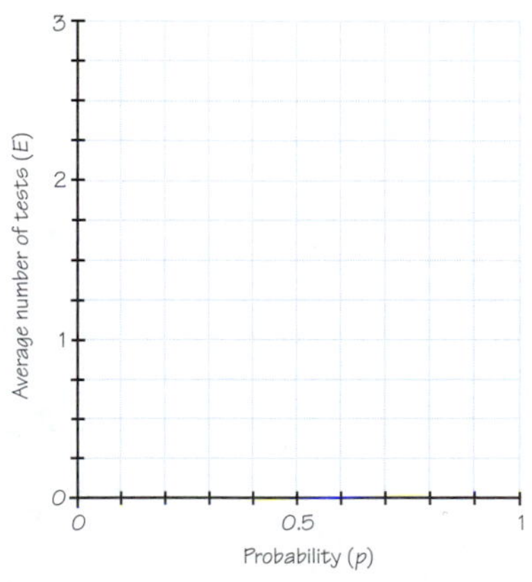

FIGURE 4.14.

Now, you will see whether the original model developed in Activity 4.2 does a good job of describing the data just generated by the DRACULA simulation.

6. Draw the line that contains the first data point (0, 1) and the last data point (1, 3) onto Figure 4.14.

7. Use the model $E = 2p + 1$ to find the predicted values of E for each value of p in Figure 4.13. Record your answers in the column labeled Predicted No. of Tests.

8. For each probability in Figure 4.13, calculate the residuals (errors) by subtracting:

   Avg. No. of Tests – Predicted No. of Tests

   Record the answers in the column labeled Residuals in Figure 4.13.

9. On a grid like **Figure 4.15** draw a residual plot—in this case a graph of errors versus the probability p.

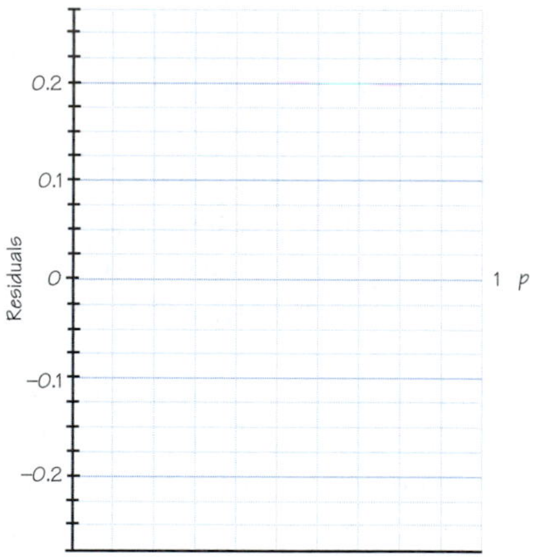

FIGURE 4.15.

CHECK THIS!

Recall that the residual is defined as the actual value minus the estimated value.

10. A *good* model will show no random pattern of regularity when examining the residuals, with about half the points above the zero line and about half below the zero line. Does your graph have those properties?

CHOOSING A REGRESSION MODEL

Since your model does not fit the data well, you must continue your search for a better model. Your calculator's statistical regression ability can fit the data to a linear, quadratic, exponential, or other kind of function. Using regression equations only guarantees that the equation is the best one of that particular *kind* of equation. Mathematical modelers must also pay attention to which kind of equation to use.

Next use your calculator to find three regression equations that fit the data. To determine which is the best fit for the data, look for the residual plot that produces a random pattern of dots.

11. a) Work with a copy of Handout 4.4. In the second column record the data for the average number of tests (E) from the second column in Figure 4.13.

    b) Enter the data from the table into your calculator. On the grid provided in Handout 4.4, make a scatter plot graph of your data.

    c) Use your calculator to find a linear regression equation for the data. Record this equation on Handout 4.4. Use the variable E for the average number of tests and the variable p for the probability. Include a sketch of the line as part of your scatter plot.

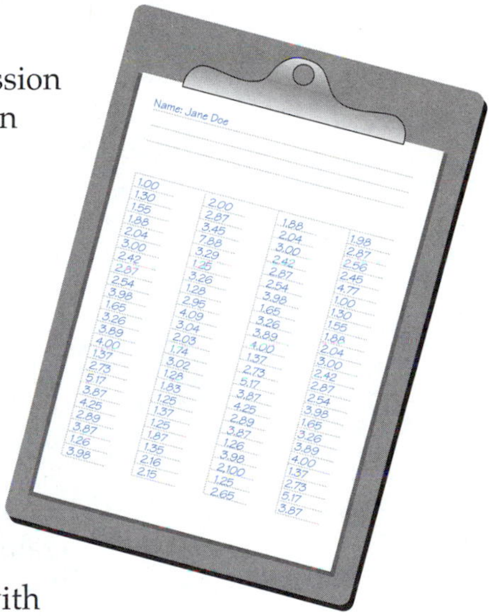

    d) Calculate the predicted values using the regression equation. Record the values in the third column of the data table in Handout 4.4.

    e) Calculate the errors (residuals) by subtracting Actual value (E) – Predicted value. Record those values in the fourth column of the data table in Handout 4.4.

    f) On the second grid provided in Handout 4.4, draw the residual plot for this regression equation.

12. a) To find a quadratic regression equation, repeat the procedure in question 11 working with Handout 4.5. In the second column record the data for the average number of tests (E) from the second column in Figure 4.13.

    b) Enter the data from the table into your calculator. On the grid provided in Handout 4.5, make a scatter plot graph of your data.

    c) Use your calculator to find a quadratic regression equation for the data. Record this equation on Handout 4.5. Use the variable E for the average number of tests and the variable p for the probability. Include a sketch of the curve as part of your scatter plot.

d) Calculate the predicted values, using the regression equation. Record the values in the third column of the data table in Handout 4.5.

e) Calculate the errors (residuals) by subtracting Actual value (E) – Predicted value. Record those values in the fourth column of the data table in Handout 4.5.

f) On the second grid provided in Handout 4.5, draw the residual plot for this regression equation.

13. a) Repeat the procedure to fit the data with a third type of equation. Work with a copy of Handout 4.6. In the second column record the data for the average number of tests (E) from the second column in Figure 4.13.

b) Enter the data from the table into your calculator. On the grid provided in Handout 4.6, make a scatter plot graph of your data.

c) Use your calculator to find a regression equation that is neither linear nor quadratic. Record this equation on Handout 4.6. Use the variable E for the average number of tests and the variable p for the probability. Include a sketch of the curve as part of your scatter plot.

d) Calculate the predicted values, using the regression equation. Record the values in the third column of the data table in Handout 4.6.

e) Calculate the errors (residuals) by subtracting Actual value (E) – Predicted value. Record those values in the fourth column of the data table in Handout 4.6.

f) On the second grid provided in Handout 4.6, draw the residual plot for this regression equation.

CONCLUSIONS

You have three regression equations that fit the data. Use your results from questions 11–13 to determine which equation is the best fit for the data.

14. Which regression equation do you think is the best model for describing the relationship between the probability (p) and the average number of tests (E) when using a paired-sample strategy to test for steroid use?

15. How did you arrive at that conclusion?

16. According to your model, when should the paired-sample strategy be applied in testing for steroid use? Explain.

17. What is the domain and range for your model?

18. a) Think back over the process used to arrive at this model. In simulating the steroid testing, you probably used a calculator program with 100 trials. If you had analyzed data that represented only ten trials instead, would it have changed the *kind* of model you chose (linear, quadratic, etc.)?

    b) Would it have changed the actual equation you recorded?

    c) Would it be a "better" model? Explain.

SUMMARY

In this activity you examined the original model $E = 2p + 1$ more carefully.

- You discovered that when you included other values of p, the probability that a person is a steroid user, the model was not a good predictor of E.
- You tried linear, quadratic, and other regression equations to model the data.
- You determined that the data is quadratic.

Since a quadratic equation is the best model for this problem, you will explore some properties of quadratic equations and their graphs in the next activity.

HOMEWORK

4.3 Testing Your Patience

1. a) In a baseball simulation game a player flips a coin (heads for the batter, and tails for the pitcher), and *then* rolls one die. Batters get a hit if the roll is either a 4 or 6. Pitchers get a strikeout when the roll is a 1. What is the probability of selecting the batter?

   b) If the batter *is* selected, what is the probability of getting a hit?

   c) What is the probability of selecting a batter *and* getting a hit?

   d) What is the probability of selecting the pitcher?

   e) If the pitcher *is* selected, what is the probability of getting a strikeout?

   f) What is the probability of selecting a pitcher and getting a strikeout?

In the baseball game, the coin toss does not affect the die roll. Because of that, you can find the probability for an event such as flipping heads *and* rolling a 4 or 6 without counting. But you need to know the probability that each event can happen independently.

2. Using your answers to question 1, explain how you can find the probability of flipping a heads **and** rolling a 4 or 6 if you know the probability of flipping a heads and the probability of rolling a 4 or 6.

3. If 20% of the population is taking steroids, what is the probability that a paired sample from two people will require only one test? In other words, what's the chance that the first person is not a drug user *and* the second person is not a drug user?

The **expected value** of an event is defined as the *sum* of the probabilistic worths for all the possible outcomes.

CHECK THIS!

Recall that the probabilistic worth of an outcome is the value or payoff of the outcome multiplied by the probability that the outcome will occur.

4. A builder is considering a contract for building homes that might make a profit of $50,000 per home. However, because of bad weather, strikes, and other unforeseen events, she might lose $10,000 per home. She estimates that there is a 30% chance of losing the money on a home. What is the builder's expected value for building the home?

The term *expected value* may be misleading. It does not mean the amount of money the builder would make on building one particular home. A better way to think about expected value is that it is the average amount she would earn per home if she built a LOT of homes.

5. Suppose you buy a $1 raffle ticket. The grand prize is $1000. Two second prizes are $300 each, and 50 consolation prizes are $20 each. There are 10,000 tickets sold in the raffle. What is the expected value of one ticket? What is the meaning of the answer?

6. a) The DRACULA calculator program produced the results for 100 trials shown in **Figure 4.16**. Find the probability for each outcome.

Number of Tests	Number of Trials	Probability of an Outcome	Probabilistic Worth
1	50		
2	19		
3	31		

FIGURE 4.16.

b) Think of the number of tests as the "payoff," and calculate the probabilistic worth of each case.

c) What is the expected value for this game? What is the meaning of the answer?

d) From the work done in the chapter so far, estimate the value for *P* (probability of steroid use) that was given to the calculator program. Explain your method or reasoning.

7. Simulations like the ones you have done are more reliable if you use a large number of trials. The **Law of Large Numbers** says that you can be more confident of a result obtained with a large number of trials. **Figure 4.17** shows the results of four simulations of a group in which half the people are using steroids ($p = 0.50$).

Number of Trials	10	100	1000	10,000
Avg. No. of Tests	1.9	2.34	2.26	2.25

FIGURE 4.17.

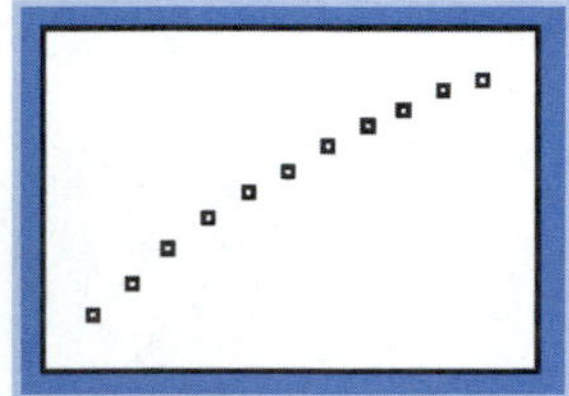
FIGURE 4.18.

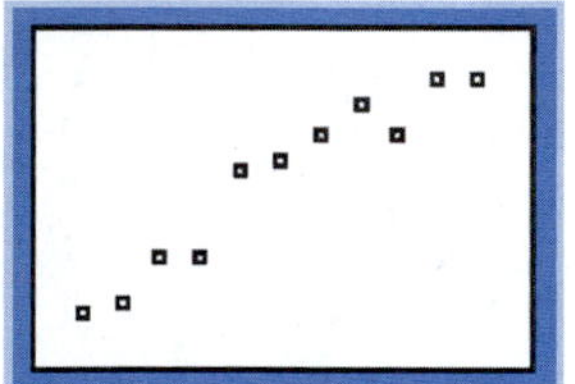
FIGURE 4.19.

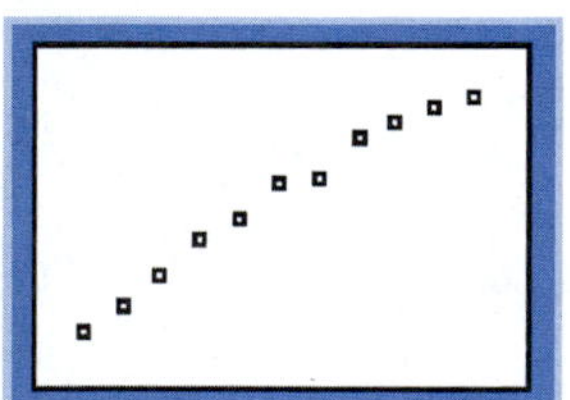
FIGURE 4.20.

a) Describe the change in the average number of tests as the number of trials increase.

b) Based on these results what is the best estimate for the average number of tests needed when $p = 0.50$?

8. a) The scatter plot graphs of three different simulations like the one in Activity 4.3 are shown in **Figures 4.18–4.20**. The only difference in the simulations is the number of trials done for each value of p. Which scatter plot do you think had the fewest trials?

b) Which do you think had the most trials?

9. a) Find the equation of the line that you think models the data in **Figure 4.21** most closely. Plot a line and then plot the residuals. Does there seem to be a pattern to the residuals? Do you think a line is a good kind of graph to fit these data?

x	−4	10	22	32	36	51	62	68	78	89	96
y	39	66	102	145	192	233	285	355	414	481	533

FIGURE 4.21.

b) Find the equation for the line that you think models the data in **Figure 4.22** most closely. Plot a line and then plot the residuals. Does there seem to be a pattern to the residuals? Do you think a line is a good kind of equation to fit these data?

x	4	6	20	30	40	47	60	70	79	92	99
y	65	58	52	46	44	32	31	22	18	14	13

FIGURE 4.22.

ACTIVITY

4.4

Automatic Quadratic

In Activity 4.3 you discovered a new model for the problem of predicting the average number of steroid tests per pair. Although your first model was a linear equation, you learned that a quadratic equation is a best fit for the data. In this activity you will find a quadratic equation that models the problem, and you will explore quadratic equations.

QUADRATICS IN THEORY

The data you gathered in Activity 4.3 suggested that a quadratic equation is a very good model for predicting the average number of steroid tests in a paired sample strategy. The results you obtained in Activity 4.3 were determined by the data you had. If your data change then your equation might change, too. You can generalize the results you obtained by going back to the original simulation for this problem in Activity 4.2.

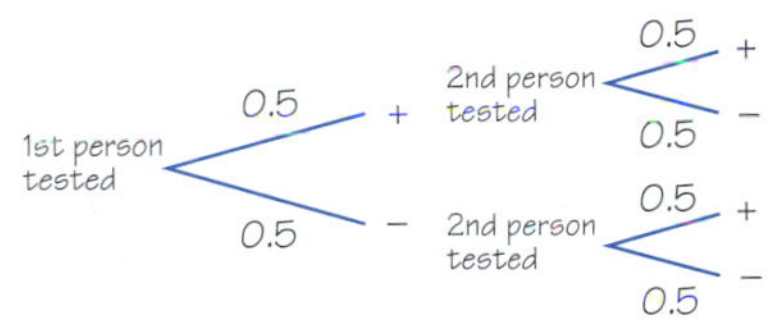

FIGURE 4.23.

In the original simulation you used a paired sample strategy, and assumed that the probability of any individual testing positive would be 50%. **Figure 4.23** shows a situation in which two people are tested one at a time from a group in which 50% of the population is using steroids. In the following question you will use a tree diagram to represent this situation.

1. a) The table in **Figure 4.24** represents the four possible outcomes of the testing process. Fill in the columns for the probability of the outcome, the number of tests each outcome requires, and the probabilistic worth of each outcome.

CHECK THIS!

Remember that you first test the pooled sample and that we are still agreeing to test the first person first!

Event	P(event)	No. of Tests	Prob. Worth
1st Person '+'; 2nd Person '+'	.25	3	.75
1st Person '+'; 2nd Person '–'	.25	3	.75
1st Person '–'; 2nd Person '+'	.25	2	.50
1st Person '–'; 2nd Person '–'	.25	1	.25

FIGURE 4.24.

2.25

b) Calculate the theoretical expected value for the situation when paired-sample testing is used in a population in which 50% of the people are taking steroids.

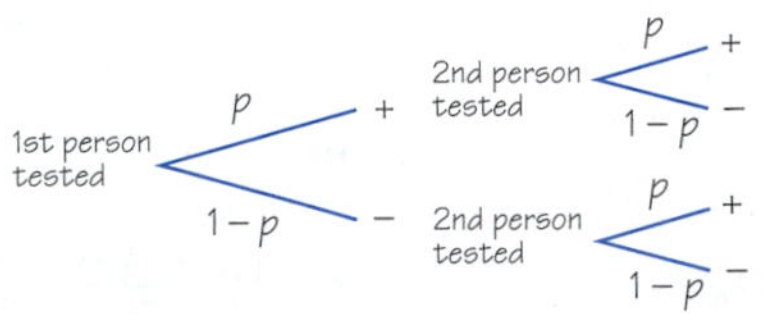

FIGURE 4.25.

c) Generalize the work done in (a) and (b). Assume that the probability that the first person takes steroids is p and the probability that the second person takes steroids is $1 - p$. Find the equation of the model. Refer to **Figure 4.25.**

2. Compare the equation you just obtained with the model that was developed using quadratic regression in Activity 4.3.

Quadratic equations like the one you obtained for your model have the general form $y = ax^2 + bx + c$. These equations model such phenomena as the path of a cannonball through the air.

QUADRATIC EQUATIONS

An equation that can be written in the form $y = ax^2 + bx + c$ is called a quadratic equation with $a \neq 0$. The equation $y = ax^2 + bx + c$ is called the general form of a quadratic equation.

CHECK THIS!

The basic quadratic equation is $y = x^2$. The graph of this equation is shown in Figure 4.26.

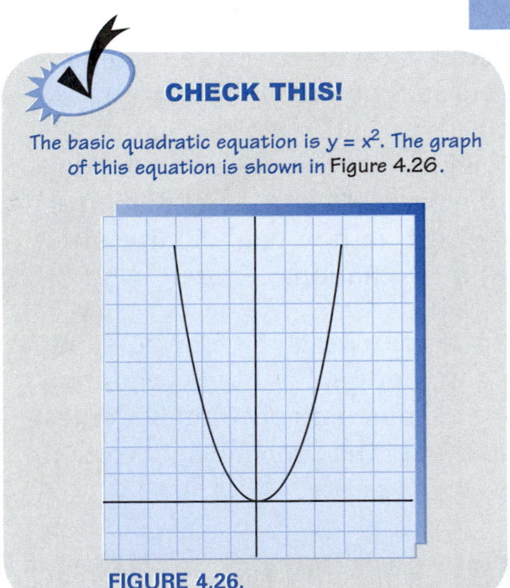

FIGURE 4.26.

Let's take a closer look at a quadratic equation.

3. a) Use your calculator and the equation $y = 0.1x^2 + 2x + 1$ to find the values for y that go with each value of x.

x	−25	−20	−15	−10	−5	0	5
$y = 0.1x^2 + 2x + 1$							

FIGURE 4.27.

b) Based on your results in (a), what WINDOW settings should be used for Xmin and Xmax to graph the equation? What setting should be used for Ymin and Ymax?

c) Describe the shape of the graph that you get. Be sure to sketch a picture of the graph as well.

d) Use the TRACE feature to find the coordinates of the point where the value of y is the least.

4. a) Use your calculator to find the values of y in the equation $y = -0.1x^2 - 2x - 1$ for the values of x in the table in **Figure 4.28**.

x	−25	−20	−15	−10	−5	0	5
$y = -0.1x^2 - 2x - 1$							

FIGURE 4.28.

b) Based on your results in (a), what WINDOW settings should be used for Xmin and Xmax to graph the equation? What setting should be used for Ymin and Ymax?

c) Describe the shape of the graph that you get. Be sure to sketch a picture of the graph as well.

d) Use the TRACE feature to find the coordinates of the point where the value of y is the greatest.

Notice that the general form of a quadratic equation $y = ax^2 + bx + c$ has three terms on the right side of the equation:

- the quadratic term ax^2
- the linear term bx
- the constant c.

The graph of a quadratic equation depends on the value of a in the term ax^2. If $a > 0$, the graph has the shape of a U. If $a < 0$, then the graph has the shape of an upside down U. The graphs of quadratic equations are used in many scientific applications because these graphs have unique properties. For example, these graphs are related to the design of satellite dishes. One important point on the graph of a quadratic equation is its lowest or highest point, called the **vertex**.

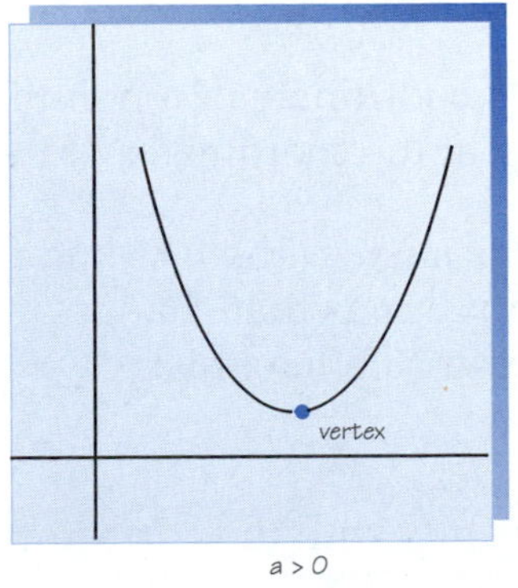

$a > 0$

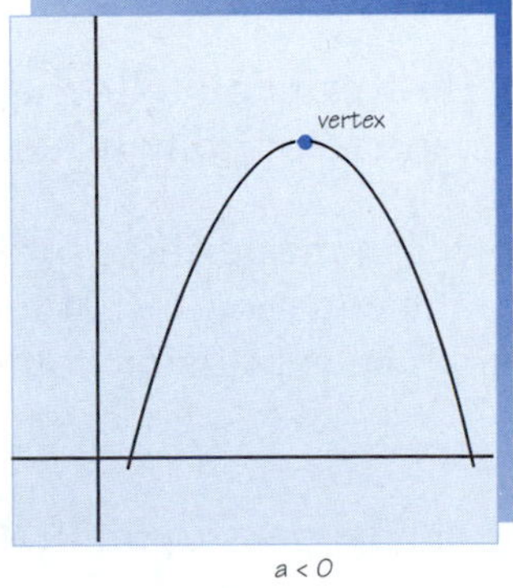

$a < 0$

FIGURE 4.29.
Vertex of the graph of the equation $ax^2 + bx + c$.

FACTORED FORM

The general form of a quadratic equation does not provide a lot of information about the graph of the equation. Other forms of a quadratic equation are more useful because they display some important features of the graph. One form of a quadratic equation is called the **factored form** and is written $a(x - r_1)(x - r_2)$.

5. a) The equation $y = -0.5(x - 1)(x - 5)$ is written in factored form. Graph this equation by plotting a table of values by hand, or by sketching the display of a graphing calculator.

   b) Where does the graph cross the x-axis?

   c) Find the vertex of the graph.

   d) Explain how each of the numbers in the equation affects the graph.

   e) Since the given equation is a quadratic, there should be a general form of the equation. What quadratic equation in general form would produce the same graph?

FACTORED FORM OF A QUADRATIC EQUATION

Equations written in the form $y = a(x - r_1)(x - r_2)$ are said to be in factored form. The numbers r_1 and r_2 are the x-coordinates of the points where the graph crosses the x-axis. These numbers are also known as x-intercepts of the graph. When $r_1 = r_2$, the graph rests on the x-axis and the equation can be written as $y = a(x - r)^2$.

   f) Take each of the x-intercept values identified in (b), and substitute them back into the original equation and evaluate the new expression. What do you notice? Explain.

   g) How can you tell whether a coordinate of a point is an x-intercept from just looking at the coordinates of the point?

The power in mastering expressions such as the factored form of a quadratic equation is that the only things that change are the numbers. The expression could be written like this:

$y =$ ____ $(x -$ ____ $)$ $(x -$ ____ $)$

and the challenge is to fill in the blanks with the correct numbers. Try it!

6. a) Find a quadratic equation in factored form that has x-intercepts –4 and +2 (there are many possible answers).

   b) Adjust your (a) answer so that the vertex is located at the point (–1, –3.6). Explain how you determined this answer.

   c) Write your answer to (b) in general form.

VERTEX FORM

Another form of a quadratic model is called **vertex form**. As the name suggests, it has to do with the location of the vertex.

VERTEX FORM OF A QUADRATIC EQUATION

If the coordinates of the vertex are (h, k), then the vertex form of a quadratic equation has the form $y = a(x - h)^2 + k$. Sometimes, the equation is written this way: $y - k = a(x - h)^2$.

7. a) The equation for one quadratic in vertex form is $y + 1 = 4(x - 2)^2$. According to the equation, where is the vertex located?

   b) For each value of x in the table in **Figure 4.30**, use the equation to calculate the corresponding value of y. Form the ordered pairs (x, y) that correspond to points on the graph.

x	0.5	1	1.5	2	2.5	3	3.5
y							
(x, y)							

FIGURE 4.30.

   c) Plot the corresponding points on a grid like **Figure 4.31**, and draw a smooth curve through the points.

   d) Use a graphing calculator to check whether your graph is correct. Be sure to set the WINDOW settings to match what was used in Figure 4.31. What equation do you enter into the calculator?

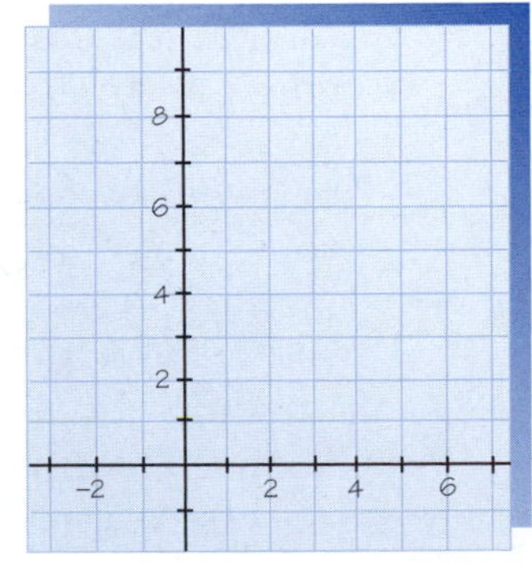

FIGURE 4.31.

e) What are the x-intercepts of $y + 1 = 4(x - 2)^2$?

f) Rewrite the vertex form of the equation in the general form.

g) Now write the equation $y + 1 = 4(x - 2)^2$ in factored form. Recall that you need to write the equation in the form

$y =$ ____ $(x -$ ____ $)(x -$ ____ $)$.

Fill in the blanks with numbers you think might work. Use your calculator to graph your resulting equation and the vertex form $y + 1 = 4(x - 2)^2$. If only one graph appears when you graph the two equations, what does that mean?

8. We have seen that *any* quadratic equation in factored form can be written in general form, but the reverse is not always the case (for real numbers). On your own paper, draw rough-sketch graphs of quadratics (or use the graphing calculator and work with a variety of general form equations) to identify what conditions are needed for the following equations.

   a) What condition is required if the factored form equation can be written as $y = a(x - r)^2$?

   b) What condition is required if a general form equation cannot be written in factored form?

SUMMARY

In this activity you determined the quadratic model for the testing problem, and you explored quadratic equations and their graphs.

- You studied the three forms of a quadratic equation: the general form, the factored form, and the vertex form.
- You used the forms of a quadratic equation to determine some properties of the graph.

This introduction to quadratics provides only some basic information about quadratic equations and their graphs. In the next activity you will learn how to find the graph of a quadratic equation easily, using a graph you already know.

HOMEWORK 4.4 Quad's Going On?

1. a) Use a tree diagram to represent testing for steroids using a paired sample strategy. Assume that the probability of testing positive in the general population is 0.6. For example, the first person in the pair who is tested has a probability of 0.6 of taking steroids.

   b) Use the tree diagram to calculate the probabilities and probabilistic worths for each condition. Then calculate the expected number of tests required for this situation.

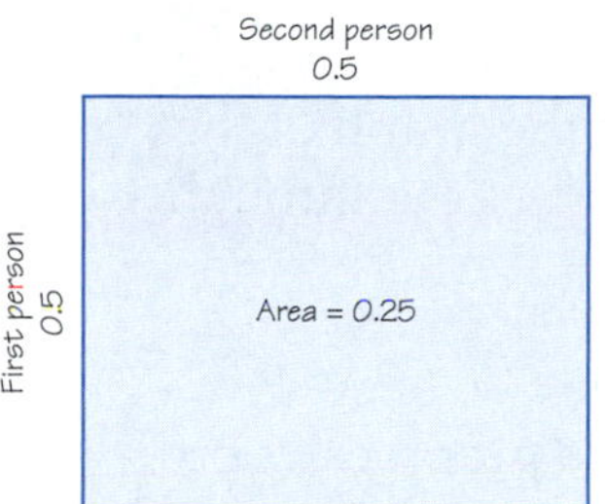

FIGURE 4.32.

Another way to represent probabilities is with the areas of geometrical figures. For example, suppose the probability of an event is 0.5 and the probability of another totally unrelated event is 0.5. The probability of both events occurring is $0.5(0.5) = 0.25$. You can represent the probability of both events using the square with area 0.25 in **Figure 4.32**.

2. a) Assume the probability of testing positive in the general population is 0.6. Which region in **Figure 4.33** represents the probability that both people in a paired sample test positive?

CHECK THIS!

Notice that the sides of the entire square in Figure 4.33 are each 1 unit long. The total area of 1 square unit represents the total probability of all possible outcomes. This square represents the probabilities of all possible outcomes of testing two people n a paired-sample strategy.

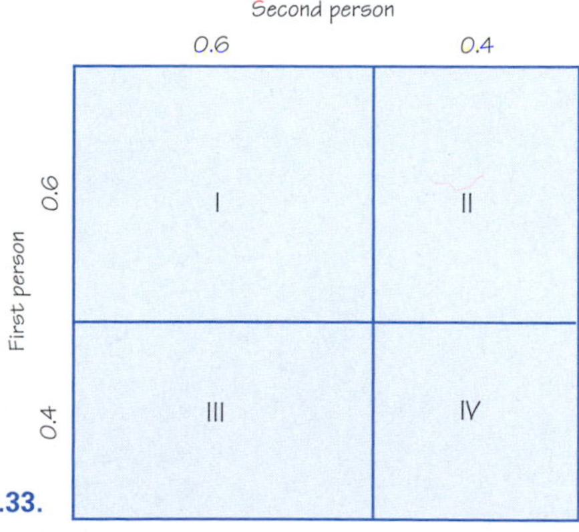

FIGURE 4.33.

   b) Which region in Figure 4.33 represents the probability that both people in a paired sample test negative?

   c) Which region in Figure 4.33 represents the probability that the first person in a paired sample tests positive and the second person tests negative?

d) Which region in Figure 4.33 represents the probability that the first person in a paired sample tests negative and the second person tests positive?

3. a) The quadratic equation $y = 2x^2 - 8$ is written in general form. Explain how to rewrite this equation in vertex form $y = a(x - h)^2 + k$ using $a = 2$ and $k = -8$. Find the vertex form of this equation.

   b) What information about the graph can you read from this form?

   c) Select some values of x and find the corresponding values of y using the equation. Which x-values will give you an idea of the shape of the graph?

   d) Using the results of (c), plot points on the graph of the equation $y = 2x^2 - 8$ and draw a smooth curve through the points to sketch the graph.

4. a) What form of a quadratic equation is $y = -0.5(x + 1)(x - 3)$ written in? What information about the graph can you read from this equation?

   b) Select some values of x and find the corresponding values of y using the equation. Which x-values will give you an idea of the shape of the graph?

   c) Using your results from (b), plot points on the graph of $y = -0.5(x + 1)(x - 3)$. Draw a smooth curve through the points to obtain a sketch of the graph.

5. a) What form of a quadratic equation is $y = x^2 + 3x - 2$ written in? What information about the graph can you read from this equation?

   b) Use your calculator to graph the equation. Use the TRACE command to estimate the x-intercepts of the graph.

   c) Use your results from (b) to write the equation in factored form.

   d) Use the TRACE command to find the vertex of the graph. Write the equation in vertex form.

6. a) What are the x-intercepts of the graph for the equation $y = 2(x - 1.5)(x + 3.5)$?

   b) Use your calculator to find the vertex of this graph.

c) How is the x-coordinate of the vertex related to the two x-intercepts? (Think of these three numbers as points on a number line.)

d) Make up any quadratic equation written in the factored form $y = \_\_\_\_ (x - \_\_\_\_)(x - \_\_\_\_)$. Choose any numbers to fill the blanks in the equation. Repeat the steps you performed in (b) and (c). What did you find?

e) Find the vertex of the graph of the equation $y = 2(x - 5)(x + 3)$ without using a calculator.

7. Change each of the following equations into the specified form.

   a) Change $y = 0.5(x + 2)^2 + 5$ into general form.

   b) Change $y = 2x^2 - 8x + 6$ into factored form.

   c) Change $y = -2(x + 2)(x - 1)$ into general form.

   d) Change $y = 1.4(x - 1)(x - 5)$ into vertex form.

   e) Change $y - 8 = -2(x + 4)^2$ into factored form.

   f) Change $y = x^2 - 2x - 15$ into vertex form.

8. Use the quadratic equation $y = 0.1x^2 + 2x + 1$ to calculate the values of y that correspond to the values for x in **Figure 4.34.** Record your answers in your table.

   b) Calculate the **first differences**, which are found by subtracting two adjacent numbers (in this case, the "bottom" number minus the "top" number). Fill in the column of first differences in Figure 4.34.

   c) Now calculate the **second differences** by subtracting adjacent numbers in the column of first differences (in the same order as before). Fill in the column of second differences in Figure 4.34. What do you notice about these values?

The pattern you found in (c) might be a conclusive way to identify quadratic models when given a table of values. However, on the basis of one example it's wise not to jump to conclusions.

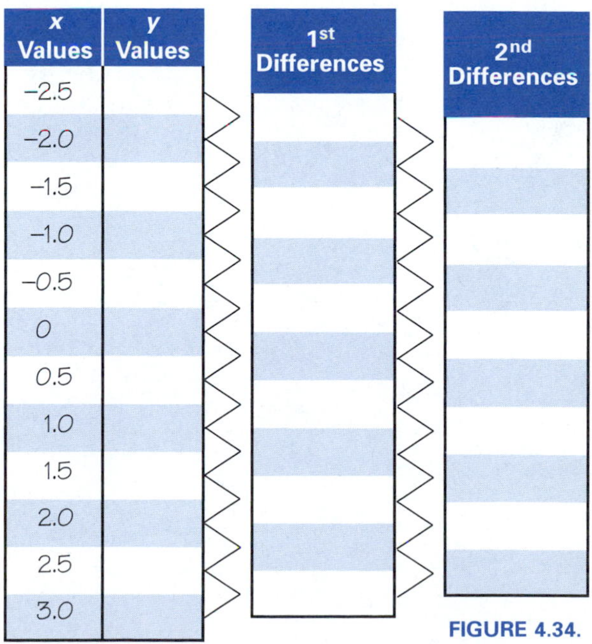

x Values	y Values
−2.5	
−2.0	
−1.5	
−1.0	
−0.5	
0	
0.5	
1.0	
1.5	
2.0	
2.5	
3.0	

FIGURE 4.34.

x	y
4.00	2.800
4.20	3.208
4.40	3.632
4.60	4.073
4.80	4.530
5.00	5.003
5.20	5.491

FIGURE 4.35.

9. Make up your own quadratic equation. List some values for x that increase by a constant amount, and use them to build a table of equation values like the one in Figure 4.32. Then calculate the first and second differences of those values. Did you get the same results?

10. Without graphing, and without using regression analysis, determine whether a quadratic model should describe the set of data in **Figure 4.35**. Explain.

ACTIVITY 4.5

Quadratics on the Move

In Activity 4.4 you discovered that the best model to predict the average number of tests when using paired samples is a quadratic equation. You explored three forms of a quadratic equation, and you learned about the graph of a quadratic equation. In this activity you will explore how a graph of a quadratic equation may be changed to form the graph of another equation.

TRANSLATING GRAPHS

One way to determine the graph of a quadratic equation is to transform another graph that you already know. Question 1 shows how this is done.

1. a) Go back to the simple quadratic equation $y = x^2$. Complete the table in **Figure 4.36**. If it is impossible to find a particular value, write 'NA' in your table.

x			-2	-1			$\frac{2}{3}$			2		3
y	-4	5			0	$\frac{1}{4}$		1	2		5	9

FIGURE 4.36.

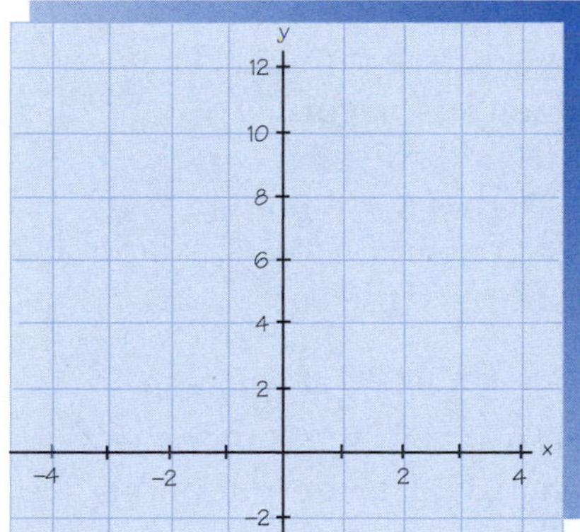

FIGURE 4.37.

b) Using the grid provided in **Figure 4.37**, sketch a graph of $y = x^2$.

c) Enter the y-values you found in (a) in the row of y-values in the table in **Figure 4.38**. Using the equation $y = (x + 3)^2$, find the corresponding x-values. Enter them in the row of x-values in the table. The x-values in the table that can be calculated should increase from left to right.

x												
y												

FIGURE 4.38.

d) How do these x-values compare to the original ones in the table in Figure 4.36?

e) Plot the graph of the equation $y = (x + 3)^2$ in the grid in Figure 4.37.

f) How do the graphs of the equations $y = x$ and $y = (x + 3)^2$ compare?

Question 1 demonstrates the first rule of the transformations of quadratic equations. (A **transformation** is a rule that changes one equation into another.)

TRANSLATING THE GRAPH OF $y = x^2$ TO THE LEFT

If the equation $y = x^2$ is changed to the equation $y = (x + k)^2$ for some positive number k, then the graph of $y = x^2$ is moved k units to the left. The shape of the graph of $y = x^2$ does not change. Only its position on the coordinate grid changes. Note: k does not represent the y-coordinate of the vertex in this expression.

2. a) Predict the change to the graph of $y = x^2$ if the equation $y = x^2$ is changed to $y = (x + 1.5)^2$. Predict what happens to the point (1, 1) on the graph of $y = x^2$.

   b) The graph of $y = x^2$ is shifted five units to the left. What is the equation of the new graph that results from this shift?

Questions 1–2 looked at what happens when the equation $y = x^2$ is changed to $y = (x + k)^2$ when k is a positive number. The next question considers what happens when k is negative.

3. a) Again start with the simple quadratic equation $y = x^2$ and consider the equation $y = (x + k)^2$ for $k = -2$. Replace k with –2 in the equation and simplify the equation.

   b) Now work with the equation $y = (x - 2)^2$. Enter the y-values you found in question 1(a) in the row for y-values in the table in **Figure 4.39**. Use the equation $y = (x - 2)^2$ to find the corresponding x-values. Enter them in the row for x-values in your table. The x-values in the table that can be calculated should increase from left to right.

FIGURE 4.39.

x												
y												

c) How do the table values for this equation compare to the other two tables?

d) The grid provided in **Figure 4.40** shows the graph of $y = x^2$. Copy this graph onto your own grid, and graph the equation $y = (x - 2)^2$.

e) How do the graphs of the equations $y = x$ and $y = (x - 2)^2$ compare?

f) How does adding or subtracting a number to the x variable (before the squaring operation) affect the location of the vertex of a quadratic?

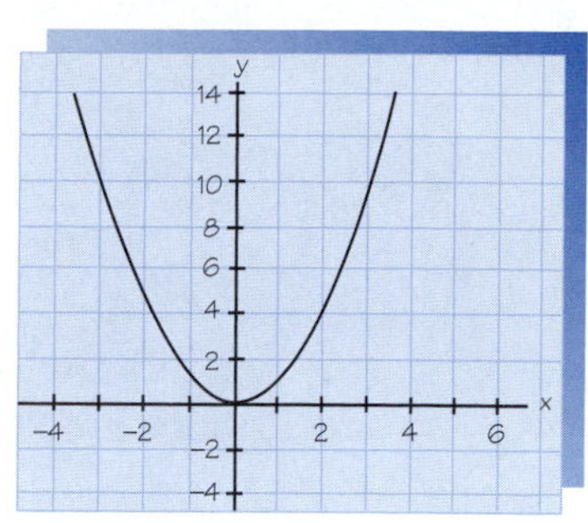

FIGURE 4.40.

Question 3 demonstrates another rule of the transformations of quadratic equations.

TRANSLATING THE GRAPH OF $y = x^2$ TO THE RIGHT

If the equation $y = x^2$ is changed to the equation $y = (x - k)^2$ for some positive number k, then the graph of $y = x^2$ is moved k units to the right. The shape of the graph of $y = x^2$ does not change. Only its position on the coordinate grid changes.

4. a) Again start with the simple quadratic equation $y = x^2$ whose graph is shown in **Figure 4.41**. Now use the equation $y = x^2 + 1$ to calculate the y-values corresponding to the x-values shown in the table in **Figure 4.42**. Enter each y-value in the last row of your table.

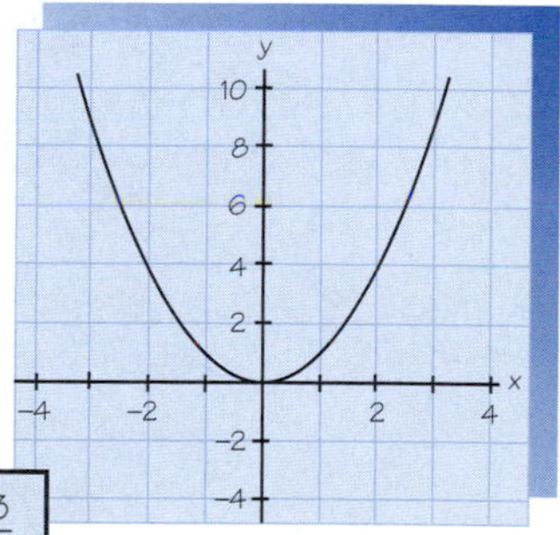

FIGURE 4.41.

x	0	1	-1	2	-2	3	-3	$\frac{1}{2}$	$\frac{3}{2}$	$-\frac{3}{2}$
$y = x^2$	0	1	1	4	4	9	9	$\frac{1}{4}$	$\frac{9}{4}$	$\frac{9}{4}$
$y = x^2 + 1$										

FIGURE 4.42.

b) How do the y-values for $y = x^2 + 1$ compare to the y-values for $y = x^2$?

c) Graph the equation on a grid like Figure 4.41.

d) How do the two graphs compare?

5. a) Now work with the equation $y = x^2 - 4$. Use the equation $y = x^2 - 4$ to calculate the y-values corresponding to the x-values shown in the table in **Figure 4.43**. Enter each y-value in the last row of your table.

FIGURE 4.43.

x	0	1	-1	2	-2	3	-3	$\frac{1}{2}$	$\frac{3}{2}$	$-\frac{3}{2}$
$y = x^2$	0	1	1	4	4	9	9	$\frac{1}{4}$	$\frac{9}{4}$	$\frac{9}{4}$
$y = x^2 - 4$										

b) How do these y-values for $y = x^2 - 4$ compare to y-values for $y = x^2$?

c) Sketch the graph of equation $y = x^2 - 4$ on the same grid with the graph of the equation $y = x^2$.

d) How does this graph compare to the other two graphs?

e) How does adding or subtracting a number to the equation (after the squaring operation) affect the location of the vertex of a quadratic?

TRANSLATING THE GRAPH OF $y = x^2$ VERTICALLY

The graph of the equation $y = x^2 + k$ is a shift vertically of the graph of $y = x^2$. If k is greater than zero, then the graph of $y = x^2$ is shifted k units up. If k is less than zero, the graph of $y = x^2$ is shifted k units down.

THE INCREDIBLE SHRINKING AND STRETCHING GRAPHS

The work done in the first part of this activity introduced **translations**, which are a type of transformation. A translation affects either the x-value or y-value by adding or subtracting. This change produces a shift or "slide" effect on the location of the graph, either horizontally or vertically. Now you will see what happens to the graph of $y = x^2$ when you multiply the x^2 term by some number.

Once again you will compare equations with the simple quadratic equation $y = x^2$. For convenience, the table of values for $y = x^2$ is shown again in **Figure 4.44**.

x	0	1	-1	2	-2	3	-3	$\frac{1}{2}$	$-\frac{1}{2}$	$\frac{3}{2}$	$-\frac{3}{2}$
y	0	1	1	4	4	9	9	$\frac{1}{4}$	$\frac{1}{4}$	$\frac{9}{4}$	$\frac{9}{4}$

FIGURE 4.44.

6. a) The table in **Figure 4.45** has the same x-values as the x-values in Figure 4.44. Working with the equation $y = 0.5x^2$, enter the corresponding y-values in your table.

x	0	1	-1	2	-2	3	-3	$\frac{1}{2}$	$-\frac{1}{2}$	$\frac{3}{2}$	$-\frac{3}{2}$
y											

FIGURE 4.45.

b) How do the table values for $y = 0.5x^2$ compare to the table of values of $y = x^2$?

c) Using a grid like **Figure 4.46**, which already contains the graph of $y = x^2$, sketch a graph of $y = 0.5x^2$.

d) How do the table values and graph for $y = 0.5x^2$ compare to the standard graph of $y = x^2$?

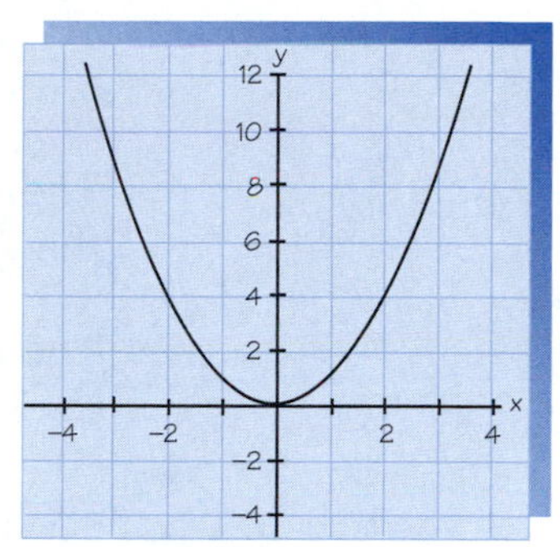

FIGURE 4.46.

7. a) Now working with the equation $y = 1.6x^2$, fill in the y-values in the table in **Figure 4.47**.

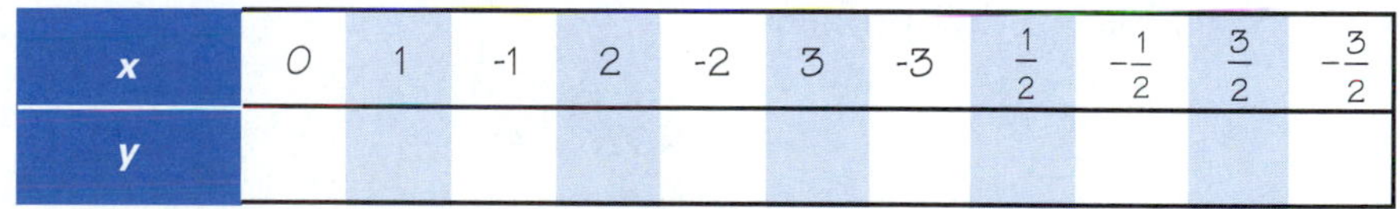

x	0	1	-1	2	-2	3	-3	$\frac{1}{2}$	$-\frac{1}{2}$	$\frac{3}{2}$	$-\frac{3}{2}$
y											

FIGURE 4.47.

b) How do the y-values in Figure 4.47 compare with the y-values of the table for $y = x^2$?

c) Sketch the graph of the quadratic equation $y = 1.6x^2$ on the grid with the graphs of $y = x^2$ and $y = 0.5x^2$.

d) How do the table values and graph for $y = 1.6x^2$ compare to those of the other two equations?

e) How does the value for the constant used in quadratic equations of the form $y = kx^2$ affect the location of the vertex of a quadratic? How does it affect the shape of the quadratic graph?

STRETCHING AND SHRINKING THE GRAPH OF $y = x^2$

The graph of $y = kx^2$ stretches or shrinks the graph of $y = x^2$ vertically. If k is greater than one, then the graph of $y = kx^2$ is stretched. If k is between zero and one, then the graph of $y = x^2$ is shrunk. (If k is less than zero, then the graph is also reflected across the x-axis).

SUMMARY

In this activity you learned about the transformations of the graph of a quadratic equation.

- The graph of $y = (x \pm k)^2$ is a horizontal translation of the graph of $y = x^2$.
- The graph of $y = x^2 \pm k$ is a vertical translation of the graph of $y = x^2$.
- The graph of $y = kx^2$ is a stretching or shrinking of the graph of $y = x^2$.

Now that you can graph quadratic equations it's time for some algebra. Activity 4.6 demonstrates how to solve a quadratic equation so that you can find a solution to the steroid testing problem.

HOMEWORK

4.5

Active Quadratics

1. For each pair of equations describe how the first graph is changed to obtain the second graph.

   a) The graph of $y = x^2$ and the graph of $y = x^2 - 0.5$.

   b) The graph of $y = x^2$ and the graph of $y = (x - 0.5)^2$.

   c) the graph of $y = x^2$ and the graph of $y = 5x^2$.

2. Each of the following graphs is the result one transformation of the graph of $y = x^2$. Identify the transformation and the equation for each graph.

a)

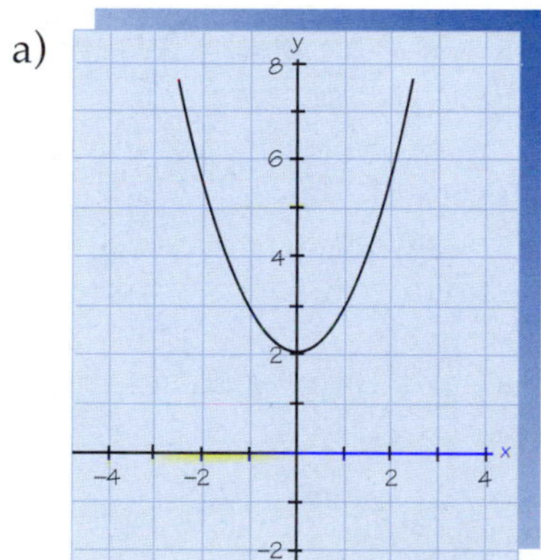

b)

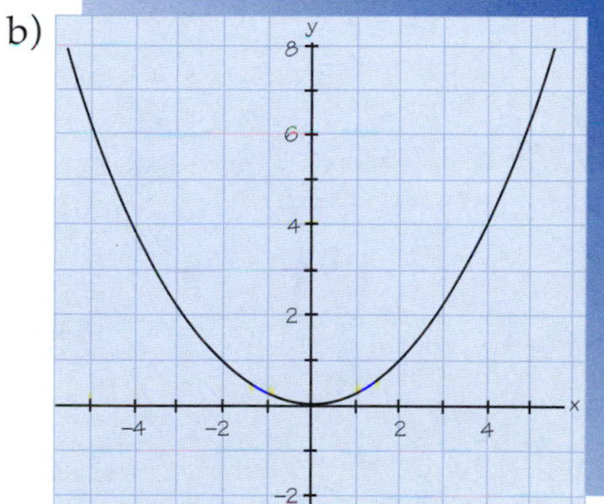

c)

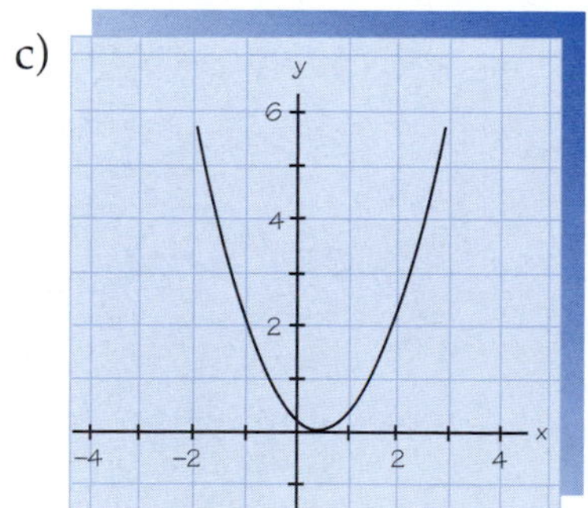

3. a) The graph of $y = x^2$ has been shifted using one of the transformations you learned in Activity 4.5. The table in **Figure 4.48** shows some y-values for the new graph. Complete the values in the figure.

FIGURE 4.48.

x	–3	–2		0		2	3
y		25	16		4		

b) What is the equation of the new graph? How was the graph of $y = x^2$ shifted?

4. a) The graph of $y = x^2$ has been shifted using one of the transformations you learned in Activity 4.5. The table in **Figure 4.49** shows some y-values for the new graph. Complete the values in the figure.

FIGURE 4.49.

x	–3		–1		1	2	
y			$-\frac{2}{3}$	0		$\frac{4}{3}$	2

b) What is the equation of the new graph? How was the graph of $y = x^2$ shifted?

The vertex form equation, $y = a(x - h)^2 + k$, may be analyzed using the transformation operations in Activity 4.5. This equation is the result of transforming the simple quadratic equation, $y = x^2$.

- First there is a stretching or shrinking of the equation $y = x^2$ by a.
- The resulting graph is translated horizontally by h units.
- The new graph is translated vertically by k units. In each translation, the only point that needs to be tracked is the vertex.

These actions are represented as an arrow diagram in **Figure 4.50**.

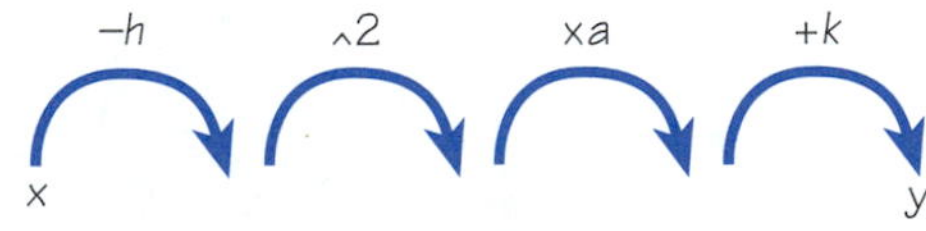

FIGURE 4.50.

5. Draw an arrow diagram to represent the order of operations for the equation $y = 3(x - 4)^2 + 7$.

6. a) The quadratic equation $y = (x/2 - 1)^2$ does not correspond to any form previously studied. What specific feature of the expression suggests that it is a *quadratic* equation?

b) Draw an arrow diagram to represent the order of operations described by this equation.

c) Use the equation to fill in the **Figure 4.51** table.

x	−3	−2	−1	0	1	2	3	4	5	6
y										

FIGURE 4.51.

d) Rewrite the equation in vertex form.

7. For each of the following situations, write a quadratic equation that describes the result of the specified transformations of the graph of $y = x^2$.

a) Stretch the graph so that the y-values of the equation $y = x^2$ are doubled.

b) Move the graph of the equation $y = x^2$ down eight units.

c) Move the graph of the equation $y = x^2$ horizontally five units.

8. a) The graph of the equation $y = x^2$ is stretched by a factor of 1.5. What is the new equation?

b) Translate the resulting graph from (a) horizontally to the right three units. What is the new equation for this graph?

c) Translate the resulting graph from (b) down two units. What is the new equation now?

d) Select two points on the graph of $y = x^2$ and determine how they were changed through these transformations.

ACTIVITY 4.6

Sounds Like Quadraphonic

In the last activity you explored some properties of quadratic equations and their graphs. Now it's time to return to the problem of steroid testing using paired samples. You will answer the questions, "When is it cost-effective to apply a strategy of pairing samples? When is it going to save money?"

CALCULATOR METHODS

Begin by recalling the equation you found for your model and then identifying the problem you wish to solve.

1. a) Write the equation you found in Activity 4.4 that models the steroid testing problem when paired-sample testing is used.

   b) What does this equation represent in terms of the problem you are considering?

2. a) How many tests would we expect per *pair* if we were testing everybody individually?

   b) If you want to save money, what condition should you place on the model?

3. a) In Chapter 1, "Modeling IS Mathematics," you used the TABLE feature of your graphing calculator to examine various values for an equation. If you use this feature to find values of the modeling equation, what settings in the Table Setup (TBLSET) Menu should you use?

   b) Enter your modeling equation into your calculator as the y_1 equation. Using the TABLE feature estimate the probability of the population for which the paired-sample strategy costs the same as testing people individually. How did you get your answer?

4. a) Another way to solve the problem is to graph the equation that models the situation. Then graph the equation that represents the condition for the problem

at the same time. Using your graphing calculator, enter the equation that represents the condition on the average number of tests as the equation y_2. What WINDOW settings are appropriate for the problem?

b) Describe in words the graph of the equation that represents the condition on the average number of tests.

c) Use the TRACE feature to find the point(s) of intersection of the two graphs, and record the x-value(s). What is the meaning of the x-value of these points, in terms of the cost of the paired sample strategy?

COMPLETING THE SQUARE

Your work with the calculator suggested that the paired sample strategy is cheaper than individual tests when the probability of steroid use is about 0.38. You obtained this result when you compared the modeling equation for average number of tests with testing two individuals separately.

Another method is to substitute that value $E = 2$ into the model, which gives you the equation $-p^2 + 3p + 1 = 2$. This equation says, "We want the expected number of tests (E) to average 2." Solving that equation means finding the value of the probability p that makes $E = 2$. Solving a quadratic equation is such an important skill that two different methods will be developed. The first method is called **completing the square**.

The goal of the method of completing the square is to rewrite a quadratic equation as an equation with a perfect square such as $(x + 4)^2$. Let's look at the graph of some equations having perfect squares.

5. Sketch a graph of each of the following equations with perfect square expressions.

   a) $y = (x + 4)^2$.

   b) $y = (x - 5)^2$.

   c) What do the graphs of these perfect square expressions have in common?

6. a) Sketch a graph of the equation $y = x^2 - 4x$.

   b) What number must be added to the expression $x^2 - 4x$ to obtain the graph of an equation with a perfect square?

c) Add the number you found in (b) to the right side of the equation in (a) and sketch a graph of the new equation.

d) Write the equation of this new graph using a perfect square.

e) Translate the graph from (c) back to your original graph by subtracting the number you added in (c) from the perfect square expression you found in (d). What is the resulting equation? (Hint: It should be in vertex form now.)

7. For each of the following, use the instructions from question 6 to rewrite them in vertex form.

   a) $y = x^2 + 8x$.

   b) $y = x^2 - 10x$.

8. Now look again at your answers to question 7.

   a) How does the linear coefficient (number in front of x) in each of the original equations compare to the number in the perfect square expression?

   b) In each equation, how does the linear coefficient compare to the number that was added (and then subtracted)?

COMPLETING THE SQUARE

To solve a quadratic equation $x^2 + bx + c = 0$ using the method of completing the square:

Step 1. Rewrite the equation in the form $x^2 + bx = c$.

Step 2. Add $\left(\frac{b}{2}\right)^2$ to both sides of the equation:

$$x^2 + bx + \left(\frac{b}{2}\right)^2 = c + \left(\frac{b}{2}\right)^2$$

Step 3. Rewrite the left-hand side as a perfect square.

$$\left(x + \frac{b}{2}\right)^2 = c + \left(\frac{b}{2}\right)^2$$

Step 4. Take the square root of each side of the equation and solve for x.

The answers to questions 6–8 are the steps that complete the square, and can be summarized in a rule:

To find out what you add (and subtract), take half of the linear coefficient and then square it. Rewrite the part that is a perfect square expression as a binomial to the second power.

This method is useful in solving quadratic equations because it changes the expression into vertex form. If a quadratic equation is written in vertex form, it can be evaluated by going through a series of steps using the traditional order of operations: parentheses, exponents, multiplication (or division), and addition (or subtraction). Solving the equation requires you to cancel those steps in reverse order.

9. Here is a quadratic equation already written in vertex form: $(x + 3)^2 + 1 = 10$.

a) Suppose you want to see if $x = 4$ is a solution to the equation. After substituting $x = 4$ into the equation, what steps do you go through, and in what order, to evaluate the expression? Do you get 20 as an answer?

b) Now start with 20 as the answer and work backwards, doing the opposite steps in the opposite order. What solutions do you get? (Hint: There are two answers!)

Solving a quadratic equation by using the method of completing the square involves two parts. You must rewrite the original equation in vertex form, and then "undo" the order of operations to isolate the variable.

Example 

Solve the quadratic equation $x^2 - 3x = 10$ by completing the square.

Solution

Step 1. The left side needs to be written in vertex form. Since $-3 \div 2 = -1.5$, and $(-1.5)^2 = 2.25$, add and subtract 2.25.

$$x^2 - 3x + 2.25 - 2.25 = 10$$

Step 2. Re-write the part that is a perfect square expression as a binomial to the second power.

$$(x^2 - 3x + 2.25) - 2.25 = 10$$
$$(x - 1.5)^2 - 2.25 = 10$$

Step 3. Solve the equation for x.

Add 2.25 to each side of the equation:

$$(x - 1.5)^2 - 2.25 + 2.25 = 10 + 2.25$$
$$(x - 1.5)^2 = 12.25$$

Take the square root of each side:

$$\sqrt{(x-1.5)^2} = \sqrt{12.25}$$
$$x - 1.5 = 3.5 \text{ or } -3.5$$

Add 1.5 to each side of the equation:

$$x = 5 \text{ or } -2$$

The two solutions to the equation $x^2 - 3x = 10$ are: $x = 5$ and $x = -2$. If the original problem was adjusted to $x^2 - 3x - 10 = 0$, the solutions would indicate where the graph crosses the x-axis and how to write the same quadratic in factored form: $(x - 5)(x - -2)$.

10. The equation that represented the break-even point in testing pooled samples is given by the quadratic equation $-p^2 + 3p + 1 = 2$. Solve that quadratic equation by "completing the square."

QUADRATIC FORMULA METHOD

The process of completing the square applies as well to the general form of the quadratic equation, $ax^2 + bx + c = 0$. By solving the quadratic equation in the general case, a general formula called **the quadratic formula** is obtained.

11. a) What must you do to the equation in $x^2 - 3x = 10$ before you use the quadratic formula to solve the equation?

    b) What are the values of a, b, and c for that equation?

    c) When you use the quadratic formula, you must calculate the expression with the radical first. What is the value of this expression for the values of a, b, and c from (b)?

    d) Verify the solutions you got in question 10 using the quadratic formula.

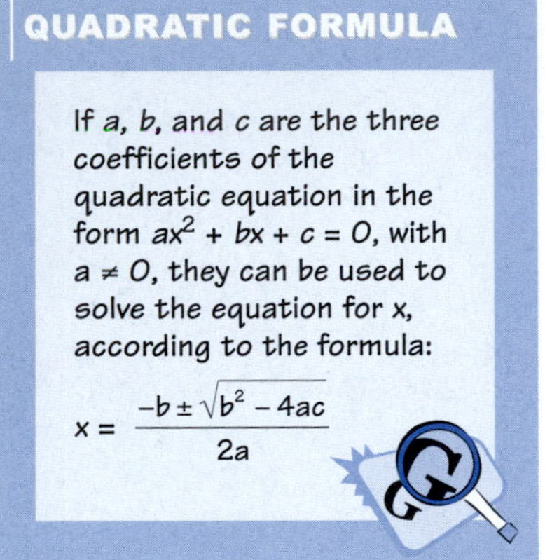

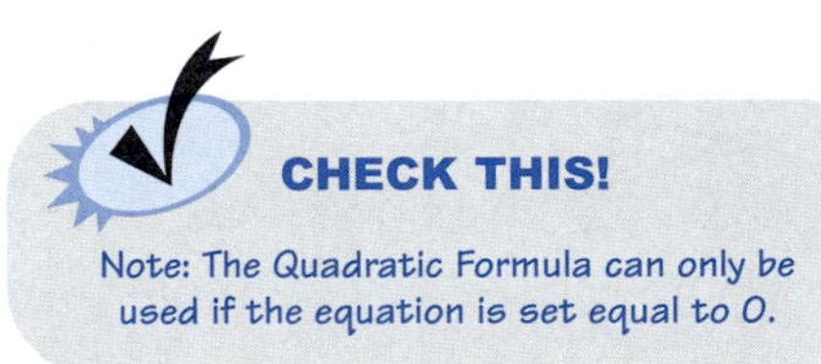

SUMMARY

In this activity you used three methods to solve the problem of when to use a paired sample strategy in testing for steroids. To solve the problem you used the following methods to find the solutions to a quadratic equation.

- You solved a quadratic equation with a calculator by graphing or using tables.
- You solved a quadratic equation by completing the square.
- You solved a quadratic equation using the quadratic formula.

HOMEWORK

4.6 Solutions in Quadruplicate

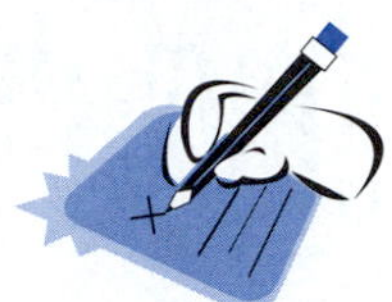

1. Identify all values of x that make the following equations true.

   a) $(x - 3)^2 = 4.$ b) $(3x - 1)^2 = 9.$

2. a) Write the expanded form for $(x - 3)^2$.

   b) Write the expanded form for $(3x - 1)^2$.

3. Complete each square, then write the squared form.

   a) $x^2 - 16x +$ _____. b) $x^2 + 7x +$ _____.

4. How do you know that $(x - 5)^2$ is not equal to $x^2 + 25$? Explain. (Diagrams may be used.)

For questions 5–7, solve by completing the square.

5. $x^2 + 4x = 12.$

6. $x^2 + 8x - 11 = 0.$

7. $5x^2 - 10x - 30 = 0.$

8. a) **Figure 4.52** is an incomplete area model. As the model is currently drawn, what algebraic expression does it represent?

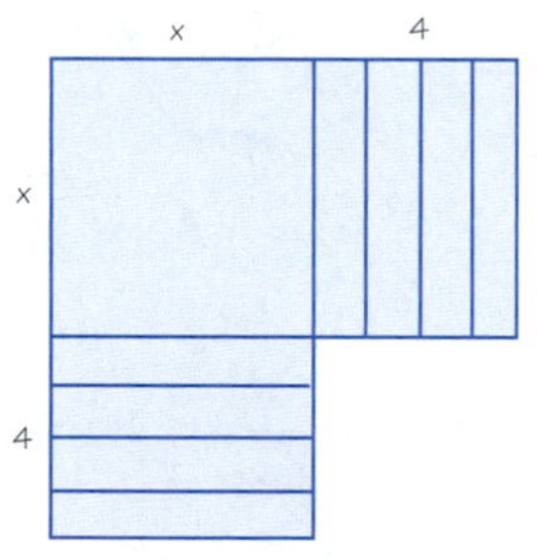

FIGURE 4.52.

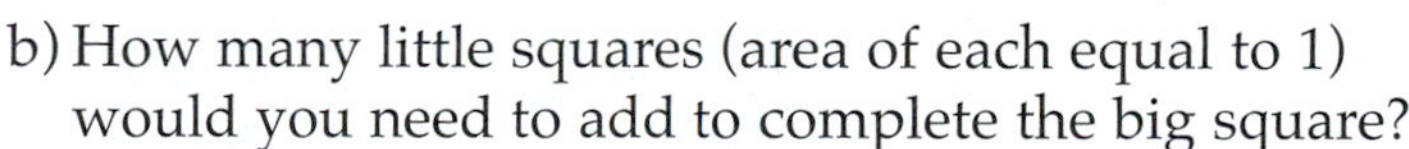

   b) How many little squares (area of each equal to 1) would you need to add to complete the big square?

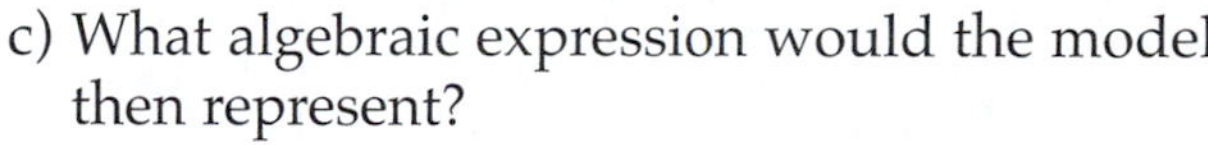

   c) What algebraic expression would the model then represent?

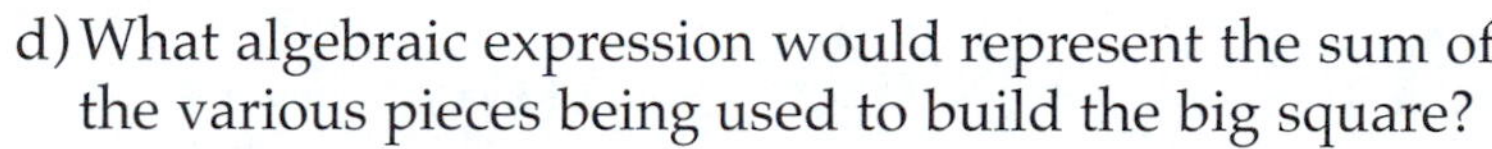

   d) What algebraic expression would represent the sum of the various pieces being used to build the big square?

Use the quadratic formula to solve the equations in questions 9–10.

9. $3x^2 + 10x + 7 = 0.$

10. $2x^2 + 5x - 9 = 0.$

Modeling Project Modeling Larger Pools

The chapter focuses on the study of quadratics because we made the assumption to test only two people in pooling samples. There was no *real* reason for deciding on two, other than to keep things simple. Now that you understand the problem of using pooled samples to test for steroid use, it's time to raise the stakes!

In this modeling project you will examine the problem of testing pools of three people. Here are some questions to consider when using three samples in the pool.

- When is it cheaper to test the people in pools?
- What happens when the pooled sample returns a positive test result?
- Does it make any difference how you proceed from a positive test result?
- Does a quadratic model still work, or does a different kind of model describe this new situation?
- If a new model is needed, what are its properties?

Don't forget the tools you've developed in this chapter for exploring the problem—tree diagrams, algebra, and solving equations on the calculator. Explore the problem and then write a summary of your findings. Have fun, and good luck!

Practice Problems

Review Exercises

1. The model developed in this chapter described the strategy of testing pairs of samples for steroid use in the population.

   a) Why not just test everyone individually? What was the logic behind pairing samples?

   b) What condition determined whether it was cheaper to apply a paired-sample strategy?

   c) What conclusion was found? In other words describe when to apply the paired-sample strategy.

2. In testing for steroid use using a paired-sample strategy, a quadratic model was developed.

   a) What is the distinction between the average number of tests needed and the expected number of tests?

   b) What is the difference between the model $E = -p^2 + 3p + 1$ and the quadratic equation $y = -x^2 + 3x + 1$ (besides the obvious change in variable names)?

3. Consider the situation in which 35% of a certain population is taking steroids.

   a) Draw an area probability model to represent this situation.

   b) Use the drawing to explain how to calculate the expected number of tests to administer in testing for steroids using a paired-sample strategy.

4. Suppose the Department of Motor Vehicles has found that 75% of people pass the test for a driver's license on the first try. Of the people who fail the first time, 80% pass the test on the second try.

   a) Draw a tree diagram to represent this situation.

   b) What is the probability that a person will need to take a third test?

5. A sweepstakes offers a first prize of $1,000,000, two second prizes of $100,000, and ten third-place prizes of $50,000. A total of 15,000,000 entries are received. It costs you 50 cents in paper and stamps to enter. If you submit only one entry for the contest, what is the expected winning?

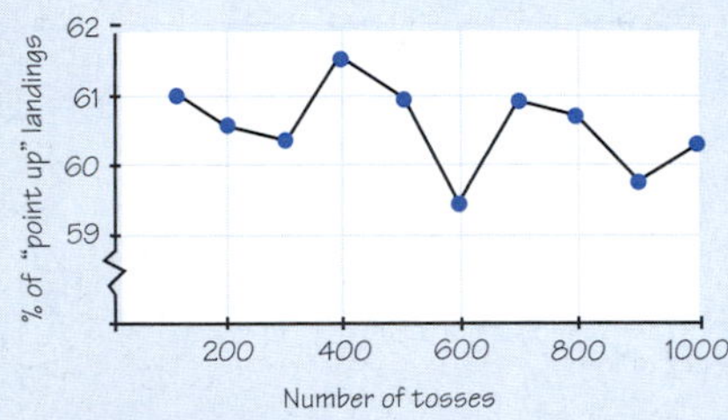

FIGURE 4.53.

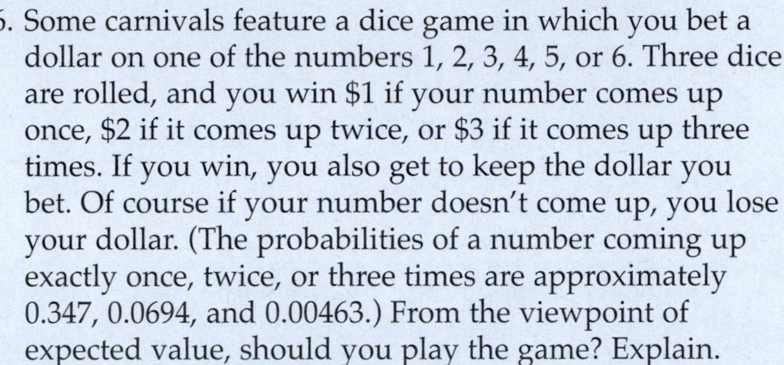

6. Some carnivals feature a dice game in which you bet a dollar on one of the numbers 1, 2, 3, 4, 5, or 6. Three dice are rolled, and you win $1 if your number comes up once, $2 if it comes up twice, or $3 if it comes up three times. If you win, you also get to keep the dollar you bet. Of course if your number doesn't come up, you lose your dollar. (The probabilities of a number coming up exactly once, twice, or three times are approximately 0.347, 0.0694, and 0.00463.) From the viewpoint of expected value, should you play the game? Explain.

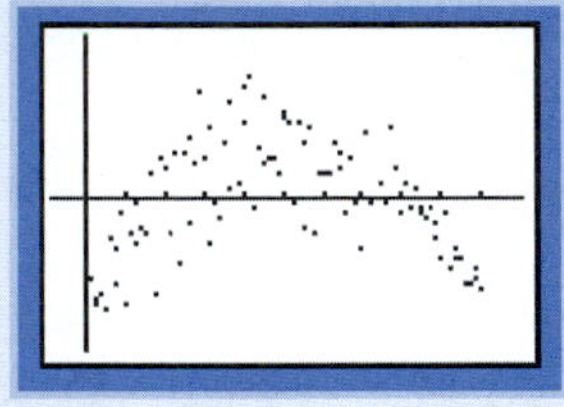
FIGURE 4.54.

7. Consider an experiment in which you flip an ordinary thumbtack, and keep track of whether it lands point up or point down.

a) Do you think it is more likely that the tack will land point up or point down?

b) A graph of the experiment is shown in **Figure 4.53**, which compares the number of tosses with the percentage of tosses that land point up. Estimate P (point up) based on the information in the graph.

FIGURE 4.55.

8. A group of students were preparing a report on the model developed for determining when to pair steroid-testing samples. They revised the DRACULA program to use values for p from 0 to 1, increasing by 0.01 each time. They carefully ran the program twice, using ten trials the first time and 100 trials the second time. They used regression to create a linear model for each data set, and made scatter plot graphs and residual graphs. They forgot to label the graphs, which are shown in **Figures 4.54–57**. Which graphs are the scatter plot graphs? Which graphs reflect doing only ten trials?

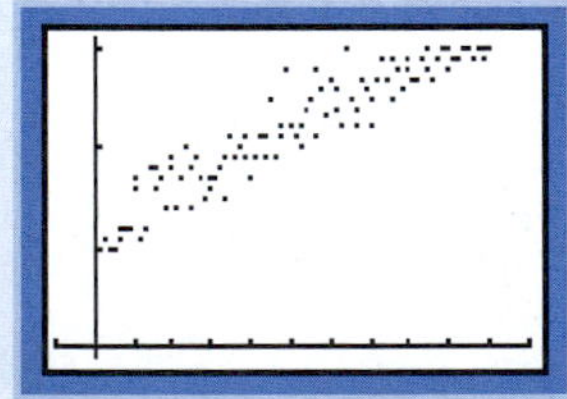
FIGURE 4.56.

9. In Chapter 8 you will investigate models that describe various kinds of motions and their properties. The table in **Figure 4.58** contains information about the distance it takes an automobile to stop while traveling at various speeds under particular weather condtitions.

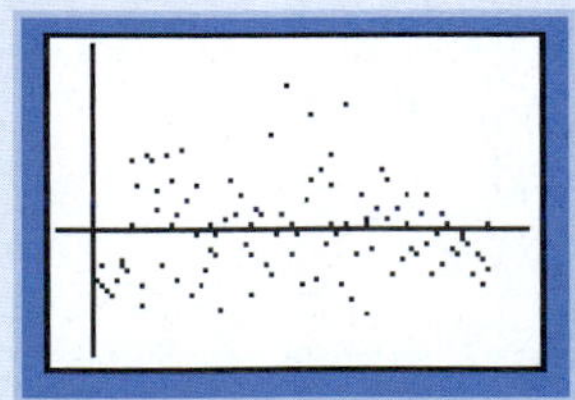
FIGURE 4.57.

Speed (mph)	20	30	40	50	60	70
Stopping Distance (ft)	42	74	116	173	248	343

FIGURE 4.58.

a) Create a mathematical model that predicts the stopping distance for any speed. Using the residuals for your model, explain why you think your model is good.

b) A police officer at an accident scene estimates that a car required 300 feet to stop. Write the equation that describes this situation. Use one of the methods you learned in this chapter to solve that equation. Describe your method.

10. Rewrite each of the following variable expressions in the specified manner:

a) $2(x + 4)(x - 3)$ as a general quadratic expression

b) $p^2 + p(1 - p) + (1 - p)p + (1 - p)^2$ in simplest form

c) $x^2 + 6x + 8$ in factored form

11. Change the following general quadratic equations into the specified form:

a) $y = 0.5x^2 + 1.5x - 5.0$ into factored form

b) $y = 2x^2 + 8x + 3$ into vertex form

12. The graph of $y = 0.5x^2 - 1.5x - 2$ is shown in **Figure 4.59**.

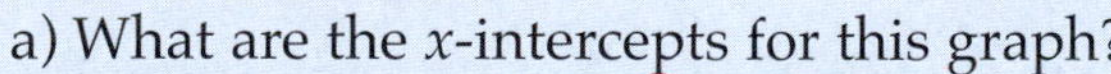

a) What are the x-intercepts for this graph?

b) What are the coordinates for the vertex of this quadratic?

c) How can you tell from looking at the original equation that the quadratic will be oriented like a "U" (and not upside down)?

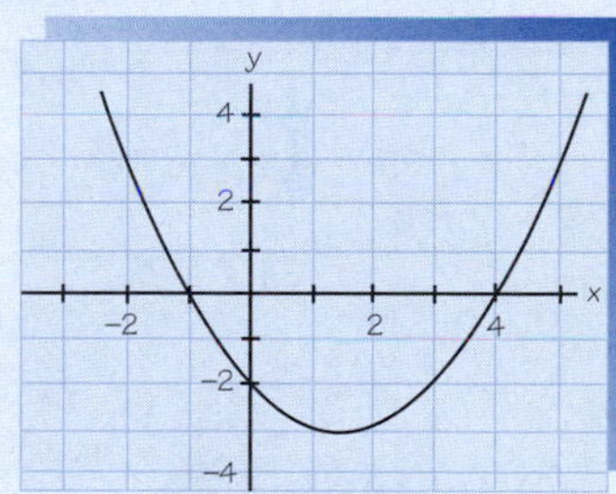

FIGURE 4.59.

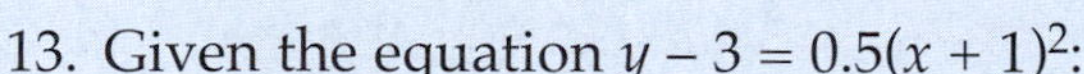

13. Given the equation $y - 3 = 0.5(x + 1)^2$:

a) Explain the role that each of the three numbers in the equation play in determining the graph of this equation.

b) Sketch the graph of the equation.

c) Stretch the graph vertically by a factor of 3, and move it four units to the right and one unit down. What equation describes the new quadratic that is produced?

14. Solve each of the following quadratic equations. Explain the method used.

a) $0.4(x + 3)(x - 2) = 0$.

b) $1.6(x + 1)^2 - 3 = 10$.

c) $2x^2 + 8x - 5 = 0$.

15. A property of quadratic relationships explored in this chapter is that second differences are always constant. You can investigate these relationships and develop ways to construct the equation from the patterns in the table of differences.

x-value	*y*-value	1st Differences	2nd Differences
0			
1			
2			
3			
4			
5			

FIGURE 4.60.

a) Substitute each value of x in the table shown in **Figure 4.60** into the general quadratic equation $y = ax^2 + bx + c$. Record the *expression* obtained for y in the column labeled "y-value" in your table.

b) Calculate first differences and second differences the same way you would find them if there were numbers in the table. Record your answers in your table.

c) Explain how the work done indicates that when working with quadratic equations, second differences of patterns are always a constant.

d) Explain how to find the values for a, b, and c from the numbers in the table.

e) Use your answer to (d) to find the equation that was used to determine the numbers shown in the table in **Figure 4.61**.

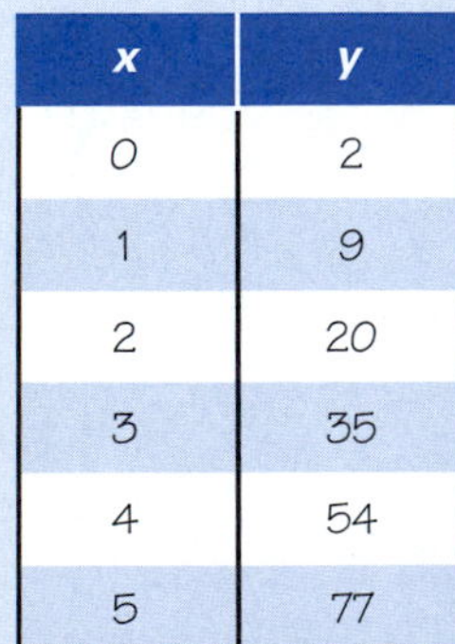

x	*y*
0	2
1	9
2	20
3	35
4	54
5	77

FIGURE 4.61.

16. A steroid tester has collected approximately the same number of samples from two groups of people. He expects the incidence of steroid use to be 8% in one group and 18% in the other. He intends to pool two samples when conducting the test. He is considering two options. One is to pool pairs of samples from each group separately. The other is to pool one sample from one group with one sample from the other group. Which is the better strategy?

4 CHAPTER REVIEW

Modeling Pooled-Sample Testing

Model Development

The problem of checking certain populations for steroid use using expensive medical tests was central to this chapter. As a way to save money the strategy of pooling samples was explored. In a general sense, when there weren't many people taking drugs in the population, it was cheaper to test people using pooled samples. In building a model to describe this situation two simplifying assumptions were made. First the problem was adjusted to consider only the pairing of two samples. Second it was assumed that the tests being used to detect steroids were 100% accurate.

Working with just two people, the process of pairing samples might require only one test if both people being tested were not taking steroids. It might require two tests if the paired sample returned a positive result, and the first individual tested showed up negative. It also might require three tests if the paired sample *and* the first individual tested both had positive results. On the other hand, it would always take two tests to check two people on an individual basis. So when few people were taking steroids, one would expect to have only one test most of the time. That explains *why* it would be cost-effective to use a paired-sample strategy when few people are using steroids. Over many tests, keeping track of whether the average was above or below 2.00 would determine if the paired-sample testing strategy should be used.

The first model considered was based on the idea that the model might be a linear relationship. A hand-drawn experiment to test whether the value halfway between extreme conditions would produce an average halfway between the range of test results proved inconclusive. With few trials a large range of average values could be expected, and a linear model could be constructed to meet the known conditions. With more trials it was found that the original linear model would not work, but perhaps a line of best-fit could still be used. After running a simulation that considered other probability situations, the data showed a definite non-linear pattern. Regression analysis was used to verify that a linear model was not appropriate. By examining residual plots, a new type of model, called a quadratic, was found to be the best choice.

Finally the problem was examined again, in a more theoretical way. Rather than relying on data collected from experiments, probability was used to describe the condition of steroid usage among the population. A calculation called the expected value gave a prediction of the long-term behavior of the testing game. Using a variable to replace a specific probability value, the same steps were repeated to produce the final quadratic model $E = -p^2 + 3p + 1$. The solution to the equation $-p^2 + 3p + 1 = 2$ was found to be around $p = 0.38$, which determined whether it would be cheaper to test in pairs or individually.

Mathematics Used

When working with a single situation, the average number of tests was calculated by adding up the number of tests and dividing by the number of trials performed. When working with data that reflected a range of possible conditions, regression analysis was used to find the best model for a particular type of equation. The model was used to generate predicted values, which were compared to the actual data. The pattern of errors (and size) was used to determine whether that type of equation was the best one to use.

Probability was introduced as a number between 0 and 1 that describes the likelihood that an outcome for an event takes place. Formally it is defined as the ratio between the number of times an outcome of interest occurs and the total number of possible outcomes that can take place. Tree diagrams were introduced as tools to help set up situations having more than one event. Probabilistic worth was defined as the product of the probability of an outcome taking place and the payoff for that outcome. Expected value was defined as the sum of all possible probabilistic worths.

Quadratics were introduced as the best model for describing the problem of pairing samples. These are equations of the general form $y = ax^2 + bx + c$, where a, b, and c are numbers. The graphs were found to have a characteristic U-shape, with symmetric halves meeting at a point called the vertex. The vertex is always the lowest or highest point on the graph of a quadratic. Another way to express the equation of a quadratic was in factored form:

$y = a(x - r_1)(x - r_2)$. When written this way, the numbers r_1 and r_2 are the roots of the quadratic equation. The third way to express the equation of a quadratic was in vertex form: either $y = a(x - h)^2 + k$ or $y - k = a(x - h)^2$. This kind of equation form reveals the coordinates of the vertex (h, k). Techniques were developed (or practiced) for changing quadratics from one form to another using algebra, arrow diagrams, and technology.

Solving a quadratic equation, in which a value for y is specified, produced the answer to the central question explored by the unit. Many methods were introduced for handling specific situations. Using a calculator the quadratic expression could be given as one equation, and the condition for the solution given as another equation. After that either graph both equations using TRACE to find the coordinates of the intersection point, or compare values from both equations using TABLE until they are the same. Equations in factored form could be solved (assuming the answer is equal to zero) by setting each factor equal to zero and solving the resulting equations. Undoing operations in the reverse order solved equations that were in vertex form. General equations could also be solved if they were set equal to zero. If solving an equation $ax^2 + bx + c = 0$, the quadratic formula was used:

$$x = \frac{-b \pm \sqrt{b^2 - 4ac}}{2a}.$$

Area Model: A representation in which the area of a geometric region represents the probability that two events happen simultaneously. The dimensions of the region have lengths that are interpreted as the individual probabilities for the events.

Binomials: Algebraic expressions involving two terms, usually with one of them containing a variable.

Completing the Square: The process of changing the general form of a quadratic to vertex form. Completing the square is used to solve quadratic equations.

Domain: A description of the possible values that can be used in a model.

Equivalent: When two expressions produce identical values and solutions for all numbers.

Event: A situation in which more than one possible outcome may take place.

Expected Value: The average result over many trials. Expected value is calculated by adding together all the probabilistic worths for all the possible outcomes.

Factored Form of a Quadratic Equation: A quadratic of the form $y = a(x - r_1)(x - r_2)$.

First Differences: A pattern of numbers produced by subtracting successive values from a data table in which the x-values are increasing by a constant amount. If first differences are constant, then the equation describing the original numbers is linear.

General Form of a Quadratic Equation: A quadratic of the form $y = ax^2 + bx + c$.

Independent Events: Two events are called independent when the likelihood of one event taking place is unaffected by whether the other event happens.

Law of Large Numbers: The average result is more likely to be close to the theoretical result if the number of trials is large than if the number of trials is small.

Negative Test Result: In medical testing, this indicates that the trait for which the test is being administered is not present.

Outcome: A specified manner in which an event actually takes place, or one particular one that is of interest.

Paired Sample Testing: This is testing the mix of two people's samples. In paired sample testing either 1, 2, or 3 tests will be required for every two people tested. If the average of the tests stays under 2.00, it will be cheaper than testing each person individually.

"Perfect" Drug Test: 100% accurate in detecting the presence of a drug, and 100% accurate in verifying that a drug is not present.

Pooled Samples: These consist of mixing more than one individual's samples. If a pooled sample provides a negative test result, then everyone who was in the sample must also be negative individually (assuming the medical test is perfect).

Positive Test Result: In medical testing this indicates that the trait for which the test is being administered has been detected.

Probabilistic Worth: The value (or payoff) of an event multiplied by the probability of the outcome happening. It measures average results for a specific outcome.

Probability: A number between 0 and 1 that assesses the likelihood of a chance outcome. The closer the probability is to 1, the more likely the outcome.

Quadratic Formula: The formula used to solve general quadratic equations of the form $ax^2 + bx + c = 0$, where $a \neq 0$. The solution, using the quadratic formula, is:

$$x = \frac{-b \pm \sqrt{b^2 - 4ac}}{2a}.$$

Quadratic Function: A function of the form $y = ax^2 + bx + c$, where $a \neq 0$. The highest power of x in a quadratic is 2.

Range: The description of the possible values that result from using a model, or the spread of numbers in a data set (from lowest to highest).

Roots of a Graph: The points where the graph crosses the x-axis.

Second Differences: A pattern of numbers produced by subtracting successive first differences. If the second differences are constant, then the equation describing the original numbers is quadratic.

Simulation: Re-enactment of the details of a situation being modeled. Both a physical activity and a computer program simulated the problem of paired sample testing.

Square Root: A factor of a number that when squared gives the original number.

√ (Square Root Symbol): The principal square root of a number. The principal square root of a positive number is the positive square root.

Transformation: The rule that changes one equation into another.

Translation: The transformation that affects either the x- or y-values of an equation by adding or subtracting a number. Translations produce a "shift" or "slide" effect on the location of the graph.

Tree Diagram: A graphic representation of a step-by-step process used when there is more than one possibility at each stage of the process.

Vertex Form Equation of a Quadratic: A quadratic of the form $y = a(x - h)^2 + k$.

Vertex of a Quadratic: A point on the graph of a quadratic where the left-side and right-side (symmetric) parts meet. The vertex also acts as either the highest or lowest point for the graph and the values that result from the equation.

***x* intercepts:** The points where the graph of an equation crosses the x-axis.

Zeroes of a Graph: The points on the graph of an equation where the y-coordinate is 0. They are also called the x-intercepts or the roots.

CHAPTER 5

Art and Perspective

Is there a connection between mathematics and beauty?

Artists are able to create magical, seemingly three-dimensional worlds on their canvases with little more than putting paint on a flat surface. They achieve these remarkable works by representing three-dimensional space on two-dimensional canvas or paper using some basic artistic principles. To produce this enchanting realism, artists use a great deal of mathematics and geometry. In fact the techniques that artists use are closely related to the science of optics—the study of light and vision.

The ability to produce the illusion of depth in painting and drawing has been an important characteristic of Western art since the 15th century. Artists of the Renaissance studied nature, color, and light to create the techniques that have been handed down to us. Italian Renaissance masters such as Filippo Brunelleschi and Leonardo da Vinci developed these techniques based on the way the human eye perceives an object. It was during this time that the precise mathematical methods of perspective drawing were invented. In this chapter you will learn how mathematics was used to create many masterpieces of Western art.

PREPARATION READING

Picture Perfect

What does mathematics have to do with creating art? An architect uses the geometry of one-point or multi-point perspective to help a client visualize what a proposed building will look like. A theater designer uses the geometric principle of convergence to create the illusion of depth and distance on a small stage. An animator guides you into and around three-dimensional objects using a two-dimensional screen. A painter blends the geometric precision of perspective with the subtle effects of shading and color to add depth to drawings.

In this chapter you answer the question, How does an artist accurately represent three-dimensional objects on a two-dimensional page? Before you begin, take a look around you and notice the lines and shapes that make up your world. Think about how you might sketch the room that you are in or the front of your school.

FIGURE 5.1. House by Jackson Barber, age 8.

Small children view the world differently than you. Their sense of depth is not well developed. Just look at a picture drawn by a young child, such as the one in **Figure 5.1**.

Notice that the sun is too large and the building appears flat. Now compare Figure 5.1 with an architect's drawing for a new high school (**Figure 5.2**).

FIGURE 5.2. Tantasqua Regional High School.

The drawing helped community citizens visualize what the completed project might look like before they voted on whether or not to fund the new school. The drawing needed to be accurate.

- What geometric principles guided the artist's creation of an accurate drawing of the future school?
- What principles gave the architect's picture its depth?

This chapter will provide some of the answers to these questions.

ACTIVITY 5.1

It's a Matter of Perspective

Artists, architects, illustrators, and animators frequently need to draw objects and scenes accurately. They use many principles that have been developed to represent three-dimensional figures in two dimensions. In this chapter you will discover some of the principles that artists use to bring their pictures to life.

PERSPECTIVE DRAWING: WHAT IS IT?

PERSPECTIVE DRAWING

Perspective drawing is the technique of representing objects from three-dimensional space on a two-dimensional sheet of paper, piece of canvas, or other flat surface. Artists use several techniques to create the impression of three-dimensional space on a flat canvas or sheet of paper.

Many artists use principles of **perspective drawing** as a guide for creating their works. An artist uses perspective to imitate the appearance of a three-dimensional object. Artists represent multiple objects of many shapes from various points of view.

Sometimes an artist violates a geometric principle of art in order to get your attention. The painting in **Figure 5.3** by William Hogarth contains several intentional mistakes.

CHECK THIS!

William Hogarth (1697–1764) was an English painter, engraver, and art theorist. Hogarth painted many wonderful portraits and pictures of everyday life. Hogarth is best known for his delightful, satirical prints. He made fun of both rich and poor Londoners in these prints.

FIGURE 5.3.
Perspective Absurdities by William Hogarth (1697–1764).

CHECK THIS!

Here are some questions you might ask your group as you examine the Hogarth painting:

- What is in the foreground of the painting?
- What is in the background of the painting?
- How do the sizes of the elements in different parts of the painting compare?
- How do the locations of the elements in different parts of the painting compare?

1. Work together with your group to identify all the real-world inaccuracies in the painting by William Hogarth (see Figure 5.3). Describe each in detail.

2. Describe what the artist needs to do to correct each of the inaccuracies you identified in question 1.

3. Study the corrections you suggested in question 2 and look for common underlying guidelines an artist might follow in order to create accurate pictures.

Perspective drawing has been called the geometry of vision. Art historians generally agree that perspective drawing was invented in Florence, Italy, in the 15th century. The Italian architect Filippo Brunelleschi is credited with developing a systematic approach to perspective drawing. Brunelleschi was the designer of the remarkable dome of the cathedral in Florence.

The painting by Georges Seurat in **Figure 5.4** demonstrates how the artist used some principles of perspective drawing to create a sense of depth (or distance) in the picture. Questions 4 and 5 will help you identify some of these principles.

FIGURE 5.4.
The Bathers by the French painter Georges Seurat (1859–1891).

4. Look at Figure 5.4. In this painting the artist tried to represent a three-dimensional subject on a two-dimensional plane (the plane of the picture).

a) Imagine that your group is in a hot-air balloon floating over the subjects in this picture. Draw a bird's-eye view of what you would see. You don't have to sketch every detail. Just draw simple outlines of some of the objects in the picture and label them. (For example, you might use ovals labeled as dog, man lying down, man wearing hat, or sailboat to indicate the objects.)

CHECK THIS!

Recall that a **plane** is a flat surface that has height and width. You can think of a sheet of paper or a piece of canvas as a plane. The **plane of the picture** is the flat surface on which the artist has created his or her work.

b) Explain what clues in the original picture helped you decide where to place particular objects in your picture.

c) Discuss how Seurat tried to create the illusion of depth.

5. Now look at **Figures 5.5** and **5.6**. In each picture the artist tried to represent a three-dimensional subject on a two-dimensional plane (the plane of the picture).

FIGURE 5.5.
Cotswold Games.

FIGURE 5.6.
The Boating Party by the American figure painter Mary Cassatt (1844–1926).

a) Imagine that your group is in a hot-air balloon floating over the subjects in these pictures. For each picture, draw a bird's-eye view of what you would see. You don't have to sketch every detail. Just draw simple outlines of some of the objects in each picture and label them. You might use ovals for the elements in the pictures just as you did in question 4.

b) Explain what clues in the original picture helped you decide where to place particular objects in your picture.

c) Discuss how the artists tried to create the illusion of depth.

SUMMARY

In this activity you examined some works of art to learn about perspective drawing.

- You examined the Hogarth painting to learn what *not* to do. The work showed you the following principles:
- Objects in the foreground that are in front of objects in the background should cover part of the distant objects. This is the **principle of overlapping**.
- Objects in the background should be smaller than objects in the foreground. This is the **principle of diminution**.
- An object that is larger than another in real life should be larger in the work of art.

These observations give you an idea of several principles of perspective drawing. As you progress through the chapter, you'll add layers of complexity to your drawings. For each new layer you'll need to extend your toolbox of drawing principles. In Activity 5.2 you will learn more about one important principle, the principle of diminution.

HOMEWORK 5.1

I've Got You Covered

One basic element, perhaps the simplest, of perspective drawing is the principle of overlapping. You use **overlapping** to show depth and distance when you place one object in front of another object in a picture.

PRINCIPLE OF OVERLAPPING

Objects that are closer to the viewer and appear to be in front of another object will hide part of the more distant object.

1. Look at Figures 5.4, 5.5, and 5.6. For each picture, describe at least one object that overlaps another. Which of the two objects is closer to the viewer?

2. Look at Figure 5.3. Describe where Hogarth violates the principle of overlapping.

Activity 5.1 presented pictures from the artist's viewpoint (see Figures 5.4, 5.5, and 5.6) and asked you to draw bird's-eye views of the scenes. Next you'll reverse the process. Questions 3 and 4 provide top views of simple objects and ask you to sketch views from the sides.

CHECK THIS!

Questions 3–5 follow a basic principle of modeling: simplify the situation and solve the simplified problem. These questions examine the principle of overlapping using cubes, a relatively simple shape.

3. **Figure 5.7** shows the top view of three cubes sitting on a table. Draw each view of what you would see if you were sitting at A, B, C, and D.

4. a) Look at the top view presented in **Figure 5.8**. Draw each view of what you would see if you were sitting at A, B, C, and D.

   b) Repeat (a) using **Figure 5.9**.

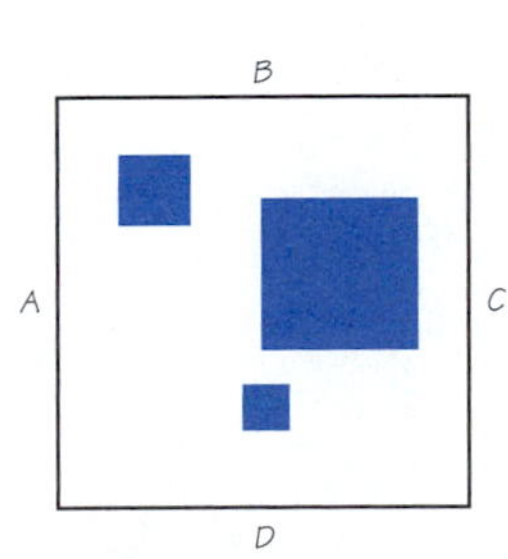

FIGURE 5.7.
Top view of three cubes on a table.

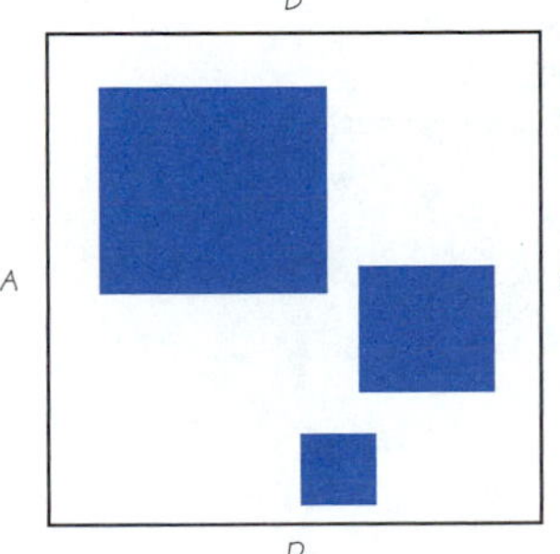

FIGURE 5.8.
Top view of three cubes.

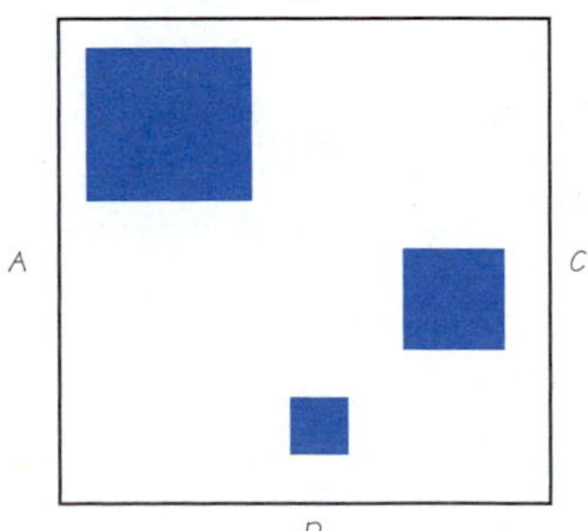

FIGURE 5.9.
Top view of three cubes.

5. Suppose that you are given a side view and asked to draw the view from the top. Sketch the top view for Arrangement 1 and Arrangement 2 in **Figure 5.10**. Which top view is more likely to be accurate? Explain your answer.

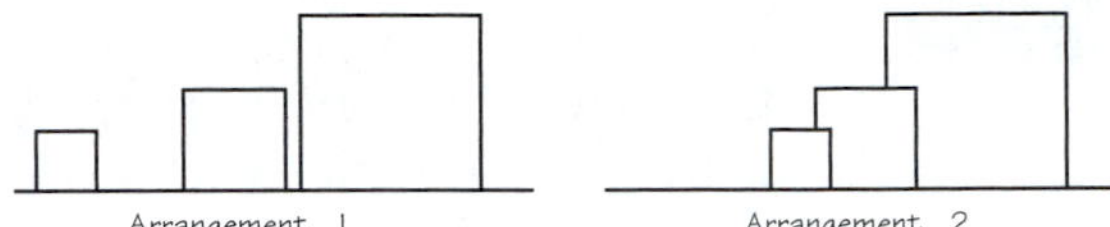

FIGURE 5.10.
Side view of two different cube arrangements.

Overlapping is one way to show distance from the viewer. However, you often need more than overlapping to show depth. Arrangement 2 demonstrates that sometimes overlapping does not show the relative distance between objects. Arrangement 1 shows that sometimes objects do not overlap. You need more principles in order to accurately represent three-dimensional scenes on paper.

6. a) Which of the people in **Figure 5.11** appears to be taller? Explain your answer.

   b) Now look at **Figure 5.12**. Does one of the people appear to be taller? Explain your answer.

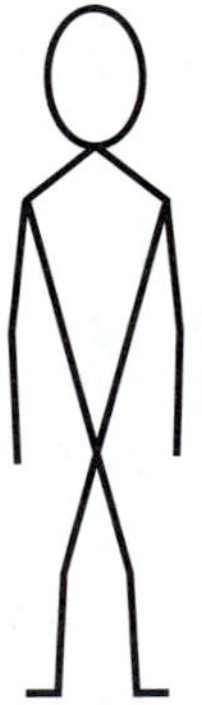

FIGURE 5.11.
Stick figures for question 6(a).

FIGURE 5.12.
Stick figures for question 6(b).

ACTIVITY 5.2

From a Distance

In Activity 5.1 you explored the principle of overlapping. This principle is used to show that one object is closer to the viewer than another. However, overlapping is not adequate for showing relative distances between objects. A new principle is needed. In this activity you will learn about the principle of diminution.

THE PRINCIPLE OF DIMINUTION

Try a quick experiment. Hold a cube directly in front of you so that all you see of the cube is the square side. Move the cube toward you. The square appears to become larger. Move the cube directly away from you and the square becomes smaller. You have just discovered the principle of **diminution**.

CHECK THIS!

The word *diminution* means the act of reducing or decreasing. This word comes from the word *diminish*.

DIMINUTION

In general, objects appear smaller as they move farther away. This basic principle of perspective drawing is called diminution.

1. Look at Figures 5.4, 5.5, and 5.6. For each picture describe specific examples of how each artist used the principle of diminution.

As you may have noted already, Seurat (see Figure 5.4) used the principle of diminution when he made the people in the distant boat smaller than the people on land. But how did he decide how much smaller? The key lies in a technique called **scaling**.

To help you see how scaling works, first you need to know what is meant by *lines of sight*.

LINES OF SIGHT

Lines of sight are imaginary lines running from your open eye to some object at which you are looking.

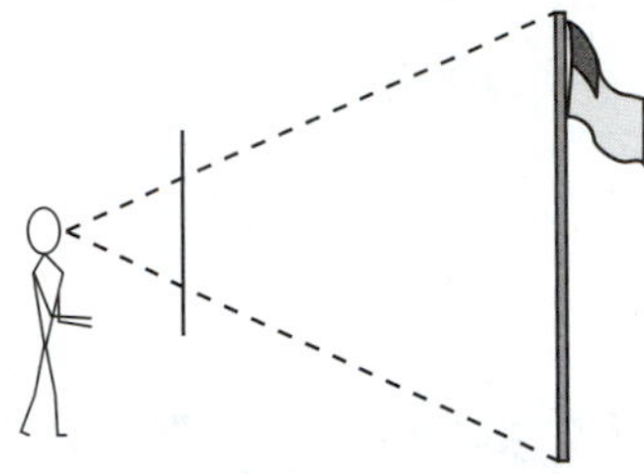

FIGURE 5.13.
Lines of sight to top and bottom of a flagpole.

Suppose you are looking at a flagpole as shown in **Figure 5.13**. The dashed lines represent the lines of sight from your eye to the top and bottom of the flagpole. Imagine placing a canvas an arm's length away. The vertical line next to the figure represents the canvas. You want to sketch the image of the flagpole as it would appear on the canvas. The lines of sight intersect the picture plane (the canvas) to indicate where the top and bottom of the flagpole should be.

PICTURE PLANE

The picture plane is the flat surface on which the picture is painted. (This is sometimes called the plane of vision or viewing plane.) You can think of the plane as a windowpane between the viewer and the scene that is shown in the painting. Many artists forget about the flat surface they are painting on. Instead, they think and feel the distances that they depict in their works as if they are seeing through a window.

2. Trace Figure 5.13 onto a piece of paper. Then imagine that the viewer steps backward. Draw your figure farther from the flagpole. Sketch new lines of sight to the top and bottom of the flagpole, and sketch another picture plane. Will your new drawing of the flagpole be larger or smaller than your first drawing?

Figure 5.13 shows one example of scaling. You can use **similar triangles** and proportions to find the proper size of each object in a picture or to draw objects at the correct size. Then you can use one or all three ways to test for similarity.

SIMILAR TRIANGLES

Two triangles are similar if and only if corresponding angles are congruent and the measures of corresponding sides are proportional. To say this in mathematical shorthand, write $\triangle ABC \approx \triangle DFE$. This means

- $\angle A$ corresponds to $\angle D$, $\angle B$ corresponds to $\angle F$, and $\angle C$ corresponds to $\angle E$.
- Side AB corresponds to DF, BC corresponds to FE, and AC corresponds to DE.

Tests for Similarity

- A-A (Angle-Angle) Similarity: Two angles of a triangle are congruent to two angles of another triangle.
- SSS (Side-Side-Side) Similarity: The measures of corresponding sides of two triangles are proportional.
- SAS (Side-Angle-Side) Similarity: The measures of two sides of a triangle are proportional to the measures of corresponding sides of another triangle *and* the included angles are congruent.

Example 1

List all the triangles in **Figure 5.14** that are similar to each other. Use the fact that corresponding sides of similar triangles are proportional to find lengths x and y.

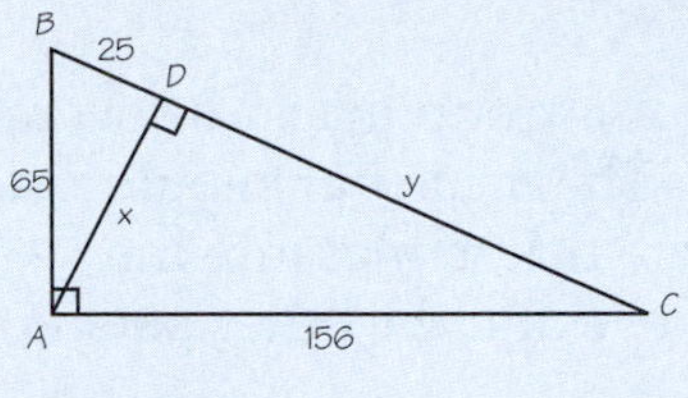

FIGURE 5.14.

Solution

Triangle ABC is similar to triangle BDA is similar to triangle DBA. To find x, use the fact that the ratios of the lengths of legs of the triangles are equal since the triangles are similar.

$$\frac{x}{25} = \frac{156}{65}$$

$$25\frac{x}{25} = 25\frac{156}{65}$$

$$x = \frac{3900}{65} = 60$$

To find y, use the fact that the ratios of the lengths of legs of the triangles are equal since the triangles are similar.

$$\frac{y}{60} = \frac{156}{65}$$

$$60\frac{y}{60} = 60\frac{156}{65}$$

$$y = \frac{9360}{65} = 144$$

SEE FOR YOURSELF

For the remainder of this activity, you will put the principle of diminution into practice. In question 3, you will hold a ruler at arm's length away and measure a distant object. You will take some other measurements and then use scaling to determine the proper height of an object in a drawing. Decide which member of your group will do each of the following:

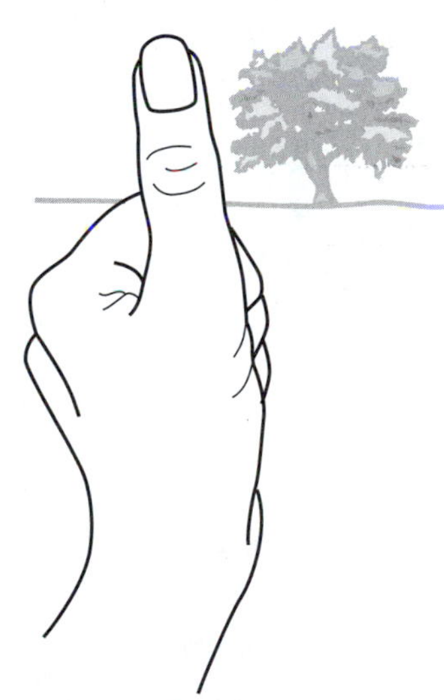

- A person to hold the ruler
- A person to stand at a distance
- A person to take measurements
- A person to record measurements.

CHECK THIS!

Artists often estimate the size of an object for a painting by holding their thumb up to a distant object. You will use a ruler instead so that you can obtain some precise measurements.

The instructions that follow are written from the ruler holder's point of view.

FIGURE 5.15.
Photograph showing how to hold the ruler.

3. a) Hold a ruler or meter stick in your hand, and extend your arm in front of you at eye level so that the ruler is vertical and your arm is extended fully (see **Figure 5.15**). The group "measurer" should carefully measure the distance from your eye to the ruler. Record this distance.

   b) Instruct a member of your group to stand several meters away from you. The measurer should use a tape measure or meter stick to measure the distance between your eye and the other person. Record this distance.

   c) Use the ruler in your extended hand to measure the apparent height of the other person (the height you would draw the person if the plane of vision coincided with the ruler). See **Figure 5.16**. Record this measurement.

FIGURE 5.16.
Measuring the apparent height of a person.

4. a) Make a sketch of a side view showing the lines of sight (see **Figure 5.17**). Use your measurements from question 3 to replace the question marks. (Note the ruler lies in the picture plane.)

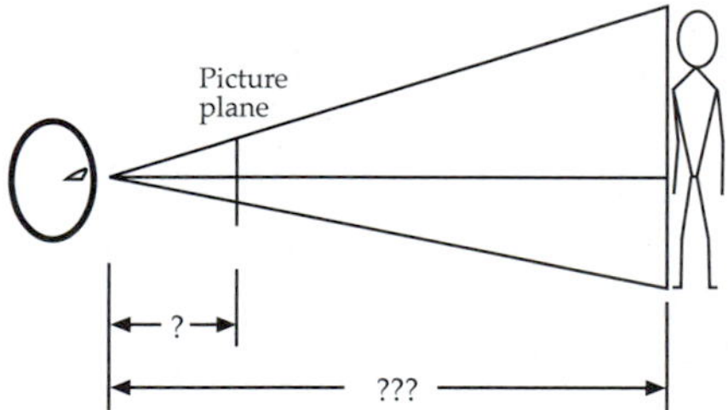

FIGURE 5.17.
Side view showing triangles.

   b) Find the ratio $\frac{\text{the distance from eye to ruler}}{\text{distance from eye to other person}}$.

   c) Find the ratio $\frac{\text{apparent height}}{h}$ where h represents the height of the other person.

   d) Use the properties of similar triangles and the results from (b) and (c) to calculate the height of the other person.

   e) Is your answer reasonable? How could you decide if your answer is reasonable?

In questions 3–4 you determined, indirectly, a person's height based on three pieces of information:

- the person's apparent height from the ruler holder's view
- the person's distance from the ruler holder's eye
- the distance between the ruler holder's eye and the picture plane.

5. a) Determine, indirectly, the height of two other objects on your school campus or in your classroom. Choose objects likely to be measured by other groups so that you can compare your answers. For each object use the Steps for Finding a Height to find a height indirectly.

   b) Discuss the reasonableness of your answers. What affects the accuracy of your answers?

   c) Suppose you make an error of 1 cm in measuring the distance from your eye to the ruler. What is the corresponding error in your height calculations?

STEPS FOR FINDING A HEIGHT

1. Use a ruler to find the object's apparent height.
2. Draw a side view (similar to Figure 5.17) showing lines of sight and similar triangles.
3. Write the resulting proportion and solve the proportion to determine the height.

If you know the height of an object, you can use this same process to find the distance to this object. For example, if you see the Washington Monument in the distance as you are driving to Washington, D.C., you can use its height (169 m) to find the distance to the center of the city. Question 6 shows how to do this calculation.

6. a) Measure the height of one member of your group.

   b) Send that person to stand an unknown distance from the ruler holder. Hold your ruler at arm's length and measure the apparent height of the person.

   c) Use your measurements and similar triangles to determine the distance between the ruler holder and the other group member. Then have the group's measurer measure the actual distance. How well did you do?

THE SCALE OF AN OBJECT

One way to compare objects in a picture with the real objects is to consider the scale of the picture. The **scale** of an object in a picture is the ratio of the image size (either height or width) to the actual size:

$$\text{scale} = \frac{\text{image size}}{\text{object size}}.$$

You can use similar triangles and proportions to find the scale of an object in a picture or to draw an object at the correct size.

7. a) Suppose a ten meter tree stands 80 meters from the place at which you are viewing a scene. What is the apparent size of the tree on a drawing that is 60 cm from your eye?

   b) Determine the scale of the ten meter tree in the drawing from question 7(a).

   c) Determine the ratio of distances (from the viewer) to the image plane and to the tree. Comment on your result.

Question 7(c) shows that you can also calculate the scale of an object using the distances from the eye to the viewing plane and from the eye to the object:

$$\text{scale} = \frac{\text{distance from eye to viewing plane}}{\text{distance from eye to object}}.$$

SUMMARY

In this activity you learned about diminution. This principle means that objects farther from the viewer are smaller than similar objects closer to the viewer.

- You described how artists use diminution in their works to indicate relative distances.
- You discovered how to use similar triangles to find the correct size of an object in a painting.

Diminution and overlapping are only two principles of perspective drawing. In the next activity you will learn another basic principle of perspective drawing.

HOMEWORK 5.2

1. Return to Hogarth's painting (Figure 5.3, Activity 5.1). Describe at least two examples in which Hogarth violated the principle of diminution.

FIGURE 5.18.
Painting illustrating diminution.

2. Describe examples of diminution in **Figure 5.18.**

3. Tamara made the sketch in **Figure 5.19** of her hand held up 2 feet in front of her eye.

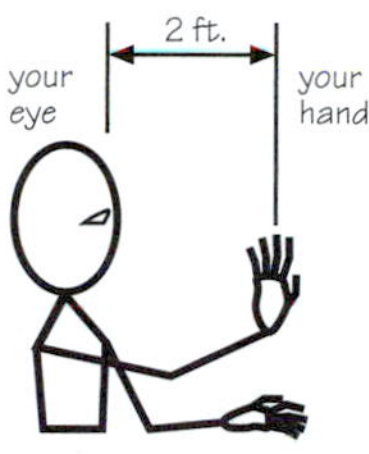

FIGURE 5.19.

a) Copy the drawing on your own paper. Draw in lines of sight from Tamara's eye past the top and bottom of her hand. Extend these lines to the right edge of your page.

b) Suppose her hand is 7 inches long and just blocks out a friend who is 5 1/2 feet tall. What is the scale of her friend in the plane of her hand? Draw her friend into your picture.

c) How far from Tamara's eye is her friend standing? Remember, the distance from her eye to her hand is 24 inches.

d) How much does the answer change if Tamara made a 1/4-inch error in her hand-size measurement?

e) How much would your answer change if Tamara made a 1/2-inch error in the distance from her eye to her hand?

4. a) The Washington Monument is 169 m high. How far would you have to stand from it to block it out with your hand? Assume your hand is 18 cm in length and the picture plane is 50 cm from your eye. Sketch the situation first. (Be careful with your units.)

b) Suppose you are drawing a tree on a piece of paper taped to an easel. The paper is 2 feet from your eye. The tree is 10 feet high and 40 feet from your eye. How tall should you make the image of the tree on the paper?

Scale is defined by the apparent and real sizes of an object, but the main control numbers affecting scale are distances. Play with these two distances:

- the distance between the object and the viewer
- the distance between the viewer and the plane of vision.

Question 5 examines this relationship between sizes and distances.

5. a) If an object is moved away from the viewer, does its apparent size increase, decrease, or remain the same? Support your answer with a drawing based on similar triangles.

   b) If the picture plane is moved closer to the viewer, does the apparent size of an object increase, decrease, or remain the same? Support your answer with a drawing based on similar triangles.

Triangles ABC and DFE in **Figure 5.20** are similar. You can prove the following corollaries (immediate consequences) about similar triangles. Explain why each is true.

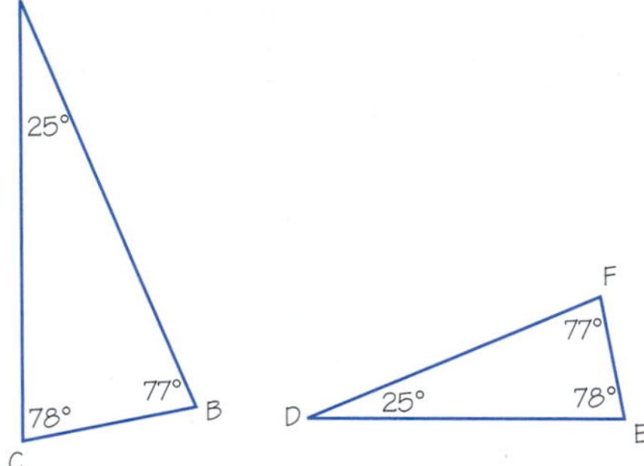

FIGURE 5.20.
Two similar triangles.

6. a) Corollary 1: If one triangle has two angles that are the same size as two angles of another triangle, the triangles are similar.

   b) Corollary 2: If an acute angle (an angle whose measure is less than 90°) of one right triangle equals an acute angle of another, the right triangles are similar.

When you hold a cube at eye level so that one face is parallel to your plane of vision, the cube looks like a square. If the cube were transparent, then you would see the otherwise invisible sides and edges of the cube (see **Figure 5.21**). The dotted lines represent the edges of the hidden faces. The edges labeled *a*, *b*, *c*, and *d* are perpendicular to the plane of vision (that is, the front face of the cube).

7. a) What do you notice about the vertical edges of the front face?

   b) How do the front and back faces compare?

   c) Why does the back face appear smaller than the front face of the cube?

   d) What do you notice about the top, bottom, and side faces?

   e) What do you notice about the edges labeled *a*, *b*, *c*, and *d* that are perpendicular to the plane of vision?

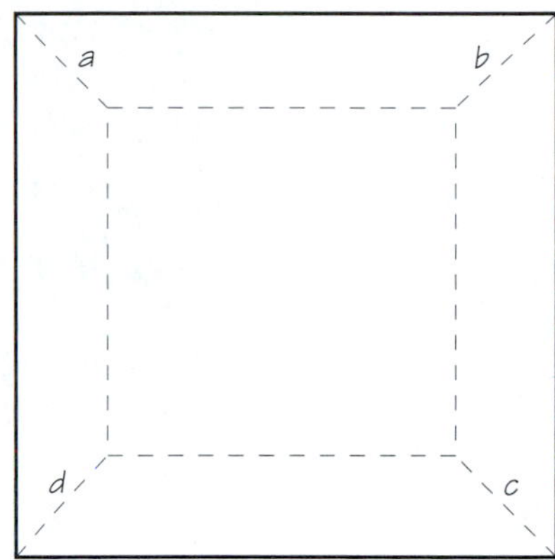

FIGURE 5.21.
Transparent cube.

ACTIVITY 5.3

When Parallel Lines Intersect

You have learned about two principles of perspective drawing, overlapping, and diminution. There are other principles that artists use to create a three-dimensional effect. This activity introduces one of them. It's called the principle of convergence, which is a primary way of showing depth.

WAY DOWN THE LINE

Figure 5.22 is a picture of a hallway that shows the principle of convergence at work. Notice that the straight lines on the floor and the lines of columns seem to be moving closer together or converging as they go off into the distance.

FIGURE 5.22.
The inside of a hallway.

You can see this effect more easily with a simpler figure. (This is another case of using the modeling strategy of starting with a simpler case.) Start with the perspective drawing of a transparent cube shown in **Figure 5.23**. (You have seen this drawing before in question 7, Homework 5.2.)

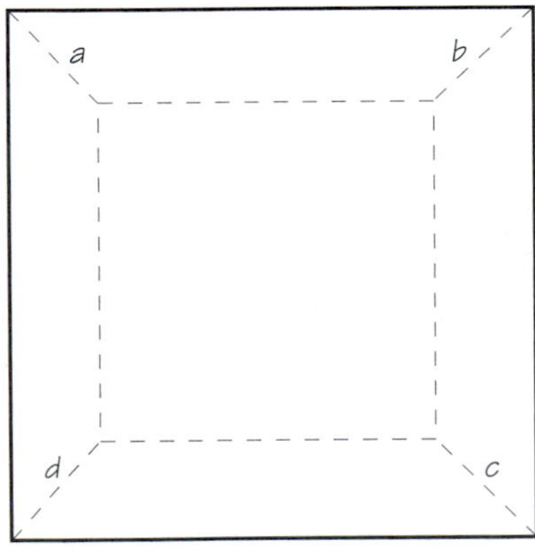

FIGURE 5.23.
Perspective drawing of transparent cube.

Since the front and back sides of the cube are parallel to the viewer's plane of vision, they are drawn as squares. In real life the front and back faces of the cube are identical. In the drawing the back face appears smaller than the front face. This is an example of the principle of diminution. The back face is farther from the viewer, so it must be drawn smaller than the front face.

Now look at the edges *c* and *d* at the bottom of the cube. In real life these edges are parallel, and they are perpendicular to the plane of vision. If you extended them beyond the back of the cube, they would never intersect with each other. In the drawing, however, as a result of diminution, edges *c* and *d* appear to get closer together as you look toward the back of the cube.

1. a) Trace a copy of Figure 5.23. Extend edges *c* and *d*. Where do they intersect?

   b) In real life the edges at the top of the cube are parallel also. However, as you look toward the back of the cube in the perspective drawing, edges *a* and *b* appear to get closer together. Extend these edges on your drawing. Where do they intersect? How is this intersection point related to the one you found in (a)?

CHECK THIS!

The key idea here is that we are looking at parallel lines that move away from the plane of vision. Since these lines move away from the viewer, they add depth to a perspective drawing. These types of lines are the focus of this whole section.

2. a) Return to the picture of a hallway in Figure 5.22. Select two of the lines that run the length of the hallway. Place a ruler or the edge of a sheet of paper along each of them. Describe where the two lines intersect.

   b) Repeat (a) using two lines that run the length of the ceiling. How does their intersection point compare to the intersection point of the floor lines?

THE PRINCIPLE OF CONVERGENCE

The narrowing of the apparent distance between parallel horizontal lines in a perspective drawing is called the principle of convergence. The single point to which the lines converge is called a vanishing point.

The lines in Figures 5.22 and 5.23 that appear to converge to a single point are examples of the **principle of convergence**.

A similar effect is visible when you look down the parallel rails of a railroad track. They too appear to converge in the distance to a vanishing point (see **Figure 5.24.**).

FIGURE 5.24.
Railroad tracks illustrate convergence.

FIGURE 5.25.
Top view of person viewing a series of railroad ties.

3. a) **Figure 5.25** is a top view of a person observing a railroad track and a series of equally spaced railroad ties. The lines of sight are drawn from the viewer to the ends of the first five ties. To simplify the picture, the rails connecting the ties are not shown. Describe the relationship between the two rails in real life.

   b) The projection of the first tie in the picture plane is almost as wide as the picture (see Figure 5.25). How does the length of the projection of the fourth tie compare with the length of the projection of the first tie?

   c) What would the projection of the railroad tie that is as far as the eye can see look like? Where is it located in the picture plane?

   d) Each tie connects the parallel rails. What does your answer to (c) tell you about the images of the rails in the picture?

CHECK THIS!

Recall that a projection is the image of a three-dimensional object projected onto a two-dimensional plane. In this case each three-dimensional railroad tie is dropped directly down onto the surface of the paper to form a rectangle.

FIGURE 5.26.
Side view observing a series of equally spaced railroad ties.

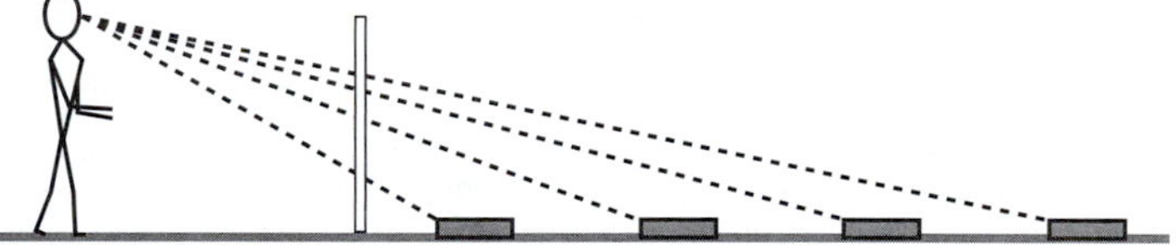

4. a) **Figure 5.26** is a side view of the same person viewing the same first four railroad ties. The projection of the first tie appears near the bottom of the picture plane. How is the location of the projection of the fourth tie different from the location of the projection of the first tie?

   b) Describe the location in the picture plane of the tie (not shown) that is farthest from the viewer. Assume the ground is level.

Questions 3 and 4 show that the apparent distance between parallel lines in a perspective drawing converges to a vanishing point. This point lies at the eye level of the viewer. Lines of sight may be drawn from a viewer to one or more objects in a scene. A line of sight that is parallel to the ground (assuming the ground is level) is said to be at the **eye level** of the viewer (see **Figure 5.27**). The eye level of the viewer plays an important role in the design of a perspective drawing.

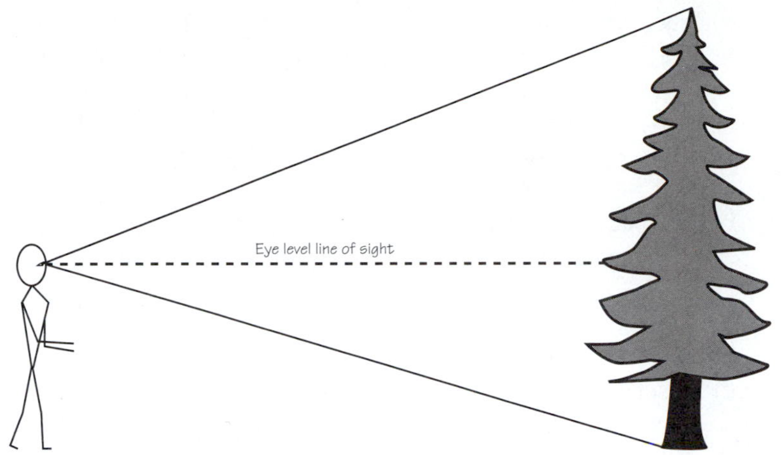

FIGURE 5.27. Sketch illustrating line of sight at the eye level of the viewer.

EYE LEVEL AND THE HORIZON

Figure 5.28 shows an incomplete picture. The artist has drawn in the **horizon**, the horizontal line at the viewer's eye level where the earth and sky appear to meet. She has added two dots at the bottom of her picture to mark the edges of a roadway that she plans to sketch.

CHECK THIS!

Recall the horizon is the farthest point on the earth that you can see. Usually the horizon line goes across the entire picture and is placed at the eye level of the viewer.

5. a) Where in the picture in Figure 5.28 do you expect to find the portion of roadway that is as far as the eye can see: above the horizon, below the horizon, or on the horizon? How large will it appear?

   b) Trace a copy of Figure 5.28. Sketch the roadway for the artist. What assumptions did you make about the viewer's location in relation to your roadway?

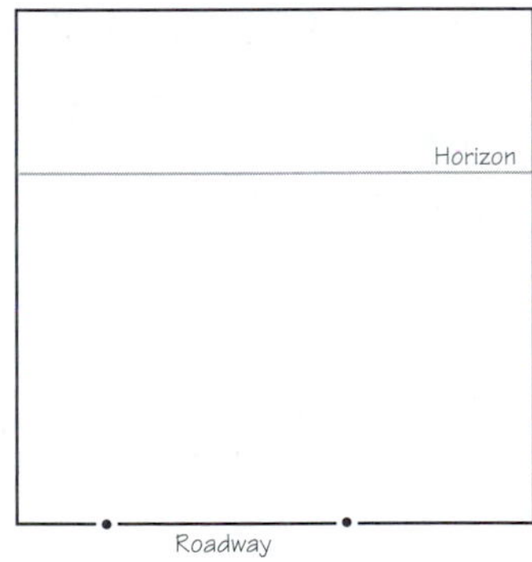

FIGURE 5.28. Incomplete picture of roadway.

In question 4 you observed objects (railroad ties) below the eye level of the artist. In the next question you will examine lines of sight to objects above the eye level of the artist. Suppose you are looking at light poles instead of railroad ties. The side view in **Figure 5.29** shows the poles are of equal height. The top of each pole is above the eye level of the viewer. The bottom of each pole is at ground level, below the eye level of the viewer.

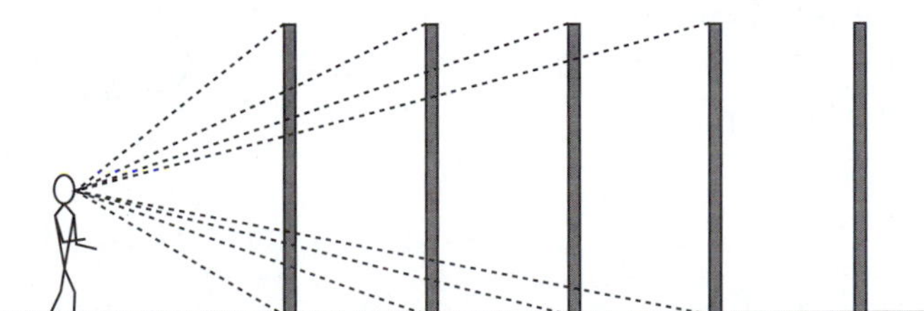

FIGURE 5.29. Side view of a series of light poles, including lines of sight.

6. a) Trace the five poles in Figure 5.29 on a sheet of paper. Draw a line connecting the tops of the five poles. Then draw a second line connecting the bottoms of the five poles (where the poles meet the ground). Describe the relationship between the two lines you have drawn.

   b) Now consider the lines of sight from the viewer to the tops and bottoms of the first four poles. Describe the difference between the projection in the picture plane of the first pole and the projection of the fourth pole.

   c) Describe the successive lines of sight to tops and bottoms of poles that are more and more distant.

   d) Where in the picture plane do you expect to find the projection of the pole that is as far as the eye can see—above the eye level of the viewer, at eye level, or below eye level? How large will it appear?

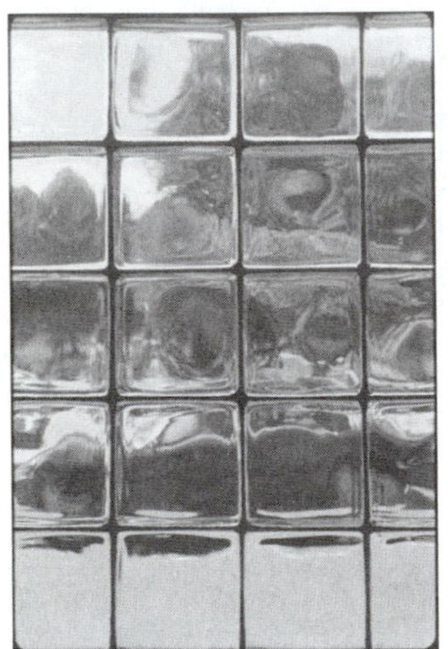

FIGURE 5.30.
Front view of window.

Questions 1–6 provided examples of perspective drawings or pictures of parallel lines. Using the principle of convergence, lines were drawn so the apparent distance between the lines narrows. This narrowing added depth to the drawings. In each case the lines were not parallel to the picture plane. For example, in question 1, the lines containing cube edges *a*, *b*, *c*, and *d* were perpendicular to the plane of vision. Next you will explore what happens when parallel lines are parallel to the picture plane.

7. a) In the front view of the window in **Figure 5.30**, which lines appear parallel? Which lines, if any, appear to converge?

   b) What role does the principle of diminution play in Figure 5.30?

   c) In the perspective view of the tiled floor in **Figure 5.31**, which lines appear parallel? Which lines, if any, appear to converge?

FIGURE 5.31.
Perspective view of a tiled floor.

d) Based on your observations thus far, when do parallel lines appear to converge? Explain how the principle of diminution supports your answer.

8. Using Handout 5.1, draw lines of convergence and identify the vanishing point of **Figure 5.32**.

FIGURE 5.32.
The Bridge at Argenteuil by Claude Monet (1840–1926).

SUMMARY

In Activity 5.2 you studied the principle of diminution. You found that objects appear smaller as their distance from the viewer increases. In this activity you have begun the study of convergence, which is a consequence of diminution.

- Parallel lines appear closer together as distance from the viewer increases. Parallel lines that move away from the plane of vision appear to meet at a vanishing point.
- The vanishing point where parallel lines appear to meet is located at the horizon.

The principle of convergence is used in many ways to draw or analyze a perspective drawing. In the next activity you will see how it is used to create a precise drawing with objects at their proper size.

HOMEWORK 5.3 Setting Your Sights

1. a) In **Figure 5.33** you see a photograph of a roadway. What do you notice about the lines that form the sides of the road in the picture?

FIGURE 5.33.
Trees lining a roadway.

b) Next, focus on the trees on the right side of the road. Suppose you drew a line connecting the treetops (assume the trees are approximately the same height) and another line connecting the tree bottoms. What would you notice about these lines?

c) Use the rulers that surround the picture to identify the location of the vanishing point for the lines joining the tops of the trees and the bottoms of the trees. Then imagine drawing a horizontal line through that vanishing point. What is the equation of this line?

d) The shoulders of the road extend beyond the white lines on both edges of the road. Pretend a ladder is placed from the edge of the shoulder to the top of each tree, forming a triangle. Describe the corresponding triangles.

e) Sketch a top view of the road. (Use circles to indicate the trees.) What should be true of the lines that form the sides of the road? What about the lines connecting the treetops and tree bottoms?

2. a) On Handout 5.2, which is the same as **Figure 5.34**, draw lines marking the sides of the road and along the treetops on either side of the road. Draw lines along the roofs of buildings on the right side of the road.

   b) What can you say about these lines? Explain.

FIGURE 5.34.
The Chemin de Sevres, Louveciennes by Alfred Sisley (1839–1899).

3. a) On Handout 5.3 showing **Figure 5.35**, find the vanishing point for this picture.

   b) Where is the vanishing point of this picture? Is the vanishing point inside or outside of the picture?

FIGURE 5.35.
The Flatiron Building.

The vanishing point for horizontal lines is at the eye level of the observer (see **Figure 5.36**). This eye level is called the **horizon line**—the line on which the earth and the sky appear to meet. The vanishing point for the railroad ties in Figure 5.36 is a point on the horizon.

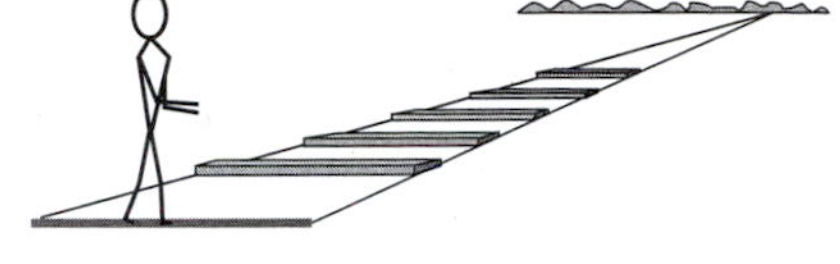

FIGURE 5.36.
The vanishing point is on the horizon, at the eye level of the viewer.

4. a) Return to Figure 5.34 from question 2. Draw the horizon line using a different color from the one used in question 2.

   b) Return to Figure 5.35 from question 3. Draw the horizon line using a different color from the one used in question 3.

5. **Figure 5.37** shows an aerial view of a scene. Draw a perspective view of the same scene with one vanishing point.

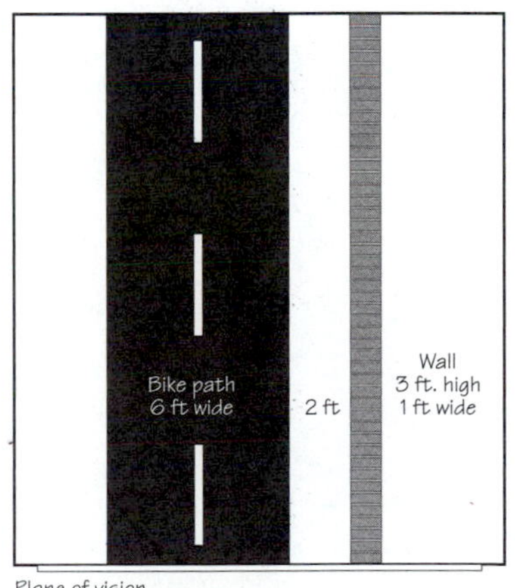

FIGURE 5.37.

Figure 5.38 shows a pair of parallel lines drawn by an architect and intersected by a line SE. A mathematical term for the line SE is transversal. A **transversal** is a line that intersects two or more parallel lines. In Figure 5.38, $\angle 2$ is to the left of the transversal and above one of the two parallel lines. Notice that $\angle 6$ is also to the left of the transversal and above the other of the two parallel lines. These two angles are **corresponding angles**.

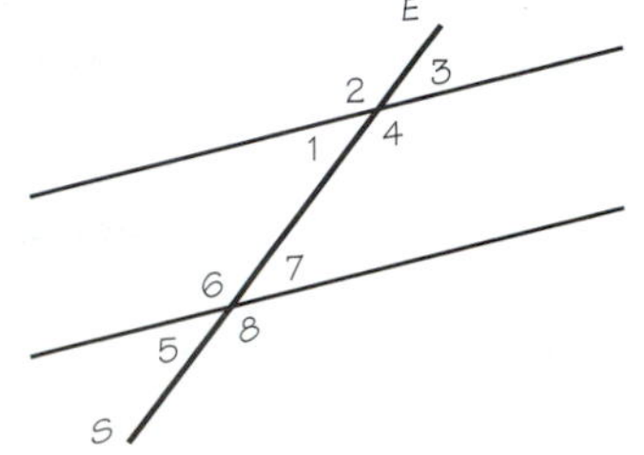

FIGURE 5.38.

6. a) What angle is in the same corresponding position as $\angle 5$?

   b) Name all the other pairs of corresponding angles in the picture.

   c) What can be said about the relationship of corresponding angles?

   d) If $\angle 1$ measures 40°, find the measures of all the other angles.

   e) Give at least one example of a picture or figure in Activity 5.3 that showed a perspective view of a series of transversals. Draw these transversals as they would appear in a picture plane that was parallel to them.

7. a) A **trapezoid** is a quadrilateral (four-sided figure) in which exactly one pair of opposite sides is parallel. Draw a trapezoid.

   b) Explain when and why some rectangles appear as trapezoids in a perspective view.

   c) Give at least one example of a picture or figure in Activity 5.3 in which a perspective drawing has a rectangle that looked like a trapezoid.

ACTIVITY

5.4

Investigating the Scale of an Object

So far, you have investigated the artistic principles of overlapping, diminution, and convergence. Given information about the size of an object and its distance from the viewer, you can use these principles to determine the proper image size and, perhaps, location in your picture.

Do the same principles work in reverse? That is, can you use a photograph or picture to determine the real-life size of the object and the real-life distance from the viewer? In this activity you will develop the investigative tools to answer these questions.

DETERMINING SCALE

Could investigators use the mathematics of perspective to draw conclusions about measurements and distances of objects in pictures? Question 1 uses **Figure 5.39** to help you find a way to determine the size of objects in a drawing without knowing how far the object is from the viewer. You will take some measurements to show how these calculations may be done.

Figure 5.39 shows a viewer, a vertical line at eye level representing a picture of the tree, and the actual tree. The line of sight at the eye level of the viewer intersects the picture plane at the horizon. Rectangle HIJK in **Figure 5.40** represents the drawing in Figure 5.39 from the ground to the *viewer's* eye level. For example, IJ represents the line of sight at eye level from the viewer to the tree.

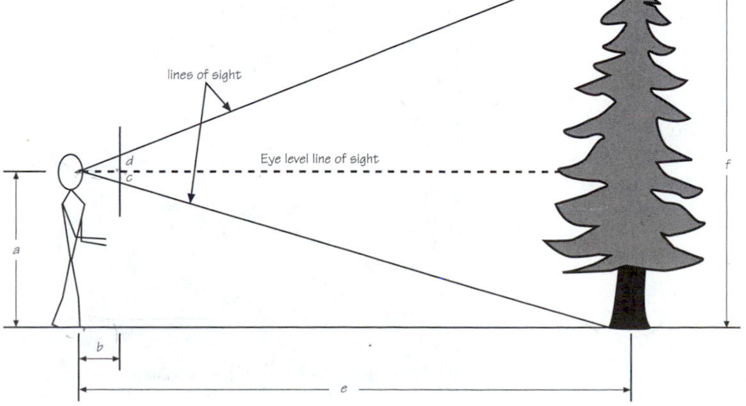

FIGURE 5.39.
Picture illustrating known and unknown measurements.

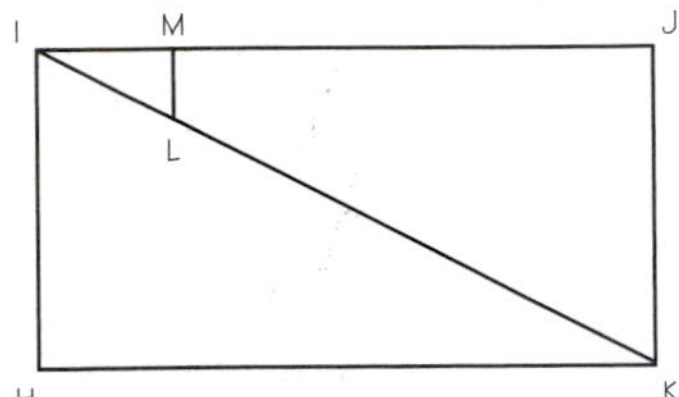

FIGURE 5.40.
Rectangle representing Figure 5.39 from eye level line of sight and below.

1. a) What does HI in Figure 5.40 represent? What about KJ? What about LM?

   b) Segment lengths a, b, c, d, e, and f are shown in Figure 5.39. What are the lengths of the segments HI, KJ, and LM in Figure 5.40?

   c) List all triangles in rectangle HIJK that are similar to triangle KHI.

   d) Consider the sketch in Figure 5.39. How are the lengths of a and c related to the lengths of b and e? Support your answer using an argument based on similar triangles.

DETERMINING THE SCALE OF A PICTURE

To determine the scale of an object in a picture, you need the following information:

- The distance between where the object touches the ground in the picture, and eye level (c in Figure 5.39)
- The distance from the ground to the viewer's eye level (a in Figure 5.39)

Then the scale of an object in the picture is the fraction c/a.

Recall that the eye level in Figure 5.39 intersects the picture plane at the horizon. If the ground in Figure 5.39 is level, the distance from the ground to the horizon line (a in Figure 5.39) represents the eye height of the viewer at that depth in the picture. That means you can determine the scale of an object in the picture using similar triangles as you did in question 1.

Next you will determine scales for two of the trees in **Figure 5.41**. Based on the scales, you will be able to determine the heights of the trees. Assume you are the same height as the person who took the photograph.

FIGURE 5.41.
Photograph showing tree-lined roadway.

2. a) Measure the distance from the ground to your eye.

   b) Select a particular tree in Figure 5.41. Carefully measure the distance from the bottom of the image of that tree up to the horizon line. (This measurement corresponds to *c* in the scale fraction.)

   c) Use the measurements from (a) and (b) to determine the scale of the selected tree in the photograph.

   d) What is the actual height of the tree?

3. a) Generalize the steps from question 2 to describe how to determine the scale of any object in an image.

   b) Use your method to determine the height of another tree in Figure 5.41.

4. Using the scale from question 2(c), determine the width of the shoulder of the road in Figure 5.41.

5. Assume the picture plane is 10 inches from the viewer. Use scale factors to estimate the distance from the photographer to each of the trees you examined in questions 2 and 3.

DRAWING EQUALLY SPACED OBJECTS

Figure 5.42 shows a picture of a rectangle with diagonals AC and BD. **Congruent** figures are similar figures of the same size. For example, two similar triangles are congruent if their proportionality ratio is 1:1. (In other words, you can place one triangle on top of the other in such a way that the two triangles coincide.) Congruent rectangles are used to determine the location of equally spaced objects.

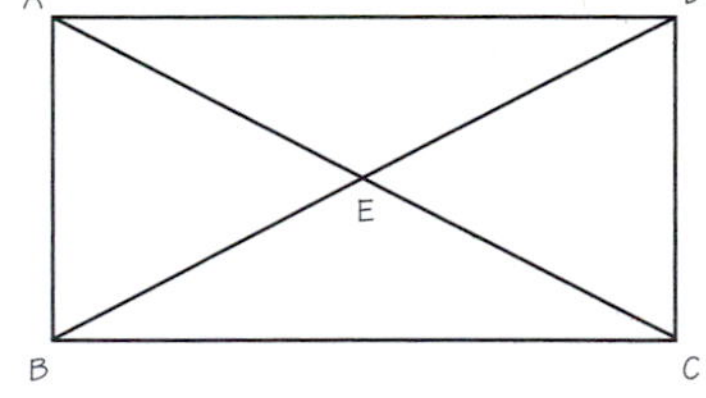

FIGURE 5.42. Rectangle with diagonals.

6. a) Which triangles in Figure 5.42 are congruent?

   b) Sarah claims that the diagonals of a rectangle bisect each other. Carefully measure the lengths of BE and ED. Then measure the lengths of AE and EC. Does Sarah appear to be correct?

   c) Give an argument based on congruent triangles that supports Sarah's claim in (b).

   d) Draw a line through E that is parallel to one of the sides of the rectangle. Explain how you know that your line divides the rectangle into two congruent rectangles.

FIGURE 5.43.
Railroad tracks.

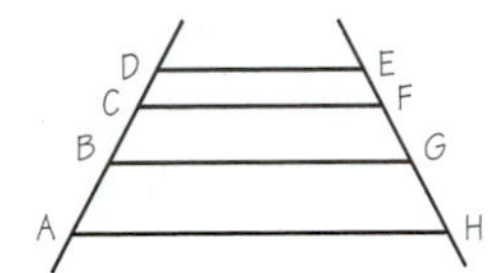

FIGURE 5.44.
Tracing of front edge of four ties.

The spacing between the railroad ties in the photograph in **Figure 5.43** appears to shorten because of diminution. However, railroad ties are usually equally spaced. In this question you will construct a method for determining whether the ties were equally spaced.

7. a) **Figure 5.44** results from laying a transparency over the photograph in Figure 5.43 and tracing the front of the second through fifth ties. Labels A–H have been added. Trace a copy of Figure 5.44 on a sheet of paper.

   b) Classify the shape of quadrilateral ADEH in your drawing from (a). What would be the shape of a top view of this section of track?

   c) Draw diagonals AF and CH of the quadralateral ACFH. Apply what you have learned about the diagonals of a rectangle in question 6 to decide if the second through fifth rails are equally spaced. Explain your method and why it works.

SUMMARY

In this activity you have used some tools of perspective drawing to estimate heights of objects in photographs and their distances from the photographer.

- You also used some principles of geometry to decide if a series of objects in a photograph were equally spaced.

In the next activity you will continue to draw and analyze more types of perspective drawings.

HOMEWORK 5.4

In Activity 5.4 you determined a method that could detect if objects were equally spaced. In this activity you will use the method to draw objects that are equally spaced.

1. a) Suppose you are drawing the image of a pole that is 10 meters tall and 30 meters away. Calculate the height of the image when the picture plane is 60 cm from the viewer. What is the scale of the object?

   b) For the image of the pole in the drawing, calculate the vertical distance from where you draw the bottom of the pole to where you draw the horizon. Assume the artist's eye height is 150 cm.

   c) Suppose the distance from the viewer to the pole is doubled to 60 meters. Calculate the length of the image, the scale of the image, and the distance from the bottom of the image to the horizon.

   d) Sketch a perspective view of two poles 10 meters in height. One is located 30 meters from the viewer and the other 60 meters from the viewer. The sketch should represent accurately your calculations in (b), (c), and (d). Be sure to include the horizon, the vanishing point, and the lines of sight in your sketch.

   e) Explain how you determined the correct location for the pole that was 60 meters away.

2. a) **Figure 5.45** shows a perspective view of two poles of equal height. Make a similar sketch on a sheet of paper. (If you wish, you may enlarge this figure or you may use the figure that you drew for question 1(e).) Explain how you can determine the correct location, in the drawing, for a pole that, in the real world, is midway between the two poles. Assume that this third pole is equal in height to the other two poles and midway between them.

   b) Use your method to add a third pole to your sketch.

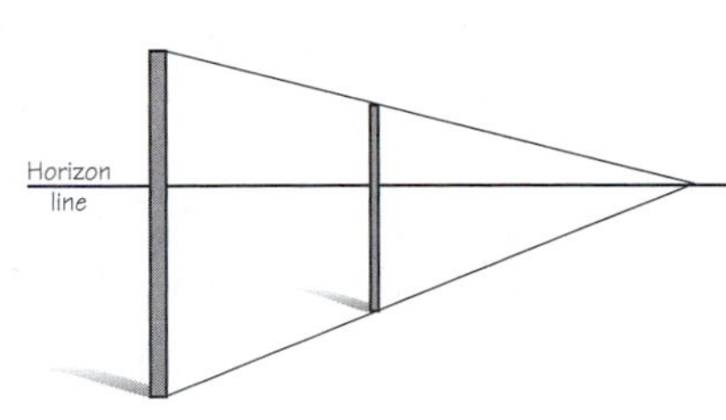

FIGURE 5.45.
Determine the midway location.

3. Suppose that you plan to draw a series of equally sized windows on a wall. Assume that the wall is parallel to the picture plane and that the windowsills are all the same height above the floor.

   a) Sketch the wall and one window so that your drawing resembles **Figure 5.46**.

FIGURE 5.46.
Wall, parallel to picture plane, with one window.

   b) Lightly sketch horizontal lines level with the top and bottom of the window and extending to the right. Connect these lines with a vertical line that represents the left side of the second window.

   c) The problem is to decide, without measuring, exactly where to place the vertical line representing the right-hand side of the second window. (Remember, you want the two windows to have equal width.) Here is one method:

   - Draw a line through the upper left-hand and lower right-hand corners of the first window.
   - Next draw a line through the upper left-hand corner of the second window parallel to your first line.
   - Mark where your second line crosses the horizontal line that is level with the windowsills. Draw a vertical line at this point.

   Use this method to complete the second window. Do your two windows appear to be congruent (the same size and shape)?

   d) In **Figure 5.47**, ABEF is a rectangle, HC and GD are parallel to AB, and AC is parallel to GE. Explain how you know that rectangles ABCH and GDEF are congruent (similar with a proportionality ratio 1:1).

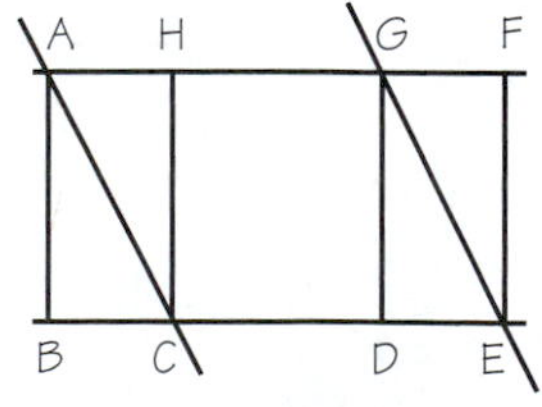

FIGURE 5.47.
Congruent rectangles.

Next you'll adapt the method you used in question 3 to sketch a perspective drawing of a hallway with a series of equally sized windows. **Figure 5.48** is an incomplete perspective drawing of a hallway with one window on the left wall. Your task will be to draw another window further down the hallway. Assume that, in reality, this window is the same size as the first window and that its top is the same height above the floor as the top of the first window.

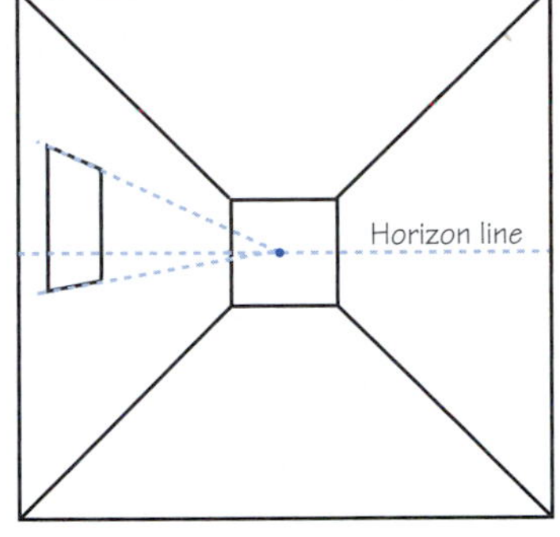

FIGURE 5.48.
Perspective drawing of a hallway with a window.

4. a) Trace a copy of Figure 5.48 on paper (or use Handout 5.4). Select where you want to place the left side of the new window. Draw a vertical line representing this side.

   b) Adapt the method that you used in question 3 to decide where to place the vertical line for the right side of the second window. Draw your window.

   c) Repeat this process and draw a third window on the left side of the hallway. How do your windows look?

   d) To add interest to your drawing, sketch a rectangular chest (or a table, if you prefer) along the right wall of the hallway. Lightly sketch the lines of sight used to draw the chest.

One application of similar figures is in enlarging (or reducing) drawings. In mathematics, this scaling transformation is usually called **dilation**. The ratio of the new lengths to the old lengths (new/old) is called the **scale factor**.

Suppose you want to double the size of triangle ABC in **Figure 5.49** (that is, the scale factor is 2). Select a point E (arbitrarily) as the center of the dilation. Extend EA through A, and mark point A′ on it so that EA′ = 2 EA. Thus, EA = AA′. Similarly, draw and extend EC so that EC′ = 2EC. Then EC = CC′. Repeat the same construction through point B.

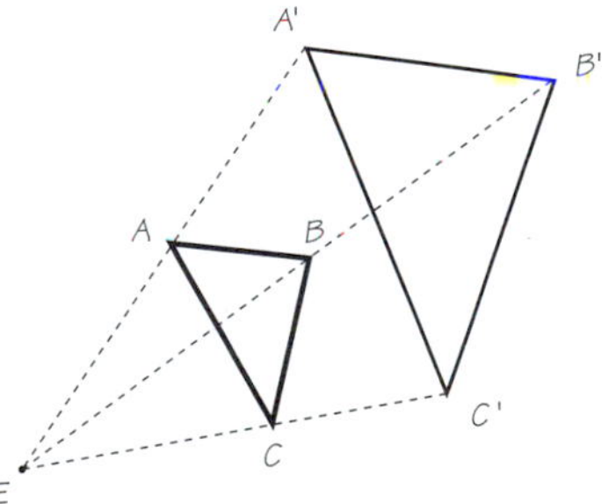

FIGURE 5.49.
Illustration of dilation that doubles lengths.

5. a) Measure the sides and angles of triangles ABC and A′B′C′ in Figure 5.49. Confirm for yourself that these triangles satisfy all the conditions for being similar and that each side of A′B′C′ is twice the length of the corresponding side of ABC.

   b) From your measurements you have found that ABC and A′B′C′ are similar. Explain why this construction guarantees similarity.

c) Describe a way to triple the size of triangle ABC. Trace triangle ABC and point E from Figure 5.49 and show that your method works.

d) Describe a way to halve the side of triangle ABC. Trace triangle ABC and point E from Figure 5.49 and show that your method works.

e) If the sides of a triangle are twice as long, respectively, as the sides of another triangle, what can you say about the relationship between the areas of the two triangles? Verify your answer mathematically.

6. a) Suppose you want to double the size of the same triangle ABC that you doubled in the last question. Now E is in a new location as shown in **Figure 5.50**. Use E to find a triangle that is double the size of triangle ABC.

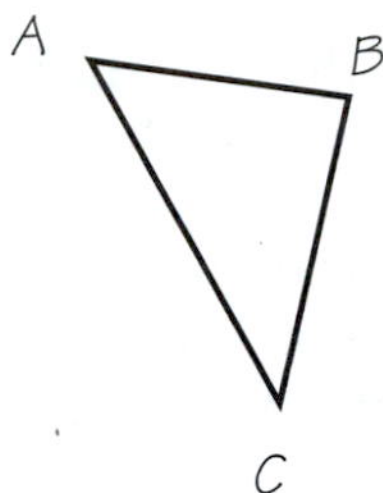

FIGURE 5.50.

● E

b) Does the construction produce a double-sized similar triangle if you place E inside the original triangle? Does it matter where you place the point E? Explain.

ACTIVITY

5.5

Vanishing Telephone Poles

In Activity 5.4 you learned how to determine if a series of ties on a railroad track are equally spaced. Then, in Homework 5.4, you drew a perspective drawing of three poles that, in the real world, were equal in height with equal spacing between them. In this activity you will modify the process used in these examples so that you can draw a series of objects that, in reality, are identical and equally spaced.

INTRODUCTION

The specific question for this activity is how do you draw a row of telephone poles? Telephone poles, like all lines that are parallel to the viewing plane, appear in the image parallel to one another. In a perspective view, the poles may diminish in size as they converge toward a vanishing point, but they remain parallel to one another. Connect the tops of the poles with a line and the bottoms of the poles with a line (see **Figure 5.51**). Notice the series of adjacent trapezoids that diminish in size. The problem is determining the spacing between the poles.

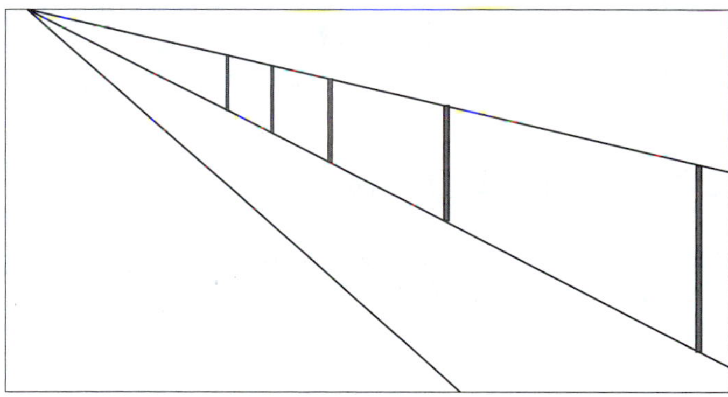

FIGURE 5.51.
Perspective view of poles with connecting lines that form trapezoids.

The key idea in drawing the poles is illustrated in **Figure 5.52**. The diagonals of a rectangle intersect at a point midway between opposite sides. A line drawn through the point of intersection and parallel to the sides of the rectangle divides the rectangle into two equal rectangles adjacent to one another.

A basic modeling strategy is used to solve this problem. Here is a simple case. The solution to the simple case will help construct a solution to the general case.

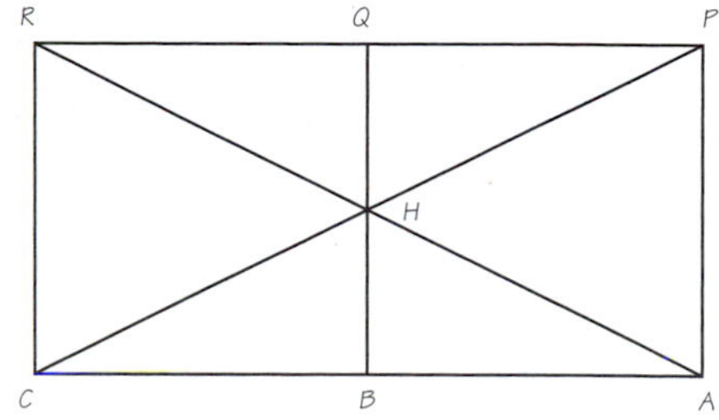

FIGURE 5.52.
Rectangle with center determined by intersecting diagonals.

PART 1: A NONPERSPECTIVE VIEW

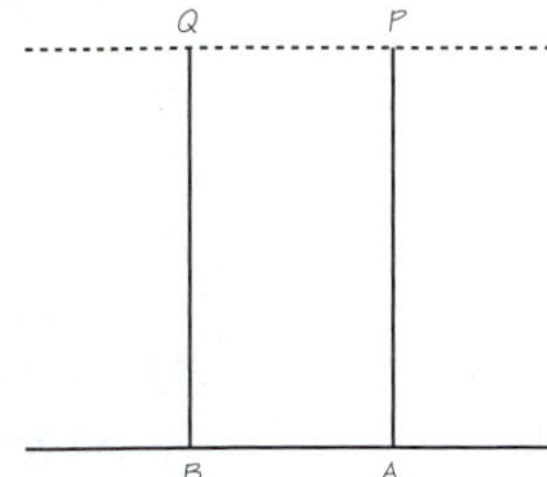

FIGURE 5.53.
Poles separated at an arbitrary distance.

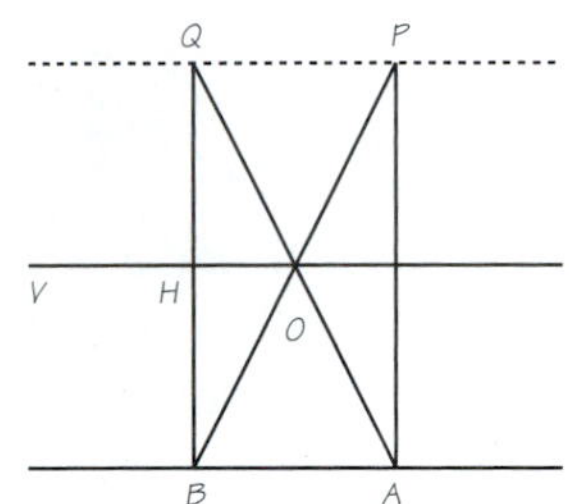

FIGURE 5.54.
View of intersecting diagonals.

The simplifying assumption is that the line of poles runs across the picture from right to left and are parallel to the picture plane.

1. To solve the separation problem in the simplified situation, begin by drawing two poles an arbitrary distance apart (see **Figure 5.53**). Label the ends of the poles A, B, P, and Q.

2. Draw diagonals QA and BP to locate the center, O, of rectangle ABQP. Draw a horizontal line VO (see **Figure 5.54**). Point H is therefore the midpoint of pole BQ.

3. Draw a line from P through H intersecting the line through A and B at a point you label as C.

4. a) Draw a line segment from C parallel to BQ and ending on PQ (extended). Call the new point R.

   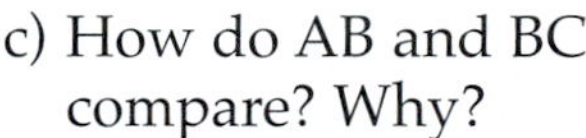

   b) How does the length of CR compare to the lengths of AP and BQ?

   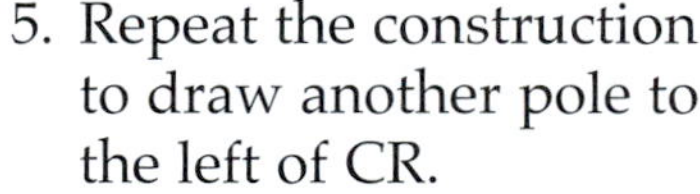

   c) How do AB and BC compare? Why?

5. Repeat the construction to draw another pole to the left of CR.

CHECK THIS!

Two ideas were crucial in your construction.

- The diagonals of a rectangle meet at a point halfway between top and bottom of the rectangle.
- Once the mid-height (the line through H) is found at *one* place, a horizontal line through that point can be used to mark that height *everywhere*.

PART 2: PERSPECTIVE VIEW

You can use a similar construction to draw a line of poles that runs at an angle to the picture plane (in other words, a perspective view).

6. Draw a horizon line and the pole closest to the viewer. Draw the second pole in perspective, making sure that the fraction of the second pole that lies below the horizon is the same as the fraction of the first pole that lies below the horizon.

CHECK THIS!

The fraction of each pole from the ground to the horizon must be the same so that the scale of each pole is constant.

7. a) Connect the bottoms of the poles with a line. Locate the vanishing point on the horizon. Label the ends of your poles A, B, P, and Q. Your picture should resemble **Figure 5.55**.

   b) Explain where the vanishing point is located and what it represents.

8. a) Draw in diagonals AQ and BP to establish the center of ABQP in correct perspective.

   b) Draw a line through the vanishing point and the center point from 8(a). In your perspective drawing, the midheights of the poles lie on this line. (Why?)

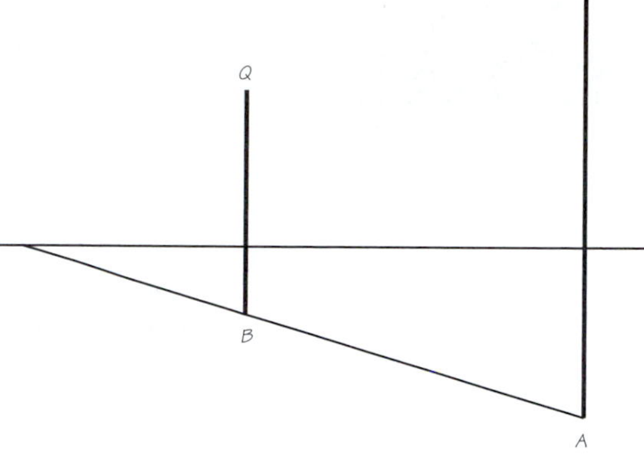

FIGURE 5.55. Perspective view of the vertical sides of the "rectangle."

9. Apply the ideas you used in Part 1 to draw the next pole in proper perspective. Explain why your method is correct.

10. Continue your process for another two or three poles.

SUMMARY

In this activity you drew a series of identical and equally spaced objects using methods you developed in Activity 5.4.

- First you drew a series parallel to the plane of vision.
- Then you drew a series in perspective.

Now that you have studied ways to draw equally spaced objects, you are ready to learn about another principle of perspective drawing. In the next activity you will study the principle of foreshortening.

HOMEWORK
5.5

Under Construction

1. Suppose you work for an architect. A client wants a plan for a series of equally spaced flagpoles along the side of a road. Repeat the telephone pole construction (Activity 5.5) on the road in **Figure 5.56**. The horizon, the vanishing point, and the first pole have been placed for you. Copy the figure on your own paper, or use Handout 5.5.

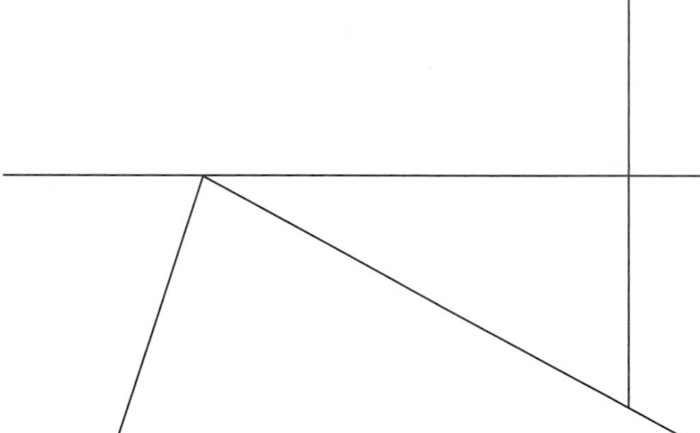

FIGURE 5.56.
Beginning view of flagpole construction.

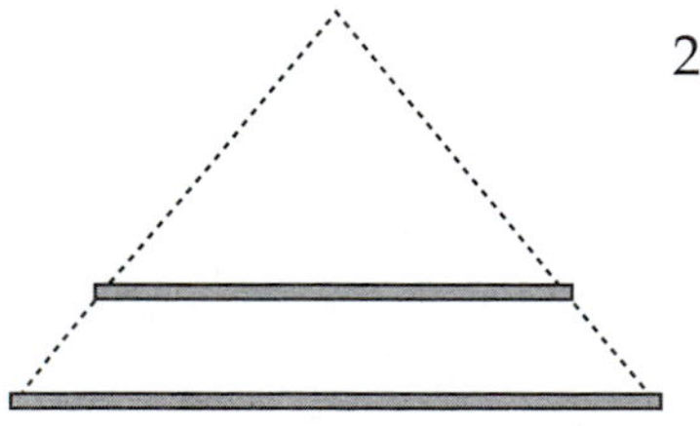

FIGURE 5.57.
Beginning view of the remodeled walkway.

2. A landscape architect wants to remodel a walkway by alternating rows of bricks with rectangular concrete panels. Construct the proper placement, in a perspective view, of the panels and lines of bricks given the placement of the first two in **Figure 5.57**. (The black rectangles represent the bricks; the white represents the concrete.) Copy this figure, or use Handout 5.6.

3. Explain why the construction method for rectangles guarantees equal spacing of poles. (Refer to questions 1–5, Activity 5.5.)

4. Suppose telephone poles are 212 feet apart in the real world. The plane of vision is 2 feet from the artist's eye. The first pole in the picture is 300 feet from the plane of vision. The artist is viewing from a point 100 feet to the left of the line connecting the poles. (See **Figure 5.58a**.) The centerline would be the artist's line of sight if he were to look in a direction parallel to the line of poles toward the horizon.

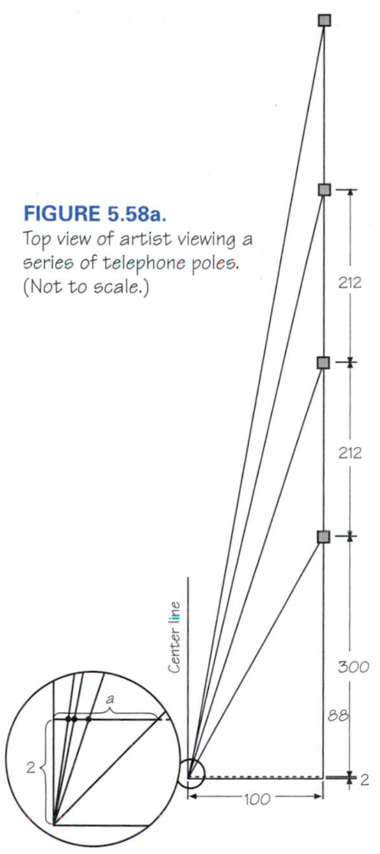

FIGURE 5.58a. Top view of artist viewing a series of telephone poles. (Not to scale.)

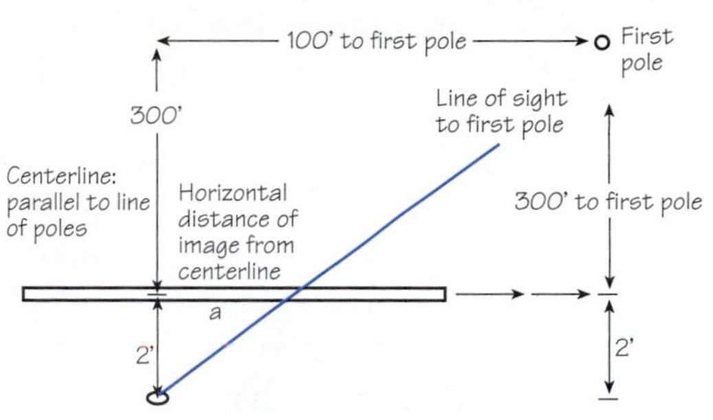

FIGURE 5.58b. Representation of picture plane and first pole.

a) **Figure 5.58b** shows a super-enlarged representation of the artist's eye, the picture plane, and the first pole. Calculate the distance *a*, the horizontal distance in the picture plane from the image of the pole to the centerline. Explain how you determined your answer.

b) What is the scale at the second pole? Calculate *b*, the horizontal distance between the image of pole 2 and the centerline.

5. Suppose the poles in question 4 are 40 feet high and the eye-level height of the artist is 5 feet. If you were to draw the first pole, how far below the horizon is the bottom of the first pole? The bottom of the second pole?

Figure 5.59 shows a sketch of a room containing three rectangular objects, a chest, a large wardrobe, and an air purification unit. The drawings of the objects are incomplete. Only the sides parallel to the viewer have been sketched.

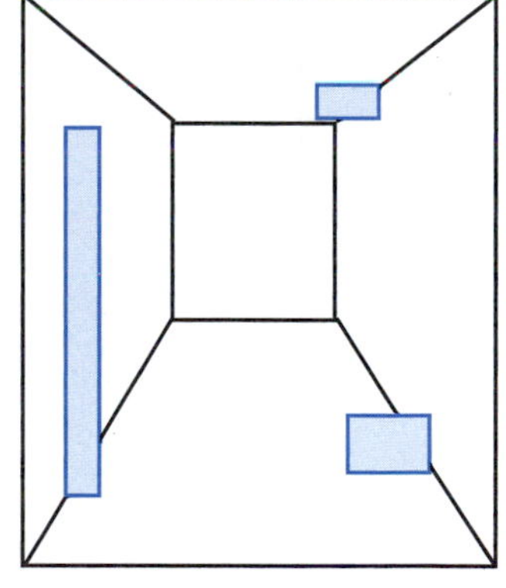

FIGURE 5.59. Incomplete drawing of room containing three objects.

6. a) Trace Figure 5.59 onto a sheet of paper. Complete the perspective drawings of the chest, wardrobe, and air purification unit so that they appear to have depth. What principles of one-point perspective did you use to guide your drawing?

b) In your completed drawing, you can see the top of one of the rectangles and the bottom of another. For the third object, you can see neither its top nor its bottom. What determines whether the viewer can see the top or the bottom of an object in a perspective drawing?

c) Under what conditions are both the top and the bottom of an object not visible to the viewer?

7. Sketch two views of cubes that are different from any you have studied in this chapter. Describe how each view is different and what makes it more challenging to draw. (If you have trouble answering this question, find a cube, such as a block or box, and look at it from different angles. Analyze what you see.)

ACTIVITY 5.6

Tilt: The Principle of Foreshortening

When parallel lines are not parallel to the plane of vision, the lines appear to intersect in the distance. This is called the principle of convergence, which you studied in Activity 5.3. Another effect that happens when the objects are not parallel to the plane of vision is called *foreshortening*. In this activity you will learn about this principle of perspective drawing.

INTRODUCTION

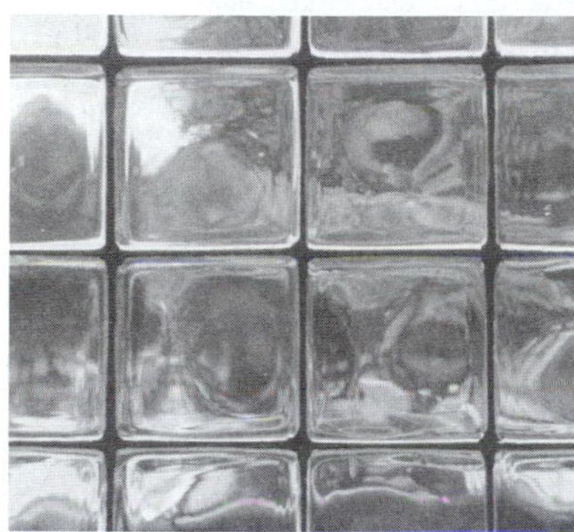

FIGURE 5.60.
Squares parallel to picture plane.

FIGURE 5.61.
Squares at an angle.

Start your investigation of this principle by comparing **Figures 5.60 and 5.61**. The windowpane in Figure 5.60 is parallel to the plane of vision (also, perpendicular to your line of sight). The individual panes look square. In contrast, the tile floor in Figure 5.61 is not parallel to the picture plane. The square-shaped tiles appear distorted. Focus on the tile that has one vertical and one horizontal edge outlined. The vertical length looks shorter than the horizontal length.

Now look at the picture of the jugglers in **Figure 5.62**. In reality the clubs are identical. Yet, depending on their angle with the picture plane, some appear shorter than others. Figures 5.61 and 5.62 provide examples of the **principle of foreshortening**.

FORESHORTENING

The principle of foreshortening occurs when lines or surfaces perpendicular to the line of sight appear increasingly shorter as they are rotated away from the observer.

FIGURE 5.62.
Jugglers, picture from stamps.

MEASURING APPARENT LENGTH

The examples you just examined introduce the following new modeling question.

How do you represent accurately three-dimensional objects tilted at an angle to a two-dimensional picture plane?

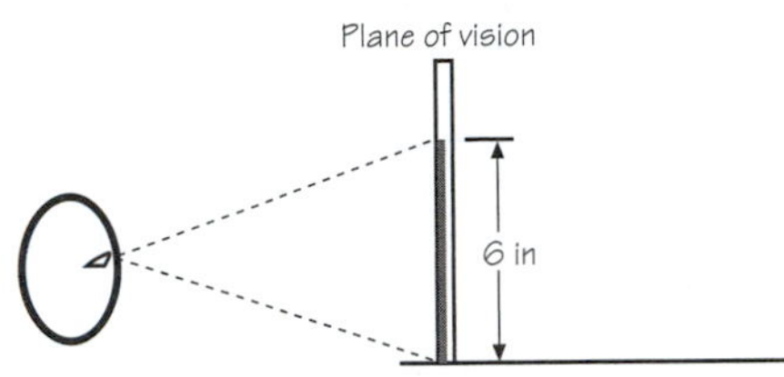

FIGURE 5.63.
Pencil in plane of vision.

Once again, use the basic modeling approach of simplifying and conquering. Assume the object is one dimensional, say a flat 6-inch pencil. That way you only have to deal with length. If you hold the pencil upright so that it coincides with your picture plane, it looks 6 inches (see **Figure 5.63**).

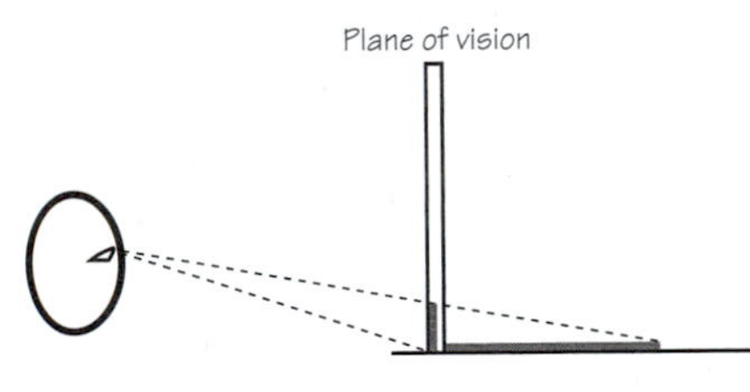

FIGURE 5.64.
Pencil lying flat on a table.

Now suppose the pencil is lying on a table so that it touches the picture plane but is perpendicular to it. In the plane of vision (the picture plane), the pencil looks shorter (see **Figure 5.64**). The question is how much shorter?

Suppose your eye level is 5 feet above the floor and a 6-inch pencil is lying on a table 3 feet high and 2 feet away from you (see **Figure 5.65**). The drawing in **Figure 5.66** is a geometrical representation of the situation. In this drawing E represents the viewer's eye and PQ, the pencil. (Note that the length of the pencil has been converted to feet.)

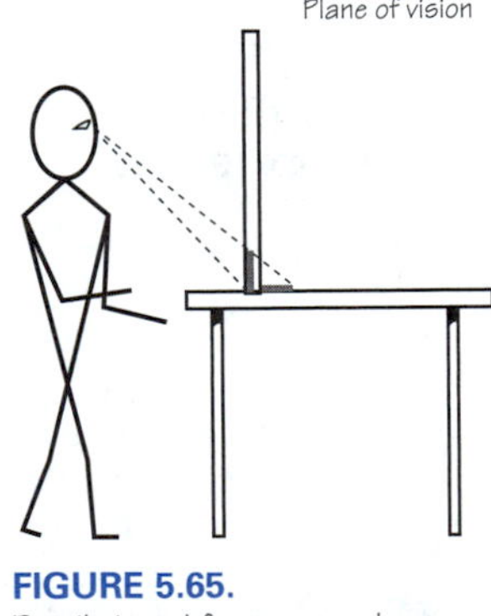

FIGURE 5.65.
Pencil viewed from an angle.

E
2
W
A
2
P 0.5 Q

FIGURE 5.66.
The geometry of viewing a pencil.

APPARENT LENGTH

We call the length of the pencil as it appears in the plane of vision the apparent length. Because of foreshortening, the apparent length of the pencil is shorter than the actual length of the pencil. In Figure 5.66, the apparent length is indicated by the vertical line WP.

1. a) Find the apparent length, WP, of the pencil. Explain how you used similar triangles to determine this length.

   b) Suppose you place two pencils end to end so the total actual length of the pencils is 12 inches. What is the apparent length of the two pencils?

   c) How long is the apparent length of the second pencil? Compare the apparent length of the second pencil to the apparent length of the first pencil (the pencil touching the picture plane).

   d) Add a third 6-inch pencil. What is the apparent length of the three pencils? How long is the apparent length of the third pencil. Compare its length to the apparent lengths of the first two pencils.

   e) Generalize your findings from (a)–(d).

2. a) Suppose you raise your eye level to 7 feet above the floor so that in the geometric representation (Figure 5.66) AE measures 4 feet. Do you think this change in viewing angle will increase, decrease, or leave unchanged the apparent length of a pencil? Explain.

   b) Confirm your answer to (a) by calculating the apparent length of a 6-inch pencil when the viewer's eye is 4 feet above the table.

   c) Would the apparent length of the pencil increase, decrease, or remain the same if you stooped so that your eye was only one foot above the table? Draw a geometric representation that supports your answer.

   d) What if you stooped so that your eye was level with the table? How would the pencil appear?

Figure 5.67 is a photograph of *Dead Christ* by Andrea Mantegna (1431–1506), an Italian painter who explored perspective thoroughly. This 15th century work is one of the first good examples of foreshortening.

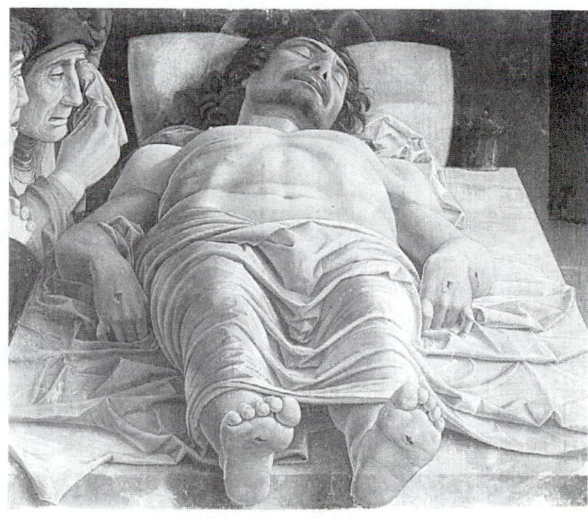

FIGURE 5.67.
Dead Christ by Andrea Mantegna (1431–1506).

3. The original dimensions of the painting are about 27 inches high by 32 inches wide. The figure in the painting was 5 feet, 10 inches (70 inches) tall, and the eye of the painter was 36 inches above the height of the platform on which the figure was lying. How far away from the platform was the painter standing if the figure's feet were touching the plane of the picture?

WHEN THE TILT IS LESS THAN 90°

So far in this activity, the object that is foreshortened has been perpendicular to the picture plane. What happens if the object is tilted away from the picture plane at an angle less than 90°? Questions 4 and 5 will provide a partial answer.

Suppose you are two feet away from the picture plane and your eye is level with the table. If a 6-inch pencil is held against the picture plane, its apparent length is also 6 inches. Now imagine tilting the pencil away from the picture plane. Label the angle between the pencil and the picture plane θ (see **Figure 5.68**). KP represents the pencil held against the picture plane and KQ the tilted pencil.

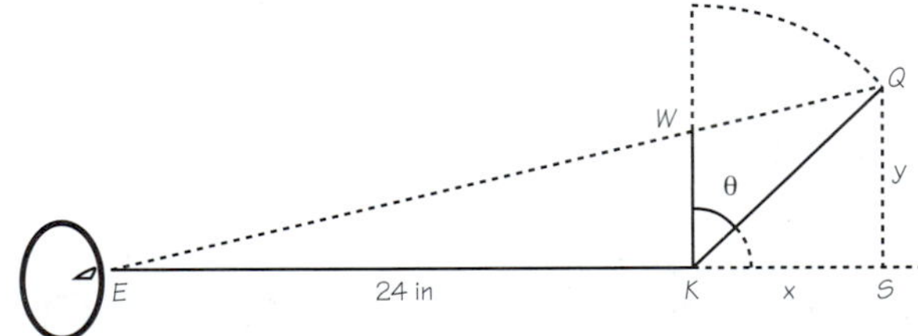

FIGURE 5.68.
Side view of a tilted pencil.

4. a) Draw geometric representations, to scale, of the situation in Figure 5.68. Each member of your group should use the same scale but vary the amount of tilt (the angle θ) for the pencil. From each drawing, measure the apparent length of the pencil.

   b) Based on your drawings in (a), what happens to the apparent length of the pencil as you increase the amount the pencil is tilted?

5. a) **Figure 5.69** shows a square with a diagonal. Suppose the length of the diagonal is 0.5 foot (6 inches). What is the measure of θ?

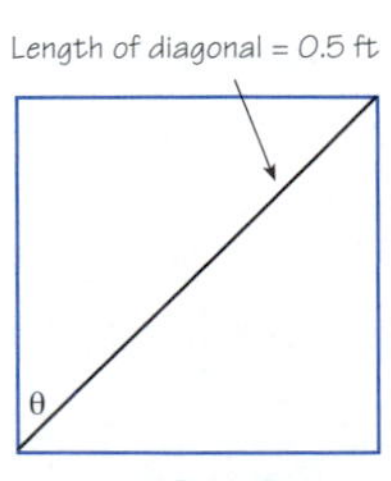

FIGURE 5.69.
Square with 1/2-foot diagonal.

b) Use the Pythagorean theorem to find the length of each side of the square.

6. a) Suppose in question 4, the 6-inch pencil is tilted 45° from the picture plane as shown in **Figure 5.70**. Identify two similar triangles in Figure 5.70.

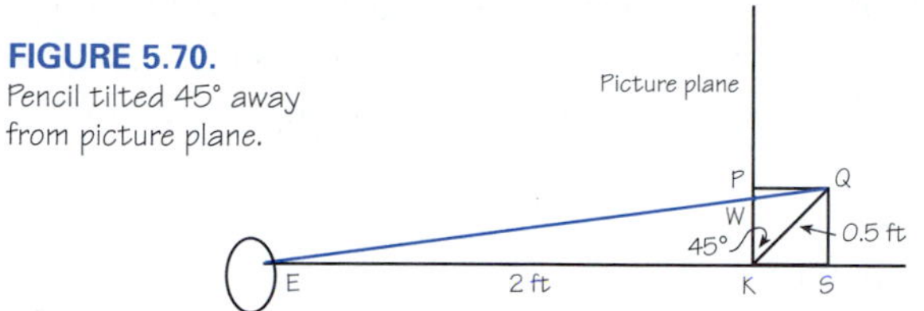

FIGURE 5.70.
Pencil tilted 45° away from picture plane.

b) Use the proportionality of corresponding sides of similar triangles, and your answer to question 5, to determine the apparent length of the pencil, KW.

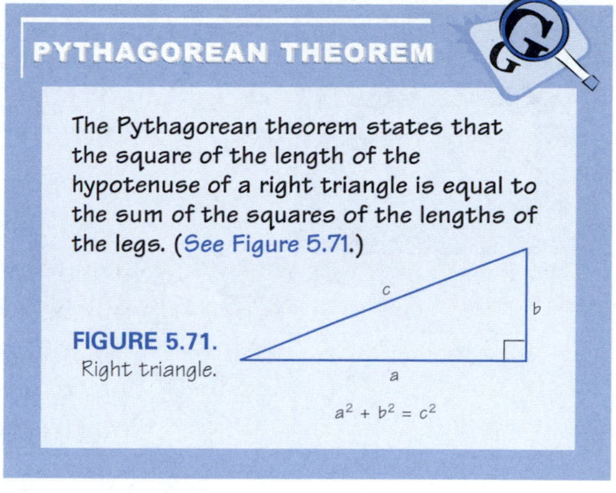

FIGURE 5.71.
Right triangle.

SUMMARY

In this activity you studied the principle of foreshortening that occurs when objects are not parallel to the plane of vision. This principle says that the farther an object is from the eye, the shorter the object appears.

- You found that the higher the eye level, the longer the apparent length of the object.
- You found the apparent length of objects that were tilted by either 90° or by certain special angles such as 45°.

To find the precise measurements of foreshortened objects that are tilted by any acute angle, you need some special ratios of triangles. These ratios will be introduced in the next activity.

HOMEWORK

5.6

A Vision of Angles

1. Suppose you are running down the beach and come upon your best friend lying flat on the sand. Your friend is 6 feet tall, and you are looking at him feet first from 8 feet away. Your eye level is at 5 feet 4 inches, as you look down on your friend.

   a) Draw a geometric representation, to scale, of this situation and label all the important distances. Be sure to include your lines of sight. State the scale you use.

   b) How long is the projection of your friend on your plane of vision if you assume that the plane of vision goes through his feet?

2. Suppose you are eye level with a flagpole that is tilted 45° away from the picture plane. The pole is 2 feet long. See **Figure 5.72** for a geometric representation of this situation.

FIGURE 5.72.

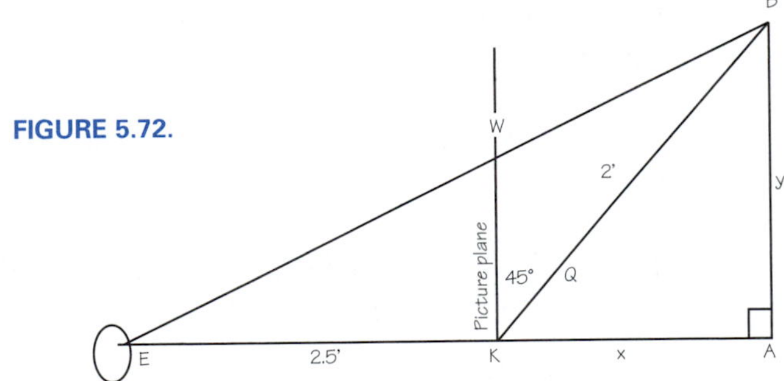

   a) Triangle KAB is a 45°-45°-90° right triangle. It is an isosceles right triangle since the measures of two angles are equal. Hence, the lengths of two sides are equal. Which two sides have the same length?

   b) Use the Pythagorean theorem to determine the unknown lengths of the sides of triangle KAB.

   c) Assume that you are 2.5 feet from the plane of vision and that the plane of vision touches the bottom of the pole. How long is the projection of the pole on your plane of vision?

3. a) The measure of angle A in **Figure 5.73** is 30°. How are triangles AMN, APQ, and ABC related? Explain.

   b) Measure the lengths of AM, AN, and MN, and find the ratios MN/AM, and AN/AM.

   c) Measure AP. Using (b) and properties of similar triangles, find the lengths of PQ and AQ without measuring them directly.

   d) Measure the length of AB. Using the properties of similar triangles, find the lengths of BC and AC without measuring them.

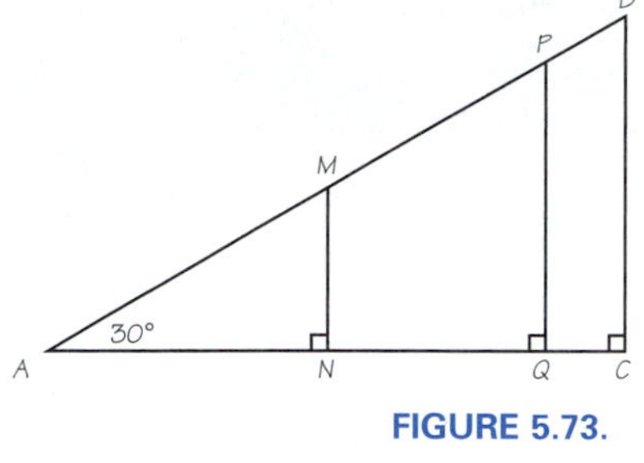

FIGURE 5.73.

4. **Figure 5.74** shows the side view of a tilted 6-inch pencil.

   - EQ is the line of sight from the eye to the tip of the pencil in the rotated position.
   - PK is the plane of vision.
   - KW represents the image of the tilted pencil.
   - KQ represents the 6-inch pencil.

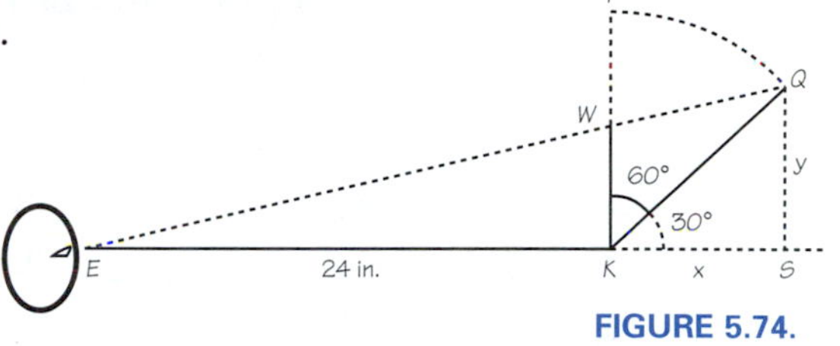

FIGURE 5.74.
Side view of tilted pencil.

   The eye level of the viewer is level with the base of the pencil. The picture plane is 24 inches from the viewer.

   a) Use your results from question 3 to estimate the lengths x and y (refer to Figure 5.74).

   b) What triangle in Figure 5.74 is similar to triangle EKW?

   c) Approximate the apparent height of the pencil, KW. Explain how you determined your answer.

5. Return to the flagpole drawing in question 2. Suppose that the pole is tilted 60° away from the picture plane instead of 45°. Assume that you are still 2.5 feet from the plane of vision and that the plane of vision touches the bottom of the pole.

   a) How long is the projection of the pole on your plane of vision?

   b) Did the increase in the amount of tilt away from the picture plane increase or decrease the length of the pole's projection on the picture plane?

ACTIVITY

5.7

Follow the Sines

In Homework 5.6, you worked with two specific right triangles: one with an acute angle of 30° and one with an acute angle of 45°. In each case you knew the size of one acute angle and the length of one side. Using that information, you were able to find the lengths of the other two sides. How can you find the measures of triangles with different acute angles? Right triangles have some special ratios that you can use to solve these problems.

THE SINE RATIO

CHECK THIS!

The word *trigonometry* is taken from the Greek words for triangle and measure. Trigonometry was used by the ancient Greeks to estimate the distances to the moon and the sun.

SINE

The sine of an angle A in a right triangle is the ratio

$$\frac{\text{length of the side opposite A}}{\text{length of the hypotenuse}}$$

The sine of A is abbreviated sin A.

The mathematical relationships between the sizes of angles and the lengths of sides in right triangles is part of what is called **trigonometry**. This field of mathematics was first developed in ancient times to aid in navigation and surveying.

There is an important relationship between an acute angle in a right triangle and the ratio of the side opposite it to the hypotenuse. This ratio is given a name: the **sine** of the angle.

In **Figure 5.75** , side k is opposite angle K, side w is opposite angle W, and side h is the hypotenuse. For the angle K, the sine of the angle K (sin K) is

$$\sin K = \frac{\text{length of HW}}{\text{length of KW}} = \frac{k}{h}.$$

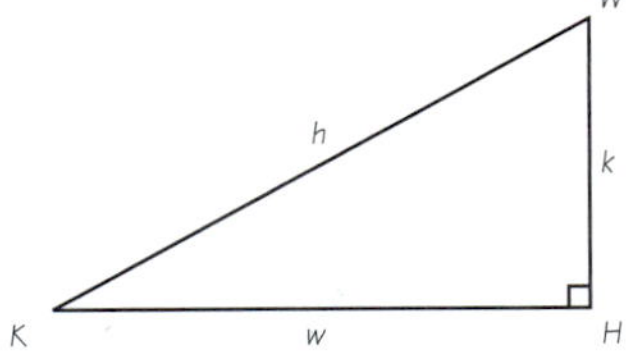

FIGURE 5.75.
Right triangle with side HW opposite angle K and hypotenuse KW.

1. What is the sine of angle T in **Figure 5.76**? The sine of angle R?

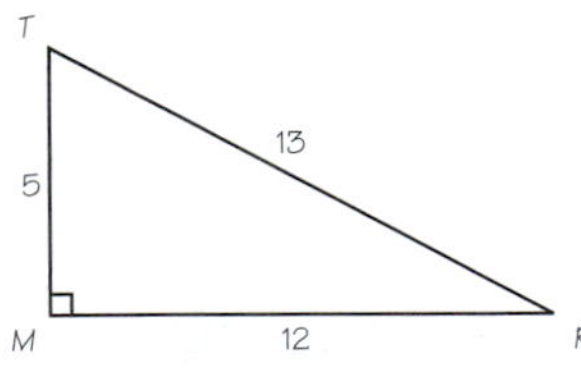

FIGURE 5.76.

In question 3 of Homework 5.6, you approximated the sine of 30°. Your value should have been close to 0.5, the actual value of sin(30°). Next you will draw a variety of triangles and use your triangles to estimate the sine of other angles.

2. Use a ruler and protractor or geometric drawing utility to draw several *right triangles*. Vary the sizes of the acute angles among the triangles you draw. Vary the sizes of the triangles. Include some triangles that are similar to each other.

   a) For each triangle, measure one acute angle, the length of the side opposite the measured angle, and the length of the hypotenuse. Record these measurements in the first three columns of a table like the one in **Figure 5.77**. Then compute the value of the sine of each angle directly from the definition of sine and your side measurements.

Angle Name	Side Opposite	Hypotenuse	Sine

FIGURE 5.77.
Table for recording estimates of the sine of an angle.

   b) Examine the sets of similar triangles among your data. Why is it possible for similiar triangles to have different sizes, but the same sine angle?

3. Use a scientific calculator or graphing calculator and check the accuracy of your answers for the sine of the angles in Figure 5.77. Make sure your calculator is in degree mode.

   a) Create a graph of the sine of an angle versus the measure of the angle. If you use your calculator rather than the data in your table and graph $y = \sin(x)$, you will need to check the MODE is set to degree measure. Restrict your domain from 0° to 90°. (To accomplish this, set Xmin = 0 and Xmax = 90.) Sketch your graph or describe its shape.

   b) Look at your graph from (a). As you increase the measure of the angle, does the value of the sine increase, decrease, or remain the same?

4. The sine of an angle relies on the ratios of lengths of two sides of only right triangles. Sometimes **helping lines** are added to create right triangles where none exist. Add a helping line to triangle RST to form two right triangles. Use the given information to find the length of side ST. (See **Figure 5.78**. Note, however, that the figure is *not* drawn to scale.)

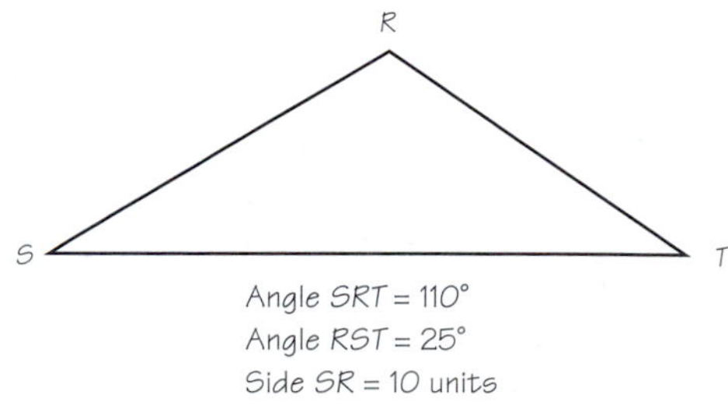

FIGURE 5.78.

COSINE AND TANGENT RATIOS

The cosine of an acute angle in a right triangle is the ratio

$$\text{cosine of A} = \frac{\text{length of the side adjacent to A}}{\text{the length of the hypotenuse}}$$

The cosine of A is abbreviated cos A.

The tangent of an acute angle A in a right triangle is the ratio

$$\text{tangent of A} = \frac{\text{length of the side opposite A}}{\text{the length of the side adjacent to A}}$$

The tangent of A is abbreviated tan A.

The sine ratio is one of three right triangle **trigonometric ratios** commonly used to solve problems. The cosine and tangent ratios are two other trigonometric ratios.

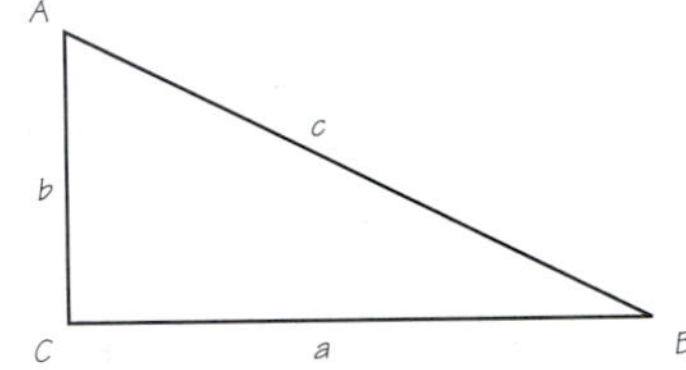

FIGURE 5.79. Right triangle.

The triangle ABC in **Figure 5.79** is a right triangle, and angle C is the right angle. Side a is opposite angle A and is adjacent to angle B. Side b is opposite angle B and adjacent to angle A. The trigonometric ratios of the acute angles A and B are

$\sin A = a/c$ $\quad \cos A = b/c$ $\quad \tan A = a/b$

$\sin B = b/c$ $\quad \cos B = a/c$ $\quad \tan B = b/a$

CHECK THIS!

Remember, the definitions of trigonometric (trig for short) ratios apply *only* to right triangles.

Angle (degrees)	Tangent (tan)	Cosine (cos)	Sine (sin)

FIGURE 5.80.

5. a) Draw five right triangles that are not similar. (You may use right triangles drawn earlier.) Measure the lengths of all three sides of each triangle. Use the Pythagorean theorem to check the accuracy of your measurements. Then use your measurements to determine the tangent and cosine ratios for the ten acute angles in your five triangles. Record your answers in a table similar to the one in **Figure 5.80**.

   b) Use your calculator's built-in trig functions to verify the accuracy of your ratios.

   c) Graph $y = \sin(x)$ and $y = \cos(x)$ using the values in your table or using a graphing calculator. (Again, make sure that MODE is set to degrees and that you restrict your domain from 0° to 90°.)

d) Look at your graphs for sine and cosine. As you increase the measure of the angle, does the value of the cosine increase, decrease, or remain the same? Compare this to what happens to the value of the sine.

e) Graph $y = \tan(x)$. What happens to the value of the tangent when the measure of the angle is close to 90°?

With the aid of trigonometric ratios, you can now solve the foreshortening problem introduced in Activity 5.6. **Figure 5.81** is a geometric representation of the tilted-pencil problem.

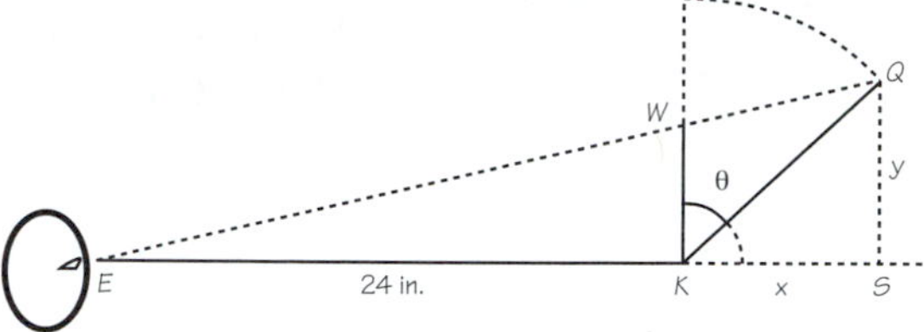

FIGURE 5.81.
Side view of a tilted pencil.

6. Suppose you are two feet away from the picture plane and your eye is level with the table. If a 6-inch pencil is held against the picture plane, its apparent length is also 6 inches. Now imagine tilting the pencil away from the picture plane so that it makes a 50° angle with the picture plane. (In other words, $\theta = 50°$.)

   a) Triangle QKS is a right triangle. What is the measure of angle QKS?

   b) Use one of the trigonometric functions defined in this activity to determine the value of y in Figure 5.81. What trig function did you use?

   c) Use one of the trigonometric functions defined in this activity to determine the value of x in Figure 5.81. What trig function did you use?

   d) Use your answers to (b) and (c) to determine the apparent length of the pencil, KW.

SUMMARY

In this activity you learned about the sine, cosine, and tangent ratios.

- Also you discovered how to use the ratios to find the length of foreshortened images in a perspective drawing. Now you have all the artistic principles and mathematical tools to create accurate perspective drawings.

Unfortunately this knowledge will not make you a world famous artist unless you also have talent!

HOMEWORK 5.7

SOH-CAH-TOA

Some students find SOH-CAH-TOA an easy way to remember the definitions of the three basic trig functions.

> **SOH-CAH-TOA**
>
> SOH: Sine is the ratio of the Opposite side to the Hypotenuse.
>
> CAH: Cosine is the ratio of the Adjacent side to the Hypotenuse.
>
> TOA: Tangent is the ratio of the Opposite side to the Adjacent side.

1. a) Draw three right triangles by hand or use a geometric drawing utility. Vary the acute angles so that the triangles are not similar to one another. Measure the lengths of the opposite sides and the hypotenuse. Write the ratio of the side opposite to the hypotenuse.

   b) Use your calculator to compare your answer with the sine of the angle. Show all your work and ratios.

2. Use the information in each of the following parts to find the lengths of all sides and the measures of all angles in the right triangle FGH (see **Figure 5.82**). In each case, angle G is the right angle.

   a) F = 37° and $f = 6$ units.

   b) F = 52° and $h = 12$ units.

   c) F = 20° and $g = 15$ units.

   d) $f = 8$ units and $g = 12$ units.

   e) H = 72° and $g = 20$ units.

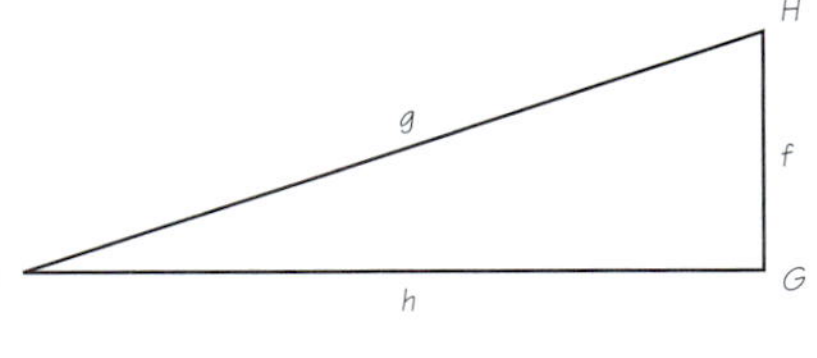

FIGURE 5.82.

3. a) Gerard places a 22-foot ladder against his house so that the end of the ladder extends 2 feet above the edge of the roof. The ladder makes an angle of 75° with the ground. Draw a geometrical representation of this situation.

b) How far is the base of the ladder from the house? Show how you determined your answer.

c) How high is the roof? Show how you determined your answer.

4. Suppose triangle ABC is a right triangle, with angle C the right angle. Use the sine, cosine, and tangent ratios to find each length from the following information. Remember that side a is opposite angle A, side b is opposite angle B, and side c is the hypotenuse.

   a) Angle A = 23°. Side c = 7.5 cm. Find angle B and the lengths a and b.

   b) Angle B is 58°. Side b is 3.84 m. Find angle A and the lengths a and c.

   c) Side a = 27 and side c = 42. Find the length b.

When the sum of two angles is 90° then these angles are called **complementary angles**. Thus, the angles measuring 32° and 58° are complements. Likewise, angles measuring 40° and 50° are complements.

5. a) Describe the relationship between the sine of an acute angle and the cosine of its complement. Investigate this relationship for several pairs of complementary angles. (Your data from question 5 in Activity 5.7 may be helpful here.)

   b) Explain the reason for the relationship you noticed in (a).

6. a) Suppose you are holding a 1-foot ruler 3 feet from your eyes. The ruler is tilted 40°. In **Figure 5.83**, $d = 3$, $l = 1$, and $\theta = 40°$. What is the measure of angle QKS?

   b) Find the value of x.

   c) Find the value y.

   d) Find the apparent length of the ruler, KW.

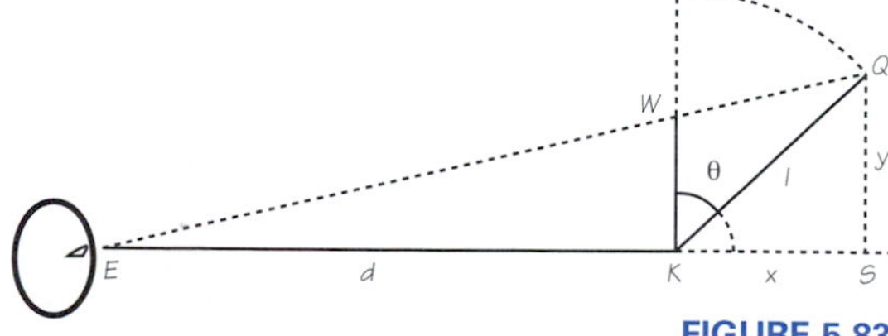

FIGURE 5.83.
Labeled side view of tilted ruler.

7. a) An animator wants to write a program to animate the image of the ruler as it rotates away from the picture plane. Write an equation for x using a trigonometric ratio of the complement of the angle θ. Use the values of d, l, and θ from question 6.

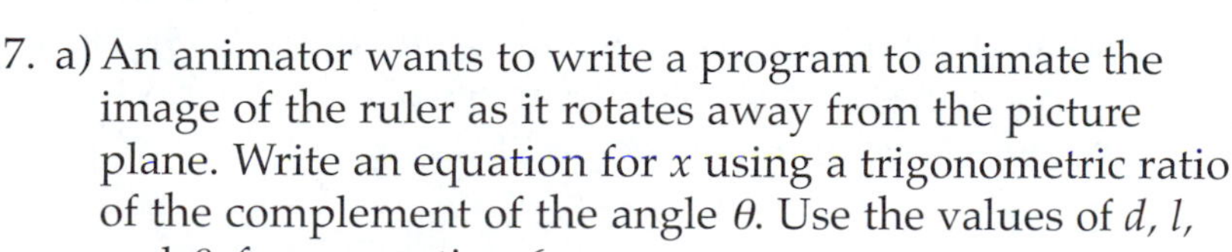

b) Write an equation for y using a trigonometric ratio of the complement of the angle θ. Use the values of d, l, and θ from question 6.

c) Write a formula that calculates the apparent length of the ruler in terms of the complement of the angle θ. Use your results from question 7(a) and (b) to write the equation.

d) Use your formula from (c) to find the apparent lengths of the ruler when $\theta = 20°$ and 80°. Show your work. Do your values seem reasonable? Explain.

e) Use your formula in (c) to find the apparent length of the ruler when $\theta = 90°$. Does your answer make sense? Explain.

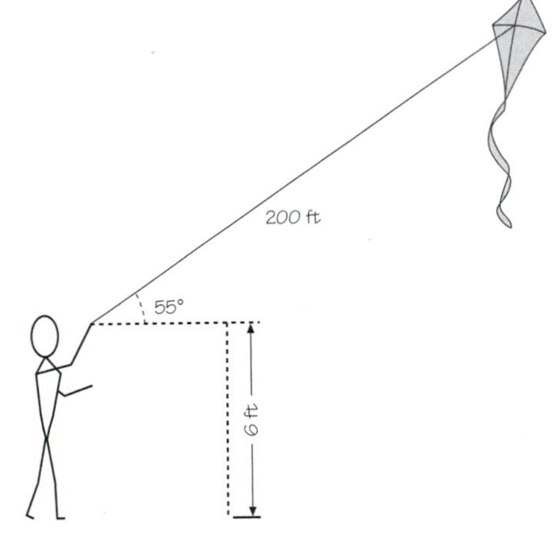

FIGURE 5.84.
Kite on string.

8. Suppose you are flying a kite on a 200-foot string. You hold the end of the string 6 feet above the ground so that the string makes an angle of 55° with the horizontal. How high is the kite flying (if you disregard any sag in the string)? (See **Figure 5.84.**) Explain how you arrived at your answer.

9. A radio tower is 30 feet tall. Miguel wants to attach cables to the top of the tower and tether the other ends to the ground to hold the tower perpendicular to the ground (right angle). He wants the angle between the ground and each cable to be 75°. Calculate the length of each cable.

10. The leaning tower of Pisa is known both for its beauty and its incline. The 184.5-foot tower makes an angle of approximately 84.5° with the ground.

a) When the sun shines, the tower casts a shadow. If the sun were shining directly overhead, how long a shadow would the tower cast? Draw a geometric representation of the situation before determining the shadow length.

b) Pisa is too far north for the sun ever to shine directly overhead. Suppose instead that the sun's rays come in at an angle of 60° with the ground (see **Figure 5.85**). How long is the shadow cast by the tower? (Hint: Draw helping lines to make two right triangles.)

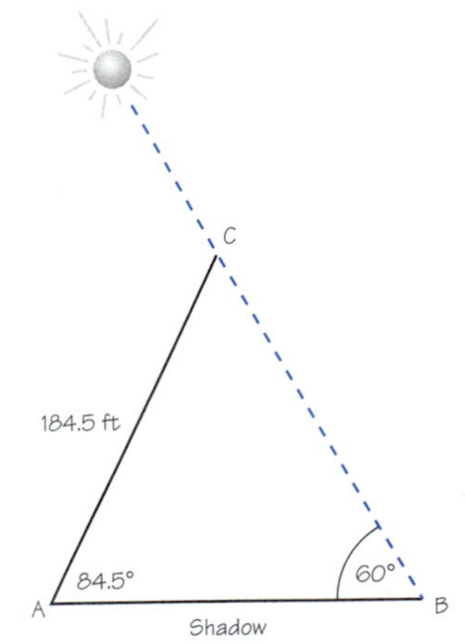

FIGURE 5.85.
Geometrical representation of leaning tower of Pisa and its shadow.

Modeling Project Video Game Highway

Suppose you are an animator working with a team of artists. Your team is working on a project to create a video game that simulates driving a car. You are given the task of determining the proper size and location for signs that appear at regular intervals along the side of your simulated highway. **Figure 5.86** is a sketch of the highway in your video game.

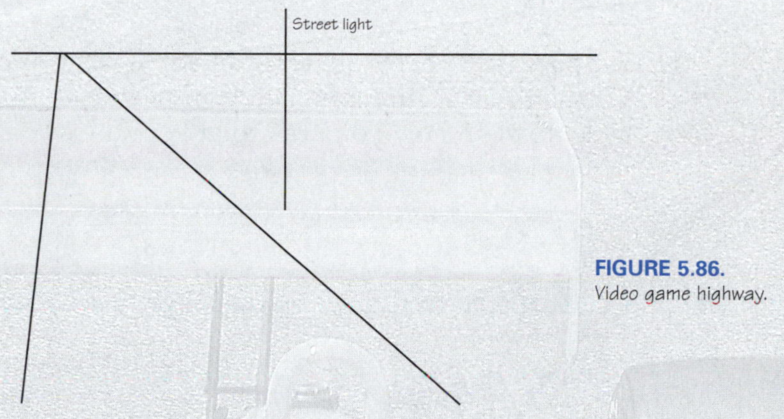

FIGURE 5.86.
Video game highway.

1. Copy the sketch onto your own paper (or use Handout 5.7). Draw three signs along the side of the highway using the principles that you have learned in this chapter. Assume that in real life the signs are 4 by 4 meter squares and are placed 50 meters apart. The streetlight seen in the picture is actually 10 meters tall. List any additional assumptions you make.

2. Explain how you determined the sizes of your signs and where to place them.

Practice Problems

Review Exercises

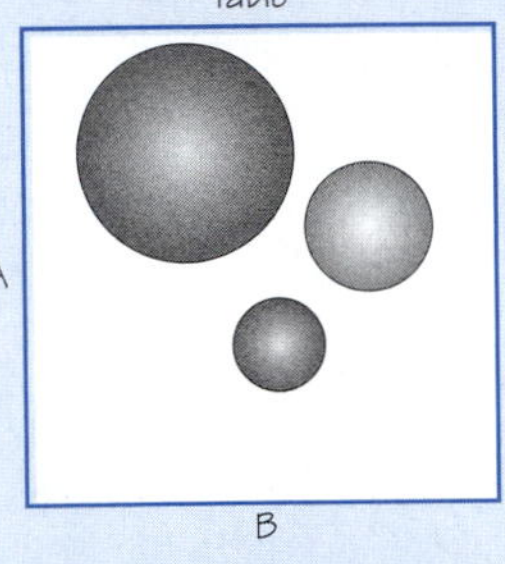

FIGURE 5.87. Three spheres arranged on a table.

1. Three spheres are arranged on a table.

   a) **Figure 5.87** shows the top view. Draw the views from A and B. What principle of perspective did you use in your drawing?

   b) Suppose the spheres are rearranged. Views from A and B are shown in **Figure 5.88**. Draw an overhead view of the table. Explain how you arrived at your answer.

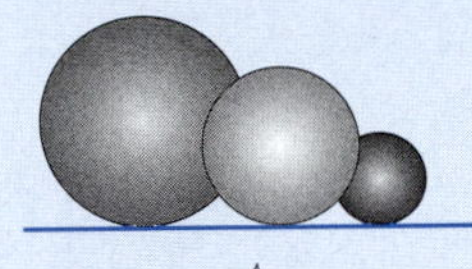

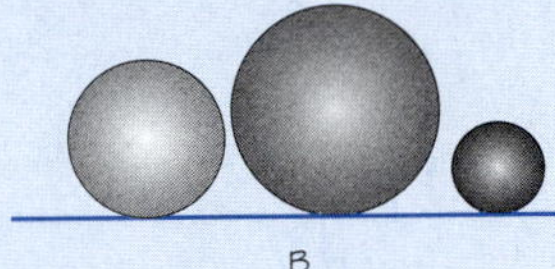

FIGURE 5.88. Side views of the rearranged spheres.

2. Use proportions to answer the following questions. In each case, suppose a ruler is held at arm's length so that the plane of vision is 27 inches from the eye of the viewer.

   a) You are 100 yards away from the goalpost on a football field. The image height of the goalpost is 1.6 inches. How high is the goalpost?

   b) Close to shore on the opposite side of a lake is a tower that is 50 feet tall. The image height of the tower is 1 inch. How wide is the lake?

   c) A certain basketball player stands 86 inches tall. You are standing under the basket and he is attempting a free throw, which means he is about 18 feet from you. How tall is his image?

3. **Figure 5.89** shows the top view of a person looking at five railroad ties.

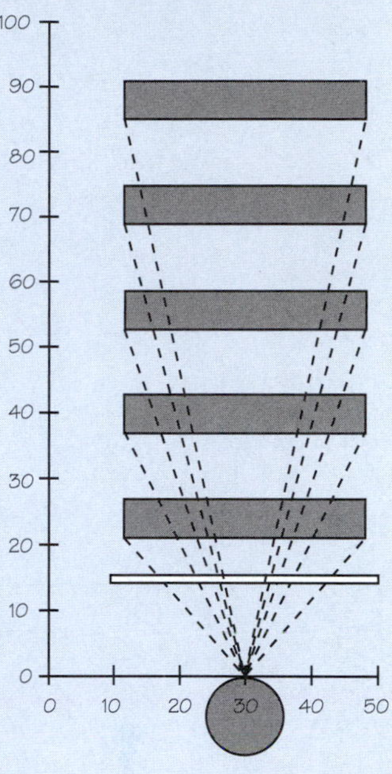

FIGURE 5.89.
Top view of person looking at railroad ties.

a) Measure the length of each projection in the picture plane. Then measure the distance between each tie and the viewer's eye. Record your measurements in a table similar to the one in **Figure 5.90**.

Tie Number	1	2	3	4	5
Distance to Viewer's Eye					
Projection Length					

FIGURE 5.90.
Table for recording railroad tie lengths and distances.

b) Based on your measurements, describe the relationship between the length of the projection of each tie and its distance from the viewer's eye.

c) Graph your data. Label the horizontal axis "Distance to Eye" and label the vertical axis "Projection Length." Describe the shape of your plot.

4. a) Copy **Figure 5.91** on your own paper and label the sides and vertices according to mathematical convention.

b) For each of the triangles in **Figure 5.92**, how long is the side opposite angle V?

c) Find the length of the leg adjacent to angle V in the triangles in Figure 5.92.

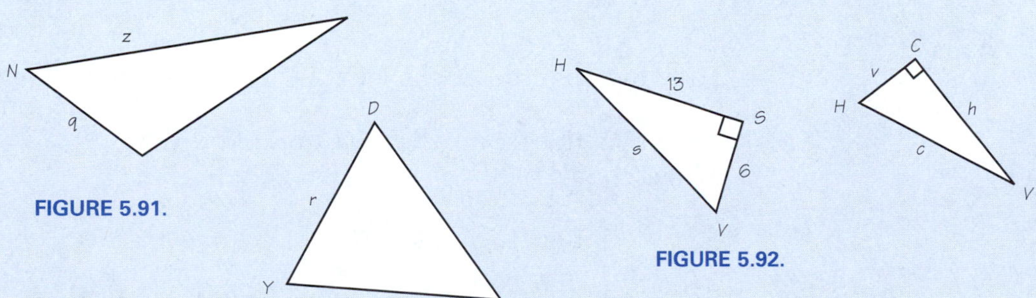

FIGURE 5.91.

FIGURE 5.92.

5. Three equal-height poles are to be placed 100 feet apart, increasing in distance from the viewer. Which drawing (see **Figure 5.93**) seems to be the best representation? Explain your answer.

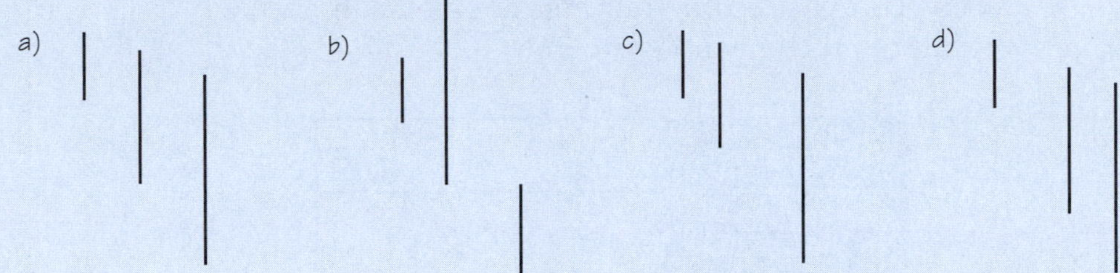

FIGURE 5.93.
Four different drawings of three poles.

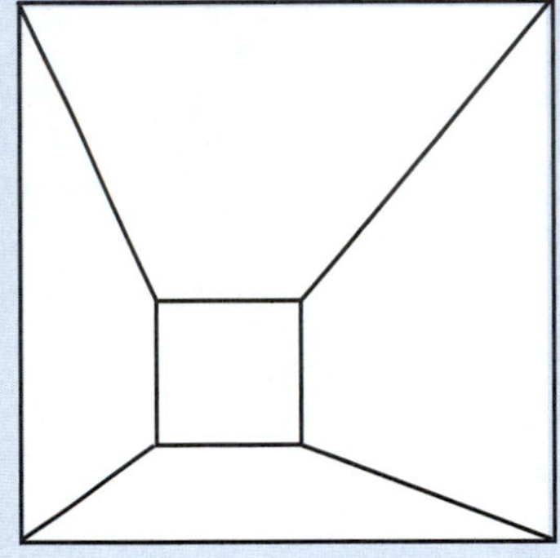

FIGURE 5.94.
Perspective view of hallway.

6. **Figure 5.94** is a perspective view of a hallway.

   a) Trace a copy of Figure 5.94 on your own paper. Determine the vanishing point and the horizon line.

   b) Raise the vanishing point and move it to the right. Redraw the hallway using the new vanishing point.

7. Recall the characteristics of trapezoids, parallelograms, rectangles, rhombuses, and squares. (See **Figure 5.95** for sample drawings of each of these quadrilaterals.)

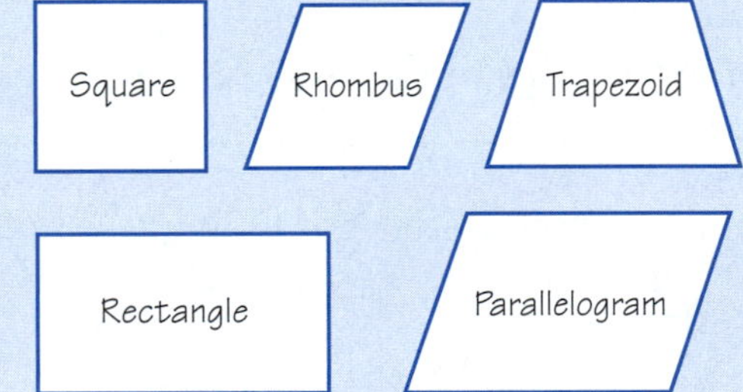

FIGURE 5.95.
Five quadrilaterals.

   a) For which figures are the lengths of opposite sides equal?

   b) For which figures are the measures of opposite angles equal?

   c) For which figures do diagonals bisect one another?

   d) For which figures are diagonals perpendicular to one another?

   e) For which figures are diagonals equal in length?

8. a) In **Figure 5.96**, ABCD is a parallelogram. List all pairs of similar triangles.

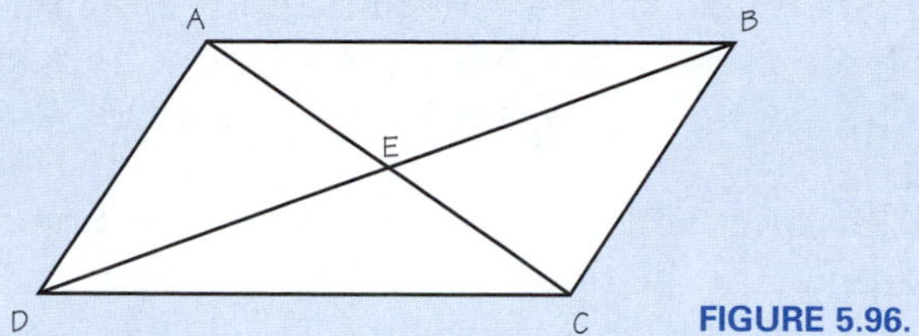

FIGURE 5.96.

b) What is the ratio of the lengths of corresponding sides in the pairs of similar triangles in the parallelogram of **Figure 5.97**?

9. In Figure 5.97, RS is parallel to HK.

a) Which angles have equal measure?

b) What is the relationship between triangles ARS and AHK? Explain.

c) **Figure 5.98** was created by adding the line segment AN to Figure 5.98. Segment AN bisects segment HK. Explain why it must also bisect RS.

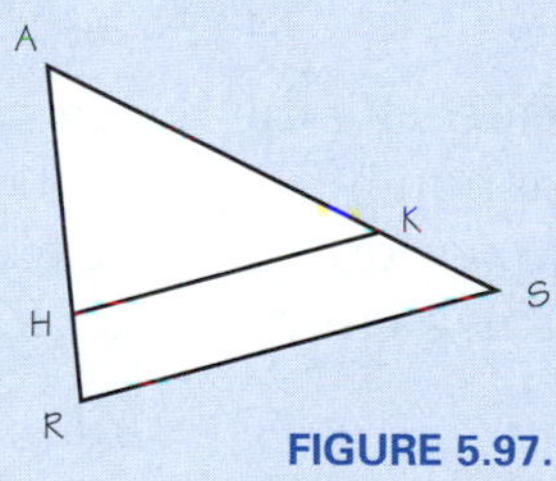

FIGURE 5.97.

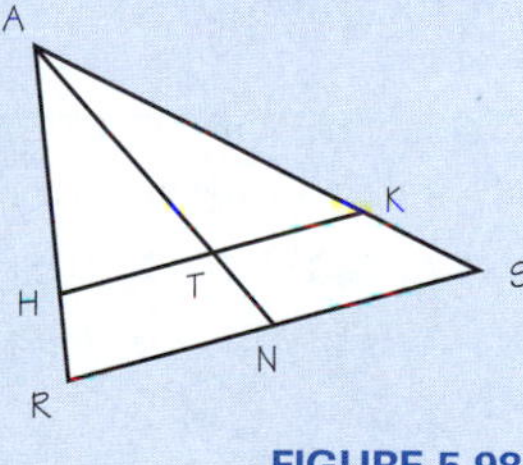

FIGURE 5.98.

10. Develop a method for creating a drawing of the front wall of your classroom at a 1/10 scale.

11. Suppose you want to draw accurately a series of poles that support a cover above a walkway. **Figure 5.99** shows a top view of the poles in relation to the artist and the plane of vision. The artist chooses a viewing position 8 feet to the left of the line of poles and 22 feet from the first pole that is to be drawn in the picture. The 9-foot poles are 15 feet apart. The distance between the artist's eye and the picture plane is 2 feet.

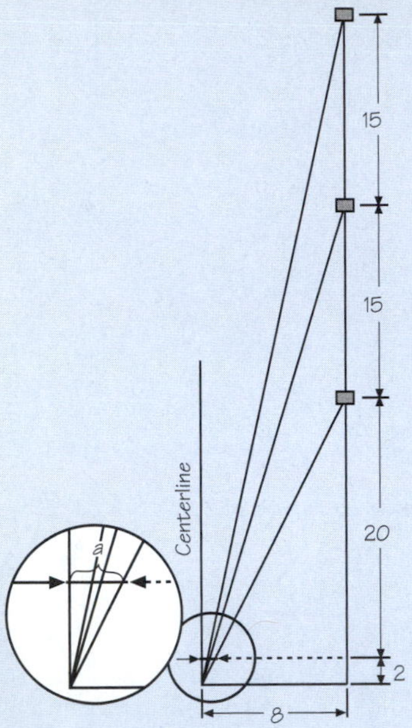

FIGURE 5.99.
Top view of hallway support poles, including artist's viewing position (not to scale).

a) The centerline is an imaginary line from the artist's eye to the vanishing point. In a top view, this line is parallel to the line of poles. Describe the relationship in a perspective drawing between (1) the centerline and (2) a line drawn through the poles at the artist's eye height. If drawn, how would the centerline appear in the picture?

b) In the drawing, what is the scale at the first pole? At the second pole? At the third pole?

c) What is the image size of Pole 1? What are the image sizes of Pole 2 and Pole 3?

d) Determine the horizontal distance *a*, the distance between the image of Pole 1 and the centerline. How can you use the scale at Pole 1 to determine this distance?

e) Determine the horizontal distance *b*, the distance between the image of Pole 2 and the imaginary centerline.

f) Determine the horizontal distance *c*, the distance between the image of Pole 3, and the imaginary centerline.

12. In this chapter you have examined image size and image placement, each separately. Your goal in this question is to determine how these two kinds of measurements are related.

    a) Use proportions (or scale) to calculate the image height of each pole based on the information in **Figure 5.100**. Then calculate the distance from the centerline to the image of each pole. Record your values in a copy of **Figure 5.101.**

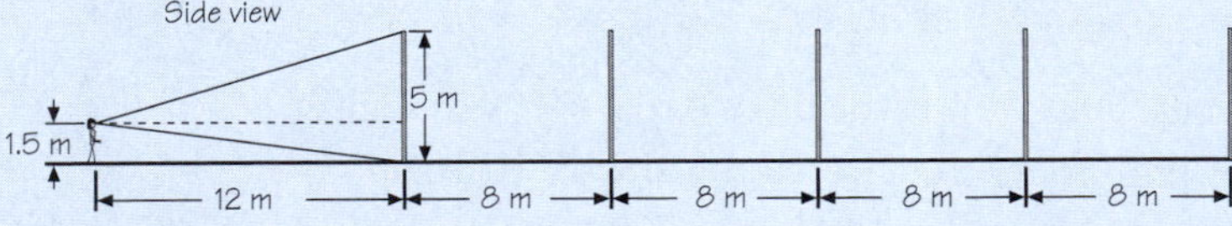

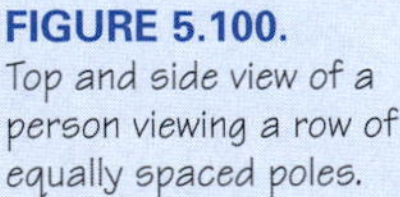

FIGURE 5.100.
Top and side view of a person viewing a row of equally spaced poles.

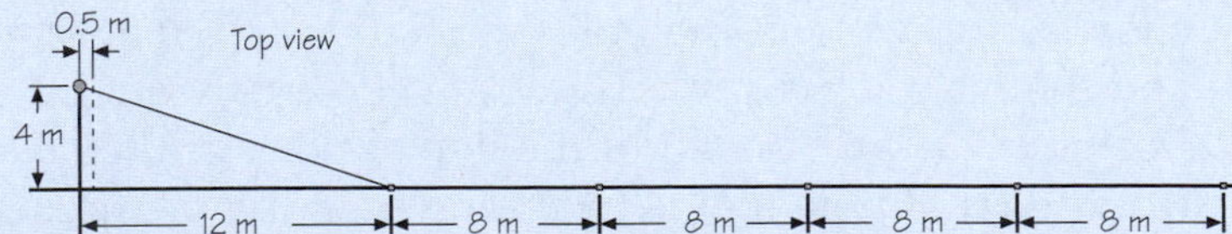

Pole Number	Picture Distance from Centerline	Image Height
1		
2		
3		
4		

FIGURE 5.101.
Table of locations and heights.

b) Graph the relationship between the image heights of successive poles and their distances from the vanishing point. What pattern, if any, do you notice? If there is a pattern, write an equation to represent the pattern.

13. Telephone poles frequently are supported by guy wires. (A guy wire is a cable from the top of the pole to the ground.) From the side, a pole with a guy wire might look like **Figure 5.102**. In perspective, it might look like **Figure 5.103**.

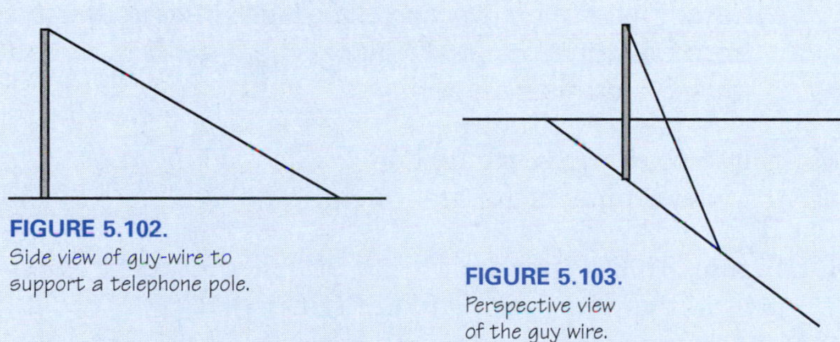

FIGURE 5.102.
Side view of guy-wire to support a telephone pole.

FIGURE 5.103.
Perspective view of the guy wire.

To protect people from accidentally running into the guy wire, the telephone company would like to hang small markers at the midpoint and at 1/4 and 3/4 points of the cable. By working with both the perspective and the non-perspective views, show how these points can be constructed.

14. a) Which angles in **Figure 5.104** are equal in measure to angle 5?

b) If the measure of angle 12 is 65°, what is the measure of angle 1? Of angle 9? Of angle 16?

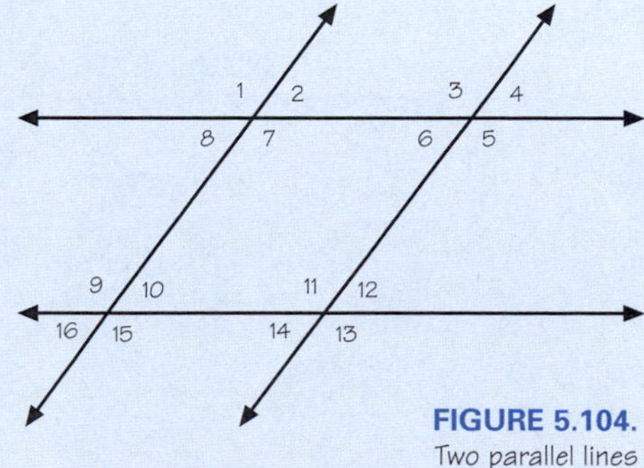

FIGURE 5.104.
Two parallel lines intersected by two parallel lines.

15. **Figure 5.105** shows a top view of a series of equally spaced cubes. Each cube has one face parallel to the plane of vision. Each individual distance is 0.75 cm. Make a perspective drawing in which you can see two sides and the top of each cube. How many vanishing points should there be?

FIGURE 5.105. Top view of a series of four cubes.

16. Suppose you have a pole 20 feet in length and your plane of vision is 1.5 feet from your eye. Your partner is holding the pole at an angle so that its base is at your eye level and in the plane of vision. The top of the pole is actually 5.18 feet above your eye level and quite a bit beyond the plane of vision.

    a) Draw a geometrical representation of the situation.

    b) How far back from the plane of vision is the pole?

    c) What is the scale for objects at the same distance away as the top of the pole?

    d) Recall that scale also can be measured using distances from the horizon line. Explain where the horizon line must be in this situation.

    e) Use your answer to (c) and the actual height of the pole to determine its apparent height in the plane of vision.

17. **Figure 5.106** shows a top view of a tiled floor. Draw a perspective view of this tiled floor.

FIGURE 5.106. Top view of tiled floor.

18. Suppose you want to use a ramp to move objects from ground level to the back of a furniture truck. The length of the ramp is 8 feet. It is difficult to move objects up the ramp if the ramp makes an angle of more than 35° with the ground. What is the maximum height for the back of the truck given the 35° restriction?

19. The angle between eye level and the line of sight to the top of a mountain is 14°. The distance from the eye to the plane of vision is 24 inches. The horizontal distance to the top of the mountain is 10 miles. Calculate the height of the image and estimate the actual height of the mountain (above the horizon).

Mathematics of Perspective Drawing

REVIEW

The three major principles of perspective drawing discussed in this chapter are diminution, convergence, and foreshortening. In this chapter each principle relied on the development of a different mathematical topic.

The mathematics of similar triangles and proportions is central to the idea of diminution. The geometry of parallel lines guides the study of convergence. Relationships between the sides and angles of right triangles, a major topic in trigonometry, are necessary to quantify the principle of foreshortening. A review of some of the key mathematical ideas behind these three principles follows.

Similar Triangles and Proportions

Observe an object moving away from you. The object appears smaller the farther away it gets. In reality the object has not changed size, but it still appears smaller. That's the principle of diminution. The mathematics of similar triangles and proportions is useful in relating size, distance, and scale—the key elements of diminution.

Look at **Figure 5.108**. Four measurements are labeled as a, b, c, and d. You can use similar triangles and proportions to find the fourth measurement when three of the other measurements are known.

- The horizontal distance a between the viewer's eye and the picture plane.
- The horizontal distance b between the viewer and the object.
- The size c, in the picture plane, of the image of the object.
- The size d of the object.

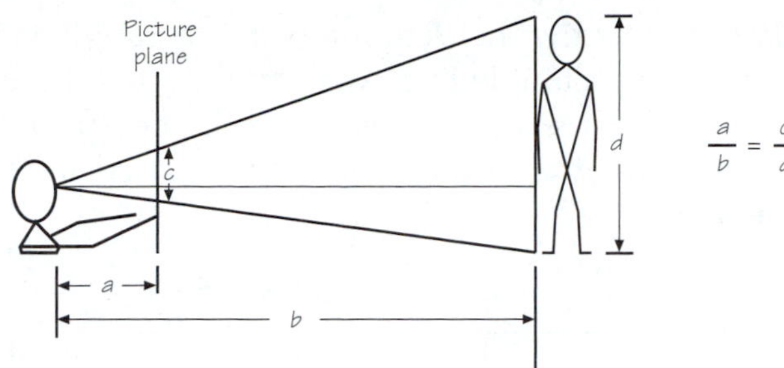

FIGURE 5.108. Proportions and perspective drawing.

From this context of Figure 5.108 comes the mathematical definition of the scale of an object in a picture:

Scale = (height in picture)/(actual height) = c/d.

From Figure 5.108 it is also clear that

Scale = (distance from eye to picture plane)/(distance from eye to object) = a/b.

When the concept of horizon is combined with scale, another ratio may be added to these representations of scale:

Scale = (distance in picture from horizon to place where object meets ground)/(eye-level height of viewer)

Proportions may also be used to determine the correct spacing, in the perspective view, between objects that are equally spaced in the real world.

Parallel Lines and Related Figures

The principle of diminution guarantees that parallel lines that are horizontal in reality and increase in distance from the viewer must appear to converge to a point at the eye level of the viewer. In a perspective drawing, parallel lines drawn into the distance appear to meet at a point, called a vanishing point. This is the principle of convergence.

One consequence of convergence is that a rectangle with two vertical sides looks like a trapezoid in the perspective view.

The diagonals of rectangles and trapezoids are key features of constructions that ensure the equal spacing of objects. The constructions rely on these two facts:

- the diagonals of a rectangle bisect one another and are equal in length
- the diagonals of a trapezoid do the same in a perspective view.

Figure 5.109 shows how the diagonals of a rectangle may be used to divide the rectangle into two equal parts. Therefore, the diagonals of a trapezoid representing the perspective view of a rectangle divide the rectangle into equal parts in the perspective view.

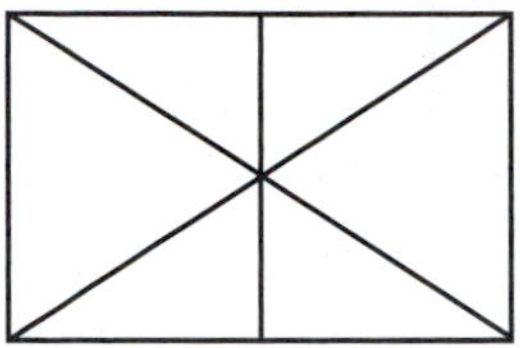

The diagonals of the rectangle divide the larger rectangle into two equal parts.

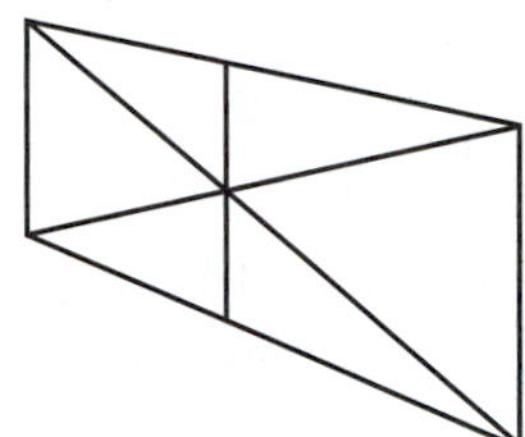

The diagonals of the rectangle in the perspective view divide the larger "rectangle" into two "equal" rectangles.

FIGURE 5.109. Diagonals bisect a rectangle in a side view and a perspective view.

▪ Right Triangle Relationships

Foreshortening results as the plane of vision intersects a smaller angle between two lines of sight when an object is tilted at an angle to the plane of vision.

Figure 5.110 shows the relationship between the foreshortened image of an object and a right triangle with horizontal and vertical sides x and y and hypotenuse defined by the object itself.

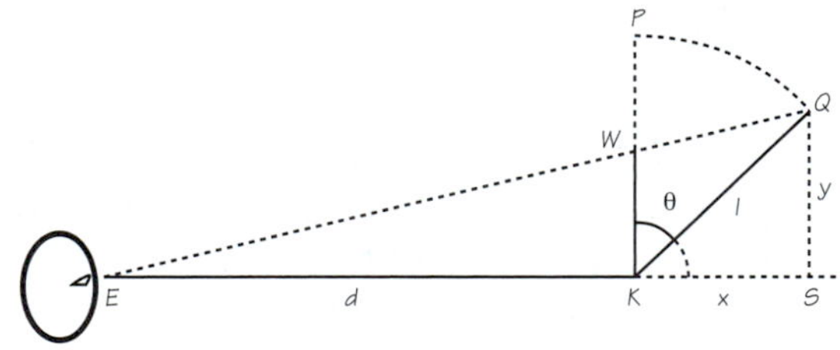

FIGURE 5.110. Side view of an object tilted at an angle from the picture plane.

The relationships between the sides and the acute angle of a right triangle are described by trigonometry. For the right triangle ABC shown in **Figure 5.111**, a is the side opposite angle A and b is the side adjacent to angle A. Angle A actually has two adjacent sides, b and c, but since triangle ABC is a right triangle, c is referred to as the hypotenuse, not as a side. Sides a and b are also called legs.

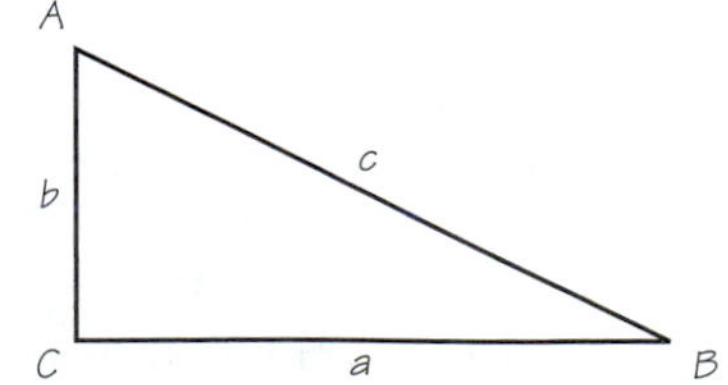

FIGURE 5.111. Right triangle ABC.

The sine ratio is one of three right triangle trigonometric ratios commonly used to solve problems. The cosine and tangent ratios are the other two. These ratios are defined by the following equations:

sin(angle) = (side opposite) / (hypotenuse)

cos(angle) = (side adjacent) / (hypotenuse)

tan(angle) = (side opposite) / (side adjacent)

For example, in triangle ABC, sin A = a/c, cos A = b/c, tan A = a/b,

sin B = b/c, cos B = a/c, and tan B = b/a.

Complementary angles: Two angles whose measures sum to 90°.

Convergence: Artistic principle asserting that lines or edges of objects that in reality are parallel appear to come together (that is, converge) as they recede from the observer.

Corresponding angles: Angles in corresponding (analogous) positions within similar figures. Corresponding angles are equal in measure.

Cosine: In a right triangle, the ratio of the length of the side adjacent to an acute angle to the length of the hypotenuse.

Dilation: A scaling transformation of the plane in which the directed distance to every point's image is exactly k times its original directed distance from some specified, fixed point. The fixed point is called the center of the dilation; k is the scale factor for the dilation. Informally, think of dilation as a stretching away from the center by a factor of k.

Diminution: The phenomenon by which an object appears smaller as its distance from the observer increases.

Foreshortening: The phenomenon that occurs when lines or surfaces perpendicular to the line of sight show their maximum length, but as they are rotated away from the observer, they appear increasingly shorter.

Helping lines: Lines added to figures to create right triangles where none exist.

Horizon line: The imaginary line on the plane of vision (or picture plane) containing the vanishing points of all horizontal (in reality, parallel to the ground), converging lines. It is always on the same level as the observer's eyes (the eye level).

Lines of sight: Any of the imaginary lines between the viewer's eye and the top and bottom of the three-dimensional object being observed.

One-point perspective: Occurs with only one family of horizontal parallel lines not parallel to the plane of vision; all the lines converge to a single vanishing point located on the horizon line.

Overlapping: A technique achieving a sense of depth and space in drawings by showing which objects are in front and which are in back.

Perspective: The technique of representing objects in three-dimensional space in a two-dimensional plane.

Plane of vision (or picture plane): An imaginary plane between the observer and the three-dimensional object(s) being viewed. The image of the object is projected onto the plane. In this unit, the picture plane is always vertical.

Pythagorean theorem: A formula relating the lengths of three sides of a right triangle: $a^2 + b^2 = c^2$, where c is the length of the hypotenuse and a and b are the lengths of the other two sides (legs).

Scale: The ratio of an image size (height or width) in a picture to the actual size of the object.

Scale factor: In a dilation, the ratio of the new lengths to the old lengths. The number by which each linear dimension of an object is multiplied in a dilation or scale drawing.

Scaling: A mathematical technique for representing depth and distance by enlarging or reducing objects represented in a drawing. The amount of enlargement or reduction depends on the scale factor.

Sine: In a right triangle, the ratio of the length of the side opposite an acute angle and the length of the hypotenuse.

Tangent: In a right triangle, the ratio of the length of the side opposite an acute angle and the length of the side adjacent to the acute angle.

Transversal: A line that intersects two or more parallel lines.

Trapezoid: A quadrilateral in which exactly one pair of opposite sides is parallel.

Trigonometric ratio: A ratio of the lengths of two sides of a right triangle.

Vanishing point: The imaginary point to which lines parallel to one another appear to converge.

6

CHAPTER

The Financial Ride of Your Life

Can mathematics make you a better consumer?

So you want to buy a car? This purchase is one of the more important financial decisions of your life, and you will probably do it several times. Since the cost of buying and maintaining a car is very high, you will want to weigh your choices carefully. You don't want to be saddled with huge bills that you cannot afford!

On the one hand, part of the decision-making has absolutely nothing to do with mathematics. Why do you need the car? Will you use it to commute to school or work? Will you use it mostly around town or on the highway? Do you want a car that you can show off to your friends? Do you prefer a sleek, stylish shape or a classic look? Do you need a luxury car or just a car to get you around?

On the other hand, there are many issues that can be analyzed using mathematics. As you will quickly learn in this chapter, buying and owning a car involves many costs. There are many people in the business, such as automobile salespeople, bank officials, auto mechanics, and insurance executives who make their living taking money from you. Modeling can help you make a good decision so that you will buy a car you can afford. Once you develop these models, they can be used to purchase any car at any time during your life. With some adjustments, these models can be used to buy other "big ticket" items, too!

PREPARATION READING

Decisions, Decisions

When buying a car, you have so much to consider.

- What make, model, style, and color of car do you want? Do you want a custom paint job?
- Do you have a choice of engine sizes? Transmission types? Tires? Safety features?
- What other possible options are available? Do you want the car to come with options installed, or add them later as time and money allow? *Can* they be installed later?
- Should you buy a new car or a used car?
- If you have an old car, should you trade it in or try and sell it yourself?
- Which dealership will give you the best deal?

- Can you pay cash for the car or do you have to finance it? How much will the payments be?
- Should you buy or lease a car?
- What kind of mileage does the car get? How soon will you have to worry about repairs?
- What about insurance? How much will that be?

There are hundreds of makes and models from car manufacturers, and the choice of "Which one?" is often as much a personal preference as it is a financial decision. In this chapter you will examine two specific models only, so the main decision has been made for you. But you will probably find that the other decisions still remain a sizeable challenge.

Our stars for this chapter, Maria and Carla, are twin sisters from Bryan, Texas. They have been thinking about owning a car for some time. They are both active girls who enjoy the outdoors with their friends. So they know that they would like to own a 4-door **SUV** (sport utility vehicle). They realize they do not need four-wheel drive, even if they could afford it. But that is where the common ground ends. Maria likes the Ford Explorer and prefers the basic model with no special features, but she will consider other options. Carla is more adventurous and thinks a loaded Nissan Xterra would be great.

You will explore these two alternatives with them, and figure out a way to determine which would be the "better" buy. In addition you will research the car of your dreams, and perform a similar analysis to see how it fares in comparison to these two choices. By the end of the chapter you will be prepared for the decision-making that goes into car ownership, and you will have a much better understanding of how much it costs to own a car. You will develop a model to compare various cars for their overall cost and worth to you. As you analyze these situations, you will come to understand that what something costs is not always what it is worth.

ACTIVITY

6.1

The Dealing Gets Rough

Since the sisters know what type of car they want, the first issue they need to analyze is price. In this activity you will investigate the costs of buying a car, including any hidden costs.

THE COSTS CONTINUE TO MOUNT

Before visiting car dealerships in her area, Maria decided to do some research. She found the inventory list of vehicles in stock on the web sites of two local dealers. She learned that there were two types of Ford Explorers: the XLS and the XLT series. Not knowing the differences between the two, she compared the prices for the 2001 cars. Her findings are summarized in **Figure 6.1**. The numbers in parentheses indicate how many of each SUV the dealer has on the lot at that price.

Dealer 1		Dealer 2	
XLS Series Price	XLT Series Price	XLS Series Price	XLT Series Price
$26,565 (5)	$29,185 (2)	$25,690 (4)	$28,490 (4)
$27,440 (1)	$29,465 (2)	$25,715 (4)	$28,515 (4)
$27,465 (6)	$29,935 (2)	$25,810 (2)	$28,620 (3)
$27,570 (2)	$30,150 (2)	$26,465 (3)	$29,440 (5)
	$30,485 (1)	$27,245 (1)	$29,770 (2)
	$30,533 (1)	$27,570 (3)	$30,415 (2)
		$27,625 (3)	$30,695 (3)
		$27,795 (4)	$31,645 (3)
		$28,270 (2)	$33,080 (3)

FIGURE 6.1. Sticker prices for Ford Explorers.

1. Use Maria's findings to answer the following questions for each dealer.

   a) What is the range of prices for a XLS series car?

   b) What is the range of prices for a XLT series car?

   c) What is the average price for a XLS? (Hint: Don't forget to consider the numbers in parentheses.)

   d) What is the average price for a XLT?

2. a) Explain why the average price for a XLS series car is different than that for a XLT series car?

   b) Explain why the price for a XLS (or a XLT) is not constant, even at the same dealership?

3. a) Which dealership has the best inventory?

   b) Which dealership has the best selection?

   c) Which dealership has the best deals?

DEALER COSTS

Maria was pretty confused by the facts she had gathered so far. She asked herself, "Why are the cars THAT expensive? What accounts for the difference in the prices?" She decided to find out exactly how automobile dealers price their cars. Her research is summarized in Handout 6.1 (for the XLS) and Handout 6.2 (for the XLT). Since Maria wants a 4-door model with 2-wheel drive, she collected data only for those models .

MSRP AND DESTINATION CHARGE

The retail price of each vehicle is called the MSRP (Manufacturer's Suggested Retail Price). By law the car maker must set this price to establish the worth of the car. The destination charge is an additional fee for transporting the car from the factory to the dealership.

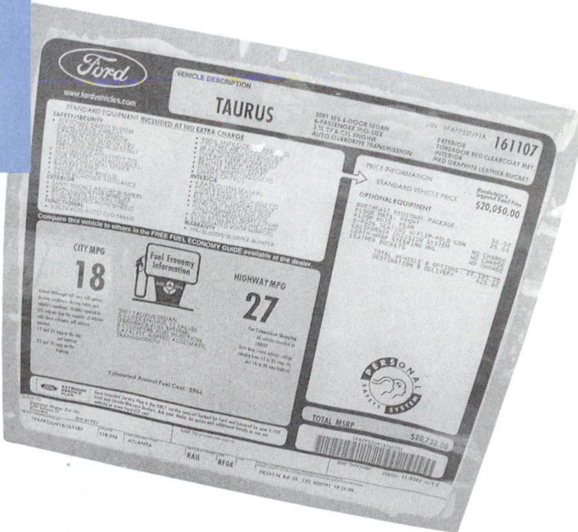

4. a) A basic model (without extra features) would have a sticker price that combines only the MSRP and the destination charge. What is the **sticker price** for a basic model 2001 Explorer 2WD XLS 4D?

   b) What is the sticker price for a basic model 2001 Explorer 2WD XLT 4D?

5. What the dealer does *not* provide is the **invoice price**, which is the price that the dealer paid for the car. Maria learned that the invoice price is $23,157 for the XLS and $25,649 for the XLT.

a) How much profit does the dealer make selling one (1) basic model of the 2001 Explorer 2WD XLS 4D?

b) How much profit does the dealer make in selling one (1) basic model of the 2001 Explorer 2WD XLT 4D?

Maria also gathered all the physical characteristics of the vehicle, warranties, engine specifications, and the choice of colors for the interior and exterior. (Note: some exterior colors may not be available with a particular interior one.) The list of standard equipment features is also provided.

For an extra cost a buyer can purchase optional equipment for a vehicle. **Accessory packages** are upgrades to the basic model that are offered as a group. For example, the "Convenience Group" package includes a cargo cover, privacy glass, keyless remote control entry system, and speed control/tilt steering wheel. These options are installed on the car, and the price for the package is added to the price for the basic model.

Some features are standard on some models and are included in the price of the car. For other models the price of those options is added to the sticker price. Examining the sticker on a car will identify which options are affecting the price.

6. a) Calculate the total price of the 2001 Explorer 2WD XLS 4D described in **Figure 6.2** using Handout 6.1.

FIGURE 6.2.

Description	Retail Price
2001 Explorer 2WD XLS 4D	
Destination Charge	
Sport Group Accessory Package	
AM/FM Stereo w/CD & Cassette	
Cloth Sports Bucket Seats	
Cast Aluminum Wheels	
Total Price:	

b) Calculate the total price of the 2001 Explorer 2WD XLT 4D described in **Figure 6.3** using Handout 6.2.

Description	Retail Price
2001 Explorer 2WD XLT 4D	
Destination Charge	
Sport Group Accessory Package	
Moon Roof	
Towing Package	
Garage Door Opener	
AM/FM Stereo w/CD & Cassette	
CD-Changer	
Restraint System	
Running Boards	
Total Price:	

FIGURE 6.3.

Other factors that affect the sticker price are the **credits**, **incentives**, and **options** that are offered to encourage customers to buy a car. Customer incentives often are cash applied to the car purchase to reduce the sticker price. Sometimes the dealer will offer to pay for an optional feature. These special offers may be made by the car dealer, or they may be programs offered nationally by the manufacturer.

7. a) What is the final sticker price for the SUV described in Figure 6.2 if the dealer agrees to "throw in" (that is, pay for) the stereo, and the car buyer is eligible for the college graduate incentive?

   b) What is the final sticker price for the SUV described in Figure 6.3 if the dealer offers the customer a $1500 incentive?

GETTING AN AGREEMENT

It is a well-known fact that car dealerships negotiate the final price of the car with the customer. Many factors affect how much a dealer will lower the sticker price of a car, including whether the customer needs a loan from the dealership, whether the customer has a car to trade in, the time of the year, or the time of the month.

8. Explain why each of the following factors might affect the final selling price for a car.

a) Whether there is a trade-in.

b) The time of the year.

c) The time of the month.

Since the dealer makes a lot of profit on options and accessory packages, there is room to negotiate a lower price. Maria did her homework and discovered that the cost of optional equipment for Ford dealers was only 85% of the retail price. (This information is available from a variety of sources.) She also learned that destination charge is the same for the dealer and the customer. It is important to know what the dealer price is for the car. Any amount above the dealer's cost is money in the dealer's pocket.

9. a) Calculate the dealer invoice price for the vehicle described in Figure 6.2.

   b) Calculate the dealer invoice price for the vehicle described in Figure 6.3.

10. A dealership near Houston is willing to reduce its *profits* on each new car (including SUVs) by 25%, because it does not offer incentives to the customer.

    a) Is it less expensive to purchase a 2001 Explorer 2WD XLS 4D from the Houston area dealer or accept the local dealer's offer described in question 7(a)?

    b) Is it less expensive to purchase a 2001 Explorer 2WD XLT 4D from the Houston area dealer or accept the local dealer's offer described in question 7(b)?

The final cost to consider is **sales tax**, which is a percentage of the selling price that is paid to the government for conducting the transaction. That amount must be included in the price that Maria must pay for buying the car. In Texas the state sales tax is 6.25%.

11. a) Assume the selling price that Maria and the car dealer agree on is $25,820.00. How much sales tax would need to be included in this case?

    b) What would be the final amount that Maria owes the dealership for the car?

SUMMARY

In this activity you learned that many factors determine the price of a car.

- You learned about how the destination charge, optional equipment, and incentives affect the price of a vehicle.
- You calculated the selling price of several different deals to figure out which one was the best.

Now that Maria has agreed on the price of the SUV, the next problem is how she will pay for it! In the next activity you will investigate the loan options that are available.

HOMEWORK

6.1 A Sticker Will Get You Quicker

1. It is hard to tell if the Figure 6.1 prices used in Activity 6.1 are correct for that car without seeing the actual stickers. But you can determine if they are *reasonable*, by considering if they are possible.

   a) Is the most expensive XLS listed in Figure 6.1 cheaper than the most expensive XLS possible based on the information contained in Handout 6.1? Explain.

   b) Is the most expensive XLT listed in Figure 6.1 cheaper than the most expensive XLT possible based on the information contained in Handout 6.2? Explain.

   c) Is the least expensive XLS listed in Figure 6.1 more expensive than the cheapest XLS possible based on the information contained in Handout 6.1? Explain.

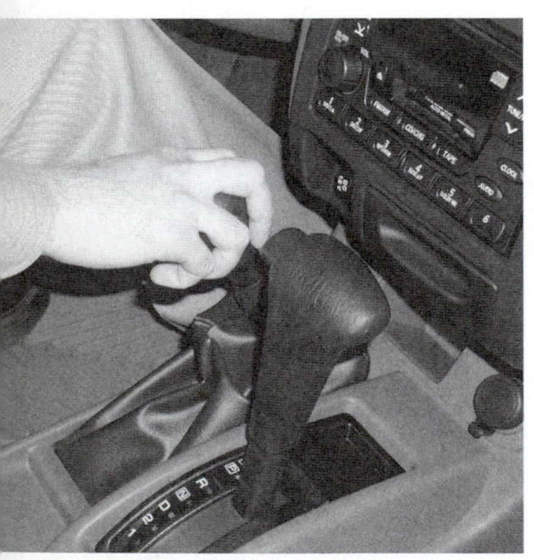

2. a) Which features are standard equipment on the XLT and are optional equipment on the XLS?

   b) Which features are standard equipment on the XLT and are not available on the XLS?

   c) Which options are available on the XLT and are not available on the XLS?

3. At one dealership Maria found the sticker information for a 2001 Explorer 4DR 4x2 XLS with Amazon green exterior and tan interior. It listed these accessories: license plate bracket, axle, 5-speed automatic O/D trans., and 4.0L SOHC V6. The dealer invoice was given as $24,020.00, and the list price was $26,080.00.

   a) In what way is the sticker information misleading or inaccurate?

   b) Why would the dealer prepare a sticker with this type of information?

   c) What is a fair selling price for this car?

4. Maria finds a 2001 Explorer 2WD XLT 4D with a V8 engine on the car lot. It has the following options: moon roof, towing package, garage door opener, AM/FM Stereo w/CD and cassette, 6-CD changer, and leather seats. She qualifies for the $1500 customer incentive, and agrees to pay full price for the car. How much will this car cost her?

Remember that Carla is shopping for a Nissan Xterra. Use Handouts 6.3–6.5 to help answer the following questions.

5. When she began shopping, Carla didn't realize there were two different series: the XE and the SE. There are also two different engine sizes, and two different types of transmissions.

   a) Can you get an automatic transmission on a 4-cylinder Xterra XE? Explain.

   b) What exactly is included in the Sports Package option? In which series is it standard equipment?

   c) Which is more expensive, the XE or the SE series? By how much?

   d) What is the price of the features that are options on one vehicle but are standard on the other vehicle?

   e) What accounts for the extra cost with the more expensive vehicle?

   f) Which is more expensive, automatic transmission or 5-speed manual transmission? By how much? Does that answer depend on whether you buy the XE or the SE series?

6. Why might the destination charge be less for the Nissan Xterra than for the Ford Explorer?

7. a) Remember that Carla wants her Xterra to be fully loaded. For now she is also thinking about high performance, and thinks it is "cool" to shift into higher gears with a manual transmission. What would be the selling price for a fully-loaded 2001 Xterra XE 2WD 4D?

   b) What would be the selling price for a fully-loaded 2001 Xterra SE 2WD 4D?

8. Carla had access to the same information source that her sister used, and found out that there is also a difference between the dealer invoice price and the retail price for the Xterra. However, there didn't seem to be a simple formula for the relationship between invoice price and retail price. The dealer invoice prices are listed in **Figure 6.4**.

Xterra XE 2WD 4D Features	Invoice Price
w/Automatic Transmission	$ 18,647
w/5-Speed Manual Transmission	$ 17,717
Protection Package	$ 83
Sport Package	$ 737
Utility Package	$ 606
Power Package	$ 1126
Floor Mats	$ 58
Microfilter, In-Cabin	$ 31
Moldings, Body Side	$ 63
Mud Guards	$ 42
6-Disc CD Changer	$ 355
Tow Hitch	$ 237

Xterra SE 2WD 4D Features	Invoice Price
w/Automatic Transmission	$ 22,161
w/5-Speed Manual Transmission	$ 21,242
Protection Package	$ 83
Floor Mats	$ 58
Grille/Taillight Guards	$ 310
Microfilter, In-Cabin	$ 31
Mirror, Inside Auto. Day/Night	$ 161
Mud Guards	$ 42
Tow Hitch	$ 237

FIGURE 6.4.
Invoice prices for Nissan Xterra.

a) How much does a fully-loaded Xterra XE with 5-speed manual transmission cost the dealer?

b) How much profit can be made by selling this car at full price?

c) How much would a fully-loaded Xterra SE with 5-speed manual transmission cost the dealer?

d) How much profit can be made by selling this car at full price?

e) With which of the two situations (a fully-loaded Xterra XE or a fully-loaded Xterra SE) is the dealer more likely to "swing a deal?" Explain.

9. Carla decides to buy the fully-loaded Xterra SE. She is a shrewd negotiator, and convinces the car dealer to lower the sticker price by 5%. How much is the *total* purchase price now, including sales tax?

10. Figure 6.1 included sticker prices for a lot of cars. Possible explanations for the differences in the various prices include the options and different dealers having different profits. But Carla came across an unusual situation at one dealership involving two identical Xterras. They had the exact same options (power brakes, power steering, air conditioning, anti-lock brakes, power windows) and same engine size. Both came with 4 doors, and they even had the same color! One of them was listed for $24,061, while the other was listed at $24,777.

    a) Were these cars from the XE series or the SE series?

    b) What could account for the difference in the car prices?

▪ Preparing for the Chapter Project

11. As part of the chapter project, you will need to select a car to research. You will need to find the following information:

    - the car's general features and specifications
    - standard equipment for that make and model
    - package accessories and other options
    - incentives that are available
    - differences between the invoice and retail prices
    - destination charges.

You might gather newspaper ads for that car from several dealers in your area for comparison, or you could visit dealerships and talk with the sales representatives. For the type of car you choose, you want to know "How much will a new car cost? What factors affect how much it will cost? How can I make the cost be as small as possible?"

ACTIVITY
6.2

Not Enough Money? No Problem!

Figuring out which car to buy is certainly an important part of buying a car. However the purchase is only the first step. You also have to pay for it! If you're like Maria you don't have an extra $25,000 ready to spend. In this activity you will learn what is required to get a car loan.

GETTING THE RIGHT LOAN

Many businesses, such as banks and other lending companies, loan money to buy a car. Getting a loan to buy a car is called **financing** the car purchase. For the privilege of getting a loan you must pay an extra amount, called the **finance charges**, in addition to the amount of the loan. When you finance your purchase, your car will cost more than you thought it would. In researching her car purchase Maria found out that the **APR** (annual percentage rate) plays a big role in financing the purchase.

CHECK THIS!

To finance a car purchase, you must qualify for the loan. The lender looks at how long you have been working, how much money you make, and your history making payments on credit cards to see whether you will be able to make the payments on the loan.

1. a) Assume Maria settles on a car whose selling price is $27,000. She considers a loan for $27,000 with an APR of 6.8%. What is 6.8% of $27,000?

   b) If an entire year went by and no payments were made, the amount you just calculated would be the **interest due**. This extra amount is charged because the loan had not been paid off completely. What is the total amount that she would owe?

Question 1 gives you an idea about APR, but the situation is not realistic. You have to make monthly payments or you lose the car. Each month you will owe some of that interest. Each monthly payment will pay for some interest *plus* repay some of the loan. As the amount you owe goes down you make the same payment, but you will owe less interest; therefore you will repay more of the loan. By the end of the **loan term**, the amount of time it takes to pay off the loan, there will be nothing left to pay. (This kind of financing will be studied in more depth in Chapter 7, "Growth: From Money to Moose.")

Handout 6.6 is a portion of a **finance table**, which provides how much the monthly payment should be, based on the amount financed and the APR for a four-year loan term. While at the car dealership, Maria discusses her options with a loan officer from the finance division.

2. a) Assume the loan from question 1 is going to be paid off in four years. How much would Maria's monthly loan payment be?

   b) How much would she pay for the loan over the entire loan term?

   c) How much did it cost her to finance the car instead of paying cash?

3. The same finance division offers a four-year loan and a five-year car loan. Handout 6.7 is a portion of another finance table for the same amounts and the same range of APR values, but assumes a loan term of five years instead.

   a) How much is Maria's monthly payment for a five-year loan?

   b) Over the entire loan term, how much would she pay?

   c) If Maria wants to keep the total amount she pays for the car as low as possible, which loan term should she take? Explain.

   d) If Maria earns *just* enough money to pay her bills each month, which loan term should she take? Explain.

One important financing decision is the length of the loan term. If Maria chooses a short-term loan, the car will be paid off more quickly and cost less money. However, the monthly payments will be higher and more difficult to manage within a budget. Longer terms will cost more money, but the payments will be easier to make.

Consumers often shop around for lower APR financing. This strategy makes sense: The lower the interest rate, the lower the amount of interest paid (and that is where the extra costs come from). Maria found the Acme Loan Company offers a slightly lower APR for the same amount of money and loan term.

4. a) Assume the amount of Maria's loan is still $27,000 and the loan term is four years, but the Acme loan APR is 6.6%. What would be her monthly payment?

   b) How much less is this payment than what she would pay by financing at the car dealership?

   c) Over the loan term, how much would she save by going with a lower financing arrangement?

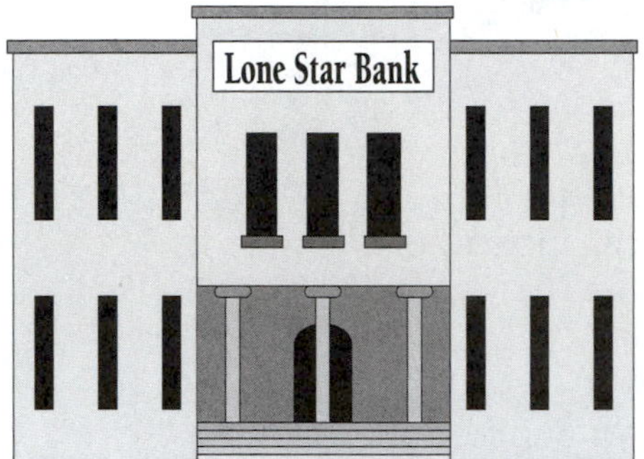

Maria has a savings account at the Lone Star Bank, and she is eligible to apply for a car loan from the bank. They are offering different financing options that depend on the loan term. Assume the amount to be financed is still $27,000.

5. a) For a four-year term, the bank is offering to loan the money at 6.4% APR. How much will the car cost in this agreement?

   b) For a five-year term, the bank is offering the money at 6.0% APR. How much will the car cost in this agreement?

   c) Which of the financing options would be better for Maria at the Lone Star Bank? Explain.

The Ford Motor Company is currently offering a customer incentive program with rates between 0.9% and 1.9% APR. All customers are eligible for this special deal (for a short time only). Since the Explorer is a popular SUV, car loans for this vehicle would have an APR of 1.9%. For a loan amount of $27,000, Maria's monthly payments would be $584.59 for a four-year loan.

6. How much would Maria save in financing the car through the manufacturer compared to the offer made by the car dealership in question 2?

It is clear that customers like Maria understand the fact that for a given loan amount and term, the lowest APR available will be the best financing option. Occasionally a customer will arrange financing at a dealership just to buy the car, and then drive their new car to a bank to finance the car at a lower rate!

When she studied the interest rate tables, Maria noticed that the monthly payment goes down if the amount being financed decreases. For car purchases, one way to reduce the loan amount is to save some money as a **down payment**. By paying some money in cash, the amount that needs to be financed is less.

7. Maria has saved $2000 over the last couple of years. She would rather not use it as a down payment, because then she would not have any extra money for emergencies. Nevertheless she is curious to know how much she would save if she *did* apply the money to the car purchase.

   a) For a selling price of $27,000, if the savings were to be used as a down payment, how much money would she need to borrow?

   b) Assume that Maria takes a loan from the car dealership at 6.8% APR financing over a five-year term. What would be the monthly payments for the amount financed?

   c) How much would she pay altogether for the car? (Hint: don't forget the down payment.)

   d) How does this arrangement compare with the answer found in question 3(b), when there was no down payment?

If a customer owns a car, it can be traded in to reduce the price of the new car. In that case the customer and dealer must agree to a cash value for the car. That cash value is then applied to the price of the new car thereby reducing the amount that needs to be borrowed. (The factors that are used to estimate the worth of a car will be explored in a later activity.)

8. Maria's current car is a Ford Taurus, which the dealer estimates is worth $5000 in trade-in value. Assume the selling price for the car is still $27,000, and the car dealership is offering 6.8% APR financing over a five-year period.

   a) If the car is traded-in and no down payment is made, what will be the monthly payments?

   b) With this financing arrangement, how much altogether will the car now cost Maria?

   c) How much would she save by trading in the car compared to no trade-in or down payment being used?

d) Maria thinks her old car is worth more than what the dealer is willing to offer her. She knows she can get at least $6000 if she sells it herself, and she may even be able to find a buyer at $7000. Should she finance the entire cost of her new car and try to sell the old one herself, or should she accept the dealer's trade-in value in the financing arrangement? Explain.

SUMMARY

In this activity you investigated how a car loan is financed.

- You found there are three factors that affect the monthly loan payments: the amount of money borrowed, the APR, and the loan term.
- You compared different loan terms and APRs to determine the total cost of each loan.
- You investigated how a down payment or a trade-in car affects the cost of a loan.

To be a smart consumer, you must compare loans to find the one that gives you the lowest total cost. Now that Maria has made arrangements for buying the car, she must determine how much money to budget to operate and maintain it. You will examine these issues in Activity 6.3.

HOMEWORK

6.2 "Hi-Fi" Nance

Carla has been considering how to finance the Nissan Xterra. Remember that she wants to buy a fully-loaded car, but isn't sure if the SE model is worth the extra money. So she must consider both the XE and the SE models. With sales tax, the list price for a fully-loaded XE is $25,095.19, and a fully-loaded SE is $26,786.69.

1. a) Use the rounded figure of $25,000 for the selling price of the Nissan Xterra XE that Carla is thinking of buying. If the car dealership is offering financing at 7.4% APR over four years, how much will the car cost Carla?

   b) If the car loan is financed over five years instead of the original term of four years, how much will she save each month in car payments?

   c) How much more will the car cost Carla if the loan is financed over five years instead of a four-year term?

2. a) Use the rounded figure of $27,000 for the selling price of the Nissan Xterra SE that Carla is also considering buying. A local bank is offering five-year loans at 7.0% APR. How much would the car cost Carla with this financing?

   b) A local savings and loan company is advertising five-year loans at 6.6% APR. How much would Carla save by borrowing the money from the savings and loan company instead of the bank?

3. Carla *really* wants the Xterra SE if she can afford it. She has been saving money for the day when she can finally get one, and has $1000 for a down payment. She also has a fairly old Nissan Sentra that the dealer will take as a $3000 trade-in towards the new car. By shopping around, the lowest financing she found was 6.4% APR for a four-year term. If she applies both the down payment and the trade-in to the car purchase and finances the rest, how much will the car finally cost Carla?

4. One tip that consumer groups suggest for cutting costs is to purchase accessory options like car stereos *after* you buy the car. Assume Maria buys the basic-model Explorer XLT. With sales tax, the selling price is around $30,000. She will use her old car's trade-in value of $3000 and finance the rest with a five-year loan at 7.0% APR. She has two choices for installing the "killer" sound system.

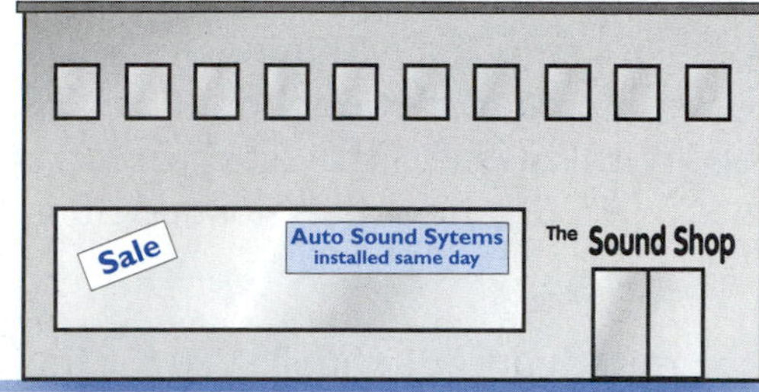

a) One option is to drive her new car to a store that sells car stereos, and buy a Mach stereo system with 6-CD Changer for $1000 (installation included). How much will that option cost her excluding the cost of the car?

b) A second option is to have the dealer install the Mach stereo system and 6-CD Changer as options. The selling price will go up by around $1000. How much extra will she pay as a result of having to borrow more money to buy the car?

c) A third option is to have the dealer install the sound system as an option, but to match the additional cost by making a $1000 down payment. How much will that option cost her above the cost of the car?

d) What assumption is the consumer group making when they recommend you buy the optional stereo *after* you buy the car?

5. In modeling the total cost of a new car purchase, there are a couple of ways to think about the calculations:

- find the selling price of the car and the total finance charge, and add them together
- find the total of all the payments made by multiplying the monthly payment by the number of months in the loan term

What is the advantage (and disadvantage) of each method of calculating the total cost?

6. Describe how to calculate each of the following situations. Either write clear directions of what you need to do or write those steps as mathematical equations. The labels in **Figure 6.5** may help you communicate your ideas.

Label	What It Stands For
NCP	Price of a New Car
DP	Down Payment
TIV	Trade-in Value
RSV	Resale Value
AF	Amount Financed
APR	Annual Percentage Rate
MP(AF, APR)	Monthly Payment (from the table) for a particular Finance Amount and APR
NP	Number of Payments
TCC	Total Cost of the Car

FIGURE 6.5.

a) Find the Total Cost of the Car if the old car is traded-in.

b) Find the Total Cost of the Car if the old car is resold.

▪ Preparing for the Chapter Project

7. As part of the chapter project you will need to know what the local lending sources are, the rates they are currently offering, the terms for their loans, and any other alternatives available for financing a car. Another factor to consider is the conditions required to qualify for a loan. You may want to interview a finance agent at a car dealership, or a loan officer at the local savings and loan or bank to find out how to qualify for a loan.

ACTIVITY

6.3

Those Nickels and Dimes Add Up

In the first two activities you determined the price of a car and how to pay for it. Now you will consider some completely new costs that must be paid to keep the car on the road. In this activity you will investigate how to calculate the costs of operating and maintaining a car.

ESTIMATING OPERATING AND MAINTENANCE COSTS

Maria spent a lot of time and energy figuring out what car to buy and how to pay for it. The reward for her efforts was her new 2001 Ford Explorer XLS. She expected to make car payments for the next five years. But after a while it occurred to her that she was still getting her wallet out, and *way* too often. She realized that she'd better model her **operating costs**—how much it costs to keep the car on the road!

Maria found the first "hand" reaching out for her money was the state government. All car owners in the state must pay a **registration and license** fee each year. For passenger vehicles in Texas this fee is based on the weight of the car. If the car weighs less than 6000 pounds, the fee is $58.50. For cars above that weight, you pay $25 plus $0.60 for every 100 pounds above 6000 pounds. In some areas of the state an additional amount is included to pay for upgrades on roads, public transportation projects, or the installation of computer systems.

1. a) The general specifications for the 2001 Ford Explorer XLS are listed in Handout 6.1. How much will Maria pay each year in registration and license fees?

   b) Over the five years that Maria will be making car payments, how much will she pay altogether for registration and license fees?

The most obvious expense Maria has is her **fuel cost**, the amount spent on gasoline. Gasoline is consumed at an average rate expressed in miles per gallon.

Each type of car tested by the Environmental Protection Agency (EPA) is given a mileage rating. This rating measures what that average rate of gasoline consumption would be. Since the consumption rate is affected by the type of driving conditions, the rating is expressed as two numbers: One rate for city traffic and one rate for freeway driving. The EPA mileage ratings for both city and highway travel for Maria's 2001 Ford Explorer XLS are also provided in the general specifications in Handout 6.1.

CHECK THIS!

Miles Per Gallon (MPG) is a fuel efficiency ratio of miles driven for each gallon of gasoline used. These ratings are only estimates of gasoline usage. Actual miles per gallon (MPG) will depend on a number of factors including driving conditions and driving style. In general, drivers who speed use more gasoline and therefore will have lower fuel efficiency.

2. In order to determine costs from mileage ratings, Maria has to make two assumptions. First, based on her normal driving habits, she figures she will drive about 8000 miles per year on freeways and 4000 miles per year in the city.

   a) Estimate how many gallons of gas she consumes per year from driving in the city.

   b) Estimate how many gallons of gas she consumes per year from driving on freeways.

   c) Besides the "break-down" on the number of miles, what assumptions must Maria make to use the EPA ratings?

3. The other assumption has to do with how much she pays for the gasoline she uses. Maria found a study from November, 2000 that said that the average pump price for regular unleaded gasoline across Texas was $1.43 per gallon. She used that estimate to price her costs.

   a) For city usage, how much will Maria's gasoline consumption cost per year?

   b) For freeway usage, how much will Maria's gasoline consumption cost per year?

   c) How much will Maria spend on gasoline consumption each year?

   d) How much will Maria spend on gasoline consumption over the five years she will be making car payments?

4. If Maria's mileage estimates were actually 6000 freeway miles and 6000 city miles per year, how much would she spend on gasoline consumption over the same five years?

Maria knew she was spending money all the time, but she didn't expect it was *that* much money! Besides gasoline there are other regular operating costs. Since these costs involve maintaining the vehicle in good condition, they are called **maintenance costs.** **Figure 6.6** shows the expenses that Maria identified.

Operating Expense	Recommended Schedule	Maintenance Cost
Change the Oil	every 3000 miles	$20/job
Replace the Tires	every 40,000 miles	$90/tire + sales tax
Replace the Battery	every three years	$70/job

FIGURE 6.6. Maintenance costs for Maria's SUV.

5. Maria wants to determine her total maintenance costs for the five-year period during which she is making car payments.

   a) How much will she spend for oil changes?

   b) How much will she spend to replace the tires?

   c) How much will she spend for batteries?

   d) How much will her total maintenance costs be for the five years?

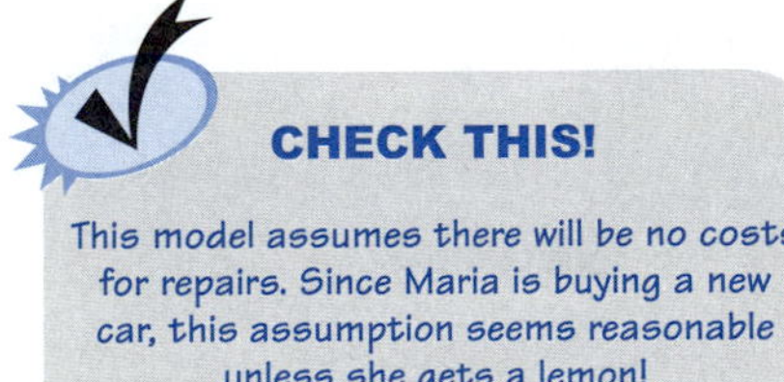

CHECK THIS!

This model assumes there will be no costs for repairs. Since Maria is buying a new car, this assumption seems reasonable unless she gets a lemon!

One obvious maintenance cost is the periodic service required by the car manufacturer as part of the new-car warranty agreement. Assume that Maria talked the car dealer into paying for those "check-ups" during contract negotiations, so she doesn't have to worry about them!

CALCULATING INSURANCE COSTS

The last major cost that Maria discovered is that *once* you own a car you must pay for **insurance.** An insurance policy is required to protect the car in case of accidents, theft, or other damage. State law requires certain types and amounts of coverage, and the banking institution that lent Maria the money requires coverage to protect their investment. There are many factors that determine the insurance cost including age, gender, and previous driving record as well as how many people are covered on the policy, and whether the driver is a "good student." The typical kinds of coverage are described in Handout 6.8.

Maria received a quote from an insurance company. The conditions and coverages she assumed are listed in **Figure 6.7**.

Conditions
17-year old living at home
No traffic violations
Drive the car 12,000 mi./year
Drive the car 11-20 mi. each way to school

Coverages	Amounts
Uninsured Motorist (bodily injury)	$50,000/$100,000
Uninsured Motorist (property damage)	$25,000 w/$250 deductible
Collision	$250 deductible

FIGURE 6.7.
Conditions and coverages of an insurance policy.

6. The quote Maria received was for six months of coverage. If paid in full at the beginning of the six-month period, it would cost her $2409.50. Also she could pay the cost in installments, but it would cost a little more for the privilege (naturally).

   a) If she continually paid off the insurance in full, how much would her insurance cost for the five years in which she makes car payments?

   b) One installment plan requires a down payment of $1105.70 and two additional payments of $798.40. If Maria used this plan, how much would her insurance cost for the five years in which she makes car payments?

   c) A second installment plan includes a down payment of $791.78, three payments of $481.84, and one payment of $479.20. If Maria used this plan, how much would her insurance cost for the five years in which she makes car payments?

Maria figures that her work is detailed enough to be a good estimate of her operating costs for the next five years. But she still has to "put it together" and figure out the total.

7. Remember that the operating costs include the registration, fuel costs, maintenance costs, and insurance costs. *All* of those costs are separate from the cost of the car, which Maria figured would total around $27,500.

a) What is the total of the operating costs for the five-year period in which Maria is making car payments?

b) Including the car payments, how much will Maria be spending on car-related costs during that five-year period?

c) How much would her **annual cost** be during that time, that is, the car-related costs each year?

d) Sometimes the cost is calculated with respect to distance traveled (rather than the time interval that has elapsed). What would her **cost per mile** be over that five-year period?

8. a) Maria plans to keep the car for an additional five years *after* the car is paid off. Do you think her average car costs will be the same when she calculates it over ten years? Explain.

   b) Some people think you should buy a new car as soon as you've finished paying off the previous one. With that strategy you never have to worry about the car getting old. How does the average cost calculation help decide whether this is a good strategy?

SUMMARY

In this activity you investigated the expenses of operating a car.

- You learned how to analyze the various categories that make up the operating costs.
- You used two methods for measuring (and comparing) costs that you can use to analyze different possibilities: The annual cost (cost per year), and the cost per mile.

Now you have a good idea about the costs of buying and operating a car. As your car ages it loses value. You must take this fact into consideration when considering your financial status. In the next activity you will learn ways to estimate the value of a used car.

HOMEWORK

6.3

Looking Down the Road

Carla is interested in comparing the cost of owning her car for four years, versus owning it for ten years. Following are the facts and assumptions upon which her model is based.

- She is buying a fully-loaded 2001 Nissan Xterra SE for $26,000.
- She is financing the entire amount over four years at an interest rate of 6.4% APR.
- She drives an average of 10,000 miles per year, with 6000 of those miles in city traffic.
- The average pump price for regular unleaded gasoline is $1.43 per gallon in Texas.
- To estimate operating costs, she uses the same maintenance schedule as Maria.
- For the same set of conditions and coverages, the insurance company quoted her this installment plan for a six-month policy: down payment of $773.51, three payments of $465.14, and one payment of $462.57.

1. First Carla must calculate the costs for owning the car for only four years. Make a table similiar to the one in **Figure 6.8** to record the answers to the following questions.

   a) Determine Carla's monthly payment, and calculate the total of the car payments.

FIGURE 6.8.

Category	4-Year Total
Purchase Cost	
Registration Cost	
Fuel Cost	
Maintenance Cost	
Insurance Cost	
4-Year Total:	

b) How much will she pay in registration costs?

c) How much will she pay in fuel costs?

d) How much will she pay in maintenance costs?

e) How much will she pay in insurance costs?

f) For the four years during which she is making car payments, what will be her total cost?

2. a) Refer to the calculations that you recorded in Figure 6.8. Over this four-year period, what is the annual cost?

   b) Over this four-year period, what is the cost per mile?

3. a) Next, Carla must calculate the costs of owning the car for a total of ten years. Determine her monthly payment, and calculate the total of the car payments. Record that in a table like the one shown in **Figure 6.9** as the purchase cost.

Category	10-Year Total
Purchase Cost	
Registration Cost	
Fuel Cost	
Maintenance Cost	
Insurance Cost	
10-Year Total:	

FIGURE 6.9.

   b) How much will she pay in registration costs?

   c) How much will she pay in fuel costs?

   d) How much will she pay in maintenance costs?

   e) How much will she pay in insurance costs?

   f) Over the entire ten years, what will be her total cost?

4. a) Refer to the calculations you recorded in Figure 6.9. Over this ten-year period, what is the annual cost?

   b) Over this ten-year period, what is the cost per mile?

5. Based on the work done in questions 1–4, is it better for Carla to keep the car for four years or ten years? Explain.

6. For each of the following cases, calculate what the fuel cost would be for Carla during the first four years. How do your answers compare to what you recorded in Figure 6.8?

   a) She still drives 10,000 miles per year, but 6000 of those miles are on freeways.

b) She drives 15,000 miles per year, with 6000 of those miles in city traffic.

c) She uses her original mileage estimates, but the car only gets 14 miles per gallon in city traffic and 20 miles per gallon on freeways.

d) She uses her original mileage estimates, but the average pump price for the gasoline is actually $1.48 per gallon.

7. In Activity 6.3 the various operating costs used two different input variables: time (in years) or distance (in miles).

   a) Which of the operating costs used the number of years as the input variable?

   b) Which of the operating costs used the number of miles traveled as the input variable?

   c) What assumption allowed Maria to convert one input variable quantity into the other?

▪ Preparing for the Chapter Project

8. As part of the chapter project, you will need to perform the same calculations that Maria did.

   - Estimate the average number of miles you figure on using your vehicle each year. How many of those miles will be driven in city traffic and how many will be on the freeway? You will need the EPA mileage ratings for your car.

   - You will want to research insurance coverages, and maybe get estimates from several companies. Interview the insurance sales person with whom your family insures their cars.

   - If you are planning on a new car purchase, find the cost for the scheduled maintenance that is needed as part of the warranty agreement.

   - If there are any additional, unusual costs that have not been anticipated, research what they would be and how much they will cost.

   - Will your model be based on the number of years you use the car, how many miles it is driven, or whether it can accommodate both inputs?

ACTIVITY 6.4

The Years Take Their Toll

Maria realizes that she spends all this money to buy and operate her wonderful Explorer, and the value of the car drops. In fact the value of the new Explorer decreases a lot quicker than she expected. She needs to understand how the value of her SUV changes. In this activity you will research the factors that determine the value of a used car.

CALCULATING DEPRECIATION

Maria has done a lot to understand the cost of owning and operating a car. But one piece of the puzzle still remains unclear to her. When she was estimating the trade-in value of her old Ford Taurus, the car dealer thought it was worth one thing and she thought it was worth another. How do people estimate the value of a car once it is no longer new?

1. a) Assume Maria is thinking of buying a new 2001 Ford Explorer XLS that is worth $25,000. She buys the car and drives it off the lot. A few days later she changes her mind and wants to return the car to the dealership. Is it still a new car, or is it a used car?

   b) Is the car still worth $25,000?

Maria is surprised to find out that the value of the car goes down as soon as the car leaves the lot. When the car is purchased and leaves the lot, it is no longer new. Maria found that the car's value had decreased by about $2000.

CHECK THIS!

It is possible to find low-mileage used cars on the lot, even for models of the current model year. If you compare the price of a low-mileage car that is less than a year old with new car sticker prices, you can find how much value a new car loses when it is bought.

2. Maria can't afford to "give away" money, so she decides to keep the car anyway. How much is Maria's Ford Explorer XLS worth now?

In Maria's research, driving the car off the lot is the only example of "losing something for nothing." However she also found that the car loses value all the time, simply by getting older. This is called **depreciation**. There are mathematical rules that determine how much value a car will lose as it ages. One method is to select a number of years over which you calculate the depreciation. Reduce the value of the car by the same equal amount each year until it has no value.

3. Maria would like to apply the idea of depreciation to her 2001 Ford Explorer XLS. She decides to use five years, the number of years she is financing the car.

   a) To calculate the depreciation on the car, should she work with the new car price ($25,000) as its starting value or the value it has when it leaves the lot? Explain.

   b) How much should the value of the car decrease over the five years?

   c) How much should the value of the car decrease each year? Explain how you arrived at that answer.

   d) Calculate how much the car will be worth each year. Record your answers in a table the same as **Figure 6.10**.

   e) Make a scatter plot graph of the value V of the car for the five years in which Maria is making payments. Sketch the graph on a grid like the one shown in **Figure 6.11**.

Year	Value of Car
0	
1	
2	
3	
4	
5	

FIGURE 6.10.

4. a) The scatter plot graph should have a familiar *form*. Why do they call this method **linear depreciation**?

   b) What is the slope of the line through the points? Why is your answer a negative number?

   c) How does the sign of the slope relate to the shape of the graph?

   d) What is the equation for the line that goes through the points?

   e) Why might the equation be useful for determining the value of the car?

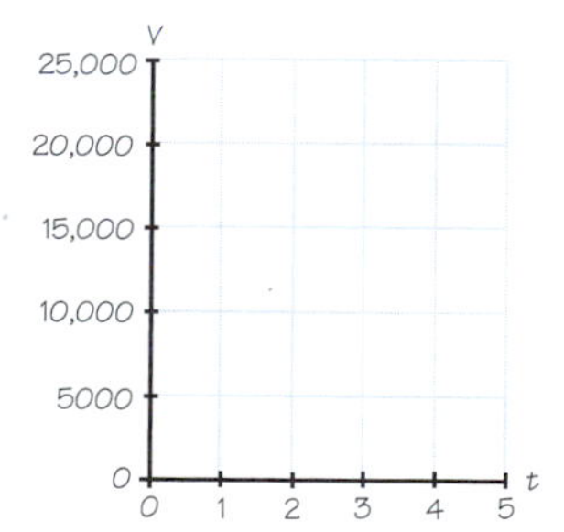

FIGURE 6.11.

Maria found that people use another method for calculating depreciation was to reduce the value of the car by a certain percentage of its value *for that year*. Using this method, as the value of the car goes down, so will the annual amount that you subtract.

Year	Value of Car
0	
1	
2	
3	
4	
5	

FIGURE 6.12.

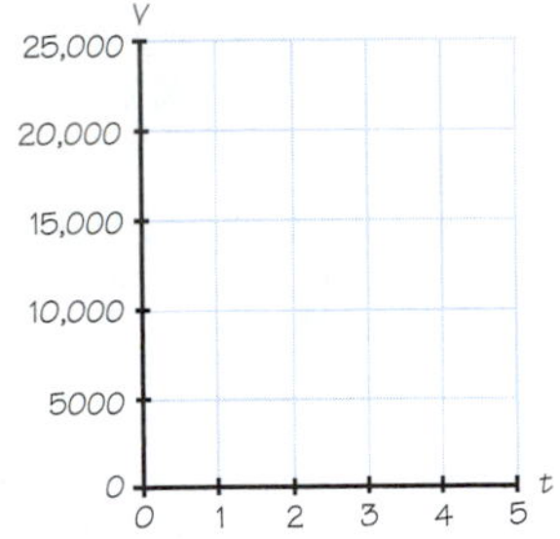

FIGURE 6.13.

5. Maria wants to see how this method would work for her 2001 Ford Explorer XLS. She decides to use 20% as the rate of depreciation for her calculations. Make a table the same as **Figure 6.12** to record your answers.

   a) Calculate the value of the car at the end of the first year by subtracting 20% of the Year 0 value. (Use the same starting value as in question 3.)

   b) Calculate the value of the car at the end of the second year by taking away 20% of the Year 1 value.

   c) Repeat this pattern for the remaining years.

   d) Make a scatter plot graph of the value V of the car for the five years in which Maria is making payments. Sketch the graph on a grid like the one shown in **Figure 6.13**.

The pattern of this scatter plot is probably new. In Chapter 7 "Growth: From Money to Moose" you will study this pattern more closely and develop an equation to describe it.

6. a) The form of the last scatter plot is different than the one that you made in Figure 6.11. Describe the general pattern in the scatter plot.

   b) When you take away 100%, you end up with nothing remaining. Explain why taking away 20% five times in a row did not give you the same result.

   c) Will this pattern ever end up with the value of the car being $0? Explain.

To further understand the effect of depreciation in modeling her costs, Maria decided that she needed data on Ford Explorers as they age. For a range of several years, she found both the trade-in value and the **retail value**, the price that the dealer would use in trying to resell the car.

CHECK THIS!

Knowing the price a dealer will ask for your car is useful when negotiating a trade-in. You can check automobile advertisements and classified ads to find the retail value of a car like yours. When you check these advertisements try to find a car with a comparable number of miles and in the same general condition.

To determine the retail and trade-in values, Maria made these assumptions: The Explorer is a basic model with a five-speed manual transmission, was driven an average of 12,000 miles per year, and is in good condition. Her findings are summarized in **Figure 6.14**. Since the XLS series was introduced in 1999, the data for previous years uses the Explorer XL values.

Year	Retail Value	Trade-In Value
2000	$21,350	$16,650
1999	$18,850	$14,325
1998	$16,475	$11,955
1997	$14,705	$10,055
1996	$12,895	$ 8445
1995	$11,265	$ 7020

FIGURE 6.14.
Retail and trade-in values of some used Explorers.

7. a) Maria found out a lot about how car dealers set used-car prices by gathering this data. How can you tell the amount of profit that a dealer will make from reselling a used car that was a trade-in?

   b) Does the amount of profit that the dealer makes depend on the age of the car? Explain.

   c) When you trade-in your used car, do you think the car dealer will let you know what the retail value is?

   d) Knowing the trade-in value, how could you estimate the retail value?

8. We can explore the relationship between trade-in value and retail value using the graphing calculator. Using statistical lists, put the trade-in values in list L1 and the retail values in list L2.

   a) If L1 is the x-list and L2 is the y-list, what will the scatter plot graph look like? Sketch a picture of your calculator screen.

   b) Use regression to find an equation of the line of best fit that predicts the retail value from the trade-in value.

   c) How does the equation verify the work you did in question 7? Explain.

Now you can investigate how the new-car calculations that form your model will be affected by depreciation. Maria assumes the new car price is $25,000, and she also assumes that the values in Figure 6.14 will describe her car as it gets older. That assumes the depreciation rate has remained constant. To keep things simple for now, also assume she pays cash for all her car purchases.

9. a) Maria buys a brand-new Ford Explorer in 1995, and trades it in when she buys another new Explorer in 2001. How much did that car actually cost her?

   b) If the original dealer cost of the 1995 Explorer was $20,700, how much has the dealer made on this car including the initial sale and the sale of the trade-in at the retail price?

SUMMARY

By this time you should have a better understanding of the problem of estimating the cost of ownership.

- You learned about how a car depreciates.
- You have investigated some methods you can use to estimate depreciation.

There is one more issue you can examine when looking for a new car. You can have the car of your dreams without actually buying it! Activity 6.5 shows you how.

HOMEWORK

6.4 Value Going Down

1. Carla was interested in examining how the value of the Nissan Xterra decreases over time. However, it is a fairly new type of car, since it was introduced first in 2000. She found the retail value was $27,415 and the trade-in value was $22,105 for a fully-loaded 2000 Nissan Xterra SX.

   a) How does the new car price for the 2001 Xterra compare with the retail and trade-in values for the year-old 2000?

   b) Do you think the selling price for a new 2001 Xterra stayed the same as the new car price for the previous year's model? Explain.

   c) What might account for any differences between the pricing for the 2000 and 2001 models?

2. With no depreciation data available on her favorite car, Carla thought it would make sense to examine a similar car that Nissan also makes, the Pathfinder SE. The information she found about retail and trade-in values for that car are provided in **Figure 6.15**.

Year	Retail Value	Trade-In Value
1999	$24,335	$19,130
1998	$22,820	$16,605
1997	$20,315	$14,390
1996	$17,795	$12,205

FIGURE 6.15. Retail and trade-in values of some used Pathfinders.

   a) Is the amount of profit the dealer makes per car the same as the profit for the Ford Explorer XLS?

   b) Carla wants to estimate the retail value for the used Pathfinders from their trade-in value. Based on the table values, will her predictions be as accurate as what Maria found for the used Ford Explorers? Explain.

   c) Using a graphing calculator, build a model that predicts the retail value from the trade-in value. What kind of model makes sense to use in this situation?

d) What is the equation for your model? How do you know the model fits the data well?

3. a) One way for Carla to estimate how her "dream car" depreciates is to go through the same calculations in Activity 6.4 that Maria did. Use the linear depreciation method over six years, and find out the value of her car each year. (The new car price is $25,211 for a fully-loaded 2001 Xterra SE; remember that it loses value as it is driven home!) Record your work in a table like the one shown in **Figure 6.16**.

Year	Used-Car Value (Linear Depreciation)
0	
1	
2	
3	
4	
5	
6	

FIGURE 6.16.

b) Use the depreciation method with a constant percentage of 15% to find out the value of her car each year. Record your work in a table like **Figure 6.17**.

Year	Used-Car Value (15% Loss Per Year)
0	
1	
2	
3	
4	
5	

FIGURE 6.17.

4. Figure 6.14 from Activity 6.4 contained data that Maria found about the depreciation of Ford Explorer XLS cars. At the time, a scatter plot graph was made to compare the two values for each year. In **Figure 6.18**, you will graph the trade-in and retail values separately to see how each type of value decreases over time.

a) Make a scatter plot of trade-in value versus the age of the vehicle on a grid like the one shown in Figure 6.18. Use round marks (dots) for each point in your graph.

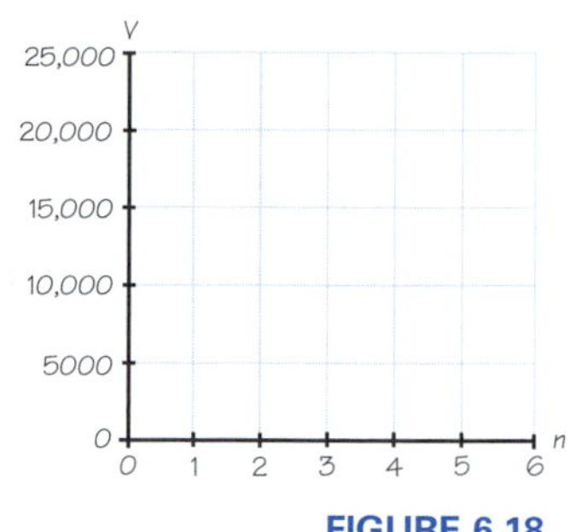

FIGURE 6.18.

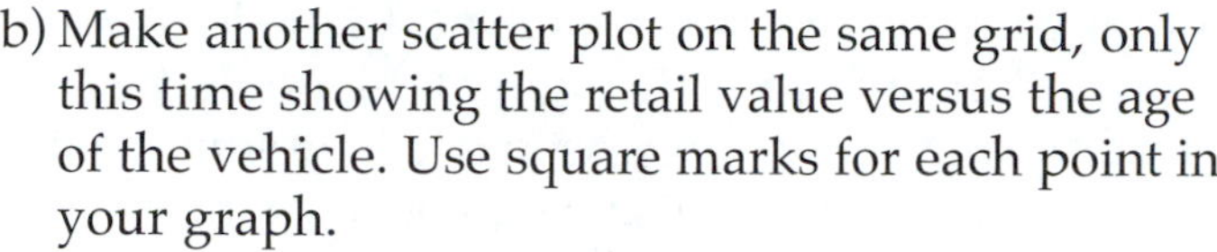

b) Make another scatter plot on the same grid, only this time showing the retail value versus the age of the vehicle. Use square marks for each point in your graph.

c) How do the two scatter plot graphs compare?

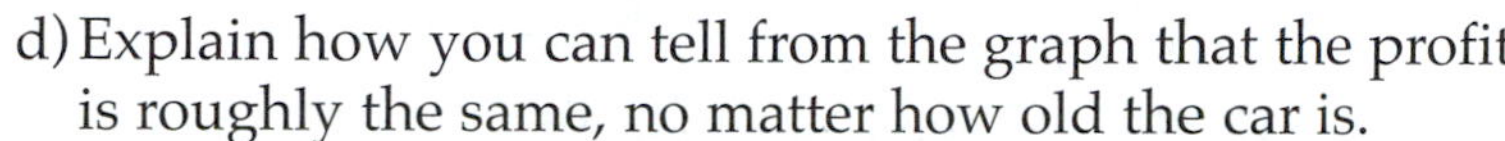

d) Explain how you can tell from the graph that the profit is roughly the same, no matter how old the car is.

e) Which method is more likely to have been used in determining these retail and trade-in values: linear depreciation or the percent-loss method? Explain.

5. In classifying used cars, the conditions in **Figure 6.19** are often used.

Label	Description
Poor	Mechanical problems or visual defects
Fair	Some mechanical problems; visual blemishes
Good	No major mechanical problems or visual blemishes
Excellent	Excellent mechanical condition; glossy paint; perfect interior

FIGURE 6.19.
Conditions of used cars.

In selling a used car yourself, you are in direct competition with the dealers. If your car is in excellent condition, you might be able to sell it at a price that is higher than the retail value. If it is in poor condition, you might not even be able to get the equivalent of the trade-in value. Aside from the factors in Figure 6.19, what other factors affect the selling price for your used car?

6. Assume that Maria buys a brand-new Ford Explorer in 1997 for $25,000 and finances that purchase at 6.8% APR over four years. After she makes the last car payment, she considers trading in her 1997 Ford Explorer for a 2001 model that is also priced at $25,000.

a) Not including operating costs, how much did the first Explorer cost Maria?

b) If the price for a new car is actually increasing over time, then the 1997 Explorer would have cost less to begin with. How would that have affected the trade-in values in Figure 6.14?

c) Assume the 1997 Explorer had cost $4000 less, and assume that each value in Figure 6.14 would be $4000 less as a result. Does it seem reasonable to assume that the trade-in values would also be $4000 less simply because the selling price is reduced? Explain.

d) Excluding operating costs, how much would the 1997 Explorer cost Maria under this scenario? How does this compare to when the selling price is kept constant?

■ Preparing for the Chapter Project

7. As part of the chapter project you are going to have to consider the effect of depreciation as a cost in owning a new car. This also affects the negotiations in buying a new car, as well as when you go to buy another car. You will need to investigate the depreciation methods that are used by car dealers, how they determine their used-car pricing, and what the difference is between the trade-in and retail values for used cars of your particular make and model.

ACTIVITY 6.5

A New Lease on Life

In the first four activities you learned all the elements related to owning a car. There is another way in which you can get a new car. In this activity you will investigate leasing a car.

LEARNING ABOUT LEASES

Even though Maria *really* wants the 2001 Ford Explorer XLS, the amount of money that it takes to own one makes a girl think twice! Fortunately she discovered a possible way to drive her dream car without having to actually buy it.

Most dealers also offer a choice of **leasing** a car. Basically, leasing is like having a long-term rental agreement. The consumer pays a fixed amount each month to the leasing company (who owns the car), and the consumer also pays for insurance and operating costs. At the end of the lease agreement, the company takes the car back. It sounds like a good deal doesn't it? As Maria has found out, it's more complicated than it seems.

1. The first thing Maria discovered about the world of leasing is that there is an entirely different set of terms for saying the same thing! For example, instead of selling price, the term in a lease agreement is the capitalized cost. This term just means the total cost of the car if someone wanted to buy it.

   a) Assume the base sticker price for her car is $25,000 and there are $2000 in options included in the deal, but the dealer will pay the destination charge. What would be the **capitalized cost**?

   b) Maria is prepared to give a down payment of $1000. She is being given a **net trade-in allowance** of $3000 for her old car. In the world of car buying, what is the net trade-in allowance called?

   c) After applying the down payment and the net trade-in allowance to the transaction, how much will her **adjusted capitalized cost** be? (This cost is the amount a buyer would finance.)

d) Assume the **lease term** is 36 months. For that interval, the dealer estimates that the **residual value** is $13,955. The residual value is the amount the dealer thinks the car will be worth at the end of the lease term. In the world of car buying, what is the residual value called?

Maria organized the information in the following formula:

capitalized cost – down payment – net trade-in allowance = adjusted capitalized cost

Both the adjusted capitalized cost and the residual value play an important role in determining the monthly payments.

MONTHLY DEPRECIATION

To calculate the monthly depreciation, subtract the residual value from the adjusted capitalized cost, and then divide by the number of months in the lease.

The monthly payment is one of the most important costs for determining how much the lease will cost. Maria found several formulas are needed to do the calculations, and they aren't too difficult to figure out. First she must find the **monthly depreciation**, which measures how much the value of the car decreases each month.

2. Using the capitalized cost for Maria's Explorer in question 1, what would be Maria's monthly depreciation?

MONTHLY INTEREST CHARGE

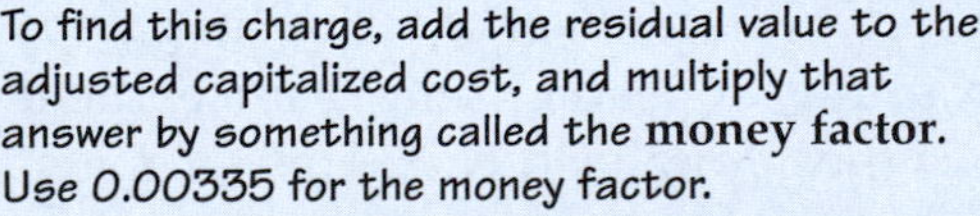

To find this charge, add the residual value to the adjusted capitalized cost, and multiply that answer by something called the **money factor**. Use 0.00335 for the money factor.

The second calculation is to find the **monthly interest charge**, which is also called the rent charge or the lease charge. This charge is like the finance charge in new car buying. Since you are not paying the total amount for the car, you have to pay for that privilege.

3. For Maria's Explorer, what would be her monthly interest charge?

4. a) A consumer must pay sales tax on both the monthly depreciation and the monthly interest charge. Assume the tax rate is 6.25%. How much sales tax would Maria pay per month?

   b) *Now* you can figure out what is the monthly payment amount by adding together the monthly depreciation, the interest charge, and the sales tax. What will be Maria's monthly payment?

c) Over the term of the lease, how much will Maria pay to the leasing company?

5. The money factor is based on the APR that would be used if the car purchase were financed. Unfortunately this factor is not calculated in the same way by all leasing companies. As a rule of thumb, if you multiply the money factor by 2400, you will get the interest rate (or a very good estimate of it).

   a) What interest rate was offered to Maria?

   b) Suppose you know the interest rate, but the dealership won't tell you what the money factor is. However, you need to know it to determine what the monthly payments will be. What should you do?

Maria was pretty pleased with herself, but she knew there must be other factors to consider. That's when she discovered the potential problems associated with leasing a car.

6. Maria learned there are other charges associated with leasing agreements. Some are **front end charges** that you negotiate when you make the deal and drive away the car. Assume that to close the deal, Maria is required to make the $1000 down payment in addition to a refundable security deposit of $500 and the first month's payment. How much money will she need to give the leasing company "up front"?

Maria wasn't worried because she will get the security deposit back if the car is in good condition, and she was going to make that down payment anyway. Sometimes a customer doesn't have the choice, or the down payment can be a fairly large amount.

There are other charges associated with a lease agreement called **back end charges**. A consumer may have to pay these charges when the car is returned to the leasing company.

7. When returning a leased car, one of the penalties is the **extra mileage** charge. Lease agreements specify a certain number of miles that the car can be driven during the term. If the mileage is greater than the miles specified in the agreement, the consumer must pay a flat fee, usually calculated on a "per-mile" basis, such as $0.10 per mile. If Maria is supposed to average 12,000 miles per year, and she actually averaged 13,500 miles per year, how much extra would Maria have to pay as a mileage charge over the term of the lease?

CHECK THIS!

Leasing can be a good option for certain people. The important thing is to read the lease contract carefully, and to understand your driving habits. Make sure to estimate how many miles you plan to drive the car.

Another of the back end charges is for **excess wear and tear**. While the terms are stated in the lease contract, the leasing company usually determines the penalty amount. Another hidden cost is the **early termination fee**, which you pay if you decide to stop leasing the car before the term is finished. It can be a large amount, depending on the contract, and it is usually charged to vehicles that are "totaled" in an accident.

8. The final decision that Maria discovered has to do with **purchase option**. When you return the car at the end of the lease, you have the choice of buying the car from the leasing company. They will provide a purchase option price, which is related to the residual value. If Maria is offered $15,000 as the purchase option price, is that a good deal for her? Explain.

Maria went through that previous example to figure out how leasing works. But she also seriously shopped around for a car to lease. Here was a particularly tempting offer:

Lease a 2001 Ford Explorer XLS, with automatic transmission, 6-cylinder engine, anti-lock brakes, AM/FM radio, air conditioning, and power steering. The MSRP was $25,810, and was being offered at a special APR rate of 0.9% for 48 months with a mileage allowance of 45,000 miles. She would have to make a $1000 down payment and also provide a security deposit of $800. The residual value would be $12,500 with a purchase option price of $11,500.

9. Maria talked the dealer down to a selling price of $25,000. Even though she thought it was worth more, she also included her old Ford Taurus in a trade-in, since the dealer offered $5000 for it.

   a) For this transaction, what is Maria's capitalized cost? Record your answer in a table like **Figure 6.20**.

Lease Conditions	Value
APR	0.9%
Term of lease	48 months
Down Payment	$1000
Security Deposit	$ 800
Mileage Allowance	45,000 miles

Leasing Figures	Amount
MSRP	$25,810.00
Capitalized Cost	
Net Trade-In Allowance	
Adjusted Capitalized Cost	
Residual Value	$12,500.00
Money Factor	

FIGURE 6.20.
Conditions and cost of leasing an Explorer.

b) What is the net trade-in allowance?

c) What is the adjusted capitalized cost?

d) Estimate the money factor for the given APR.

10. a) Maria needs to know what her monthly payments are going to be. Assume the sales tax rate is 6.25%. Calculate the monthly depreciation. Record that value in a table like **Figure 6.21**.

Leasing Calculations	Amount
Monthly Depreciation	
Monthly Interest Charge	
Monthly Sales Tax	
Monthly Payment	

Lease Charges	Amount
Total of Payments	
Excess Mileage Charge	
Excess Wear & Tear Charge	$ 1000.00
Total Charges	

FIGURE 6.21.

b) Calculate the monthly interest charge.

c) Calculate the monthly sales tax.

d) Calculate the monthly payment.

11. In order to compare other kinds of financing, Maria needs to know how much the leasing agreement actually cost her. Assume she drove the car 52,000 miles during the four years, the surcharge for extra miles is $0.12 per mile, and she was charged an additional $1000 for excess wear and tear. Enter the following calculations in your table.

Leasing Figures

	MSRP Amount			
	$15,000	$20,000	$25,000	$30,000
Capitalized Cost				
Net Trade-in Allowance				
Adjusted Capitalized Cost				
Residual Value				
Money Factor				

a) Calculate the total of the payments she made to the leasing company.

b) Calculate the excess mileage charge.

c) Calculate the total cost of leasing the car for four years.

d) Calculate the total cost of owning the car for four years. (Ignore operating costs for now.)

Well, that covers the cost of leasing a 2001 Ford Explorer XLS for four years. Now she can compare this cost with the cost of buying the same kind of car.

12. Maria considers buying a new Explorer XLS with the same selling price ($25,000), down payment ($1000), trade-in allowance ($5000), and finance rate (0.9% APR). Her monthly payments over four years would be $403.15.

    a) How much would it cost her to finance the car for four years?

    b) Recall that the trade-in value for a 1997 Explorer XLS was $10,055. Assuming the 2001 model depreciates in the same way, what was the actual cost of owning the 2001 Explorer for the four years?

SUMMARY

In this activity you examined an alternative way to obtain a new car.

- You learned about all the costs involved in leasing a car.
- You studied the differences between front end charges and back end charges.
- You compared leasing to buying, which are the two basic consumer strategies for obtaining a car.

Since car ownership involves many variables and costs, developing a *model* for the problem of determining the cost of ownership helps you analyze the problem. Now it's your turn. In the chapter project you will analyze the costs associated with the car of your dreams.

HOMEWORK 6.5

The "Lease"r Option

1. a) Having completed Activity 6.5, you have some idea of how leasing works. List some reasons why leasing is popular among consumers.

   b) List some reasons why leasing might not be a good thing.

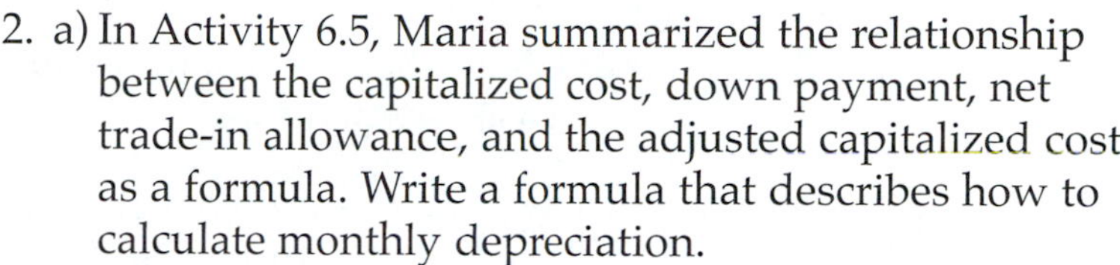

2. a) In Activity 6.5, Maria summarized the relationship between the capitalized cost, down payment, net trade-in allowance, and the adjusted capitalized cost as a formula. Write a formula that describes how to calculate monthly depreciation.

   b) Write a formula that describes how to calculate the monthly interest charge.

   c) Write a formula that describes how to calculate the monthly sales tax.

   d) Write a formula that describes how to calculate the monthly payment.

3. For this problem, refer back to the work done in Activity 6.5 questions 9–11.

   a) If you multiply the monthly depreciation by the lease term, what are you calculating? Do the calculation to find this amount.

   b) What is the average of the adjusted capitalized cost and the residual value? How did you get that answer?

   c) If you multiply the answer to (b) by the APR (as a decimal), you will be calculating the extra amount that you owe at the end of the year from financing some amount. Dividing that number by 12 shows you the amount you owe each month. What answer do you get? Where have you seen that number before?

4. Carla also found dealers that were willing to lease 2001 Nissa Xterra SE's. One dealer offered her the following terms: MSRP of $25,211; 36 month lease at 6.4% APR; 10,000 miles per year; excess mileage surcharge $0.18/mile; residual value of $14,500.

a) Under those terms, how much would Carla's monthly lease payments be?

b) Excluding operating costs, how much would the car cost Carla to drive for the three years?

5. Carla really liked the car and talked the dealer down to a selling price of $24,500 and extending the lease term to 48 months. (The residual value was adjusted to $12,200.) She agreed to use $1000 of her cash savings as a down payment and applied the $3000 trade-in value of her old Nissan Sentra.

   a) Under these terms, how much would Carla's monthly lease payments be?

   b) Assume Carla *really* liked driving the car, and drove 53,000 miles during the lease term. Assume she was also penalized $2100 for excess wear and tear on the car when she returned it. Excluding operating costs, how much would the car cost Carla to drive for the four years?

   c) If she had *bought* the car, and financed what she needed at the same APR, how much would the car have cost Carla during those four years? (Use the residual value to estimate the worth of the car after the lease term has expired.)

   d) Was leasing the car a good option for Carla? Explain.

6. Carla decided that shopping around for the best lease option made sense. She checked with a couple of car dealers, using the same assumptions: Capitalized cost of $24,500, down payment of $2000, with financing of 6.9% APR. Her information is collected in **Figure 6.22**.

	Leasing Company A			Leasing Company B		
Lease Term	**36 mo.**	**48 mo.**	**60 mo.**	**36 mo.**	**48 mo.**	**60 mo.**
Monthly Payment	$ 402	$ 344	$ 316	$ 474	$ 412	$ 373
Money Factor	0.00308	0.00312	0.00319	0.00436	0.00431	0.00427
Mileage Penalty	$0.17/mi.	$0.17/mi.	$0.17/mi.	$0.15/mi.	$0.15/mi.	$0.15/mi.
Purchase Option	$12,849	$12,102	$10,857	$11,747	$10,486	$ 9225
Residual Value	$12,699	$11,952	$10,707	$11,597	$10,337	$ 9076

FIGURE 6.22.

a) Which company is making more money from monthly interest charges? Explain.

b) Which company is making more money from back end penalty charges? Explain.

c) Which company is making more money from reselling leased vehicles as used cars? Explain.

d) If Carla must select between the two companies, which one should she choose? Why?

e) For the company selected in (d), which lease terms cost her the least amount? Why?

7. Carla made a curious discovery about how the APR affects the monthly payments for leasing versus buying. She assumed the adjusted capitalized cost and selling price were both $24,500, a $2000 down payment, and a 48-month leasing (or financing) term. She assumed the residual value was the worth of the car at the end of the lease term, which is $12,000. (Ignore operational costs.) The curious part is shown in **Figure 6.23**.

	APR	0%	2%	4%	6%
Leasing	Monthly Payment	$ 218.75	$ 262.97	$ 293.52	$ 324.06
	Total Cost	$10,500.00	$14,622.56	$16,088.96	$17,554.88
Buying	Monthly Payment	$ 468.75	$ 488.14	$ 508.03	$ 528.41
	Total Cost	$12,500.00	$13,430.72	$14,385.44	$15,363.68

FIGURE 6.23.

a) Verify that the total cost for leasing the car at 4% APR is $16,088.96. Show all your work.

b) Explain how the monthly payments are determined when there is 0% APR financing.

c) How does this information help Carla decide whether to lease or buy a new car? Explain.

At the end of Activity 6.5, Maria had compared the total cost of leasing a new Ford Explorer for four years with the cost of purchasing the car and financing it for the same amount of time. Her calculations showed that for the low APR rate she was offered, the two strategies would cost her just about the same. By leasing a car she would spend a lot of money and wouldn't even have a car to show for it.

8. Maria has several options available now. Exploring each one will allow you to answer her question about which is the best. For each choice, assume that operating costs are about the same, so that we can focus on the cost of buying versus leasing.

   a) She could lease another Ford Explorer in four years. Assume the same leasing arrangements exist at that time. How much will it cost her to lease a new Explorer four years in the future for another four years?

   b) How much will her total leasing costs be over the entire eight-year period?

   c) She could buy the one she originally leased for the purchase option price, and finance it over the next four years at the same APR. Recall that the purchase option price was $11,500. Financing that amount would mean monthly payments of $244.01. Assume that after eight years, the resale value of a 2001 Ford Explorer XLS is only $5000. How much will it cost Maria to keep the car for another four years?

   d) How much did the "lease and buy" option cost her over the entire eight-year period?

   e) Had Maria just bought the car, financed it over four years, and kept it for a total of eight years, how much would it cost her? (Hint: the value of the car is only $5000 at the end of the eight years.)

Preparing for the Chapter Project

9. As part of the chapter project, you need to know what leasing plans are available for the car of *your* dreams. Find out all the details you can about the financing arrangements, and which term lengths are available. You may want to interview a representative from a leasing company. Compare their terms with what a dealer will offer through the manufacturer. The more information you gather, the better prepared you will be to make the best decision for you.

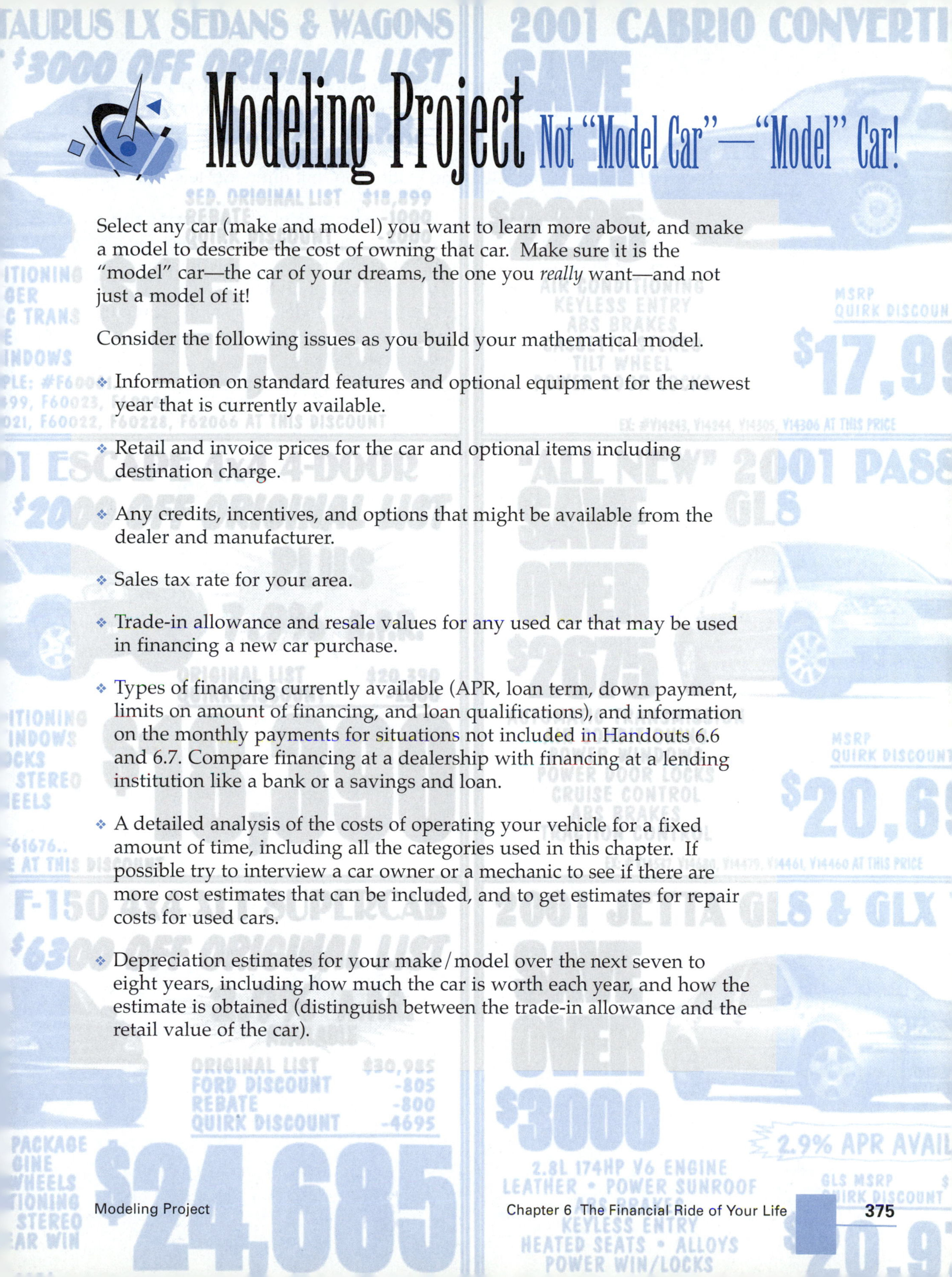

Modeling Project Not "Model Car"—"Model" Car!

Select any car (make and model) you want to learn more about, and make a model to describe the cost of owning that car. Make sure it is the "model" car—the car of your dreams, the one you *really* want—and not just a model of it!

Consider the following issues as you build your mathematical model.

- Information on standard features and optional equipment for the newest year that is currently available.
- Retail and invoice prices for the car and optional items including destination charge.
- Any credits, incentives, and options that might be available from the dealer and manufacturer.
- Sales tax rate for your area.
- Trade-in allowance and resale values for any used car that may be used in financing a new car purchase.
- Types of financing currently available (APR, loan term, down payment, limits on amount of financing, and loan qualifications), and information on the monthly payments for situations not included in Handouts 6.6 and 6.7. Compare financing at a dealership with financing at a lending institution like a bank or a savings and loan.
- A detailed analysis of the costs of operating your vehicle for a fixed amount of time, including all the categories used in this chapter. If possible try to interview a car owner or a mechanic to see if there are more cost estimates that can be included, and to get estimates for repair costs for used cars.
- Depreciation estimates for your make/model over the next seven to eight years, including how much the car is worth each year, and how the estimate is obtained (distinguish between the trade-in allowance and the retail value of the car).

- Types of leasing arrangements currently available (APR, lease term, front and back end restrictions, residual value, money factor, and purchase option price for a particular adjusted capitalized cost). Compare financing at a dealership with financing from a lending institution that offers car leasing.

At the very least you should test your model by comparing the cost of buying the car with the strategy of leasing it. Also you should test your model by comparing the cost of ownership when you buy a new car with the cost of ownership when you buy a used car of the same make and model. If possible, compare your "dream car" with another make and model that is similar to it (same type of vehicle, same price range).

Practice Problems

Review Exercises

1. As a comparison exercise, Maria considers buying a 2001 Ford Explorer 2WD XLT 4D, with a V-8 engine, moon roof, Mach radio system with 6-CD changer, running boards, and leather seats.

   a) What will the car dealer list as the sticker price?

   b) Maria and the car dealer are willing to agree to a selling price that is 95% of the sticker price. She is also eligible for the college graduate incentive. What will be the total price of the vehicle, including sales tax?

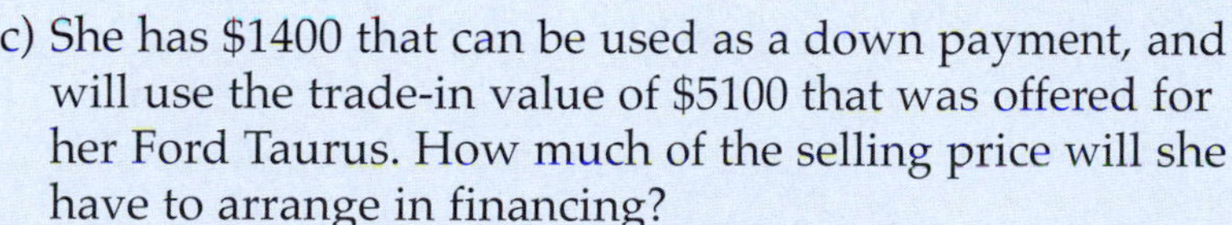

   c) She has $1400 that can be used as a down payment, and will use the trade-in value of $5100 that was offered for her Ford Taurus. How much of the selling price will she have to arrange in financing?

   d) If she arranges for 6.8% APR financing over five years, what will be her monthly payments?

2. Maria figures on keeping the car for seven years, and then selling it as a used vehicle. During that time, she plans on regular maintenance to keep it in good working condition. She averages 10,000 miles per year driving the car, with 3000 of that on the freeways.

   a) How much will the vehicle registration cost during those seven years?

   b) Assuming that over the seven years the pump price for gasoline is an average of $1.50 per gallon, how much will her fuel costs be?

   c) Using the same recommended maintenance schedule as in Activity 6.3, how much will her oil, tires, and battery cost her during the seven years?

d) Maria decides to use the same insurance carrier as her parents (Maria lives at home). The insurance agent quotes the following payment plan for the coverage she needs: Six months coverage with a down payment of $328 and four installments of $173. Assuming she continues living at home, is covered on her parents' plan, and the prices stay fixed, how much will her insurance cost her for the entire seven years?

e) How much will her operating costs be for the entire seven-year period?

3. Maria knows that the value of the car is based on the MSRP and the options that were included on the car. She estimates that the value will go down by $3000 when it becomes used. However, because this is a "fancier" Explorer than the ones she had been considering, Maria figures the car will depreciate at a rate of around 12% per year after that.

   a) With those assumptions, what will be the used-car trade-in value after seven years?

   b) Maria figures the car is in good condition, and she can sell it for $1000 over the trade-in value, rounded to the nearest thousand dollars. Assuming a buyer matches her selling price, how much will she get back when she resells the vehicle?

4. Maria would like to know what the cost of owning the Ford Explorer XLT really was for that seven-year period.

   a) How much did she actually spend when you consider purchasing and financing expenses, operating costs, and she recovers the value of the used car from reselling it?

   b) What was Maria's cost per mile?

   c) What was Maria's annual cost?

5. At the car dealer, Maria saw a similar 2001 Ford Explorer XLT for lease. (It didn't have the V-8, the upgraded sound system, or the moon roof.) The terms of the lease were 36 months, with 6.0% APR financing, a residual value of $17,000, and a purchase option of $16,500.

a) Maria and the dealer agree to a capitalized cost that is 90% of the MSRP, destination charge, and available options. For this car, what would be the capitalized cost?

b) Maria would like to use the $1400 down payment and the $5100 trade-in value on her used Ford Taurus in the leasing contract. What will be the adjusted capitalized cost for the car?

c) How much will the monthly lease payments cost?

d) How much did Maria spend altogether in the lease contract for the first three years?

6. At the end of the lease term, Maria decides to exercise the purchase option, financing the entire amount of the used car at 6.4% APR over four years.

a) How much is the monthly payment for the used-car loan?

b) How much will Maria spend in car payments over the four years?

c) If the trade-in value for this car is estimated as $10,000 at the end of the seven years, and she sells it at that time, how much did owning the car for four years cost her?

7. Assume that Maria's operating costs are the same as they were for the scenario in which she bought the Explorer XLT.

a) What would be her total costs for the "lease and buy" scenario over the entire seven-year period?

b) What would be her annual cost for the "lease and buy" scenario?

c) What is her cost per mile for the "lease and buy" scenario?

8. The two strategies being compared are:

- buy the new 2001 Ford Explorer XLT with a lot of options and finance most of it at a fairly high APR over five years
- lease a similar car (although slightly less expensive) for three years at a low APR; buy the car from the leasing company at the end of the term, and finance the used-car purchase at a mid-range APR over four years.

a) Are the two scenarios a "fair" comparison? Are the conditions for calculating the costs reasonably the same? Explain.

b) Which has the lower cost of ownership? Explain.

c) Out of all the details provided for consideration in these questions, what was the most critical factor in determining which strategy had the lower cost of ownership? Explain.

9. In comparing the "cost of ownership," two calculations have been used consistently: The average cost and the cost per mile. In answering the following questions, assume some number values, if necessary.

   a) If a new car is purchased and then parked in a garage for a couple of years, how would the average cost for a seven-year period compare to a similar new car that is purchased and used more regularly?

   b) If a new car is purchased and used quite heavily, how would the cost per mile over seven years compare to a similar new car that is purchased and used less often?

   c) If a new car is purchased and used quite heavily, how would the annual cost compare to a similar new car that is purchased and used less often?

REVIEW

Modeling the Cost of Automobiles

Model Development

In this chapter the true cost of owning and operating a vehicle was explored. There are many decisions to make when selecting a car, and all of them act as parameters for the model. Choosing the best plan is complicated, since every decision affects every other one. For example, if you go for a high-performance engine, expect the fuel consumption and insurance costs to reflect that decision. However, if you want to keep the costs as small as possible, then best can be defined in clear, measurable terms as "how much money is spent."

Assuming there are no finance charges, operating costs, or depreciation, the problem reduces to considering what decisions to make when buying a new car. This situation was considered in Activity 6.1. Parameters for that part of the model included the make/model of the car, year, any options installed on the car, and available credits, incentives, and options. Negotiating is not included in the mathematical model, but is a *major* part of the car-buying process. Anyone seriously interested in minimizing the cost of owning a car is advised to know as much as possible about the "game" of car pricing and the models that dealers use for bargaining with the customer.

Few people have enough money available to purchase a new car with cash. So the second part of the model deals with the options available to the customer to pay for the car (Activity 6.2). Included in that part of the model as parameters are the down payment, the trade-in allowance, and the financing arrangement. Financing any part of the car purchase means that you will pay more for the car, but the cost can be spread out over a longer period of time. The financing rate, or APR, and the loan term will both affect how much the monthly payment is, and how much the car will cost.

Most people don't buy a car to display in their garage. They need transportation and are willing to pay for this convenience. That "willingness to pay" are the costs calculated in Activity 6.3 that are associated with operating and maintaining the vehicle after it is purchased. While none of those costs (registration, fuel, servicing, insurance) were as big as the actual price of the car, they add up to a lot of money, especially over longer time periods. Any good model for estimating the cost of owning a car must consider whether the car gets good mileage, how much gasoline costs, and other day-to-day expenses.

The hidden cost of owning a car is the depreciation as a result of old age. Depreciation is covered in Activity 6.4. No one would buy a used car for the same price as a new car, not when they could get something new for the same price. To compare cars of different ages, an adjustment is made in the price of the car—the lower the price, the more likely someone will decide they don't need new that badly! The trade-in value and the retail value are measures of the worth of a used car, and can be estimated reasonably well. Popular models do not depreciate as quickly because of the demand for them, and sometimes figures like \$10,049 are lowered to \$9999, simply to say that it is less than ten thousand. So there isn't a hard and fast rule for pricing used cars. The value of a used car is important to monitor in a cost model, because it represents a potential for cash instead of having to pay it out all the time!

The final consideration in the model is also quite complicated, but is a lot like trying to decide between a pickup basketball game or batting practice on a warm spring day. In this case the games are "Should I buy?" and "Should I lease?" If you don't know how to play then you will lose your shirt. The costs associated with leasing options are all expressed in dollar amounts so that you can compare strategies as you did in Activity 6.5. A good modeler will include the lease option as part of the model, simply because under certain circumstances, it *is* better to go with a lease!

In developing a model for determining the cost of owning and operating a vehicle, issues come up. Determining what is best is like comparing apples and oranges, so to speak. Standardized answers like annual cost and cost per mile are a way of comparing possible buying or leasing strategies. However, they are just numbers obtained from a calculation. A good modeler will ask whether the answer makes sense for that situation. Also a lot of subjective evaluation goes into working with people and machines, such as when someone decides that their used car is worth an extra \$1000 simply because it is purple! A model of this type will be more difficult to give precise predictions, but it is okay as long as you take that into consideration.

▪ Mathematics Used

Most of the mathematics that goes into automobile transactions involves regular arithmetic computations, especially finding percents. The most complicated mathematical task in this chapter was working with the financing conditions to determine a monthly payment. In the automotive world, those are found in tables that every car dealer and banking institution use. Interest is one of the topics explored in Chapter 7, "Growth: From Money to Moose." Depreciation, another topic from this chapter, is related to the same patterns of growth and will be explored in Chapter 7 also.

The remaining calculations for this chapter were either finding totals by adding, finding totals by multiplying, working with rates, or determining what was still left. Calculations related to leasing transactions use technical formulas, but once you know the formulas, the mathematics isn't hard. The hardest thing was keeping track of what to do when, and to which quantity. That is as good a reason for building a mathematical model as anything!

Accessory package: Upgrades to the basic model that are offered as a set and are already installed on the car.

Adjusted capitalized cost: Amount of the value of the car that is being financed in a lease agreement.

Annual costs: Amount of car-related costs on a per year basis.

Annual Percentage Rate (APR): Interest rate on which any compounding is based, expressed as a yearly percentage rate.

Back end charges: Charges that are incurred when you return a leased vehicle at the end of the term.

Capitalized cost: Agreed-upon selling price of the vehicle, when negotiating a lease.

Cost per mile: Amount of car-related costs on a "per mile" basis.

Credit: Special offer made to the customer to lower the selling price for a new car.

Depreciation: Determination of the value (worth) of a car as it ages.

Destination charge: A fee included in the sticker price to cover the cost of transporting the car from the factory to the dealership.

Down payment: Money given to the car dealer to lower the amount that needs to be financed.

Early termination fee: Penalty charge paid for ending a lease agreement before the term is completed.

Excess wear and tear: Penalty charge paid for returning a lease car in worse condition than what was specified in the agreement.

Extra mileage: Penalty charge paid for mileage above the amount specified in the lease agreement.

Finance charge: Amount of interest paid as part of the monthly payment of a loan.

Finance companies: Companies that loan you money to buy a car and you pay them back in monthly installments.

Finance table: Table of monthly payments based on the amount financed, APR, and length of the loan term.

Front end charges: Charges that are incurred when you negotiate a lease agreement.

Fuel cost: The amount of money spent on any gasoline that is used.

Incentive: Special offer made by the car dealer or manufacturer to promote sales.

Insurance cost: The amount of money paid to protect a car in case of accidents or other damage.

Interest due: Extra amount of money owed because the loan had not been completely paid off.

Invoice price: Price the car dealer paid the manufacturer for a car.

Lease: Contract agreement in which a fixed amount is paid each month for the use of a car.

Lease term: Number of months that is the duration of the leasing agreement.

Linear depreciation: Depreciation method in which the value of a car goes down by a fixed amount (new car value ÷ number of years over which to be depreciated) each year.

Loan term: The amount of time it takes to pay off the loan.

Maintenance costs: Costs of keeping a car in good operating condition.

Mileage allowance: The number of miles specified in a lease agreement.

Money factor: Number used to calculate monthly interest charge; it is based on the APR percentage.

Monthly depreciation: How much the value of a car goes down each month of a leasing agreement.

Monthly interest charge: How much extra the customer must pay each month for financing a lease agreement.

MSRP: (Manufacturer's Suggested Retail Price) the price that a manufacturer sets to establish the worth of a car.

Net trade-in allowance: Credit provided in a leasing agreement for the value of a used car that is included in the negotiation.

Operating costs: How much it costs to operate and maintain a car.

Options: Extra features that are included on a particular car.

Purchase option: Quoted selling price for buying a leased vehicle at the end of the lease term.

Registration/license fee: Tax paid directly to the government to operate the vehicle.

Residual value: Amount the dealer thinks a car is worth at the end of a lease term.

Retail value: Price the dealer uses in trying to resell a used car.

Sales tax: A percentage of the selling price paid to the government for conducting the transaction.

Sticker price: Manufacturer's suggested price that is written up on a sheet (sticker) and placed in the window of a new car.

SUV: Sport Utility Vehicle, such as a Ford Explorer.

Trade-in: Car that is given to the dealer in return for a discount in the cost of a new car.

7 CHAPTER Growth: From Money to Moose

Why do some companies and people succeed financially when others fail?

There's an old saying, "Money makes money, and the money that money makes makes more money." This means you invest money to make more money. Financial planners are in the business of "growing" your money. They advise people on ways to get the maximum earnings on the money they invest.

Part of a financial planner's job is to evaluate the financial needs of a client. A planner will examine all of a client's finances before suggesting some investments. For example, a planner will ask, "What is the purpose of the investments, and how soon is the money needed?" Some important goals include:

- saving for college
- improving the client's lifestyle by increasing earnings
- paying debts
- planning for retirement

Financial planners use mathematical models to analyze different types of investments. In this chapter you will investigate many of the models that financial planners use. (You will see that many of these models may be used to analyze problems that have nothing to do with money.) You will use the models to evaluate some investments, and can even get started on your retirement planning. It's never too soon to start thinking about these issues!

PREPARATION READING

The Nature of Change

Just about everything in our world changes over time. Sometimes the change happens quickly, such as growing six inches over a summer or a rise of 600 points in the stock market in one day. Other changes occur *very* slowly—for example, the erosion of a mountain peak or the "wobble" of the Earth's north magnetic pole. Often, just the fact that something changes at all can alert officials to serious problems. Imagine being the scientist recording earth movements near a volcano. You would sleep better at night knowing that nothing is changing!

Bank accounts and wildlife populations do not seem to have anything in common. However, patterns of growth are determined by *how* quantities change over time. Those patterns can be used to classify their behavior. If the pattern of growth is the same for both money and moose, the modeling approach is to try and understand the pattern rather than learn about each application separately. The underlying pattern (represented as a table, order of operations, or even an equation) can then be used to describe both real-world problem settings. The mathematical model will have similar features because the pattern of growth is the same.

In this chapter various kinds of patterns for creating change in quantities will be explored. Most of the models focus on personal finance, especially banking. The content of "money matters" provides a context in which all of the growth patterns can be found and more easily understood. More importantly, probably you will tackle some of these problems during your lifetime. The challenge is to understand what kind of growth pattern a quantity is undergoing, and how those patterns of growth produce the models.

You will also find some problems from other contexts in which the quantities are changing. Once you can determine the type of growth pattern, you can apply the models to a host of other situations, such as wildlife populations, radioactive decay, depletion of non-renewable resources, and harvesting of renewable ones.

ACTIVITY

7.1 "Add-On" Models

When it comes to finances, there are many types of accounts, and a variety of programs for handling your money. Financial advisors always suggest that people keep a *little* money easy to reach, in case of an emergency. For interested customers, banks offer a free safe-deposit box, where money can be safely kept under lock and key. This forms the beginning point for studying how things can change.

MODELING SAVINGS USING ARITHMETIC SEQUENCES

1. a) Assume you put $200 in a safe deposit box for a rainy day, and leave it there. How much money would be in the safe deposit box after one week? Ten weeks? What assumptions are you making about the situation?

   b) How much does the amount of money change each week?

Unfortunately, putting your money in a safe deposit box will not earn you any interest. Since the easiest way to make money is to let your money earn it, let's see what happens when the assumptions are changed slightly.

To encourage customers to deposit money in a savings account, the Lone Star Bank promotes the following New Saver's Account:

For every $1000 deposited for an entire year, the bank will pay the customer $50 in interest.

CHECK THIS!

The assumption of the New Saver's Account is that interest is paid at a rate of 5% per year and only on the principal, or amount deposited. For each $100 you deposit, the interest you will receive is 5% of $100 or $5 each year that the money is in the account. When interest is calculated only on the principal, it is called **simple interest.**

2. Suppose you deposit $1000 in a Lone Star Bank New Saver's Account. (That is, the principal is $1000.) Use the table in **Figure 7.1** to determine the **balance** (how much money is in the account) at the end of each year.

Year Number	Interest	Total Balance
0	$0	$1000
1		
2		
3		
4		
5		

FIGURE 7.1.

SEQUENCE

A sequence is a list of numbers written in a certain order and often generated by some rule. Each number in the sequence is called a term in the sequence.

CHECK THIS!

Often the first term of a sequence is numbered term 0 when the sequence involves time. For example, in the sequence of account balances, the opening balance ($1000) can be considered term 0. This term refers to the amount of the balance when you opened the account.

The list of account balances in Figure 7.1 is an example of what is called a sequence.

Another example of a sequence is the list of numbers 10, 20, 30, 40, 50. The **term number**, or **index**, is used to "name" each term of the sequence according to its place in the sequence. The following are the term numbers for each term of this sequence:

term number: 0 1 2 3 4

term: 10, 20, 30, 40, 50

For example, 40 is the third term of the sequence or is term 3.

3. Return to the New Saver's Account you studied in question 2. Enter the total balance for each year from Figure 7.1 into a table like the one in **Figure 7.2**. Calculate the change in the balance each year and enter it in the right-hand column.

Number of Years (n)	Account Balance (A)	Change in Amount Each Year (ΔA)
0		----
1		
2		
3		
4		
5		

FIGURE 7.2.

The numbers representing the change in amount are also called the **first differences** for the sequence A. The sequence of first differences is determined by subtracting amounts at the beginning and end of each time interval:

(amount at the end of the time interval) – (amount at the beginning of the time interval).

4. a) Use the data that you gathered for Figure 7.2 to write ordered pairs (n, A), where n is the number of years and A is the account balance. Graph the ordered pairs on a coordinate grid like the one in **Figure 7.3.**

   b) Using the data in Figure 7.2, write ordered pairs $(n, \Delta A)$ where n is the number of years and ΔA is the first difference. Graph the ordered pairs on a coordinate grid like the one in **Figure 7.4.**

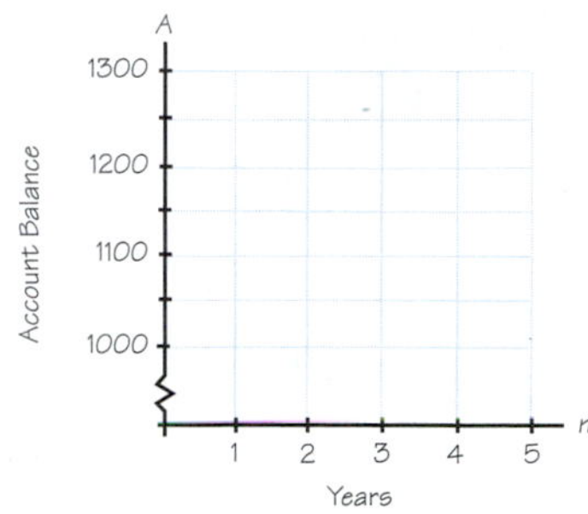

FIGURE 7.3.

The sequence of account balances is an example of an **additive model**, since the sequence is formed by repeated addition. This kind of sequence is called an **arithmetic sequence.** It is formed by specifying term 0 (for the opening balance), and repeatedly adding a fixed amount. The number you add is called the **constant difference.** Since any two terms in succession can be subtracted to produce this value, the constant difference is represented by the letter d.

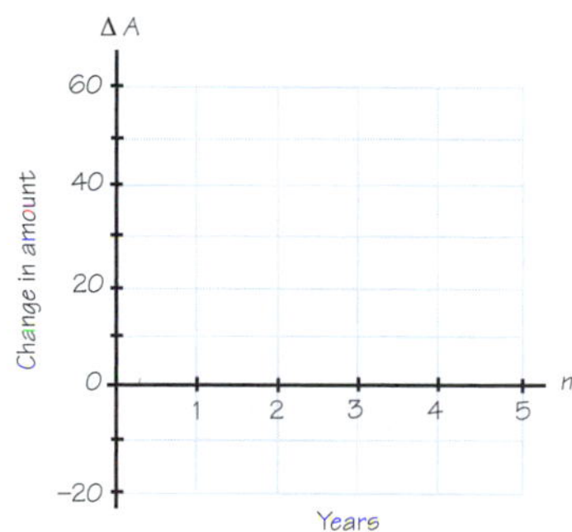

FIGURE 7.4.

5. a) Referring to the work done in questions 3 and 4, how can you tell whether a sequence of numbers is arithmetic by looking at the numbers?

   b) How can you tell whether a sequence of numbers is arithmetic by looking at the first difference values?

   c) How can you tell whether a sequence of numbers is arithmetic by looking at the graph of sequence values versus term numbers?

   d) How can you tell whether a sequence of numbers is arithmetic by looking at the graph of first difference values versus term numbers?

The terms of a sequence can be represented by a letter and using term numbers as a **subscript**. For example, the zero term is represented by A_0, the first term is A_1, the second term is A_2, etc. For the sequence of account balances,

A_0 is the opening balance

A_1 is the balance at the end of the first year

A_2 is the balance at the end of the second year, etc.

6. Calculate A_{20}, the account balance at the end of 20 years. Explain how you found that answer. Use "home-screen iteration" on a graphing calculator to check your answer.

DESCRIBING AN ARITHMETIC SEQUENCE

One way to represent the pattern of an arithmetic sequence is to write the starting or initial value, A_0, and then add the common difference as many times as necessary to obtain A_n. **Figure 7.5** shows the order of operations of the sequence viewed in this way:

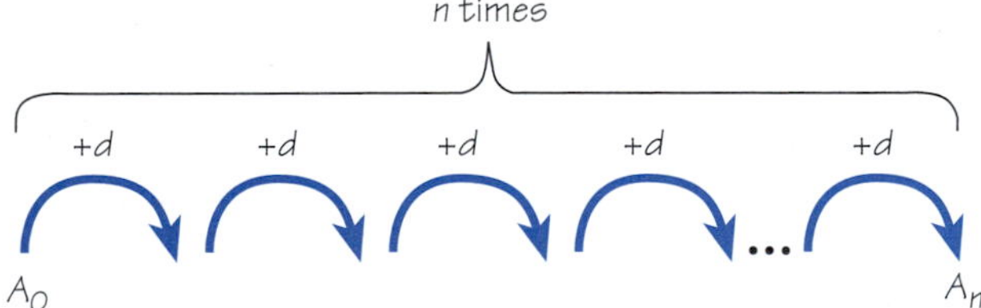

FIGURE 7.5.

The amount of change caused by adding d so many times can be quickly calculated by multiplying: $\Delta A = d + d + d + \ldots + d = d \cdot n$. The closed-form equation describing this process is:

$$A_n = A_0 + d \cdot n.$$

This equation is called a **closed-form equation** because it can be used to calculate a future value of the account from the opening balance.

7. a) Refer back to the New Saver's Account described in question 2. What is the closed-form equation that describes the ending balance for *any* number of years, n?

   b) Explain how you would use your equation to determine the balance after 15 years.

CLOSED-FORM EQUATIONS

Equations that allow you to find the value of one variable given the value of the other variable are called closed-form equations. The equations $C = 2p + 4$ and $x = 3 + 5t$ are examples of closed-form equations.

c) How does the equation help explain why the graph of the account balance versus the number of years forms a straight line?

d) What is the slope and y-intercept for the line that goes through the points you plotted in question 4(a)?

e) In the general case, what would be the slope and y-intercept of the line that goes through the graph of an arithmetic sequence?

CHECK THIS!

The word *recursive* means that a rule or formula is being applied over and over to generate a sequence of numbers. Recursive processes occur frequently in many branches of mathematics.

Notice that the pattern of any two terms of an arithmetic sequence is exactly the same. In going from any term to the *next* one, you always add d, the common difference. This pattern may be represented using a **recursive** description. **Figure 7.6** illustrates that method using A_0 as the starting or initial value.

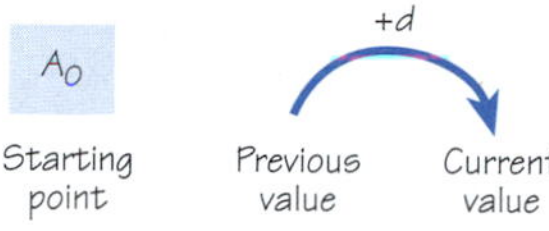

FIGURE 7.6.

Traditionally, the current value is represented by A_n, and represents the value after the recursive pattern has been applied n times. Since the previous value is the one *just before* A_n, it is called A_{n-1}.

8. a) Refer back to the New Saver's Account described in question 3. Use the table in **Figure 7.7** to find the previous and current values for each year.

Number of Years (n)	Previous Value (A_{n-1})	Current Value (A_n)
1		
2		
3		
4		

FIGURE 7.7.

b) Make a graph for this data for the New Saver's Account. Form ordered pairs (A_{n-1}, A_n) and plot them on a grid like the one in **Figure 7.8**.

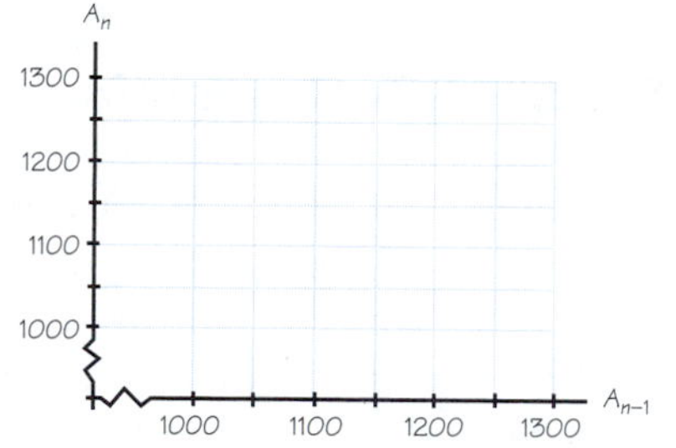

FIGURE 7.8.

c) What is the slope of the line that goes through the points plotted in Figure 7.8? Use the patterns in your table to explain why the slope *has* to be that value.

d) What is the y-intercept for the line that goes through the points plotted in Figure 7.8? Use the patterns in the table to explain why the y-intercept has to have that value.

e) Why was there no row in Figure 7.7 that corresponds to $n = 0$?

The rule for an arithmetic sequence may be written using a recursive equation. A **recursive equation** has both an initial value *and* a rule that calculates one term in terms of an earlier one. The recursive equation can be written this way:

$$A_n = A_{n-1} + d$$

$$A_0 = \_\_\_\_\_\_$$

This equation cannot predict a specific value until all the values that precede it are calculated. A recursive equation is useful in spreadsheets and performs home-screen iteration on a calculator. For mathematical modelers a recursive equation provides another way to view the same growth process, and an alternative way to study a problem. You should become comfortable with listing sequence values using both closed-form and recursive equations.

9. Use the SEQ mode capability of a graphing calculator to check the work done in analyzing the Lone Star Bank savings account plan.

   a) Enter the closed-form equation, and use the calculator to check your work in questions 3 and 4.

   b) Enter the recursive equation, and use the calculator to check your work in question 8.

SEQUENCES IN THE WILD

To appreciate the mathematical power of arithmetic sequences, consider a completely different situation involving the growth of a wildlife population. **Migration** is the term given to the movement of animals from one place to another. Migration affects a wildlife population in the same way that simple interest affects an account balance.

10. In 1988, Adirondack State Park, in upstate New York, had an estimated moose population of about 17 moose. The park is not fenced in, and moose migrate into the park from neighboring regions. Assume that the only change in population is due to migration, and that the migration rate is two moose per year (on average).

    a) How many moose would be in the park in the year 2000 (12 years later)?

    b) What is the closed-form equation that describes this migration model? Use the variables P (for population) and t (for the time interval).

    c) What is the recursive equation describing this situation?

SUMMARY

In this activity you learned how to use arithmetic sequences to model simple interest.

- You studied some arithmetic sequences, and you found the common difference for an arithmetic sequence.
- You used a closed-form equation and a recursive equation to describe a sequence.
- You learned that arithmetic sequences may be used to model other applications that are not related to finance.

Not all sequences are arithmetic sequences. In the next activity you will discover that another Lone Star Bank savings plan describes a new type of sequence.

HOMEWORK 7.1

An Adder Set of Problems

1. a) A cookie jar acts like a safe-deposit box. Suppose you put \$20 into the cookie jar, and do not add any more money later. The weekly balance of this account may be considered a sequence. Describe in words the rule for this sequence.

   b) Make a table like the one in **Figure 7.9** and fill in the second column with the amount of money that would be in the cookie jar after the specified number of weeks.

Number of Weeks (n)	Total Amount of Money (A)	Change in Amount Each Week (ΔA)
0		
1		
2		
3		
4		

FIGURE 7.9.

   c) Calculate how much the amount of money *changes* from one week to the next and record those answers in the third column of your table.

   d) Use the grids like the ones in **Figure 7.10** to make a graph of the total amount of money available each week, and how much the amount of money *changes* each week.

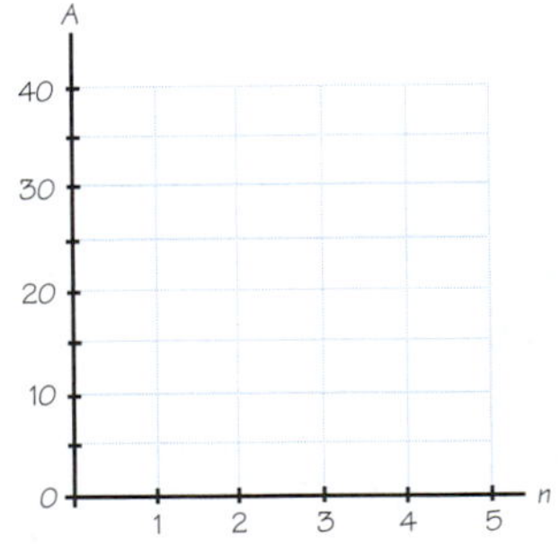

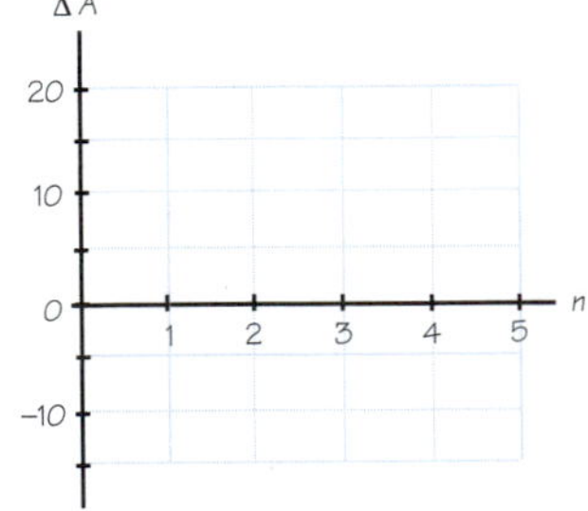

FIGURE 7.10.

2. a) The sequence of account balances in question 1 is an example of a **constant sequence.** Based on your work in question 1, describe how you could tell when you are working with a constant sequence, from using *only* the table of actual sequence values.

   b) Describe how you could tell when you are working with a constant sequence, from using *only* the table of first difference values.

   c) Describe how you could tell when you are working with a constant sequence, from using *only* the graph of sequence values versus the term numbers.

   d) Describe how you could tell when you are working with a constant sequence, from using *only* the graph of first difference values versus the term numbers.

3. Auntie Em uses a cookie jar to keep money for emergencies. On Monday she put a $10 bill in it, but decided a week later that it wasn't enough, so she put another $10 into the jar.

   a) Make a sketch of the graph of the total amount of money, M, versus the number of days that have elapsed, n.

   b) How is the pattern of numbers describing the change in the amount of money each day different than what was discovered in question 1?

   c) Write an equation for the sequence describing the amount of money in the cookie jar for any given number of days. (Actually, you will probably need to use two equations.) Be sure to include a description of the values of n.

4. Each of the following contains some values of an arithmetic sequence. In each case, write the equation that describes the sequence.

   a) The values in **Figure 7.11**.

   b) The values in **Figure 7.12**.

n	B_n
1	23
2	27

FIGURE 7.11.

n	B_n
1	31.0
4	35.5

FIGURE 7.12.

5. Each of the following situations involves an arithmetic sequence. Use the information provided to answer the question. Explain how you found your answer.

   a) Find the tenth term, if the zero term is 128 and the common difference is –8.

   b) Find the zero term, if the eighth term is 82 and the common difference is 5.

   c) Find the common difference, if the fifth term is 33 and the ninth term is 57.

   d) Find the term number that corresponds to a value of 125, if the first term is 71 and the common difference is 3.

Sometimes it is useful to be able to describe sequences with term numbers starting at 1 instead of zero. However, some adjustments must be made on the way the equation is formed. Question 6 involves such a situation.

6. A sequence has a common difference of 5, and starts with 12 as the value for B_1. The table in **Figure 7.13** shows how you would calculate the sequence term knowing the common difference and the value for B_1.

FIGURE 7.13. Calculating a term knowing B_1 and the common difference.

Number of Weeks (*n*)	Calculation of Sequence Value from Knowing '*d*' and B_1	No. of Times '*d*' was added	Sequence Value (B_n)
1	(given information)	0	12
2	12 + 5	1	17
3	12 + 5 + 5	2	22
4	12 + 5 + 5 + 5	3	27

   a) Explain how you would calculate the fifteenth term in this sequence from just knowing the common difference and the first term.

   b) What is the relationship between the term number, n, and the number of times the common difference, d, was added?

   c) Write an equation that uses the term number to calculate any value for this sequence, from *just* knowing the common difference was 5 and the first term was 12. (Check your equation with one of the lines from Figure 7.13.)

d) In general, if d and B_1 are known, what equation would describe the sequence values, B_n, in terms of the term number, n? (Your answer should be an adjustment on the formula developed in Activity 7.1 to account for the sequence starting with the first term.)

7. a) At a bookstore the first pencil you buy costs 25 cents. After that each additional pencil costs only 17 cents. Write a recursive equation describing the total amount charged for buying some pencils (not including sales tax).

   b) Write the corresponding closed-form equation to describe this sequence. Use it to determine how much 503 pencils would cost.

8. In addition to the cookie jar, Auntie Em has checking and savings accounts at the local bank, and does her banking through an ATM machine. She currently has $1500 in her checking account, and $6800 in her savings account. She wants to transfer some money from savings to checking but the machine restricts the amount to $200 per day.

   a) Write the next four terms for the sequence C_n (the daily checking account balance) and the sequence S_n (the daily savings account balance). Assume that she visits the bank every day and transfers the maximum allowed.

   b) Write a closed-form equation for the sequence C_n.

   c) Write a recursive equation for the sequence S_n.

   d) How many days will it take for the checking account balance to be greater than the savings account balance?

In Activity 7.1 the savings account plan earned $50 in interest for every $1000 kept in the account for an entire year. The underlying assumption was that the interest was paid at the rate of 5% per year. This kind of savings account pays **simple interest**, because the interest calculation is based on using the principal only.

9. a) Use the 5% interest rate to find the interest earned when $1000 is deposited for one year.

   b) How much interest would the bank have paid for one year, at that same rate, if only $600 was put into the account? Explain.

c) At that same rate, if \$600 were put into the account for five years, how much interest would the bank have to pay? Explain.

d) Suppose the \$600 was put into an account for five years, but the simple interest rate was only 4% per year. How much interest would the bank have to pay? At the end of the five years, what would be the balance in the account?

10. Generalize the work done in question 9 into equations that model simple interest calculations. Let P represent the principal, or amount deposited in the account. Let t be the number of years the money is put into the account, and let r be the simple interest rate (expressed as the decimal equivalent of a percent).

    a) Write an expression that would calculate the amount of interest, I, that the bank must pay.

    b) Write an expression that calculates the balance, B, in the bank at the end of that time period. (Assume the interest is deposited in the savings account.)

11. If you deposit \$1250 into a savings account that pays at a simple interest rate of 6.2%, what will be the balance after three years?

12. The Lone Star Bank features a Christmas savings account that pays a 6% simple interest rate per year, but the account can only stay open for one year. If you open the account with \$500, and deposit an additional \$300 half a year later, how much money will the account have when it closes?

13. a) The migration problem in Activity 7.1 was based on observations. In 1988, the estimate for the moose population in the park was between 15 and 20 moose. In 1993, the population was estimated at somewhere between 25 and 30 moose. Explain how the migration rate of two moose/year that was used in Activity 7.1 was derived from these observations.

    b) The only reason to use the *average* migration rate is because it predicts "typical" results from a model. Concern about overpopulation might lead a mathematical modeler to use the *largest* migration rate, instead. Based on the observations, what would be the largest rate?

c) Estimate the largest possible moose population in the park in the year 2000 based on the actual observations. (We *still* assume that the only change in population is due to migration.)

14. The "Six Flags Over Texas" amusement park is capable of admitting around 50 people a minute at peak times. At 10:00 A.M. when the peak time begins, there are already 1400 customers in the park. How long will it take for the attendance to reach 5000 people?

ACTIVITY

7.2

The Times Are A' Changin'

Lone Star Bank's New Saver's Account described in Activity 7.1, may interest some people. The idea of getting something for nothing should be appealing. However, many experienced financial planners realize that there is another form of interest that makes you even more money. In this activity you will learn about another form of interest, called compound interest.

COMPOUND INTEREST AND GEOMETRIC SEQUENCES

Looking a little deeper into the assumptions of a New Saver's Account raises an interesting question. Suppose you have $1000 in a New Saver's Account. At the end of the year, according to the bank's plan, you should be paid $50. The bank doesn't exactly hand over this money; they credit your account $50. So shouldn't the principal change for the next year? Let's see what happens with an account that bases their calculations on including interest with the principal.

1. Assume that you deposit $1000 in a Lone Star Bank Growth Savings Account. For this type of account, the bank pays 5% interest on the balance you have at the beginning of *each* year, instead of just the principal.

   a) Make a table like the one in **Figure 7.14**, and fill in the first three columns. Calculate the amount of interest earned each year, and include this interest as part of the ending balance for the year. Use this balance as the beginning balance for the next year.

Year	Beginning Balance	Interest Earned	Ending Balance	Ratio Between Consecutive Beginning Balances
0	$1000.00			------
1				
2				
3				

FIGURE 7.14.
Growth Savings Account table.

b) Are you earning more, less, or the same amount as you would with a New Saver's Account? Explain.

When interest is paid on both the principal and earned interest, it is called compound interest. Since the interest on the Growth Savings Account is calculated once each year, we say the interest is **compounded annually**.

2. a) The list of account balances in your Growth Savings table is a sequence. Is this sequence an arithmetic sequence? Explain.

   b) For each pair of consecutive beginning balances in the table, find the ratio

$$\frac{\text{current year's balance}}{\text{previous year's balance}}.$$

   Record the answers in the fourth column of your table. What do you notice about these ratios?

   c) What is the meaning of the ratios you found in (b)?

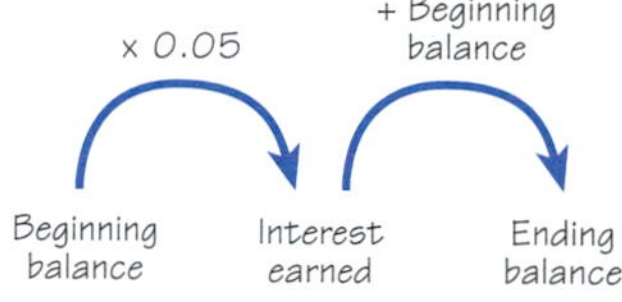

FIGURE 7.15.

The growth pattern exhibited in question 1 may be represented by a recursive description. The steps of the calculations that filled in a row of the table are shown in **Figure 7.15**.

There is another way to find the ending balance shown in Figure 7.15 that reveals more about this growth pattern. The key is that the beginning balance is used twice in the calculation.

3. a) What percent of the beginning balance in your Growth Savings Account table is the interest?

   b) What percent of the beginning balance is the beginning balance? (Trick question!)

   c) When the two percents in (a) and (b) are combined, what percent of the beginning balance is the ending balance? How does that answer compare to the ratio that was recorded in the fourth column of your table?

   d) Using the percent in (c), write an equation that finds the ending balance from the beginning balance.

   e) Write an expression for the ending balance after 15 years, then calculate the answer. (Look for an easy way to do it!)

The balances in the Growth Savings Account exhibit a different kind of growth pattern, called a **multiplicative model**. In this model a beginning value is multiplied repeatedly by another number, called the **growth factor,** b. The sequence of numbers that is produced is called a **geometric sequence**.

> **GEOMETRIC SEQUENCE**
>
> A geometric sequence is a sequence of numbers that has the property that the ratio between two successive terms is constant (rather than the difference). The constant ratio is called the growth factor, b.

Figure 7.16 shows both the recursive and closed forms for a geometric sequence.

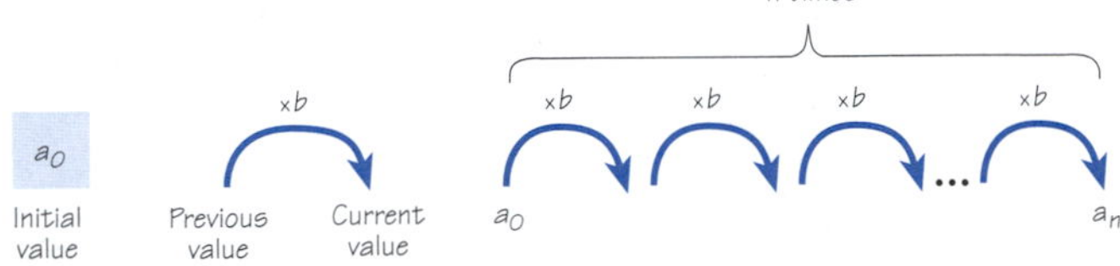

FIGURE 7.16.

The recursive form of a geometric sequence reminds us that it is a multiplicative model. This form describes what goes on from one term to the next, but is awkward for calculations. In the closed form, the repeated multiplication is done all at once by using an exponent:

$a_n = a_0 \cdot (b)^n$, where a_0 is the initial value, b is the growth factor and $n = 0, 1, 2, 3, \ldots$

On a calculator that same expression is obtained by writing $a_0 * b\wedge n$.

4. Use a graphing calculator to check the following calculations.

   a) The values in Figure 7.14 using the TABLE feature and a recursive-form equation in SEQ mode.

   b) The values in Figure 7.14 using the TABLE feature and a closed-form equation in SEQ mode.

   c) The calculation made in question 3(e) using home-screen iteration.

   d) The calculation made in question 3(e) with a single expression issued in home-screen iteration.

5. Elba Jay starts a Growth Savings Account at a new branch of the Lone Star Bank that is having a grand opening special. This branch will pay 6% interest per year on *any* money that is kept in the bank whether it is principal, additional deposits, or any interest earned. She opens the account with $1600.

a) To calculate the amount of interest earned during the year in a single step, what multiplier (rate) would you use?

b) How much interest would be earned in the first year?

c) To calculate the balance at the end of the year in a single step, what multiplier (growth factor) would you use?

d) What would be the balance at the end of the year?

COMPOUNDING PERIOD

This particular Lone Star Bank branch has an interesting plan for attracting new customers. Not only does it pay interest on all the money in the account, but the bank decides to calculate the interest earned every six months instead of at the end of the year. This account pays interest that is compounded **semiannually** (twice a year).

6. a) How would you adjust the interest rate in question 5 to calculate the interest over half a year?

   b) How would you adjust the growth factor to calculate the balance at the end of half a year? (Watch out!)

   c) Use your new rates to calculate the interest and the balance for this account, each as a single multiplication step. Make a table like the one in **Figure 7.17** to record your work.

Term	Beginning Balance	Interest Earned	Ending Balance
1st 6 months	$1600		
2nd 6 months			

FIGURE 7.17.

   d) Use your results in questions 5 and 6 to determine how compounding interest twice a year compares with calculating interest just at the end of the year (assuming the same yearly interest rate).

Question 6 shows that the number of times per year the interest is compounded makes an important difference. An account containing $1600 compounded semiannually generates more interest than the same account compounded annually. The number of times per year (n) that the interest is compounded is called the **compounding period. Figure 7.18** shows some common compounding periods.

Number of Times per Year	Interest is Compounded
$n = 1$	annually (once a year)
$n = 2$	semiannually (twice a year)
$n = 4$	quarterly (4 times a year)
$n = 12$	monthly (12 times a year)
$n = 52$	weekly (52 times a year)
$n = 360$	daily (360 times a year)

FIGURE 7.18.
Common compounding periods.

7. a) A single calculation can determine the account balance at the end of one year when $1600 is deposited into a savings account that pays 6% annually, but is compounded twice per year. Write an expression for this calculation.

CHECK THIS!

There aren't exactly 52 weeks or 360 days in a year. Bankers use those numbers out of convenience, and for accounts with small balances, the interest calculation comes out the same when rounded to the nearest penny.

b) Evaluate your expression to make sure that the balance is correct.

c) Write a similar expression to calculate the balance at the end of 10 years. Then evaluate your expression.

d) Use home-screen iteration to check your answer to (c). Record the appropriate sequence values that are displayed on the calculator screen in the second column of a table like the one in **Figure 7.19**. (Remember: The account is being compounded twice per year.)

Year (t)	Ending Balance(A)	Interest Earned During NextYear	$\frac{\text{Next Year's Interest}}{\text{This Year's Ending Balance}}$
0	$1600.00		
1			
2			
3			
4			
5			
6			
7			
8			
9			
10			

FIGURE 7.19.

e) Calculate the interest earned during the next year (first differences), and record the values in the third column of your table. What do you notice about the pattern of these values?

f) For each year calculate the ratio: (Interest earned next year) ÷ (Balance at end of this year). Record the values in the last column of your table. What do you notice about the pattern of these values?

When the Growth Savings Account is compounded annually, the growth factor is 1.06. When the Growth Savings Account is compounded semiannually, the annual growth factor is about 1.0609. For interest that is compounded more than once per year, the actual rate of growth over one year is called the **annual percentage yield (APY)**. This rate of growth always is slightly greater than the stated interest rate over one year, called the **nominal rate**, since any interest earned also earns interest.

CHECK THIS!

APR, or Annual Percentage Rate, used in Chapter 6 to determine monthly payments in automobile financing is similar to the annual percentage yield of a savings account.

GROWTH RATES IN BANKS AND FORESTS

Graphs can provide a different way to view the growth patterns formed by accounts that compound interest.

8. a) Write ordered pairs (t, A) using the calculations you made in Figure 7.19.

   b) Use a grid like the one in **Figure 7.20** to plot these ordered pairs.

   c) Describe the shape of the graph formed by the points you plotted.

   d) Write ordered pairs of the form (previous year's balance, current year's balance) for years 1 through 10. Plot the corresponding points on a grid like the one in **Figure 7.21**.

   e) Find the equation of the line going through the points you plotted.

   f) How is the slope of the line related to the growth pattern of the sequence of account balances?

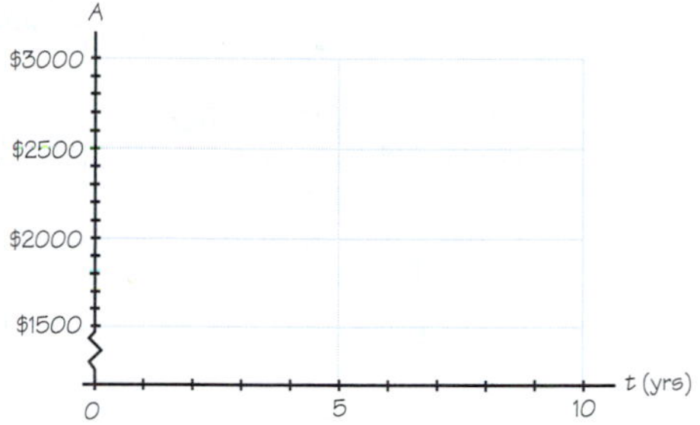

FIGURE 7.20.

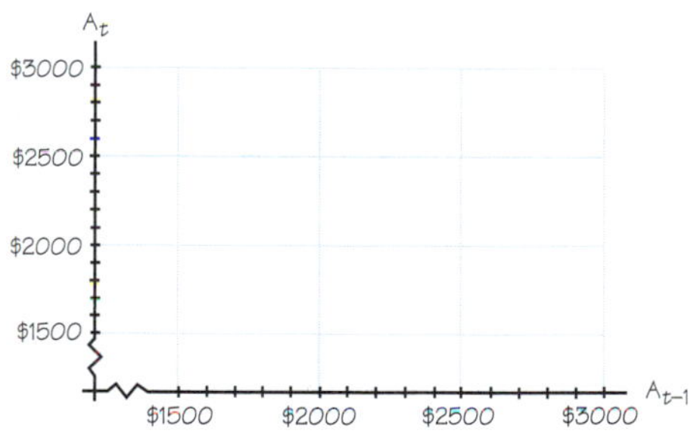

FIGURE 7.21.

Notice that the graph of the growth of an account that is compounded shows a significantly different pattern. When the balances are graphed with respect to time (Figure 7.20), the overall pattern is a slightly curved upward shape. This shape is caused by that little extra money earned by the interest on the interest.

In general, the ending balance of an account that earns compound interest can be calculated according to the following equation for a geometric sequence:

$$A = a_0 \cdot \left(1 + \frac{r}{n}\right)^{nt},$$

where r is the yearly interest rate (expressed as a decimal), n the number of times the interest is compounded per year, and t the number of years. The amount a_0 is the initial amount invested.

9. Competing banks offer similar services, but sometimes they vary how many times the interest is compounded per year. Assume the initial deposit is still $1600, and the yearly interest rate is kept at 6%.

   a) Calculate the ending balance and the interest earned for each compounding period in the banking plans in **Figure 7.22** after one year and after five years. Record your answers in a table like Figure 7.22.

Number of Times Compounded (n)	After 1 Year		After 5 Years	
	Ending Balance	Interest Earned	Ending Balance	Interest Earned
4				
6				
12				

FIGURE 7.22. Comparing interest for different compounding periods.

   b) Explain how you can calculate the interest earned from knowing the initial deposit in the savings account and the ending balance after some time period.

   c) Write out the four interest payments made during the first year when the interest is calculated quarterly. Does the total equal the amount recorded in the table as the interest earned after one year?

   d) What kind of sequence is formed by the interest payments during each compounding period? Write an equation to describe the sequence.

When considering growth due to **reproduction**, wildlife populations exhibit something similar to the growth found in bank accounts that earn compound interest. Populations grow over time. Babies eventually have their own babies, just as interest is earned by money that was deposited into an account as interest earned. It is reasonable to express that growth as a percent of the total population. This growth rate will not be a constant, but good models can be built to project population totals using an average growth rate.

10. The moose population in Adirondack State Park was estimated to be around 27 moose in 1993. It is expected that the population will grow by around 13% annually, on average.

   a) Use P to represent the moose population at any time, and let t represent the number of years. Write a equation that describes a reproduction model for the moose in the park.

   b) Use your moose model to predict the size of the moose population in the park in the year 2000.

SUMMARY

In this activity you studied compound interest.

- You learned that the balances of an account that earns compound interest form a geometric sequence.
- You discovered that the amount of interest earned depends on the interest rate and the compounding period.
- You found that geometric sequences can be used to explore other types of growth, such as population growth.

So far you have learned about some important models for analyzing finances. In the next activity you will explore a unique mathematical relation that is an important model in analyzing finances.

HOMEWORK

7.2

The Multiplication Factor

1. a) Identify each of the following sequences as arithmetic or geometric (or neither). In each case explain how you determined the answer.

   a) 5, 8, 11, 14, 17, …

   b) 5, 10, 20, 40, 80, …

   c) The pattern of dots in the left-hand graph in **Figure 7.23**.

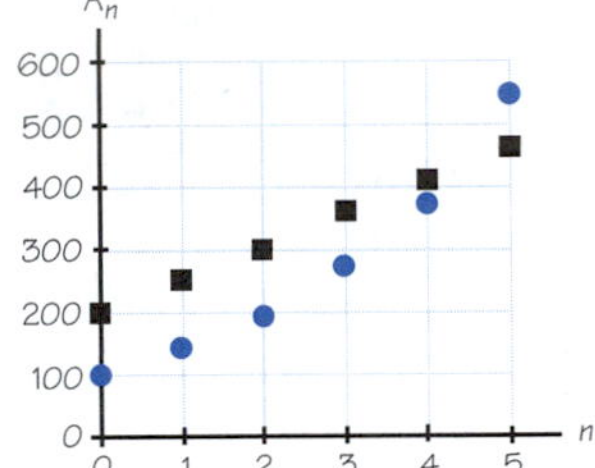

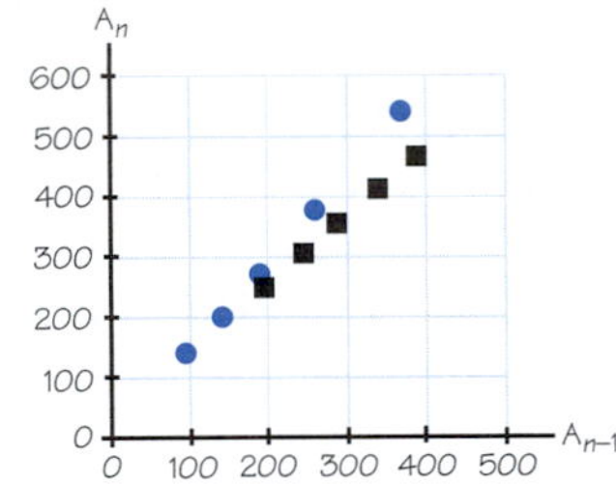

FIGURE 7.23.

   d) The pattern of squares in the left-hand graph of Figure 7.23.

   e) The pattern of dots in right-hand graph of Figure 7.23.

   f) The pattern of squares in the right-hand graph of Figure 7.23.

2. For each of the following geometric sequences, find the zero, first, second and third terms (a_0, a_1, a_2, and a_3).

   a) $a_n = 1.25 \cdot a_{n-1}$; $a_0 = 400$.

   b) $a_n = 1400 \cdot (1.05)^n$.

   c) Initial value: $a_0 = \$25{,}000$; annual percentage rate: $r = 12\%$; number of times compounded per year: $n = 6$.

3. a) Given the following model for calculating the balance in a savings account that earns interest compounded quarterly,

$A_n = 15000 \cdot \left(1 + \frac{0.06}{4}\right)^n$, what is the nominal interest rate for this account?

b) What growth factor would be used to make the sequence of yearly ending balances?

c) Which term number corresponds to a time interval of two years, nine months?

d) How much interest is earned during the entire second year?

4. For each of the following geometric sequences identify:

   - the growth factor
   - the nominal rate
   - the annual percentage yield (APY)

   All sequence values have been rounded to three decimal places.

   a) The sequence represented in **Figure 7.24**.

Number of Years	0	1	2	3
Term Number	0	1	2	3
Sequence Value	200.0	220.0	242.0	266.2

FIGURE 7.24.

   b) The sequence of account balances for $2500 deposited in a savings account that pays 6% interest annually and is compounded monthly.

   c) The sequence represented in **Figure 7.25**.

Number of Years	0		1		2	
Term Number	0	1	2	3	4	5
Sequence Value	120.000	122.400	124.848	127.345	129.892	132.490

FIGURE 7.25.

5. Write an equation and a recursive formula for each of the following geometric sequences:

   a) The sequence represented in Figure 7.24 (question 4(a)).

b) The sequence of account balances for $2500 deposited in a savings account that pays 6% interest annually and is compounded monthly (question 4(b)).

c) The sequence represented in Figure 7.25 (question 4(c)).

6. Suppose you deposit $1800 in an account with a nominal interest rate of 8% per year. What is the balance at the end of one year if the interest paid is:

   a) simple interest?

   b) compounded annually?

   c) compounded quarterly?

   d) compounded daily?

7. Suppose you invest $250 in an account that pays 4.5% interest compounded quarterly. After 30 months, how much is in your account?

8. a) The amount of money that needs to be invested *now*, in order to reach a specific amount at a certain time in the future is called the **present value**. If you know the interest rate, the compounding period, and the amount expected at a future time, explain how you can find the present value of a certain account.

   b) If a savings bond matures in three years and will pay $10,000 at that time, what is its present value? (Assume the bond has an interest rate of 6% compounded annually.)

9. Suppose that a new parent will need $15,000 to pay for a year of college for a child 18 years in the future. Right now, they can buy a certificate of deposit whose interest rate, 10% compounded quarterly, is guaranteed for that period. How much would they need to deposit?

10. Auntie Em deposits $500 in a savings account that earns 8% per year, compounded semiannually. **Figure 7.26** shows the steps for calculating the balance at the end of a year.

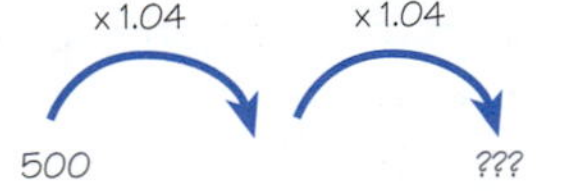

FIGURE 7.26.

a) The calculation is shown in the right side of Figure 7.26. It can be entered into the calculator in a similar fashion: 500*1.04^2. In what *order* are these calculations done?

b) Auntie Em decides that she would like to have $600 at the end of the year from her $500 investment (still compounded semiannually). Draw an arrow diagram that starts with the growth factor (now, an unknown), and goes through the two steps required to get $600 at the end of the year. What are those two steps?

c) To find the growth factor needed, retrace your steps *backward* through the arrow diagram. What are the two steps that undo the ones recorded in (b)?

d) What growth factor would be needed to have $600 at the end of the year?

e) What nominal rate should Auntie Em be considering?

11. Each of the following gives an investment in a savings account earning compound interest, and the final balance after a certain time period. Find the nominal rate necessary to provide those earnings.

a) Investment: $900; Final balance: $1000; $n = 2$ compounds/year; $t = 1$ year.

b) Investment: $1100; Final balance: $1250; $n = 1$ compound/year; $t = 3$ years.

c) Investment: $2000; Final balance: $2250; $n = 6$ compounds/year; $t = 1$ year.

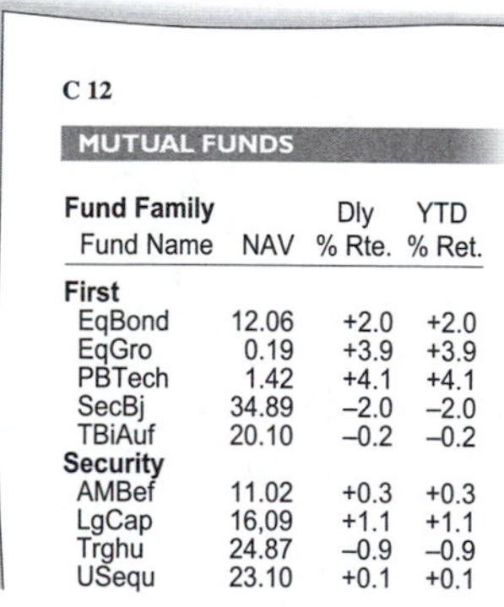

C 12

MUTUAL FUNDS

Fund Family Fund Name	NAV	Dly % Rte.	YTD % Ret.
First			
EqBond	12.06	+2.0	+2.0
EqGro	0.19	+3.9	+3.9
PBTech	1.42	+4.1	+4.1
SecBj	34.89	–2.0	–2.0
TBiAuf	20.10	–0.2	–0.2
Security			
AMBef	11.02	+0.3	+0.3
LgCap	16,09	+1.1	+1.1
Trghu	24.87	–0.9	–0.9
USequ	23.10	+0.1	+0.1

12. Money market funds change investments regularly to adjust to trends in the market. Usually these accounts report earnings to investors each month. Suppose that the monthly statement from one such fund reports a beginning balance of $7373.93, and a closing balance of $7416.59, over 28 days.

a) What is the average daily growth factor?

b) What is the average daily percentage increase?

c) Based on this performance what would be the annual percentage yield (APY)?

13. a) In Activity 7.2, the moose model was used to predict the population of moose in Adirondack State Park in the year 2000. Use the same starting population of 27 moose in 1993, and assume the growth rate is 10% per year, on average. What would be the population for the year 2000 based on this set of assumptions?

    b) What would the yearly percentage increase have been if the same population reached 75 moose in the year 2000?

14. a) **Figure 7.27** shows the official census figures for Texas for the last three censuses. Was the population growth rate over the ten-year period from 1990 to 2000 the same as the rate during the ten-year period from 1980 to 1990? Explain.

FIGURE 7.27.

Year	1980	1990	2000
Population	14,229,191	16,986,510	20,903,994

    b) Find the annual percentage increase over the twenty-year period from 1980 to 2000.

ACTIVITY

7.3

Fortunes Won—and Lost

You have explored what happens when a fixed amount of money is put into an account that earns interest. In the last activity you learned that compounding the interest earns a little more money. Now you will take this process one step further to discover another form of interest called continuous compounding.

COMPOUNDING AND COMPOUNDING AND COMPOUNDING

An interesting question that hasn't been asked yet is:

"How much money can be made by shortening the compounding period or increasing the number of times the interest is calculated?"

Let's make some calculations to see what happens when the compounding period gets smaller and smaller.

1. Suppose you put $1 into a savings account that draws compound interest at a rate of 100% per year. You want to know how the number of times the interest is compounded per year affects the balance in the account at the end of the year.

Compounding Period (years)	Number of Times the Balance is Compounded	Ending Balance (r = 100%)	Ending Balance (r = 10%)
1			
1/2			
1/4			
1/12			
1/365			
1/10,000			
1/100,000			
1/1,000,000			

FIGURE 7.28.
Compounding $1 for smaller and smaller compounding periods.

a) For each compounding period in **Figure 7.28**, calculate the number of times the balance is compounded. Record your answers in the second column of your table.

b) Use the compound interest formula to determine the ending balance for each period if the interest rate is 100%. Record your answers to five decimal places in the third column of your table.

c) Based on the pattern of the balances in the third column, do you think it's possible to continually gain more interest simply by shortening the compounding period?

d) Repeat the work done in (a) using an annual interest rate of 10% per year. Record the answers to five decimals in the fourth column of your table. Describe the pattern in those ending balances.

2. a) Suppose that x is the number of times the interest is compounded in Figure 7.28. What is the corresponding expression for the length of the compounding period in years?

b) If x is the number of times the interest is compounded, write an expression for the ending balance if $r = 100\%$ (column 3 in Figure 7.28).

In the expression $\left(1+\frac{1}{x}\right)^x$, as the value of x increases ($x \geq 1$), an interesting "tug-of-war" happens between the base $\left(1+\frac{1}{x}\right)$ and the exponent x. The base $\left(1+\frac{1}{x}\right)$ starts out with a value of 2 (when $x = 1$) and shrinks toward a value of 1. As that happens, the exponent x (column 2 in Figure 7.28) increases so that the base is multiplied by itself a greater number of times.

You might think that $\left(1+\frac{1}{x}\right)^x$ continues to increase forever. Since the base is greater than 1, multiplying by the base more times will make the value even bigger. You also might think that $\left(1+\frac{1}{x}\right)^x$ decreases to 1 since the base is getting close to 1, and everyone knows that $1 \cdot 1 \cdot 1 \cdot 1 \cdot 1 \cdot \ldots \cdot 1$ is still equal to 1! It turns out that $\left(1+\frac{1}{x}\right)^x$ gets closer and closer to a value that is so special in mathematics that it has its own name (and a special calculator button):

$$\left(1+\frac{1}{x}\right)^x \xrightarrow{\text{as } x \text{ becomes really big}} \approx 2.718281828 = e$$

Hundreds of years ago mathematicians discovered that e is the uppermost value of the growth of compounded income. Using a percentage rate r in the compound interest calculation, there is also a predictable answer:

$$\left(1+\frac{r}{x}\right)^x \xrightarrow{\text{as } x \text{ becomes really big}} = e^r$$

CHECK THIS!

Here are some characteristics of an exponential expression a^x that you can use as you examine the exponential expression.

- The larger the exponent x is, the larger the answer. For example, $2^3 = 8$, whereas $2^5 = 32$.
- The larger the base a is, the larger the answer. For example, $2^3 = 8$ and $3^3 = 27$.

3. a) Refer back to the work done in Figure 7.28. Using your calculator, find the value for $e^{0.10}$. How does this result compare with the values you calculated for the last column of Figure 7.28?

   b) Using your result from (a), write an expression for the growth factor using e, if $r = 10\%$ and the number of times the interest is compounded increases to a very large number.

   c) If \$250, rather than just \$1, had been deposited in an account that earns 10% interest, what would the values in the last column be approaching?

   d) If \$250 was deposited at 20% interest, compounded as many times as possible, what would be the balance at the end of one year?

In banking terms, this process of increasing the number of times the interest is compounded is called **continuous compounding**. The mathematical formula for this type of compounding is:

$A = Pe^{rt}$, where t is the number of years, r is the interest rate expressed as a decimal, P the principal deposit, and A the account balance.

Continuous compounding is like spending every moment of every day calculating the interest that was *just* earned (and crediting the account with that miniscule amount), only to do it again in the next moment. It is a continuous model, where the other compound interest situations are discrete.

EXPONENTIAL EQUATIONS AND THEIR GRAPHS

The formula for continuous compounding is an example of an entire *family* of mathematical equations called **exponentials**.

4. Set your calculator on 'Func' mode, with these WINDOW settings: Xmin = –3; Xmax = 6; Ymin = –5; Ymax = 50.

   a) Sketch a graph of each of the following equations: $y = 10(1.10)^x$; $y = 10(1.5)^x$; $y = 10e^x$; and $y = 10(1.25)^x$ using a grid like the one in **Figure 7.29**. (Remember: The * is the multiplication symbol, and the ^ is the exponent operator on your calculator!)

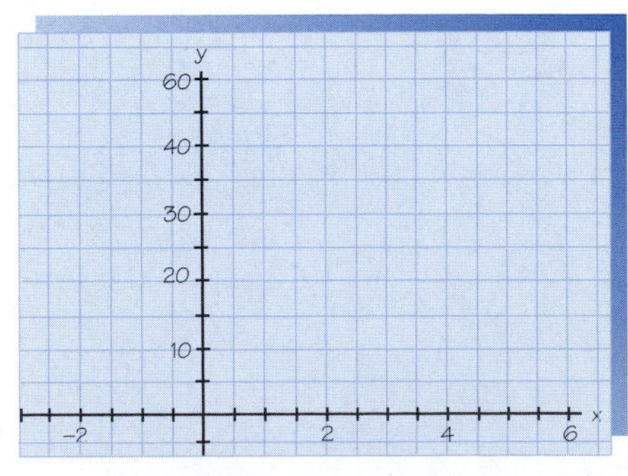

FIGURE 7.29.

b) What do the graphs have in common?

c) What do the four equations have in common?

d) What general features or similarities do these exponential graphs share?

5. a) Sketch a graph of each of the following exponential equations: $y = 12(1.12)^x$; $y = 10(1.12)^x$; $y = 8(1.12)^x$; and $y = 15(1.12)^x$ using a grid like the one in **Figure 7.30**.

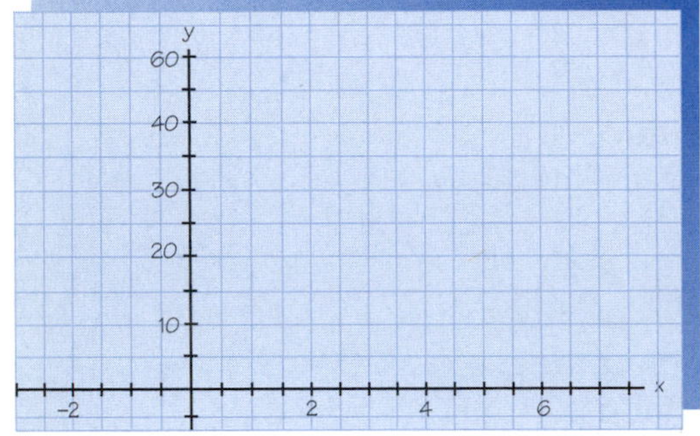

FIGURE 7.30.

b) Using any of the four graphs, trace over the portion that goes from $y = 10$ to $y = 15$. Then lay the tracing down on the other three graphs *in the same vertical position* [10, 15]. What do you notice?

c) How does the value 1.12 in these exponential equations affect the graphs of the equations?

Exponential equations are expressions that have the exponent as an independent variable. These equations can be written as: $y = ab^x$ (using a growth factor) or $y = a(1 + r)^x$ (using a relative rate). In either case the equations contain two "control" numbers:

- The starting value, a, works like the y-intercept of a line (since it is the value of y when $x = 0$). This value determines where the graph crosses the y-axis.
- The base is written either as a single number b using the growth factor, or an expression $(1 + r)$ where r is the relative rate of growth. The base determines how much change the graph experiences as x increases. Just like the growth pattern for geometric sequences, the amount of change keeps increasing (in fact, Δy is proportional to y). Hence, the shape of the graph is curved.

NEGATIVE GROWTH

Growth is not a one-way street where the values are always increasing. In mathematics we can talk about negative growth, where the values are decreasing. Some examples of negative growth include a decreasing rate of smoking among teenagers, or a decreasing unemployment rate like during the 1990's when the economic conditions were good.

6. a) Consider the following exponential equation: $y = 100(0.8)^x$. Calculate the value of y that corresponds to the x-values in **Figure 7.31**.

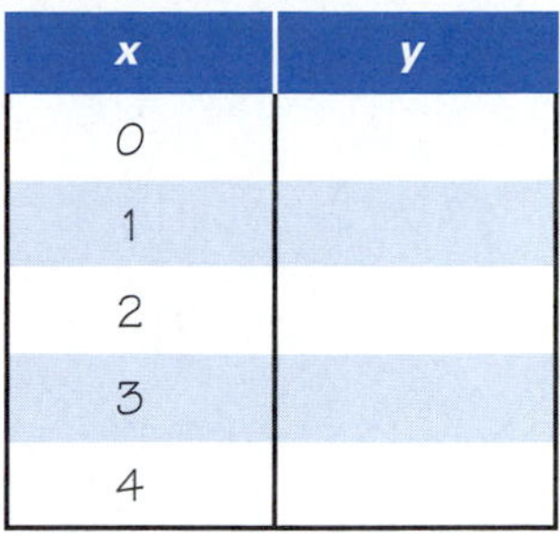

x	y
0	
1	
2	
3	
4	

FIGURE 7.31.

b) Make ordered pairs of the form (x, y) and plot the corresponding points on a grid like the one in **Figure 7.32**.

c) How is your graph different from the ones in Figures 7.29 and 7.30?

d) In the equation what control number is causing this type of pattern?

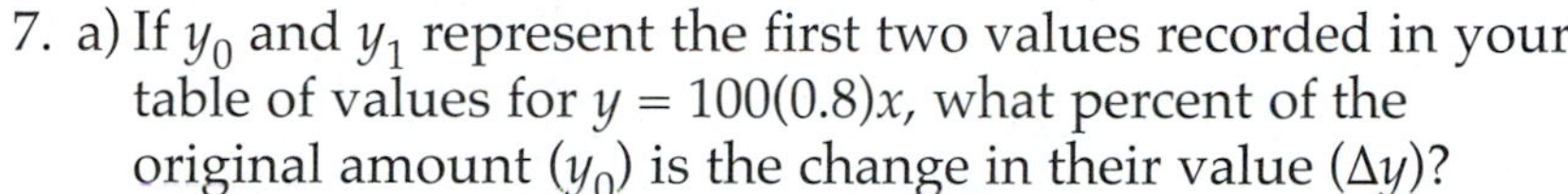

7. a) If y_0 and y_1 represent the first two values recorded in your table of values for $y = 100(0.8)x$, what percent of the original amount (y_0) is the change in their value (Δy)?

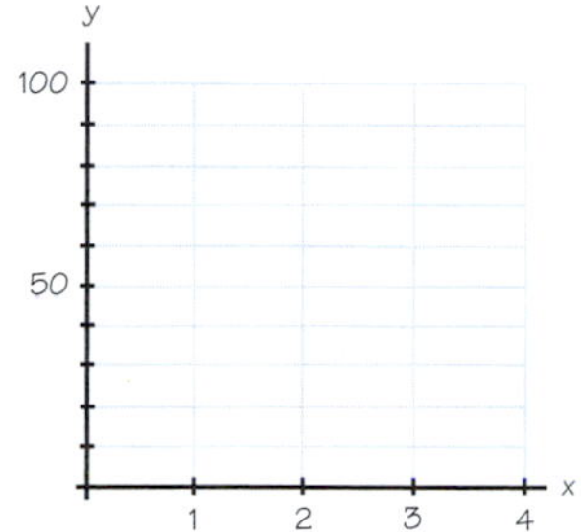

FIGURE 7.32.

b) Think of your answer to (a) as an "interest" rate r. What is the rate r in decimal form for this set of values?

c) Use your value of r to rewrite the equation using a relative rate.

d) If the table were extended to larger values for x, what pattern would be produced by the values for y?

8. Using a grid like the one in **Figure 7.33**, sketch the curve that represents the continuous graph for the equation $y = 100(0.8)^x$. Then verify your work by displaying the same equation on a graphing calculator.

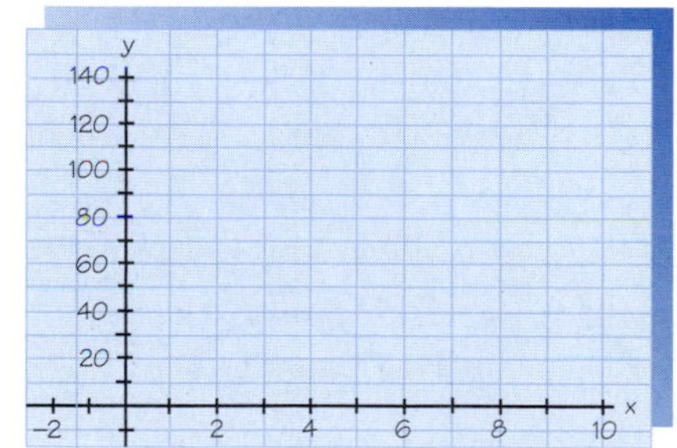

FIGURE 7.33.

Multiplicative models, such as geometric sequences or exponential equations, can describe quantities that *decrease* in proportion to the original amount. The type of pattern that suggests that "growth" is actually going down is called **decay**.

Decay models have the following properties:

- The growth factor b in the equation $y = a \cdot b^x$ is less than one.
- The percent change of successive values generated by these models is negative.

The graph of a decay model has the same curved shape as the graph of a growth model. The main difference is that the graph decreases from left to right and approaches the horizontal axis as the exponent (independent variable) gets large.

LOGARITHMS

The activities in this chapter demonstrate that understanding the behavior of exponentials is a key to financial success. Often the length of time needed for the investment is what you want to know. To answer a question about the length of time, you need to solve an exponential equation. Since the variable for time is an exponent in many recent formulas, we need a new way to solve these equations.

9. Assume that you have \$500 in a bank savings account that is compounded continuously at a rate of 11.33% per year. You wish to know how long it will take for the account to reach \$800.

   a) What is the growth factor that corresponds to this interest rate?

   b) What is the equation that describes the account balance over time?

   c) One way to find the time t it takes to reach \$800 is to build a table. You can make a table of values of t and A either by hand or by using the TABLE feature of a graphing calculator. Then see how long it takes to reach the desired balance of \$800. How many years are needed to the nearest 0.1 year?

   d) Another approach uses the graphing calculator to graph the equation recorded in (b) as $Y1$, and the condition for the problem (\$800) as $Y2$. Then determine the intersection point of the two graphs by using either the TRACE or INTERSECT features. Using this method, how many years are needed?

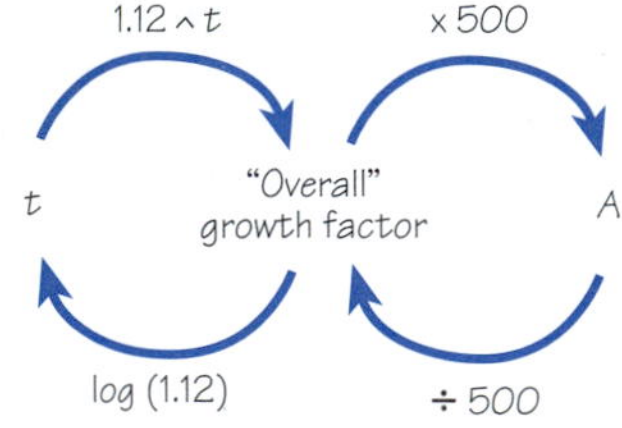

FIGURE 7.34.

There is an algebraic approach to solving this same equation. **Figure 7.34** shows the steps involved in evaluating (or solving) the exponential equation: $A = 500 \cdot (1.12)^t$.

To evaluate $A = 500 \cdot (1.12)^t$ (going from left to right in Figure 7.34), you find the value of base 1.12 raised to some power (the value of t). For any time period, that number is the growth factor from the beginning to the end of that time period. The next step uses the multiplier to determine the final account balance from the balance at the beginning of the time period.

Working backward to find t given A (right to left in Figure 7.34), divide the final account balance A by the beginning balance to obtain the same overall growth factor. Then reverse the process of raising the base to some power. The result is called a **logarithm** and can be found by using the log key on your calculator. The last step in determining this investment time can be found as:

$$t_{\text{investment}} = \frac{\log(\text{overall growth factor})}{\log(\text{yearly growth factor})}.$$

Be careful to put ending parentheses on each log calculation, and don't forget to divide.

10. Calculate the time it would take the investment in question 9 to grow from \$500 to \$1000. (This is called its **doubling time**.)

One of the places where decay occurs in the natural world is with radioactive materials. These are unstable substances that give off nuclear particles at a constant rate to become a different element. Because the rate is constant, the amount of material that decays in a time period is proportional to how much is present at the beginning of the time interval. The exponential models that describe continuously compounded interest can keep track of how much of a radioactive sample remains over time.

H Hydrogen																	He Helium
Li Lithium	Be Beryllium											B Boron	C Carbon	N Nitrogen	O Oxygen	Fl Fluorine	Ne Neon
Na Sodium	Mg Magnesium											Al Aluminum	Si Silicon	P Phosphorus	S Sulfur	Cl Chlorine	Ar Argon
K Potassium	Ca Calcium	Sc Scandium	Ti Titanium	V Vanadium	Cr Chromium	Mn Manganese	Fe Iron	Co Cobalt	Ni Nickel	Cu Copper	Zn Zinc	Ga Gallium	Ge Germanium	As Arsenic	Se Selenium	Br Bromine	Kr Krypton
Rb Rubidium	Sr Strontium	Y Yttrium	Zr Zirconium	Nb Niobium	Mo Molybdenum	Te Technetium	Ru Ruthenium	Rh Rhodium	Pd Palladium	Ag Silver	Cd Cadmium	In Indium	Sn Tin	Sb Antimony	Te Tellurium	I Iodine	Xe Xenon
Cs Cesium	Ba Barium	See Lanthanides *	Hf Hafnium	Ta Tantalum	W Tungsten	Re Rhenium	Os Osmium	Ir Iridium	Pt Platinum	Au Gold	Hg Mercury	Tl Thallium	Pb Lead	Bi Bismuth	Po Polonium	At Astatine	Ra Radon
Fr Francium	Ra Radium	See Actinides **	Rf-Ku (Rutherfordium Kurchatovium)	Ha Hahnium													

* Lanthanides	La Lanthanum	Ce Cerium	Pr Praseodymium	Nd Neodymium	Pm Promethium	Sm Samarium	Eu Europium	Gd Gadolinium	Tb Terbium	Dy Dysprosium	Ho Holmium	Er Erbium	Tm Thulium	Yb Ytterbium	Lu Lutetium
** Actinides	Ac Actinium	Th Thorium	Pa Protactinium	U Uranium	Np Neptunium	Pu Plutonium	Am Americium	Cm Curium	Bk Berkelium	Cf Californium	Es Einsteinium	Fm Fermium	Md Mendelevium	No Nobelium	Lr Lawrencium

11. a) One type of radioactive substance, called cobalt-60, decays fairly quickly. For any sample of colbalt-60, 12.4% of it will break down in a year. Assume that you have 5.00 grams of cobalt-60. After three years how much will remain?

    b) How long will it take for exactly half of the cobalt-60 to decay? (This is called the **half-life** for that substance.)

SUMMARY

In this activity you studied a form of interest called continuous compounding.

- You discovered that continuously compounding \$1 at an interest rate of 100% produced a special value called $e \approx 2.71828$.
- You learned that the formula for continuous compounding is $A = Pe^{rt}$.
- You explored negative growth and studied applications of exponential models in the natural world.

Now that you have learned some options for earning interest, it is time to put them to work. In the next activity you will explore what happens when money is invested regularly into the same account.

HOMEWORK

7.3

Life Has its Ups and Downs

A sequence is defined recursively according to the rule: $b_{n+1} = b_n + \frac{1}{n!}$, with $b_1 = 1$. The '!' is called the **factorial** symbol when it is used in mathematics. The value of $n!$ (read "n factorial") is calculated by multiplying all the counting numbers less than, or equal to, n. For example, $6! = 6 \cdot 5 \cdot 4 \cdot 3 \cdot 2 \cdot 1 = 720$.

1. a) Use a table like the one in **Figure 7.35** to calculate the next six terms of this sequence.

n	Current Term b_n	Value of $1/n$	Next Term b_{n+1}
1			
2			
3			
4			
5			
6			
7			

FIGURE 7.35.

b) What value does the sequence appear to approach?

2. Suppose $1 is deposited in an account that pays 10% interest compounded annually. The plan is to keep the money in the account for just the first term. Later, the owner decides to keep the money in the account for the second term as well.

a) We would like to know how the individual growth factors compare with the overall one. Fill in a table like the one in **Figure 7.36** with the growth factors that correspond to the specified term lengths.

1st Term Length	Growth Factor	2nd Term Length	Growth Factor	Total Term Length	Overall Growth Factor
2 years		5 years		7 years	
4 years		3 years		7 years	

FIGURE 7.36.
Calculating individual growth factors.

b) How can you determine the overall growth factor if you know only the individual ones? Test your idea on the two examples in Figure 7.36. Does it work?

c) What do you notice about the overall growth factor in each case? Does that make sense in this problem?

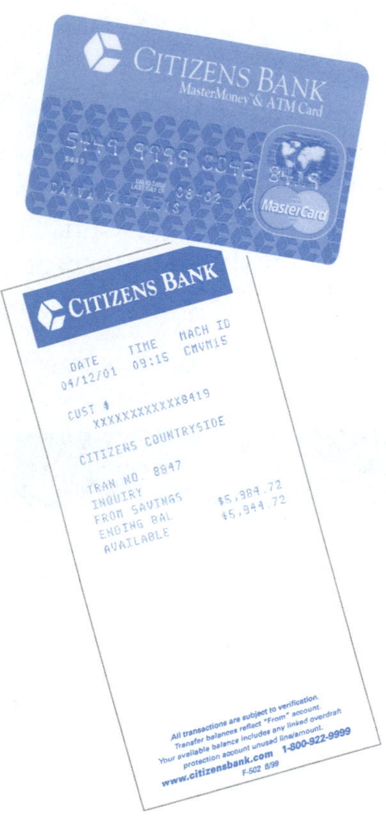

3. a) The situation represented by the first row in Figure 7.36 can be expressed using exponential notation as $(1.10)^2 \cdot (1.10)^5 = 1.10^7$. Express the second situation in the same fashion. Then use your calculator to see if the left side of each equation equals the right side. Is it true?

b) Based on your answer to (a) when multiplying two exponential expressions *that have the same base*, how can you get the exponent for the answer from knowing the exponents of the two expressions? (This is one of the Laws of Exponents.)

4. Suppose you deposit \$1200 in a bank savings account at an annual rate of 4%. How much interest do you receive after one year:

a) if the bank compounds continuously?

b) if the bank compounds daily using a 360-day year (called the "360 over 360" method)?

c) if the bank compounds daily using a 365-day year (called the "365 over 365" method)?

d) Why might a bank use the "360 over 360" method, anyway?

5. The exponential equation $A_t = A_0 \cdot b^t$ has four quantities in it where A_0 is the amount initially invested, b the yearly growth factor, and t the number of years. Given any three of them you can solve for the missing one. In each of the following, three quantities are given. Describe how to solve for the missing item, and then find its value.

a) $A_0 = \$800$; $b = 1.08$; $t = 5.2$ years

b) $A_t = \$1200$; $b = 1.12$; $t = 3.8$ years

c) $A_0 = \$1000$; $A_t = \$1100$; $t = 5$ years

d) $A_0 = \$1500$; $A_t = \$2500$; $b = 1.05$

6. The following exponential equations were displayed using a computer drawing utility: $y = 4(1.25)^x$; $y = 10(1.25)^x$; $y = 12(1.10)^x$; and $y = 10(0.75)^x$. The graphs are shown in **Figure 7.37**. Unfortunately the curves were not labeled and the axes don't show any scale. Explain how you can identify the graph of each equation.

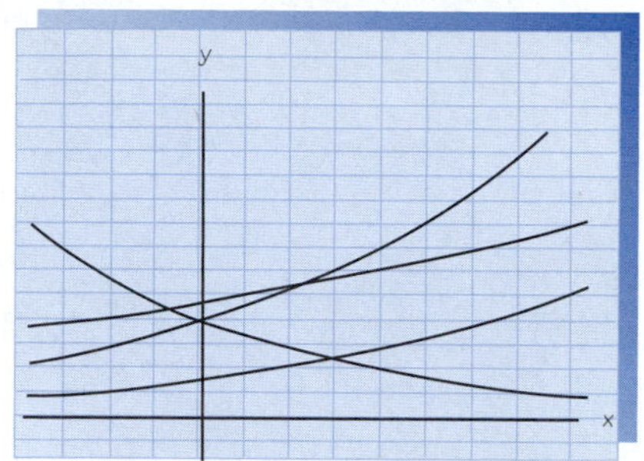

FIGURE 7.37.

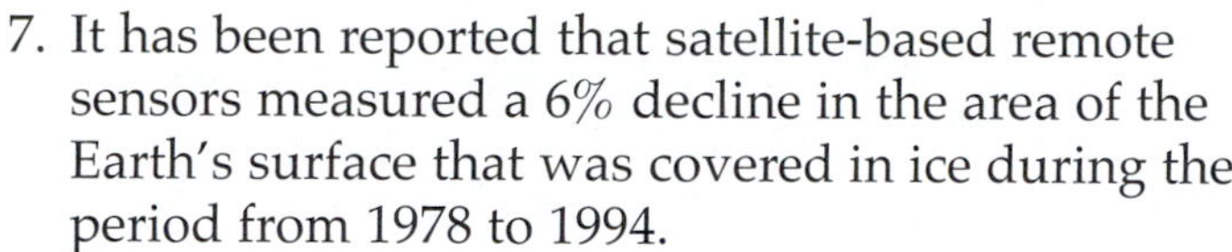

7. It has been reported that satellite-based remote sensors measured a 6% decline in the area of the Earth's surface that was covered in ice during the period from 1978 to 1994.

   a) What is the overall growth factor over the 16 years?

   b) Find the corresponding annual percent of decrease in the ice coverage.

   c) What is the annual growth factor?

   d) Using 1978 as year zero, and considering the amount of ice coverage to be 100% at that time, write an equation for the percentage of ice coverage remaining after t years.

   e) Graph your equation over a large domain (several centuries). If the model holds true well into the future, what is the long-term result of this decline in the ice coverage?

8. In 1996, a magazine article reported that the buffalo population in Yellowstone National Park had grown from 25 to 4000 during the period from 1900 to 1996.

   a) Find the annual growth factor for the buffalo population over that time period. Explain what it means.

   b) Write an equation to predict the population of buffalo in the park over any time period. (Use the year 1900 to represent $t = 0$.)

   c) Use the model developed in (b) to predict how many buffalo will be in the park in 2010.

During **inflation** prices go up. Therefore you have to spend more money to buy the same things. But you can also think of the money you *have* as not being able to get the same amount as before. In this interpretation the value of a dollar is decreasing.

9. Suppose there is a constant 2% annual inflation for the next four years as a result of economic policy changes in 2001.

   a) How much will you have to pay in 2005 for an item that currently costs $100?

   b) What is the cost of an item now that will cost $100 in four years. (That is, how much will $100 be worth then?)

Depreciation is an economic term that describes the decreasing value of some piece of equipment, like a car, as it ages. It was explored in Chapter 6 as one of the costs of owning a car.

10. Suppose you buy a car at the beginning of 1996 for $10,000, and there is no inflation over the next decade. Assume that after ten years, the worth of the car, based on the resale market, is only $2000.

    a) If the depreciation is linear, then how much does the value of the car go down each year?

    b) What will the car be worth at the beginning of 2001?

    c) If the depreciation is exponential, what is the rate of depreciation each year?

    d) At that rate what will the car be worth at the beginning of 2001?

11. The 2000 Census determined the official population figure used to allocate representatives to Congress. As of April 1, 2000, the population of the United States was around 281,422,000. That was a 13.2% increase over the 1990 census figure.

    a) What was the yearly percentage increase in the population?

    b) At that same rate of increase, what would the population of the U.S. be on April 1, 2003?

    c) If the population increases at a yearly rate of 1.00% instead, how much of an effect would that have on the population estimate from (b)?

12. Suppose you have a bank account with a balance of $5432.10 at the beginning of the year and $5632.10 at the end of the year. Your bank advertises "continuous compounding" but actually compounds continuously over each 24-hour day and posts interest to accounts daily.

a) What was the annual percentage yield for this account?

b) What was the nominal rate that the calculations were based upon?

c) What effect does the banking practice have on your account balance when compared to true continuous compounding?

13. a) A bank offers continuous compounding at the annual rate of 4.5%. Calculate the doubling time for a principal deposit of $1500.

    b) Calculate the doubling time if the principal deposit is $2500.

    c) What do you think determines the doubling time? Explain.

Carbon-14 is present in all living things, and undergoes radioactive decay. The half-life of carbon-14 is around 5580 years. The relative amount of carbon-14 that still remains in a fossil can be used to determine the age of a fossil.

14. Suppose that a fossil contains 0.027 grams of carbon-14; had it been alive it would have had 0.126 grams (the rest decayed). How long ago did the creature die?

ACTIVITY 7.4

"Sum" Kind of Savings

The savings accounts we have explored are like planting seeds. After putting the principal into the bank, you simply sit around and watch the money make money. The assumption is that the *only* money you actually put into the account is the principal. That works well for savings, but that isn't the only kind of investment strategy available to a smart investor. In this activity you will investigate other types of savings strategies.

ACCUMULATING SIMPLE INTEREST—ARITHMETIC SERIES

Does it make a difference if a series of deposits are made over time or the same total invested all at once? Let's start investigating that question by considering a savings account that pays simple interest.

CHECK THIS!

Simple interest is based on receiving interest only on the principal and no compounding takes place. For example if you deposit $100 at 5% simple interest:

Year	Principal	Interest Amount	Total
1	$100.00	$5.00	$105.00
2	$105.00	$5.00	$110.00
3	$110.00	$5.00	$115.00

1. a) Suppose $2500 is deposited into a bank account that earns 5% per year (simple interest) for six years. How much interest is earned each year?

   b) What is the total interest earned?

2. Suppose instead that $500 is deposited in the account at the beginning of each year. The account still earns 5% simple interest each year.

   a) Fill in a table like the one in Figure 7.38 with the amount of interest earned each year from this account.

Year (*n*)	Beginning Balance	Deposit Made	Total of all Deposits (Amount that Earns Interest)	Interest Earned during Year	Ending Balance
1	$0	$500			
2					
3					
4					
5					
6					

FIGURE

This strategy doesn't earn as much as before, since all the money isn't in the account for the entire six-year period. But all the payments that are made earn interest, and over time this interest becomes a noticeable amount. There is also a short-cut to calculating the total interest earned.

b) What is the total interest earned in years one and six? In years two and five? In years three and four? How could you calculate the total for all six years more quickly than adding all six numbers together?

Notice that each yearly interest is a multiple of 25, since 5% of 500 is 25. The sequence of interest payments forms an arithmetic sequence. Each term is the product of 25 and the year number, n. The *total* interest earned each year is also a multiple of 25. Since the total interest earned is the sum of the terms of an arithmetic sequence, it is called an **arithmetic series**. The total interest that has been paid through any number of years, n, can be found by multiplying 25 by the expression $n \cdot (n + 1) \div 2$.

c) Verify that the expression $25[n \cdot (n + 1) \div 2]$ calculates the total interest earned for all six years.

d) The balance after each year is simply the total amount deposited (which can be calculated from knowing how many years have elapsed) plus the total interest earned. Write an expression for the account balance at the end of any number of years, n. Check to make sure your expression correctly predicts the balance at the end of six years.

ACCUMULATING COMPOUND INTEREST—GEOMETRIC SERIES

The model for finding an **accumulation** in question 2 requires making fixed-amount payments on a regular schedule into a savings account that pays simple interest. In reality most banks pay compound interest instead. The extra interest earned by the interest forces us back to the drawing board.

3. a) Assume that the $500 is put in a saving account that earns 5% compounded annually at the beginning of each year for the same six-year time period. Fill in a table like the one in **Figure 7.39**. For each year calculate the amount in the account that earns interest (after the deposit is made), the amount of interest earned, and the ending balance.

Year (n)	Beginning Balance	Deposit Made	Total Amount Earning Interest	Interest Earned during Year	Ending Balance
1	0	$500	$500		
2		$500			
3		$500			
4		$500			
5		$500			
6		$500			

FIGURE 7.39. The value of an account earning 5% compounded annually when you deposit $500 each year.

CHECK THIS!

Compound interest receives interest on both the principal and any interest previously earned. It is compounded during a specified time period usually daily, monthly, or yearly. For example, if you deposit $100 at 5% interest compounded annually:

Year	Principal	Total
1	$100.00	$105.00
2	$105.00	$110.25
3	$110.25	$115.76

b) What two-step calculation finds the beginning balance for year two from the beginning balance in year one?

c) Does that same process work for finding the beginning balance for year three? Write a recursive equation for the sequence of beginning balances.

d) Use home-screen iteration to check the sequence of beginning balances.

- Type 0 then press <ENTER>.
- Next type (Ans+500)*1.05.
- Each time you press <ENTER> after that, you should get another value in the sequence.

CHECK THIS!

You obtain Ans by pressing the 2nd key, and then the (–) key.

e) How does the expression in (d) relate to the recursive equation you identified in (b)?

4. a) What two-step calculation finds the total amount earning interest for year two from the total amount earning interest in year one?

b) Does that same process work for finding the next value in *that* sequence? Write the recursive equation for this sequence.

c) Use home-screen iteration to check the calculations of the amount earning interest. What number do you enter first? What is the next command to give the calculator?

5. a) What two-step calculation calculates the interest earned during year two from the previous year's earned interest?

   b) Does that process work for finding the next value in the sequence of interest earned? Write the recursive equation for this sequence.

   c) Use home-screen iteration to check the calculations of the interest earned. What number do you enter first? Why? What is the next command to give the calculator?

MIXED GROWTH

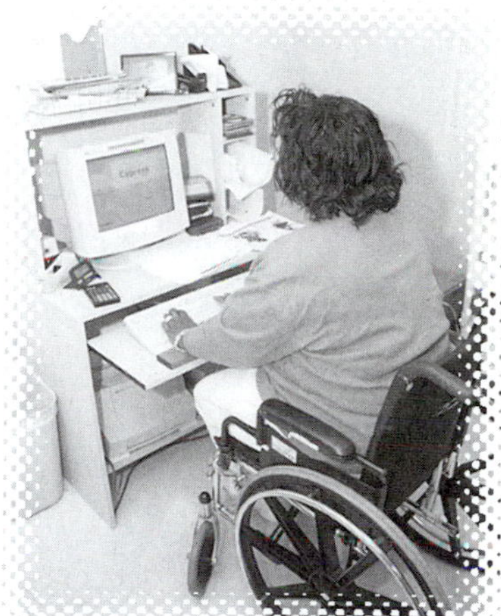

Making regular deposits into a bank account that earns compound interest creates what is called a **mixed growth** sequence. It can be described recursively as a repetitive two-step calculation of multiplying and adding (in either order, but the answer is affected by which is done first). Because the basic process is two arithmetic operations, it is often easy to "crank out" terms of such a sequence. However, finding the total of the various terms is a little more complicated.

6. Assume again that the $500 is put in a bank that earns 5% compounded annually at the beginning of each year for the same six-year time period. Only this time think of each deposit as a separate account with deposits made at different times, and different amounts of interest earned based on how long the money is in the account.

   a) Fill in a table like the one in **Figure 7.40**. Use the number of years each deposit is in the account to calculate the ending balance and interest that portion of the account earns.

Year (*n*)	Deposit Made	Number of Years Deposit Is in Account	Ending Balance	Total Interest Earned by Deposit
1	$500			
2	$500			
3	$500			
4	$500			
5	$500			
6	$500			

FIGURE 7.40.

   b) Write expressions that calculate the ending balance for each deposit.

c) Write *one* expression that calculates the ending balance at the end of the six-year period from all the deposits made. Verify that the expression yields the same result you recorded in (a).

d) Each term of the expression in (c) has two common factors, since every payment earns interest for at least one year. Simplify that expression using the Distributive Property so that the common factors are in front of a set of parentheses that group what is left of your original expression.

Geometric sequences can be formed by choosing any number as a base and raising it to various exponent powers, as long as the exponents are consecutive integers. When those sequence values are added together, that is called a **geometric series**. There is a shortcut to finding the sum of a geometric series when the first term is 1 and the common ratio is b:

$$1 + b^1 + b^2 + b^3 + \ldots + b^{n-1} = \frac{b^n - 1}{b - 1}.$$

7. a) Use the formula for summing a geometric series to help you find the sum of the expression you wrote in 6(d).

   b) Use the same formula to find the general formula for this type of savings plan. Assume the following three conditions.

      - Suppose at the beginning of each time period you put d dollars into an account.
      - Suppose the account has a growth factor of b during that time period.
      - Suppose there are a total of n time periods.

   c) To calculate the total interest, adjust the formula developed in (b) by subtracting an expression for the total of all the payments made into the account. Check to see if this new formula correctly predicts the total interest earned after six years.

One of the powerful features of mathematics is that a neat formula that required a lot of work to derive can be used to solve many problems of the same type. (If you are lucky, it can even describe situations that aren't from the same context, but *do* have the same underlying process!) The **savings formula** is one of those "friendly" formulas.

8. Now assume the money is deposited into a bank account that earns 5% compounded quarterly, and the deposits are made at the beginning of each quarter in payments of $125.

   a) Calculate the balance at the end of six years.

   b) If you wanted the account balance to reach $5000 in that same six-year period, how much would you have to deposit each quarter?

SAVINGS FORMULA

If you deposit d dollars per period (with the deposits at the beginning of the period) at an interest rate of r per period, the amount A accumulated is found by the savings formula:

$$A = d(1 + r)\left[\frac{(1+r)^n - 1}{r}\right].$$

When the deposits are made at the end of each period the formula is even simpler:

$$A = d\left[\frac{(1+r)^n - 1}{r}\right].$$

SINKING FUNDS, LOANS, AND OTHER APPLICATIONS

Savings plans that accumulate a fixed sum by a certain date are called **sinking funds**, often for the purpose of paying off a loan all at once at the end of the term.

CHECK THIS!

Money borrowed works like money invested; an interest rate is established at the time of the loan, and the total payment is based on that interest rate and how long it takes to pay off the loan.

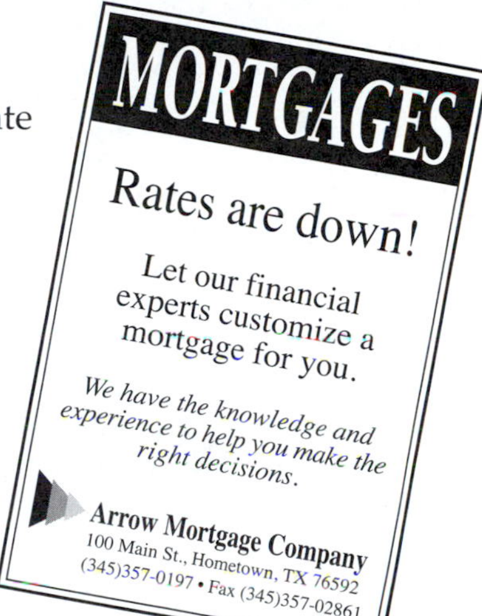

9. a) Suppose some payments are deposited in a bank account in order to pay back a loan. The loan amount is $5000 at 5% compounded quarterly, and the term is for six years. In order to pay it off by the end of six years what is the total amount that is owed from the loan?

   b) If you want the account balance to reach that amount in the same six-year period, how much would you have to deposit each quarter?

Loan payments pay off any interest earned during the compounding period *first*. Any remaining money is applied toward reducing the actual debt, or the principal on the loan.

10. Suppose that payments of $246.88 are made at the end of each quarter on a loan that was made with an interest rate of 5% compounded quarterly.

a) When the balance on the loan is \$6000, how much interest is due at the end of the first compounding period? How much is actually paid on the principal?

b) When the balance on the loan is only \$2000, you are making the same loan payments. How much interest must you pay now? How much is paid on the principal now?

Loans are usually set up to be paid back in equal installments with payments made on a regular basis. The payments **amortize** the loan; a portion of any payment is applied to the interest earned during that compounding period. The rest of the payment reduces the balance of the loan a little. During the next compounding period the amount owed is slightly less, so the interest you owe is a little less as well. Thus more of the payment is applied to reducing the loan balance.

If the principal on the loan is P, the interest rate per compounding period is r, the payment made at the end of each period is d, and the number of loan payments n, then the **amortization formula** is:

$$P(1+r)^n = d\left[\frac{(1+r)^n - 1}{r}\right].$$

The savings formula can be used in situations that don't have anything to do with money (or saving). Some resources, like petroleum, are **non-renewable** resources; when you use them up they are gone.

11. Current estimates of the remaining recoverable oil in the United States put the total around 200 billion barrels. One estimate of the oil consumption in the United States right now is 18.5 million barrels per day. (We would like to consider what happens if our resources alone are used to meet our demand, so assume that no oil is imported from other countries.)

a) At that consumption rate, how many barrels of oil are consumed per year?

b) How many years can the oil supply last if the consumption rate remains constant?

c) Over the last 25 years the rate at which the petroleum is consumed has gone up by 0.5% per year, on average. If that rate of increase remains approximately the same, how many barrels will our country consume next year? The year after that?

Think of this oil depletion situation as a savings account where every year you are depositing the amount consumed this year *plus* a little more each additional year (that extra due to the increased demand is like the compound interest being earned).

12. a) Use the savings formula to write an equation that describes the total consumption of oil for any number of years in the future, n, with this year corresponding to $n = 0$. For simplicity, assume that the oil is consumed all at once and at the end of the year.

    b) Use the fact that the total U.S. resources are estimated at 200 billion barrels to determine how many years the oil supply can last if the consumption rate continues to increase by 0.5% per year.

    c) One source estimates that the consumption rate will increase at a rate of 1–2% per year in future years. Using the mid-range value, how many years can the oil supply last if the consumption rate increases at a rate of 1.5% per year?

SUMMARY

In this activity you studied accounts in which you make fixed deposits at regular time intervals.

- You learned that formulas for summing sequences may be used to model the account balances generated by these types of deposits.
- You examined mixed growth sequence patterns and applied that mathematics to savings plans, loans, and non-renewable natural resources.

Now you have the models to study retirement accounts and other accounts where withdrawals are made. Using these models, you can start thinking about retirement even before you've even started your career.

HOMEWORK 7.4 "Sum-mation" Arguments

1. It would be nice to know where the formula for summing an arithmetic series comes from.

   a) Think about the problem: $1 + 2 + 3 + 4 + 5 + 6 + 7 + 8$. If you add the first and last numbers together, what sum do you get? The second and second to last number? The third and third from the end? The two middle numbers?

   b) How many pairs of numbers can be formed from this list? Use the results of (a) to find the sum quickly by *multiplying*. What numbers do you multiply together? What is the sum?

   c) The formula for adding up counting numbers starting from 1 (and going up to n) is:

   $$1 + 2 + 3 + \ldots + (n - 1) + n = \frac{n(n+1)}{2}.$$

   In the problem in 1(a), what is the value for n? How do the numbers you used in (b) relate to the formula for summing the counting numbers?

   d) **Figure 7.41** shows another way to view the problem of adding up the counting numbers from 1 to 8:

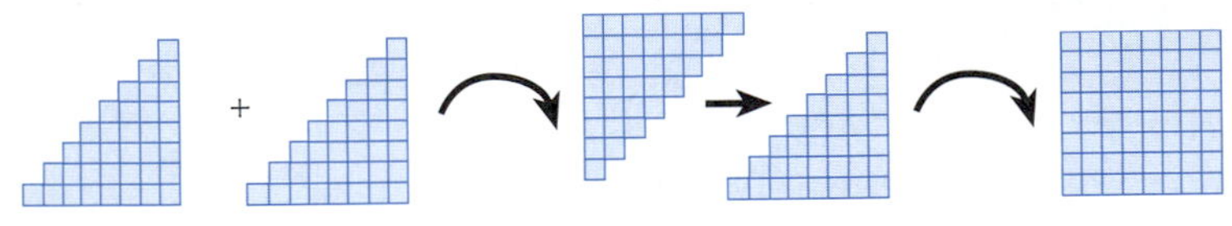

FIGURE 7.41.

How does this way of viewing the problem help explain how the formula works?

A "slick trick" for adding terms from an arithmetic sequence is to list the numbers in order, then list them in reverse order on the next row, as shown:

$$\begin{array}{c} 1 + 2 + 3 + 4 + 5 + 6 + 7 + 8 \\ 8 + 7 + 6 + 5 + 4 + 3 + 2 + 1 \end{array}$$

Start adding pairs vertically, and discover that each sum is the same. To get the final answer, multiply the number that you get as that sum by the number of pairs you have and then divide by 2 (since you listed the numbers twice). In this case, the sum would be $(9)(8) = 72 \div 2 = 36$.

2. a) Suppose the problem is changed to add these numbers: $3 + 4 + 5 + 6 + 7 + 8$. Explain how to find the sum of these numbers easily.

   b) Suppose the original problem is changed to: $5 + 8 + 11 + 14 + 17 + 20 + 23 + 26$. Find the sum of these numbers easily.

3. Each of the following has terms of an arithmetic sequence being added together. In each case find the indicated sum without actually adding the numbers together:

   a) $1 + 2 + 3 + 4 + ... + 24$

   b) $12 + 13 + 14 + 15 + ... + 48$

   c) $4 + 9 + 14 + 19 + 24 + ... + 74$

   d) The first 25 terms of the sequence: $S_n = 2 \cdot n + 13$, where $n = 1, 2, 3, ...$

4. The first five terms of a mixed-growth sequence are provided in **Figure 7.42**.

n	S_n	1st Differences
0	12	
1	20	
2	32	
3	50	
4	77	

FIGURE 7.42.

   a) Calculate the first differences for the sequence, S_n. Record your answers in the third column of Figure 7.42. Is the S_n an arithmetic sequence? Explain.

   b) Calculate the ratio between successive values of S_n for several terms. Is S_n a geometric sequence? Explain.

   c) Is the sequence of first differences arithmetic? If so, what is the common difference?

   d) Is the sequence of first differences geometric? If so, what is the common ratio?

   e) Mixed-growth sequences are produced by using a two-step rule of multiplying and adding (or subtracting) in going from one term to the next. Use the work in (c) and (d) to figure out the rule that was used. Write the recursive equation for this mixed-growth sequence.

5. Each of the following are examples of mixed-growth sequences.

   a) Find the 15th term of the sequence: $A_n = 1.5 \cdot A_{n-1} + 125$; $A_0 = 500$.

b) Find the 23rd term of the sequence: $A_n = (A_{n-1} + 60) \cdot 1.2$; $A_0 = 80$.

c) Find the ending balance after one year for a savings account in which \$1200 principal is deposited, an additional \$200 is deposited on the last day of each month, and interest is earned at the rate of 8% per year compounded monthly.

n	Series Expression	Value
0	1	
1	$1 + 2^1$	
2	$1 + 2^1 + 2^2$	
3	$1 + 2^1 + 2^2 + 2^3$	
4	$1 + 2^1 + 2^2 + 2^3 + 2^4$	
5	$1 + 2^1 + 2^2 + 2^3 + 2^4 + 2^5$	

FIGURE 7.43.

d) Find the 18th term of this sequence: 11, 13.1, 15.41, 17.951, 20.7461, ..., where $S_0 = 11$.

6. It would be nice to know where the formula for summing a geometric series comes from.

a) The table in **Figure 7.43** lists several examples of a geometric series formed by summing terms from a geometric sequence that uses a growth factor of 2.

For each value of n (from 1 to 6), find the value of the series expression. Record your answer in the third column of Figure 7.43. What do you notice about the pattern formed by the values?

b) Use the pattern you identified in (a) to predict the value of this expression:

$$1 + 2^1 + 2^2 + 2^3 + 2^4 + 2^5 + 2^6 + 2^7 + 2^8 + 2^9 + 2^{10} + 2^{11}.$$

Then verify your prediction using the formula for summing a geometric sequence.

There is a different "slick trick" for finding the sum of a geometric sequence easily. Start off listing the various terms, just like before. On the line above, multiply each term by the base and list those answers above and to the right of the term. In this particular case it will look like this:

$$\begin{aligned} 2^1 + 2^2 + 2^3 + 2^4 + \ldots + 2^{10} + 2^{11} + 2^{12} \\ 1 + 2^1 + 2^2 + 2^3 + 2^4 + \ldots + 2^{10} + 2^{11} \end{aligned}$$

Then just subtract the second line from the first line (good things happen!).

c) After subtracting the two lines, what will remain as a result of doing all this "work?"

d) If the variable A represents the original problem (bottom line), what expression represents the top line? What expression results from subtracting the bottom line from the top line?

e) Since the results of (c) and (d) must be equal, set the two answers equal to each other and solve the equation for A. Before working out the actual answer, what does the resulting expression look like?

f) **Figure 7.44** contains two columns of algebraic expressions. Using the Distributive Property, multiply every term in the first factor by *each* term in the second factor. Add "like" terms to finalize your answer. Record your final answer in the third column of Figure 7.44.

1st Term	2nd Term	(1st Term) · (2nd Term)
$x - 1$	1	
$x - 1$	$1 + x$	
$x - 1$	$1 + x + x^2$	
$x - 1$	$1 + x + x^2 + x^3$	

FIGURE 7.44.

g) What do you notice about the pattern produced by multiplying the two expressions together? By that same pattern what should $(x - 1)(1 + x^1 + x^2 + x^3 + \ldots + x^n)$ equal?

h) Explain how the pattern described in (d) relates to the formula for summing a geometric sequence.

7. Find the sum of each of the following geometric series using the summation formula:

a) $1 + 1.5^1 + 1.5^2 + 1.5^3 + 1.5^4 + 1.5^5 + 1.5^6 + 1.5^7 + 1.5^8$

b) $90 + 90(0.75)^1 + 90(0.75)^2 + 90(0.75)^3 + 90(0.75)^4 + 90(0.75)^5$

c) $10(1.2)^1 + 10(1.2)^2 + 10(1.2)^3 + 10(1.2)^4$ (Caution: this one starts with the first term, not the zero term, but every term in the expression has two factors in common!)

8. If you buy a home by taking a 30-year mortgage for \$80,000, at an annual interest rate of 8% compounded monthly:

a) How much will the payments be at the end of each month?

b) How much will you end up paying for your house?

9. A couple wishes to save for the college education of its child. They want to make regular deposits at the end of each month into the account, which earns 6.5% interest per year compounded daily.

a) Using the fact that a year has around 360 days (360 over

360 method), find the rate of increase for the account for one month.

b) If they want to have $100,000 available when the child turns 18, how much should they save each month?

10. You decide to buy a new car. After making a down payment you need to borrow $12,000 from a finance company. You check the interest rates offered by the car dealership, the local banks, and your credit union. The best rate you can find is 7.7% compounded monthly over 48 months. What will be your monthly payment?

11. Suppose you begin retirement with $1 million in savings and you don't trust banks or the stock market so you put your money in the old "mattress account." It costs you $50,000 in living expenses each year and you pay for everything with cash out of the mattress.

    a) If there is no inflation, how long will your nest egg last?

    b) If there is constant 5% per year inflation it will cost more to live each year. How long will your savings last then?

12. In 1990 the known global oil reserves totaled 917 billion barrels. Consumption, which had been 53.4 million barrels per day in 1983, rose an average of about 1.7% per year through 1990, when the consumption was about 60 million barrels per day.

    a) Predict the year that the world's oil reserves will run out if the consumption remains a constant 60 million barrels per day after 1990. (Use 1990 as time $t = 0$).

    b) If the oil consumption continues to increase at the rate of 1.7% per year, predict the year that the world's oil reserves will run out.

    c) What other assumptions (besides the consumption rate) do these predictions take into consideration?

ACTIVITY

7.5

A Balancing Act

You have seen how various decisions affect how much money you make, and how much money you will have. These decisions include the type of interest an investment earns, and whether to make one lump-sum deposit or lots of smaller ones on a regular basis. In this activity you will learn how to manage your money once it has been made.

FINDING THE RIGHT BALANCE

The accumulation of wealth is only half the battle—you need to know what to do with it once you have it. Whether as a gold miner or a day trader on the stock market, history is full of stories about people who made fortunes only to lose them afterwards. The difference between keeping your money or squandering it away is in understanding how it changes over time.

Suppose you are close to retirement age and have been saving money your entire adult life. You have managed to put enough money in a savings plan so you will get $100,000 when you retire. Not wanting to risk losing your life savings you put the money in an account that earns 5% per year, compounded monthly. At the end of each year you will withdraw $20,000 to use on living expenses.

1. a) Determine the account balance at the end of each year; record your work in a table like the one in **Figure 7.45**.

Year	Beginning Balance	Interest Earned	Withdrawal Amount	Ending Balance
1	$100,000			
2				
3				
4				

FIGURE 7.45.
The value of an account that earns 5% compounded monthly when you withdraw $20,000 each year.

b) What was the change in the account balance after the first year? What is the change between the beginning and ending balances of the second year?

c) How can you tell that your money was already in trouble from just examining those two figures?

d) The recursive calculation for the ending balance is $B_{t-1} \cdot (1 + 0.05/12)^{12} - 20000$. Use home-screen iteration to determine how many years it will take for the account to run out of funds.

2. a) Suppose the $100,000 in question 1 is deposited in an account that has an interest rate that earns 6% compounded quarterly. If only $10,000 is withdrawn at the end of each year would the account run out of funds? If so, how many years would it take?

   b) If the interest rate were changed to 10% compounded bi-monthly, and the amount withdrawn at the end of the year reduced to only $8000, would the account run out of funds? If so, how many years would it take?

The last two questions show that a financial planner must consider the following factors for each plan:

- the principal
- the interest rate
- the number of compounding periods per year
- the amount withdrawn each year

All these factors affect whether the account grows or decays, and how quickly.

Since money doesn't grow on trees it is unlikely that the principal can be changed, except to move money from one account to another. Furthermore you cannot change the interest and the number of compounding periods the banks set for their accounts. You can choose a different account but those conditions are still dictated to you. However, you have total control over how much money to withdraw from the account. Spend too much and you'll go broke!

3. a) Start with the same amount of accumulated wealth ($100,000), and invest it in an account that pays 5% interest rate, compounded annually. This time take out only $5000 at the end of each year. Determine the account balance at the end of each year using the table in **Figure 7.46**.

Year	Beginning Balance	Interest Earned	Withdrawal Amount	Ending Balance
1	$100,000			
2				
3				

FIGURE 7.46. The value of an account that earns 5% compounded annually when you withdraw $5000 each year.

b) What makes this situation different than the previous ones? If a little less was withdrawn each year, how would the situation change?

c) If the account was changed to pay 5% interest, but compounded monthly, how much could you take out of the account each year and still have the same results as in (a)?

4. a) Suppose the interest rate was changed to 6% compounded quarterly, and you wanted to be able to withdraw $10,000 at the end of each year. Explain how to determine the amount to deposit as principal and still have the same results as in question 3(a)?

b) To the nearest $10,000, find the principal that produces the same results as in question 3(a).

BRINGING THINGS INTO BALANCE

Certain mixed-growth sequences are a combination of "increasing" and "decreasing." Some of these sequences are generated by an expression that is the difference of two terms:

A term with a growth factor greater than one – some constant amount.

Other sequences are generated by an expression that is the sum of two terms:

A term with a growth factor that is less than one + some constant amount.

Under the right conditions a sequence will approach **equilibrium**, a stable condition in which the amount increasing and the amount decreasing balance each other and no more change takes place. **Figure 7.47** illustrates several ways of thinking about equilibrium:

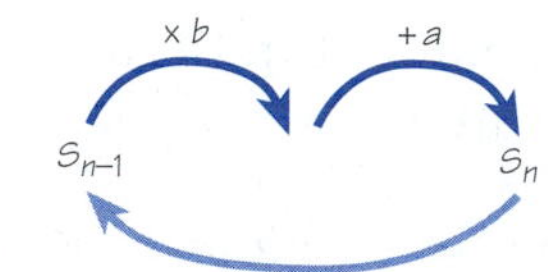

FIGURE 7.47.

"It either IS a constant sequence, or it gets closer and closer to becoming one ..."

n	S_n
:	:
48	1245.97
49	1245.99
50	1246
51	1246
:	1246

5. As protection against hard times you decide to put $10,000 into your mattress. The money loses value at the rate of 2% per year. For example, at the end of the first year the $10,000 is worth ($10,000)(0.98) = $9800. You would like to have approximately the same purchasing power, so you decide to add an extra $300 at the end of each year (a little more than what you lost can't hurt!).

   a) Use **Figure 7.48** to determine the worth of the money at the end of each year. Also find the change in the worth of the money that takes place from one year to the next.

Year	Previous Worth	Amount Devalued	Amount Deposited	Current Worth	Change in Worth (ΔW)
0	-----	-----	-----	$10,000	-----
1	$10,000				
2					
3					
4					

FIGURE 7.48. Depositing $300 each year in an account that loses 2% of its value each year.

   b) Is the worth of the account going up or down?

   c) Is the change in how much the account is worth going up or down?

   d) What kind of sequence is the change in the worth of the money, ΔW? What recursive formula describes that sequence?

CHECK THIS!

Recall that inflation means that prices are increasing. Thus you must spend more than $10 to buy items that used to cost $10. In other words, the value of a ten dollar bill decreases because it cannot buy the same amount of things it used to buy.

For any time period, you can think of the current worth as being an accumulation of (or adding together) the changes in the worth of the money ΔW and the principal.

6. a) Does the sum of the numbers for the ΔW column in question 5(a), added to the principal, give you the current worth at the end of year four?

   b) Write an expression for the sum of the numbers in the ΔW column after *any* number of years.

   c) Adjust the expression you got in (b) to predict the current worth of the $100,000 after any number of years.

   d) Use the expression in (c) to calculate the current worth after 20 years.

e) Repeat (c) for 100 years and for 500 years.

f) Does this sequence approach equilibrium? If it does, what is the highest value that the sequence will have? (Hint: You could use home-screen iteration, but it will take a while. It will be easier to repeat the calculations using the expression from (c) with much longer time intervals!)

That probably seems like a lot of work to find out whether the terms of a mixed-growth sequence approach equilibrium, and to find the equilibrium value. Notice however, the sequence of ΔW values approached equilibrium instead of just "being" at equilibrium. This sequence also provided some insights on how the change in growth behaves. Also you were able to write an expression describing the ending balance. Hang in there; there are short-cut approaches!

To make the next point more "graphic," consider a more extreme situation than the one described in question 5.

7. Suppose the money is losing value at a rate of 10% per year. (Let's hope this is not *really* the case!) Start with $1000 in principal, and add an extra $500 at the end of each year.

   a) For each year in **Figure 7.49** find these values:

   - the previous year's actual worth
   - the current year's actual worth (including the extra money that is added)
   - what the current year's worth *would* be if the process was already at equilibrium, based on the previous year's value.

Year	Actual Worth for the Previous Year	Actual Worth for the Current Year	Equilibrium Worth the Current Year
1	$1000.00		
2			
3			
4			

FIGURE 7.49.
Calculating equilibrium worth.

   b) For each year consider the difference of the values:

   Actual Worth For the Current Year — Equilibrium Worth For the Current Year

   Do these values form a geometric sequence? If so, find the growth factor.

c) Based on your answers to (b), are the sequences of actual worth for the current year and equilibrium worth getting closer together?

8. a) What is the recursive formula that describes the current year's actual worth from the previous year's actual worth?

   b) What is the recursive formula that describes the equilibrium worth for the current year from the previous year's actual worth?

   c) Using either SEQ or FUNC mode on your graphing calculator, display the graph of the equations at the same time. Adjust your WINDOW to be able to identify the actual equilibrium value. What WINDOW settings did you use? Make a sketch of the calculator display.

   d) Is the mixed-growth sequence for current worth going to approach equilibrium? What feature of the graphs can tell you this?

   e) What is the equilibrium value that the sequence is approaching?

One short-cut method to determine whether a sequence approaches equilibrium is to look at graphs of the equation of the actual sequence and the equation of a hypothetical equilibrium sequence. If equilibrium is to occur, the growth sequence needs to approach equilibrium. So the graph that describes the growth sequence must approach the graph describing the equilibrium sequence. Since these recursive graphs are linear, the picture should show two lines intersecting at a point. The point of intersection represents the condition that the sequence has reached equilibrium, and its coordinates yield the equilibrium value.

9. Each equation identified in question 8 represents a condition on the problem, and both conditions must be true at the same time for the sequence to be at equilibrium. In mathematics this is called a system of equations, and there are standard algebraic methods for solving such a system.

   a) Since both equations must have the same y-value, set each of the equations from question 8 equal to each other. (This is using the substitution method.) What new equation is produced?

b) Solve that equation for x. What equilibrium value do you obtain? (Note: Since the sequence is at equilibrium, the x- and y-coordinates will be the same.)

c) Determine the equilibrium value for the problem described in question 5 using algebra.

When a mixed-growth sequence is at equilibrium, you may not know what the value is. But you *do* know that the next value must be the same as the previous one. **Figure 7.50** illustrates this:

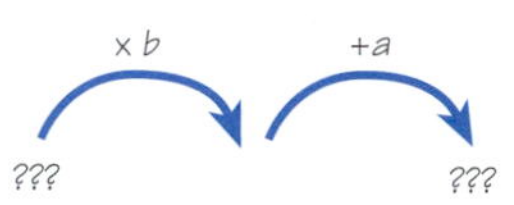

FIGURE 7.50.

Since the *answer* after the calculations must be the same as the starting number, the equation to solve for finding the equilibrium value is always set up in the following way: $bx + a = x$, where the first x is the previous value and the second x is the current value. This assumes that there *is* an equilibrium condition for the sequence.

Patterns of mixed-growth have important implications for managing **renewable resources**—ones that can be restocked, or replenished, over time.

10. You hold the timber rights to a tract of land that contains an estimated 60,000 trees. Because of the geography, cutting down trees will make it harder to find more trees to cut down. As a result you can only harvest 20% of the available forest each year. Being an environmentalist (as well as an amateur mathematician), you *know* you will run out of trees by simply cutting them down, and that is *not* okay. So you plant 8000 trees each year to replace the ones cut down.

a) Will you eventually run out of trees? Explain.

b) Will the size of the forest ever stabilize (reach equilibrium)? If so how many trees will be on the property eventually?

c) Suppose a forestry biologist estimates that the available land could support a maximum of 75,000 trees. How many trees would you need to plant each year if the harvest rate were kept at 20% per year?

d) Corporate management decides to use this property at a point in the future when the cost of wood will be much greater. So they lower the harvest rate to 10% per year, which will maintain the property by removing the older growth. If they still want the land to have 75,000 trees, how many trees would need to be planted each year?

SUMMARY

In this activity you studied patterns that are formed by repeating two-step processes of multiplying and adding (or subtracting). The mixed-growth sequences that are formed combine the properties of both arithmetic and geometric sequence patterns. You found:

- Equilibrium takes place when the difference between successive terms of a mixed-growth sequence get continually smaller. Eventually the pattern becomes a constant sequence.
- Equilibrium value can be found in a variety of ways:
  - using home-screen iteration
  - graphing the recursive equation of the actual sequence with the recursive equilibrium equation ($y = x$), and looking for the point of intersection
  - solving the equation formed by setting the two recursive equations equal to each other using algebra.
- Equlibrium is a useful condition to achieve (or maintain) when managing resources in which a portion is being consumed and hopefully replaced.

HOMEWORK 7.5

Getting There is Half the Fun

1. Each of the following *looks* like a mixed-growth sequence. List the first four terms, including the initial value, and describe the sequence.

   a) $S_n = S_{n-1} \cdot 0 + 20$; $S_0 = 0$.

   b) $S_n = S_{n-1} \cdot 1 + 10$; $S_0 = 350$.

   c) $S_n = S_{n-1} \cdot 1.04 + 0$; $S_0 = 100$.

   d) $S_n = S_{n-1} \cdot 1 + 0$; $S_0 = 240$.

An equilibrium value is called **stable** if you can start at other initial values below (or above) it, and have the sequence still approach equilibrium; otherwise it is called **unstable**.

2. If you start with the given initial value, each of the following mixed growth sequences is at equilibrium. In each case decide if it is stable or unstable.

   a) $S_n = S_{n-1} \cdot 0.90 + 200$; $S_0 = 2000$.

   b) $S_n = S_{n-1} \cdot 1.25 - 500$; $S_0 = 2000$.

   c) $S_n = S_{n-1} \cdot 1.08 - 160$; $S_0 = 2000$.

   d) $S_n = S_{n-1} \cdot 0.75 + 500$; $S_0 = 2000$.

   e) $S_n = S_{n-1} \cdot 0.95 + 100$; $S_0 = 2000$.

3. Based on the work done in question 2, what condition do the stable mixed growth sequences have in common? Explain.

4. For each of the following mixed-growth sequences, determine the equilibrium value.

   a) $S_n = S_{n-1} \cdot 0.84 + 500$.

   b) $S_n = S_{n-1} \cdot 1.12 - 200$.

   c) $S_n = S_{n-1} \cdot 0.97 + 150$.

5. An account is started with a principal amount of \$50,000 earning 6% per year compounded monthly, but \$100 is withdrawn from the bank at the end of each month.

   a) Calculate the previous and current balance for each month; record your answers in the second and third columns of the **Figure 7.51**.

FIGURE 7.51.

Month	Previous Balance	Current Balance	Change in Balance (ΔB)
0	-----	\$50,000.00	-----
1			
2			
3			
4			
5			
6			

   b) Calculate the change in the account balance from one month to the next; record your answers in the fourth column of Figure 7.51.

   c) Write an expression for the sum of the numbers that make up the change in the account balance after any number of months. Then adjust the expression to calculate the current balance after any number of months.

   d) Use the expression you developed in (c) to find the current balance after five years (60 months).

6. You have seen that the sum of the geometric series $1 + 0.8 + 0.8^2 + 0.8^3 + 0.8^4 + \ldots + 0.8^n$ is determined by the following formula:

$$S = \left[\frac{(0.8^n - 1)}{0.8 - 1}\right]$$

for any number of terms, n.

   a) For the given values of n, calculate the value of the expression 0.8^n, the numerator of the fraction, and the value of the fraction itself. Record your answers in a table like **Figure 7.52**.

n	Value of 0.8^n	Value of $(0.8^n - 1)$	Value of Entire Fraction
1			
5			
10			
20			
50			
100			
500			

FIGURE 7.52.

b) What happens to the value of the expression 0.8^n as n gets very large? Why does that happen? How does that affect the value of the entire numerator?

c) Explain why the value of the fraction must approach a "largest" value in this case. How does this help explain the behavior of mixed-growth sequences as they approach equilibrium?

d) If you know the value for n is really big, how could you determine the sum of all those terms quickly? For any growth factor, b, which is less than 1, what rule would find the sum of the first million (or billion) terms?

The pattern in a mixed-growth sequence that approaches equilibrium can be visualized in a different way using a **web diagram**.

7. Work with the recursive sequence: $S_n = S_{n-1} \cdot 0.6 + 8$; $S_0 = 4$.

   a) Calculate the previous and current values for the sequence; record your work to the nearest 1/10th in a table like **Figure 7.53**.

n	Previous Value S_{n-1}	Current Value S_n
1	4	
2		
3		
4		
5		
6		

FIGURE 7.53.

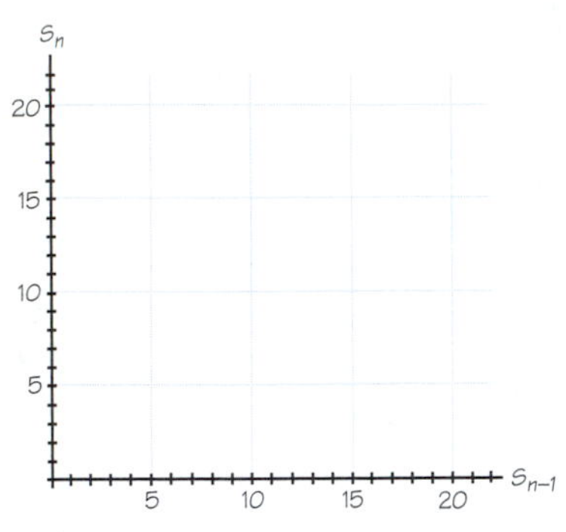

FIGURE 7. 54.

b) For this next part, copy the grid provided in **Figure 7.54**. Start by making a dot on the x-axis at 4 that corresponds to the first value for S_{n-1}. Then follow this rule: Draw a vertical line until the y-coordinate matches the S_n value, then draw a horizontal line until the x-coordinate matches the next S_{n-1} value. Repeat this process until you run out of values in the table.

c) Draw the line that goes through the top of each vertical line. What is the equation for that line?

d) Draw the line that goes through the right endpoint of each horizontal line. What is the equation for that line?

e) What does the intersection point represent for this problem? How does the web diagram help visualize that process?

8. You have an accumulation of $420,000 in your retirement account that is earning 6% per year compounded monthly.

   a) What is the largest amount that you can withdraw at the end of each month without causing the account balance to decrease?

   b) If you can find an account that is paying 6% interest per year but is compounded *continuously*, how much can you withdraw now (without affecting the account balance)?

   c) If you wanted to live off of the withdrawals you make from your original 6% retirement account (without reducing the balance), and needed $3000 per month in income, how much should you have accumulated before retiring?

   d) Even if you didn't save enough you can still accomplish your financial goals by finding an account that compounds monthly but pays a higher interest rate. What rate would allow you to take out $3000 per month without reducing the balance?

9. A trout farm is estimated to have around 5000 fish when it opens for business. Each day approximately 2% of the fish currently in the lake are caught. At the end of each week the lake is restocked with 1200 additional fish.

   a) How many fish are caught by the end of the first week?

   b) Will the lake run out of fish? Explain.

c) How many fish will be in the lake at the end of the summer (20 weeks later)?

d) If the business were to stay open all year, how many fish would eventually be in the lake?

10. Suppose there are 115 of a particular animal species in a wildlife preserve and the population is growing naturally at a rate of 11% per year. In order to manage the growth, wildlife biologists in the preserve decide to remove 15 of the animals at the end of each year and relocate them to other areas in an effort to further protect their survival.

    a) How many animals will be in the preserve in ten years? How many will there be in 20 years?

    b) Calculate the equilibrium value for this situation. How would you use that answer to improve the efforts of the wildlife biologists?

11. Suppose a professional baseball player signs a new contract for $25 million per year to be paid at the end of each year for the next ten years. Assume that the team's payroll account earns 0.5% per month.

    a) How much money would the team need to set aside on January 1 in order to pay the player's salary out of interest earnings *only*? (In other words, they would still have the principal at the end of the ten years.)

    b) How much money would the team need to set aside on January 1 in order to pay the player's salary out of the money in the account (interest plus withdrawals)? (In other words, they would have a balance of zero dollars at the end of the ten years.)

Modeling Project: Money for Life

One *very* real mathematics problem that faces all people in our society is managing the money earned in income and planning for your retirement when that cash flow suddenly stops. Do you buy a house or rent, and at what point? Do you keep your money secure and easily accessed, or do you put it in an account that pays higher interest, but with restrictions on its use?

The real world of finances is incredibly complicated with interest rates that change almost daily, speculative investments that might lose all value, and a multitude of costs—some changing with the seasons, others constant, and a few that seem to come from nowhere (and always at the worst possible time). Even the decision to raise a family has serious implications for your finances. As good mathematical modelers, it is better to keep the problem reasonable and work on making it more realistic in the next version of the model.

For that reason, this Modeling Project is going to include only the following assumptions:

- You just finished school, have no outstanding debts, and $10,000 in a bank savings account.
- You just took a job at a firm where you will remain for the rest of your professional career. The starting salary is $32,000 and you receive pay raises every two years in the amount of $2500.
- You are currently 25 years old and can work for up to 40 years. After that you have to retire.
- The selling price for a typical house is $180,000 and is going up by 2% per year (the value of the house goes up with it at the same rate).
- You need at least 20% of the selling price of a house as a down payment, and the closing costs are 5% of the selling price (on average). The down payment must come from assets but the closing costs can be financed with the house mortgage, which is financed at 6.5% per year but compounded monthly over 30 years.
- You will be in the 15% tax bracket until your salary goes above $44,000; then it will be 20% until your salary goes above $100,000. (Note: This is greatly simplified from the tax schedule provided by the Internal Revenue Service.)
- You currently need $1400 a month in living expenses including the monthly rent on your apartment, which is currently $900 but goes up by $100 every two years. The other expenses (and your lifestyle) go up by 3% per year.
- Your savings account pays 5% interest compounded monthly.
- Certificates of deposit (CDs) pay 7% interest compounded quarterly but you have to keep the money in the bank for two full years.
- You may put up to $2000 in an Individual Retirement Account (IRA) every year, which pays 10% per year. Any contribution to an IRA is not taxed until the money is taken out of the account. However, you may not take out any of the money until you retire.
- Plan right now for your financial future. You know the mathematics and you can explore this problem with spreadsheets as another option. What investments will you make? At what point in time will you make them? When you retire what will your worth be? Try and make as much money as you can—but remember, *"Money doesn't buy happiness!"*

Practice Problems

Review Exercises

1. Identify each of the following as being an arithmetic, geometric, or mixed-growth sequence. If it is arithmetic or geometric, provide the closed-form equation for the sequence. In each case find the twentieth term of the sequence.

   a) $A_n = A_{n-1} \cdot 0.9 + 20$; $A_0 = 48$.

   b) 12, 18, 27, 40.5, 60.75, …, where A_0 is 12.

   c) 24.60, 27.80, 31.00, 34.20, 37.40, …, where A_0 is 24.60.

   d) $A_n = A_{n-1} + 12.50$; $A_0 = 37.50$.

   e) $A_n = A_{n-1} \cdot 1.25$; $A_0 = 160$.

   f) 5, 10, 16, 23.2, 31.84, 42.208, …, where A_0 is 5.

2. A principal amount of \$3500 is deposited in an account that earns 4.5% interest. Calculate the ending balance after five years if the interest is:

   a) simple interest

   b) compounded annually

   c) compounded monthly

   d) compounded daily, using the 360 over 360 method

   e) compounded continuously

3. One critical skill in working with the models introduced in this chapter was determining an overall growth factor (multiplier) from a rate of increase (or vice versa) for a specified time period. Find the missing values in **Figure 7.55**.

APR	Compounding Condition	Time Interval	Growth Factor	APY
5%	annually	3 years		
5%	bi-monthly	6 mos.		
5%	monthly	2.5 years		
5%	continuously	1/4 year		
	bi-monthly	1 year	1.10	
	quarterly	1.5 years	1.25	
	monthly	5 years	2.00	
	semiannually	4 years		8%
	bi-monthly	3 years		10%

FIGURE 7.55.

4. Each of the following graphs describes an arithmetic, geometric, or mixed-growth sequence pattern. In each case identify the kind of sequence that the graph represents, and explain how you can tell the type of sequence from the behavior of that particular graph. (Caution: Pay attention to the axis labels.)

   a) the sequence described by graph 1 in **Figure 7.56**

   b) the sequence described by graph 2 in Figure 7.56

   c) the sequence described by graph 1 in **Figure 7.57**

   d) the sequence described by graph 2 in Figure 7.57

   e) the sequence described by graph 3 in Figure 7.57

   f) the sequence described by the graph in **Figure 7.58**

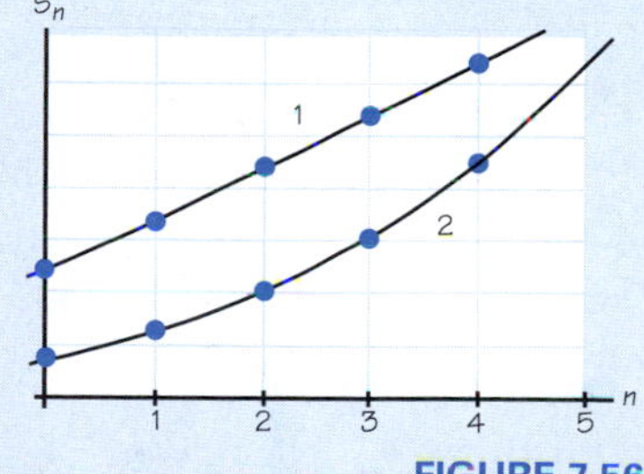

FIGURE 7.56.

FIGURE 7.57.

5. Each of the following involves working with an arithmetic sequence $A_n = A_0 + d \cdot n$.

   a) Calculate the twentieth term if the first term is 53 and the common difference is 4.

   b) Calculate the common difference if the twelfth term is 88 and the twenty-second term is 148.

   c) Calculate the term number associated with a sequence value of 153 when $A_0 = 41$ and $d = 4$.

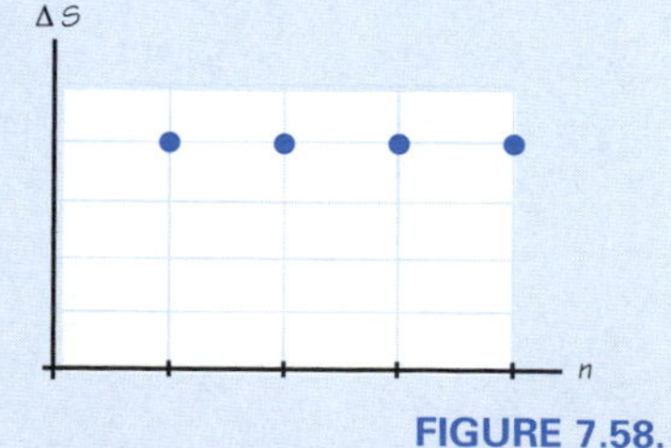

FIGURE 7.58.

6. Each of the following involves working with a geometric sequence of the form: $A_n = A_0 \cdot b^n$.

   a) Calculate the starting value A_0, knowing that the growth factor is 1.12 and the fifteenth term of the sequence is approximately 263.

   b) Calculate the growth factor knowing that the sequence starts with $A_0 = 180$, and the twelfth term of the sequence has a value of 48.

   c) Calculate the term number associated with a sequence value of approximately 817 when $A_0 = 110$ and the common ratio $b = 1.2$.

7. **Figure 7.59** shows an exponential graph; the equation is not known, but it *does* go through the points (4, 6) and (6, 10).

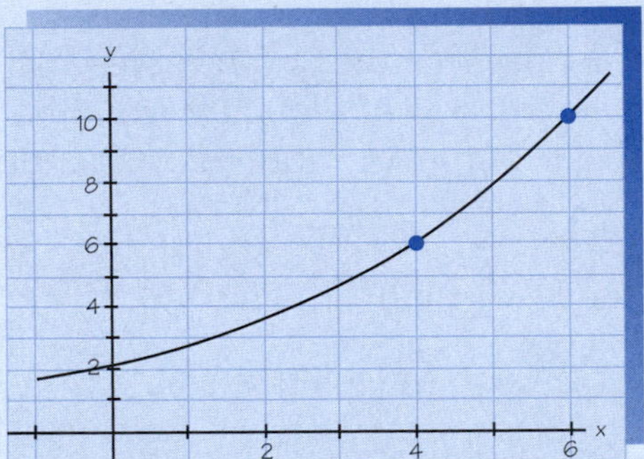

FIGURE 7.59.

   a) Find the growth factor for the exponential equation. (Hint: Think about geometric sequences; the fourth term has a value of 6, and the sixth term has a value of 10.)

   b) Find the initial value for the exponential equation. What method did you use to get your answer? Does it match the graphical behavior in Figure 7.59? Explain.

   c) What is the equation that corresponds to this exponential graph? Verify that the two points satisfy the equation.

8. In Activity 7.3 one important property of exponential equations that was briefly mentioned is summarized in the statement: "Δy is proportional to y." It would be nice to verify that statement for a couple of examples.

   a) Use the equations $y = 12(1.25)^x$ and $y = 25(0.90)^x$ to find the column of y-values in **Figure 7.60**. Then calculate the Δy values (amount of change in going from one line to the next) and record those answers in the table.

$y = 12(1.25)^x$		
x	y	Δy
0		—
1		
2		
3		
4		
5		
6		

$y = 25(0.90)^x$		
x	y	Δy
0		—
1		
2		
3		
4		
5		
6		

FIGURE 7.60.

b) For the equation $y = 12(1.25)^x$, is Δy proportional to y? In other words, is there a direct variation relationship between the two quantities? Explain.

c) For the equation $y = 25(0.90)^x$, is Δy proportional to y? Explain.

d) What is the relationship between the constant of proportionality (or slope of the graph of the direct variation) and the original problem?

9. For each of the following problems, find their sum.

   a) 7 + 7 + 7 + 7 + 7 + 7 + 7 + 7 + 7 + 7 + 7 + 7 + 7 + 7 + 7 + 7 + 7 + 7 + 7

   b) 10 + 13 + 16 + 19 + … + 82

   c) $8 + 8(1.2) + 8(1.2)^2 + 8(1.2)^3 + \ldots + 8(1.2)^{10}$

10. The following is the recursive equation for a mixed growth sequence: $A_n = A_{n-1} \cdot 0.80 + 12$; $A_0 = 10$.

    a) How can you tell from looking at the equation that the sequence approaches equilibrium?

    b) What is the equilibrium value for the sequence?

    c) If the value for A_0 could be changed, what starting value would produce a *constant* sequence? Explain.

11. Suppose the authorities at Adirondack State Park had counted 27 moose in 1988 and 42 moose in 1993.

    a) If you assume that the growth is caused by migration, what would be the migration rate (average number of moose migrating each year)?

b) At that same migration rate, how many moose would be in the park in the year 2013?

c) If you assume that the growth is caused by reproduction, what would be the annual percentage increase for the moose population?

d) At that same reproduction rate, how many moose would be in the park in the year 2013?

e) How many years would it take for the population of moose to reach 84 moose (double the original amount)?

12. If you invest \$5000 into a savings plan that pays 6% interest compounded quarterly, and you wish to have \$9000 after five years, how much should you make in payments at the end of each quarter? (Hint: How much will the principal earn? The rest needs to come from the savings plan accumulation.)

13. You wish to buy a house that is selling for \$220,000. Closing costs will be an extra \$5600. You are going to pay 20% of the total cost of the house immediately as the down payment, and you want to finance the rest over 30 years through a mortgage company that charges 7.6% interest per year compounded monthly.

    a) What is the total cost for the house?

    b) How much are you going to need for the down payment?

    c) What will your monthly payments be?

14. As regional director of the Fish and Game Department, you estimate that there are 6000 deer in the area, and their population grows at a rate of 9% per year. To best manage this valuable resource you would like to keep the population size reasonably constant. How many deer should you allow to be harvested (killed by hunting parties) next year?

7 CHAPTER REVIEW

Modeling Growth

Model Development

In this chapter several types of growth models have been introduced, along with mathematical "machinery" to describe them. In most cases the growth pattern was described in two different ways—recursively (in which the previous value in the pattern is used to determine the current one) and closed-form (where the term number, or index, is used to determine the value in the pattern).

The first type of growth model was called "additive," since it involved adding a constant number to generate a new term. Informally, the rule describing an additive process recursively would be: "To get the next value, add the number to the previous value." That rule can be stated more formally: $A_{\text{next}} = A_{\text{previous}} + d$, where d is the constant number being added each time. To complete the description, the rule must also specify a starting point or value, A_{start}. In closed form the same additive model would look like this: $A = A_{\text{start}} + d \cdot n$, where n represents the number of times the constant value was added. The other basic type of growth model was called "multiplicative," since it involved multiplying by a constant number to generate a new term. Informally, the multiplicative process can be described recursively as: "To get the next value multiply the number by the previous value." That rule can also be stated formally: $A_{\text{next}} = A_{\text{previous}} \cdot b$, where b is the constant number being multiplied together, and should have a starting value specified as well. In closed form a multiplicative model would have this form: $A = A_{\text{start}} \cdot b^n$. Examples of additive models were simple interest and migration, while compound interest and reproduction were used to illustrate multiplicative growth.

Mathematics Used

The mathematics needed to work with growth models was conveniently handled by marking each possible value with an index, or subscript, and creating a sequence. Arithmetic sequences manage additive growth models, and the equations were slightly altered to include the index: $A_n = A_{n-1} + d$ (with A_0 as the starting value), or $A_n = A_0 + d \cdot n$. Constant sequences were a special case of arithmetic sequences in which $d = 0$. The common difference could be found by subtracting two values and dividing by the difference in their term values. Geometric sequences describe multiplicative growth models with the equations changed to look like this: $A_n = A_{n-1} \cdot b$ (with A_0 as the starting value), or $A_n = A_0 \cdot b^n$. The growth factor, b, could be found by dividing two terms and taking a radical, using the difference in the term values as the root specified. Finally mixed-growth sequences were described recursively, involving both adding and multiplying in either order. Equations describing mixed-growth sequences in which multiplication came first had the form: $A_n = A_{n-1} \cdot b + d$. When addition came first, the equation had the form: $A_n = (A_{n-1} + d) \cdot b$.

Exponential functions were also introduced in this chapter to work with the situation in which interest is compounded *continuously*. This is a class of mathematical equations of the form: $y = a \cdot b^x$, with $a \neq 0$ and $b > 0$. Exponentials are the continuous model that has the same pattern of a geometric sequence, for a particular value of b. The graphs of exponentials move away from the x-axis slowly and eventually look like one side of a hill that gets increasingly steep. The amount of curve depends on whether the value for b is close to 1, and the orientation for the graph on whether b is greater than 1 (or less than 1). If a value for the equation is known, the exponent which produced that value can be determined. Divide the value by a, then divide the log of that answer by the log of b. Exponentials that have a growth factor $b < 1$ were distinguished from the others, and are said to decay instead.

For certain situations, such as savings plans and amortized loans, it was necessary to add up the various terms of a sequence. When the sequence is arithmetic, the sum is called an arithmetic series. The formula for an arithmetic series is based on adding up the counting numbers from one to n: $1 + 2 + 3 + \ldots + n = n \cdot (n + 1)/2$. The following formula shows how to add up terms of an arithmetic sequence:

$$a_0 + a_1 + a_2 + \ldots + a_n = (a_0 + a_n) \cdot \frac{n+1}{2}.$$

When the sequence is geometric, the sum is called a geometric series. The formula for a geometric series is based on adding up numbers which are consecutive powers of the same base: $1 + b + b^2 + b^3 + \ldots + b^n = (b^{n+1} - 1) / (b - 1)$. The following formula shows how to add the terms of a geometric sequence:

$$a + a \cdot b^1 + a \cdot b^2 + a \cdot b^3 + \ldots + a \cdot b^n = a \cdot \frac{(b^{n+1} - 1)}{b-1}.$$

The savings formula was an equation describing how much money would be in a bank account if regular payments of d dollars were made at the end of each compounding period, and the account had a constant interest rate of r per period. According to that formula, the amount A that was accumulated (both deposits and interest earned) would be:

$$A = d \cdot \frac{(1+r)^n - 1}{r}.$$

The amortization formula was an equation that could determine the monthly payments d that are necessary to repay an amortized loan at a fixed rate of r per period, when the principal P was to be paid off in n installments. The amortization formula is:

$$P \cdot (1 + r)^n = d \cdot \frac{(1+r)^n - 1}{r}.$$

Finally a special condition for certain types of mixed-growth sequences was studied. If one of the two steps in the recursive calculation causes the quantity to go up while the other causes it to go down, it is possible to find a situation in which the two growths balance each other and the quantity will remain constant. This condition is called equilibrium. When the growth factor $b > 1$, equilibrium will take place only when the starting value is a particular amount. When $b < 1$, the value will approach the equilibrium condition as long as the amount being added isn't too big. Equilibrium turned out to be of interest in retirement planning and as an ideal goal for managing renewable resources.

Accumulation: Combination of deposits made on a regular basis and the interest that those deposits make.

Additive Model: One that features adding the same number repetitively to some initial value.

Amortization: Process in which a debt is repaid in regular installments.

Amortization Formula: Formula for situations involving installment loans:

$$P(1+r)^n = d\,\frac{(1+r)^n+1}{r},$$

where P is the principal, r is the effective interest rate per compounding period, d is the payment amount made at the end of each period, and n is the number of payments required to pay off the loan.

Annual Percentage Yield (APY): Actual percentage increase in the account, expressed as a yearly percentage rate.

Arithmetic Sequence: Sequence formed by specifying an initial value, and generating each successive term by using a rule that continually adds a fixed amount to the previous term.

Arithmetic Series: Accumulation of an arithmetic sequence.

Balance: Amount remaining in a savings account, or remaining to be repaid on a loan.

Base: Number that is raised to a power in an exponential expression.

Bi-monthly: Term used to describe six compounding periods per year.

Closed-form equation: An equation that finds the value of one variable given the value of another variable.

Compound Interest: Method of paying interest on both the principal amount and the interest previously earned.

Compounding Period: Time interval between interest calculations.

Constant Difference: The fixed amount being added to successive terms in an arithmetic sequence.

Constant Sequence: Sequence in which all terms are the same.

Continuous Compounding: Payment of interest based on using the limiting rate from compounding as frequently as possible.

Decay: Negative exponential growth; corresponds to a growth factor in which $0 < b < 1$.

Doubling Time: Time it takes for a quantity that is growing exponentially to become twice as much as before.

Equilibrium: Condition on a mixed-growth sequence growth, in which the positive and negative changes balance out, and the values for the sequence remain constant for higher term numbers.

Exponent: Number that indicates how many times the base is multiplied.

Exponential: Growth that is proportional to the size (or amount) of the quantity.

First Differences: Difference between successive values in a sequence.

Geometric Sequence: Sequence of numbers that can be described recursively by a multiplication operation, or is characterized by a repetitive multiplicative process.

Geometric Series: A sum of consecutive terms (either some or all of them) of a geometric sequence.

Growth Factor: Amount by which an exponential quantity is multiplied in one recursive period.

Half-Life: Time it takes for a quantity that is decaying exponentially to become half as much as before.

Home-Screen Iteration: Feature of a calculator that allows you to repeat a process (many times) simply by pressing the ENTER or RETURN key.

Index: Number used to identify specific terms of a sequence.

Inflation: Rate at which the cost of goods increases.

Installment: Amount paid on a regular basis to repay a loan.

Migration: Process by which the population (wildlife or people) changes by moving from one location to another.

Mixed Growth: Sequence of numbers that can be described recursively by a combination of one multiplication and one addition operation.

Multiplicative Model: One that features multiplying the same number repetitively by some initial value.

Nominal Interest Rate: The stated interest rate on an investment without including compounding.

Present Value: The value of money today that is to be received in the future.

Principal: Initial deposit or balance on a savings account or loan.

Quarterly: Four times per year.

Recursive: Description of change during a time interval, based upon the previous value.

Recursive Equation: An initial value and a rule that calculates a term in terms of an earlier term.

Renewable Resources: Resources, such as trees and fish populations, that can be restocked or replenished over time.

Reproduction: Process by which the population (wildlife or people) increases by births.

Savings Formula: Formula for the amount in an account in which regular deposits are made:

$$A = d\frac{(1+r)^n - 1}{r},$$

where A is the amount accumulated, d is the amount deposited at the end of each compounding period, r is the nominal interest rate over the compounding period, and n the number of compounding periods (and payments).

Savings Plan: Savings account with compound interest, in which deposits are made regularly for the purpose of accumulating money.

Semiannual: Term used to describe two compounding periods per year.

SEQ Mode: Calculator setting that allows for working with sequences.

Sequence: A list of values, ordered by using an index.

Simple Interest: Method of paying interest on only the initial balance in an account, and not on any interest that is earned.

Sinking Fund: Savings plan to accumulate a fixed sum by a certain date, often for the purpose of paying off a loan at the end of the term.

Stable: Condition in which starting at an initial value near the equilibrium value will allow the sequence to get closer still to equilibrium.

Subscript: Method of identifying different values, but using the same variable name; in sequences, the subscript would be the term number, and written to the right and down from (and slightly smaller than) the variable name.

Term Number: Another name for an index; used to identify specific sequence values.

Unstable: Condition in which starting at an initial value near the equilibrium value will cause the sequence to "move away" from equilibrium.

Web Diagram: Recursive graph technique for showing the patterns of change in a mixed-growth sequence through the use of vertical and horizontal line segments.

Chapter 8: Modeling Motion

How do functions help make action-packed movies and television?

When you see a stunt person in a movie jumping from a car that is going off a cliff, you probably don't think about mathematics. Believe it or not, a lot of mathematics is used to prepare for such a hair-raising fall.

To grab your interest and entertain you, a movie stunt must have at least some element of danger. A successful and thrilling movie stunt takes a huge amount of planning and precise measurements. Think about that person jumping out of a car that is about to sail off a cliff. When should the jump happen? Where should the jump happen? How should the fall be broken to prevent injury? These are some of the many questions that must be answered as the professional stunt designers plan the scene.

Mathematics cannot solve all these problems, but you can use mathematics to model the motion of the car so that the stunt person knows when to jump. This type of stunt is one that you will investigate in this chapter. Also you will study some models of motion and use the models to plan some realistic stunts.

PREPARATION READING

It's Show Time!

You've probably watched scenes like these in the movies:

> *The star, with no other route of escape, jumps off a rooftop, miraculously landing in the back of a passing pickup truck.*
>
> *During a chase, the star speeds down the street on a motorcycle, crossing an intersection and narrowly missing a truck.*

Auto thrill shows also present many stunts using speeding cars and trucks. Jeff Lattimore's specialty in Chittwood's Thrill Show is "The Leap for Life." Jeff has performed it successfully for years. In his stunt, Jeff climbs a ladder and stands on an eight-foot stool. Then the ladder is removed, leaving Jeff stranded just as a car comes speeding toward him. Jeff jumps a moment before impact. The car hits the stool and snaps it out from under him. Jeff lands safely on the ground.

Many stunts are very complicated and can require a lot of equipment. Many involve cars and trucks traveling at high speed. To entertain the audience, a stunt must be exciting and have an element of danger. Since lives are at risk, trial-and-error alone would be a poor method for planning these stunts. Just one mistake could result in loss of life. Mathematical calculations and an understanding of the laws of physics are often an important part of planning successful stunts.

In this chapter you'll plan several of the stunts described in this reading. You'll get a chance to test small-scale versions of some of these stunts in your classroom.

ACTIVITY

8.1

Walking the Walk

The Preparation Reading described several types of stunts. In this chapter you'll study each one. You'll begin with the design of a two-vehicle, near-collision stunt. In order to avoid a collision, it's important to be able to measure precisely the position and speed of each vehicle. In this activity you will study all about speed.

MEASURING AVERAGE VELOCITY

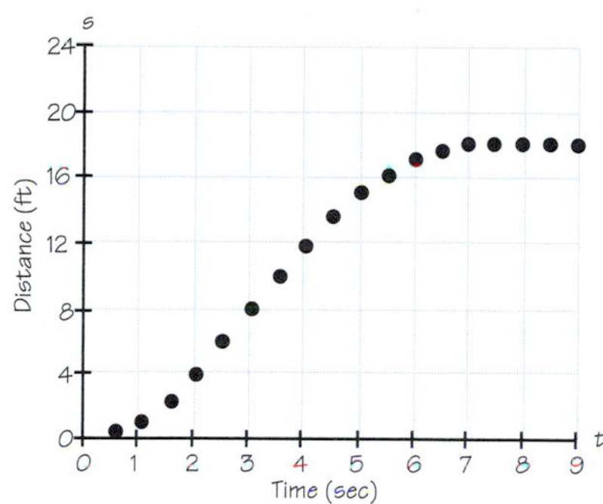

FIGURE 8.1.
Distance-versus-time graph for a toy car.

Imagine that a battery-operated, toy car is moving away from you along a straight line across the floor of your classroom. **Figure 8.1** shows the graph of the distance between you and your car during the nine-second trip.

The rate of change of distance with respect to time is called **velocity**. Velocity is usually measured in feet per second (ft/sec) or in miles per hour (mi/hr or mph). For example, suppose the velocity of a car traveling on Interstate 40 near Amarillo, Texas is 70 miles per hour. If the car's velocity stayed constant for an entire hour, the car would travel 70 miles in an hour.

We can calculate the **average velocity** of the car over the interval from time 1 to time 2 by measuring the distance the car travels during that time.

Notice that the formula for average velocity is just the formula for the slope of the line joining the points (time 1, distance 1) and (time 2, distance 2) shown in **Figure 8.2**.

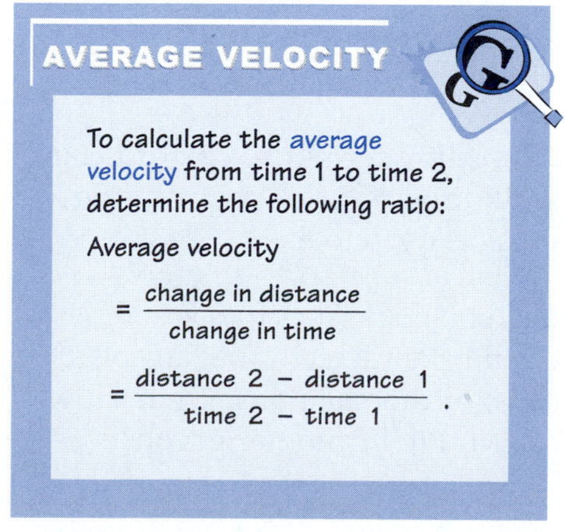

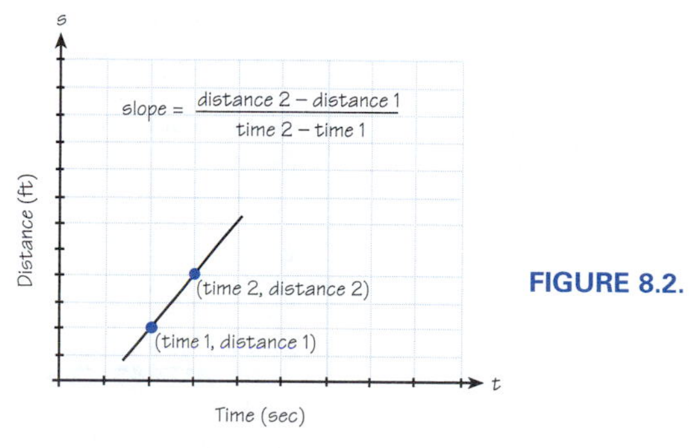

FIGURE 8.2.

1. a) The graph in Figure 8.1 provides a lot of information about the car's velocity during its nine-second trip. What is the average velocity of the car from $t = 0$ to $t = 9$? (Be sure to include the units for the velocity.)

   b) What is the average velocity over the two-second time interval from $t = 0$ to $t = 2$?

   c) What is the average velocity from $t = 2$ to $t = 4$?

   d) Is the car traveling faster during the first two seconds of its trip or during the second two seconds? How could you tell from the graph in Figure 8.1?

2. a) Find a two-second interval over which the car appears to be traveling at a constant velocity. What is this velocity?

   b) Can you find more than one such two-second interval during which the car's velocity appears to be constant?

3. Describe in words what is happening to the toy car's velocity during its nine-second trip.

MOTION GRAPHS

For the remainder of this activity you will analyze data created by a student who walked in front of a motion detector. Using a motion detector to track a student walking is no different than using the motion detector to track the motion of a toy car. The advantage of tracking the walk of another student is that you can easily imagine how you would have to walk in order to produce a similar distance-versus-time graph.

4. Cara walked in a straight line in front of a motion detector. The motion detector tracked her walk by recording her distance from the motion detector every 0.1 second for six seconds. Cara's distance-versus-time graph appears in **Figure 8.3.**

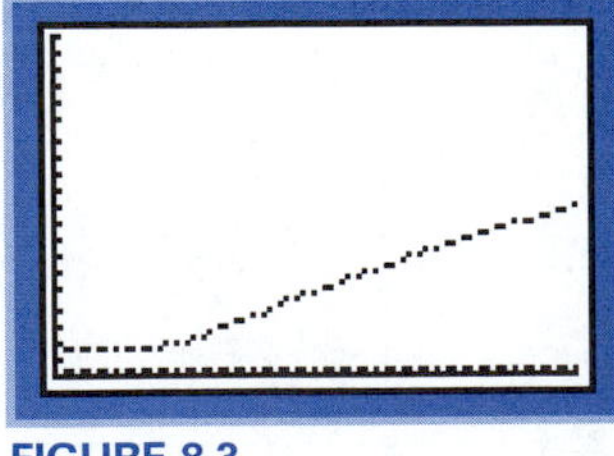

FIGURE 8.3.
Cara's distance-versus-time graph.

   a) According to Figure 8.3, did Cara walk toward or away from the motion detector?

   b) Did she begin moving as soon as the motion detector began collecting data, or did she pause first? How can you tell from her graph?

c) After she began moving, did she walk at approximately a constant velocity, or did she speed up or slow down? Explain how you decided.

5. a) A portion of Cara's distance-time data appears in **Figure 8. 4**. What was Cara's average velocity from $t = 0$ to $t = 0.5$? Be sure to specify the units for velocity.

Time (sec)	0	0.5	1.0	1.5	2.0	2.5	3.0	3.5	4.0	4.5	5.0	5.5	6.0
Distance (ft)	1.53	1.53	1.62	2.01	3.01	3.94	4.87	5.91	6.77	7.60	8.51	9.19	9.94

FIGURE 8.4. A portion of the distance-time data from Cara's walk.

b) What was Cara's average velocity from $t = 0.5$ to $t = 1.0$?

In previous chapters you have used the regression feature of your calculator to find a model that fits some data. For example, in Chapter 4, using your calculator, you discovered that a quadratic equation was the best model for the paired-sample testing strategy.

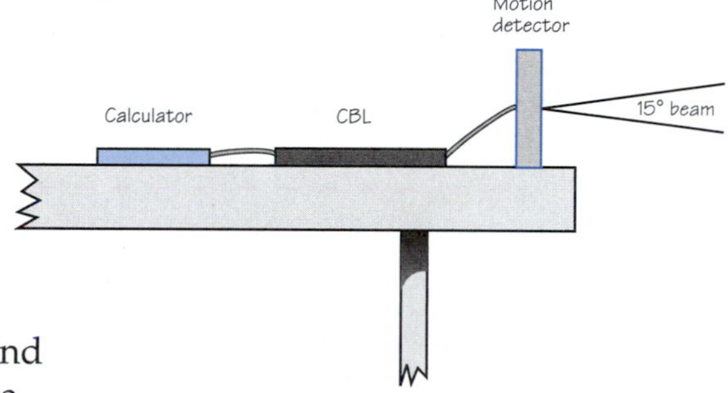

6. Get out your calculator so that you can find a model that fits Cara's distance-time data.

   a) Enter the data from Figure 8.4 corresponding to times from $t = 1$ to $t = 6$ into your calculator. Determine a linear model that describes the relationship between distance and time. Write the equation of your model.

   b) Make a scatter plot of the data that you entered into your calculator. In the same viewing screen, graph your equation from (a).

   c) How well does your linear equation describe these data? Explain.

   d) Assume that Cara's walk can be described by your equation from (a). What, if anything, does the slope of the graph tell you about Cara's walk?

   e) What, if anything, does the y-intercept of the graph tell you?

7. **Figure 8.5** shows distance-versus-time graphs of the motions of four people who walked in front of a motion detector. Write a description of how each person walked to produce each graph. Include details such as

   - where to start
   - which direction to move
   - how fast to walk
   - any other necessary instructions.

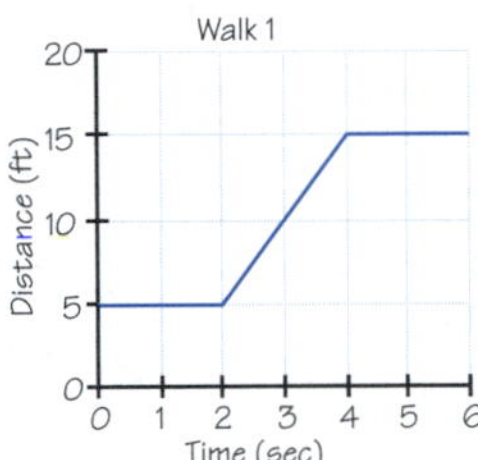

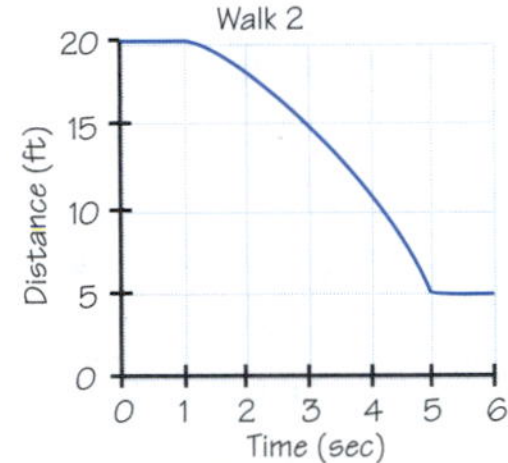

FIGURE 8.5. Distance-versus-time graphs for four walks.

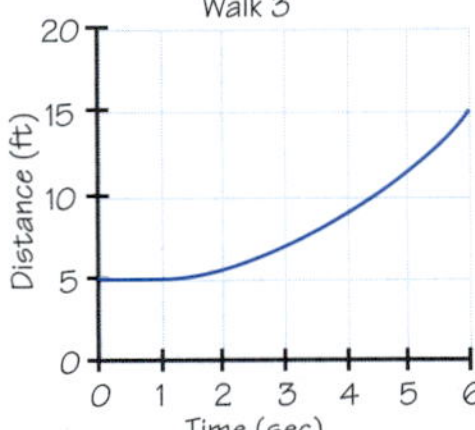

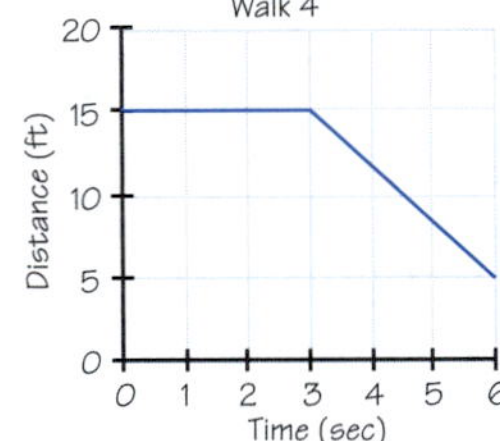

Imagine that your group used the directions you wrote in question 7 to reproduce the four walks graphed in Figure 8.5. The graphs in **Figure 8.6** show four possible graphs based on these directions.

Walk 1

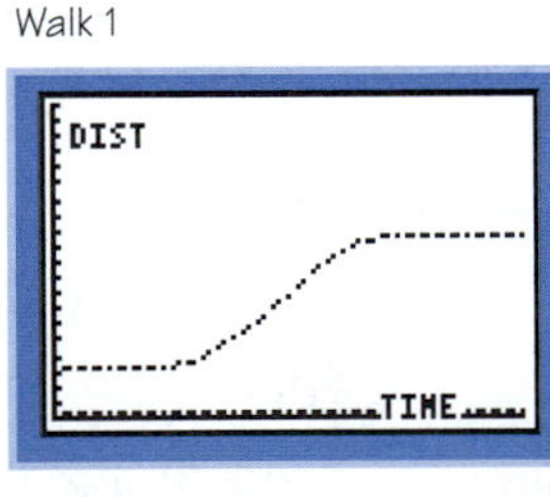

Walk 2

Walk 3

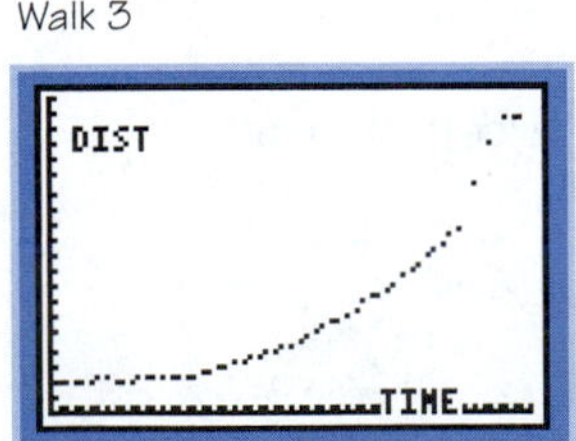

Walk 4

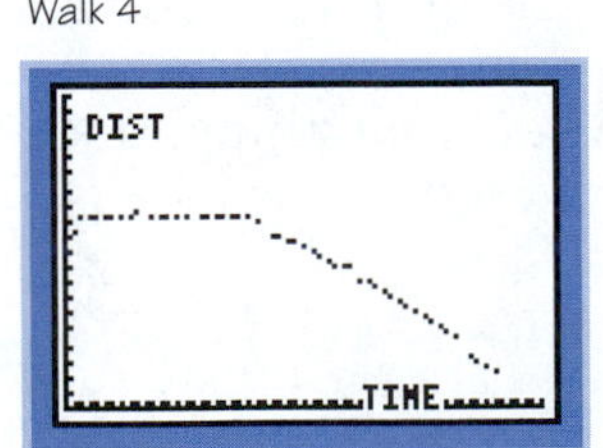

FIGURE 8.6. Distance-versus-time graphs corresponding to four walks. The calculator window settings are the same: [0, 6] x [0, 20].

8. a) Identify any differences between the graphs of Walk 1 in Figure 8.5 and Walk 1 in Figure 8.6.

   b) Identify any differences between the graphs of Walk 2 in Figure 8.5 and Walk 2 in Figure 8.6.

   c) Identify any differences between the graphs of Walk 3 in Figure 8.5 and Walk 3 in Figure 8.6.

   d) Identify any differences between the graphs of Walk 4 in Figure 8.5 and Walk 4 in Figure 8.6.

9. If motion detector equipment is available, complete the following experiments.

   a) Members of your group should take turns following the "walking instructions" for each of the graphs in Figure 8.5. Use the motion detector to record data from each walk. Make sketches of your actual graphs.

   b) How do your graphs compare with the graphs in Figure 8.5?

SUMMARY

In this activity you studied some distance-versus-time graphs to investigate some motions.

- You learned how to calculate average velocity.
- You discovered the slope of the graph of a line can represent velocity.

Using your knowledge about linear motion, you can design a stunt that can be modeled by a linear equation. In the next activity you will design a near collision using a model that you create.

HOMEWORK 8.1

Interpreting Motion Graphs

1. The first graph in **Figure 8.7** was recorded when Jasmyne walked in front of a motion detector. The program used to collect the data recorded time in seconds and distance in feet. The TRACE function of her graphing calculator was used to find the points corresponding to $x = 0.9$, $x = 2.9$, and $x = 4.9$ seconds in the other three graphs.

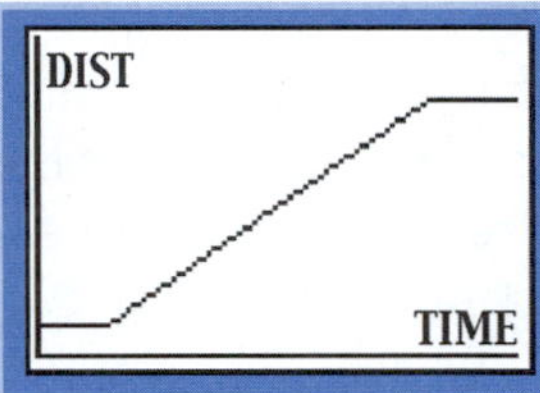

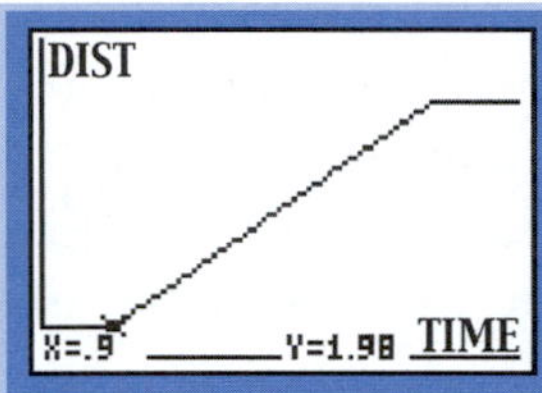

FIGURE 8.7.
Graph of Jasmyne's walk.

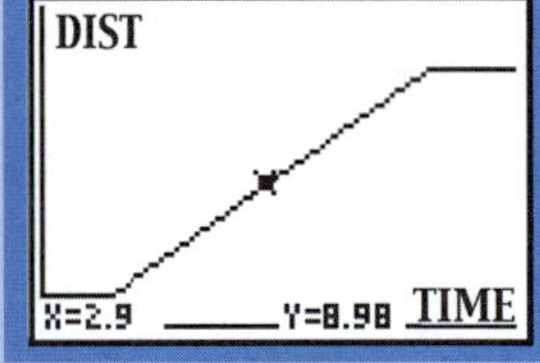

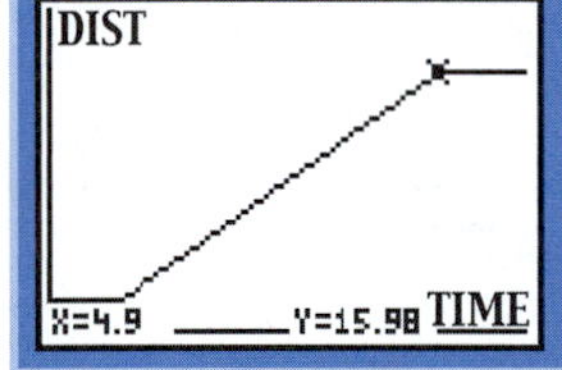

a) How far was Jasmyne from the motion detector when the program began running?

b) How many seconds did Jasmyne stand still before moving?

c) How much time did Jasmyne spend walking?

d) How far did she walk?

2. a) During the time that Jasmyne was walking, did she walk at a constant speed, increase her speed, or decrease her speed?

   b) During the time that Jasmyne was walking, how fast did she walk?

   c) Write an equation that models the relationship between distance, d, and time, t, during the period that Jasmyne was moving. For what t-values does this model make sense?

d) Using the equation from (c), does the value of d corresponding to $t = 0$ have some meaning in terms of Jasmyne's walk? Explain.

e) Does the slope of the line associated with the equation from (c) have some meaning in terms of Jasmyne's walk? Explain.

3. The two graphs in **Figures 8.8 and 8.9** were recorded by a motion detector when a student walked in front of it. Assuming the scales are the same for both graphs, in which walk was the student walking faster? Explain your answer.

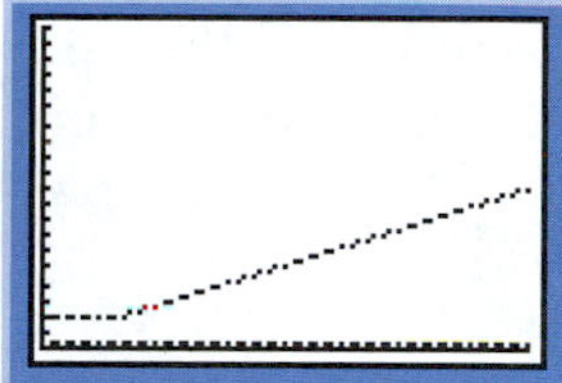

FIGURE 8.8.
Student walk 1.

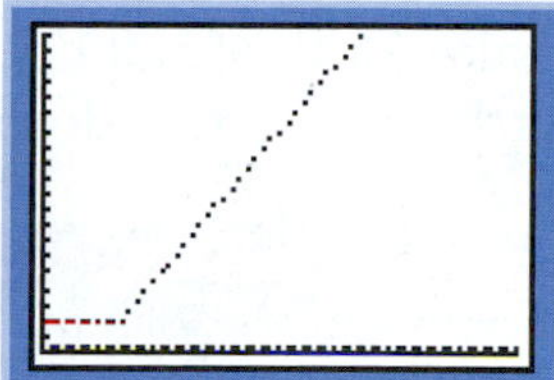

FIGURE 8.9.
Student walk 2.

To get ready for the near-collision stunt in the next activity, you will apply what you have learned from your walks to the motion of toy cars in the following questions.

4. a) A toy car moves along a straight line in front of a motion detector. The motion produces a distance-versus-time graph that is a line. What does this tell you about the toy car's velocity?

   b) How could you determine the toy car's velocity?

   c) Suppose that the distance-versus-time graph has a negative slope. What does that tell you about the car's motion?

5. Suppose that a toy car is traveling along a straight line in front of a motion detector. The motion produces a distance-versus-time graph that is curved rather than straight. What does this tell you about the car's velocity? Explain.

6. a) Recall that the relationship between distance (d), rate (r), and time (t) is $d = rt$. Imagine you are driving at a constant rate of 30 mph. How far will you drive in a half hour?

b) If you drive at a constant rate of 30 mph, how far will you travel in two hours?

c) If you plan to take a break after driving 45 miles, how long will you be driving before you take a break?

7. a) Jared's favorite battery-operated toy car moves at approximately 1.3 ft/sec. His car is a red sports car, nine inches long and five inches wide. Convert the length of Jared's car into feet.

   b) Jared places the front of his car at the zero marker. (See **Figure 8.10.**) Then he switches the car on. How long will it take for the front of the car to reach the three-foot marker? (Use two decimal places in your answer.)

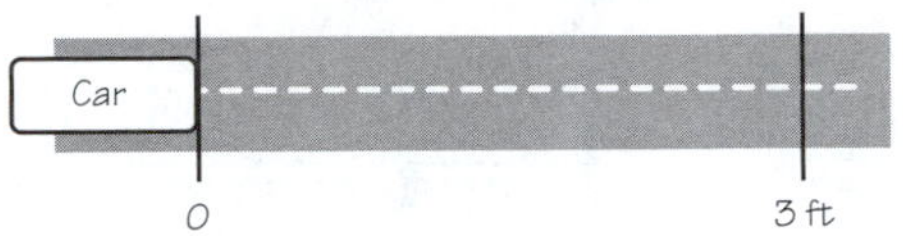

FIGURE 8.10.
Toy car at zero marker.

   c) How much longer will it take before the back of the car crosses the three-foot marker?

8. Jared decided to plan a near-collision stunt using two battery-operated cars. He uses his favorite red sports car and a green racer that is seven inches wide and 12 inches long. The green racer moves a little faster than the red sport car at approximately 1.36 ft/sec. **Figure 8.11** shows the roadway that Jared has set up for his stunt.

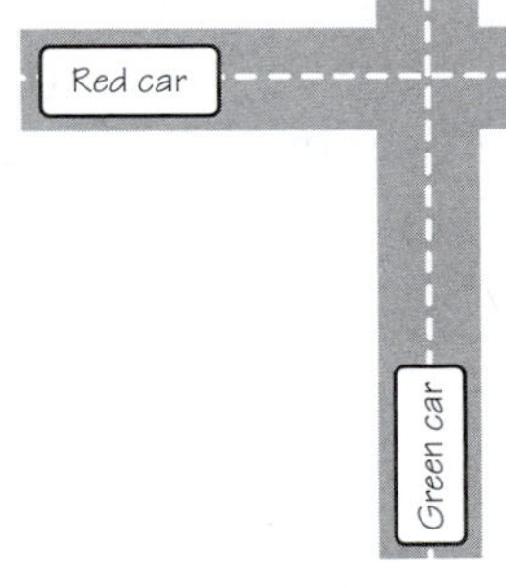

FIGURE 8.11.
Roadway for near-collision stunt.

Jared arranges his cars so that the front of the red sports car is three feet from the center of the intersection and the front of the green racer is four feet from the center of the intersection. Jared makes sure that each car is in the center of the road. He and a friend start each car at the same time.

a) Which car will reach the intersection first?

b) Will the cars hit or miss each other?

ACTIVITY

8.2

I Miss You Nearly

Now it's time to use what you have learned to design a stunt. Recall the description of the motorcycle-truck, near-collision stunt in the Preparation Reading. In this activity you'll first plan and then stage a scale version of this stunt.

PLANNING THE NEAR COLLISION

The Intersection. The first part of the plan is to describe the intersection where the vehicles will have their "near miss." Assume that the roads on which the vehicles will drive are straight. Also assume the two roads intersect at right angles.

The Vehicles. Next, either use two battery-operated vehicles (cars, trucks, robots) or select the data from *two* of the vehicles described in **Figure 8.12**. The distance-time data in Figure 8.12 was collected by running each toy vehicle in front of a motion detector.

Blue monster truck: length = 8 inches; width = 7 inches

Time (sec)	0.2	0.4	0.6	0.8	1.0	1.2	1.4	1.6	1.8	2.0	2.2	2.4
Distance (ft)	1.56	1.67	1.79	1.90	2.01	2.10	2.19	2.30	2.40	2.50	2.61	2.72

Red fire chief's car: length = 12.5 inches; width = 5.25 inches

Time (sec)	0.2	0.4	0.6	0.8	1.0	1.2	1.4	1.6	1.8	2.0	2.2	2.4
Distance (ft)	1.44	1.58	1.84	2.07	2.33	2.57	2.82	3.04	3.28	3.52	3.76	4.00

FIGURE 8.12. Motion detector readings from four battery-operated toy vehicles.

Purple sports car: length = 10 inches; width = 6 inches

Time (sec)	0.2	0.4	0.6	0.8	1.0	1.2	1.4	1.6	1.8	2.0	2.2	2.4
Distance (ft)	4.62	4.82	5.02	5.25	5.47	5.68	5.89	6.12	6.32	6.50	6.73	6.94

Green pro wrestler's car: length = 7.75 inches; width = 4.75 inches

Time (sec)	0.2	0.4	0.6	0.8	1.0	1.2	1.4	1.6	1.8	2.0	2.2	2.4
Distance (ft)	2.48	2.64	2.80	2.96	3.10	3.25	3.41	3.56	3.70	3.85	4.00	4.15

1. a) Describe the two toy vehicles you will be using for your stunt. (Be sure to include the dimensions of the vehicles in your description.)

   b) Use the distance and time data for your two toy vehicles to sketch a distance-versus-time scatter plot of the data.

   c) Use your calculator to find a model for the data for each vehicle. Use your model to determine how fast your vehicles move and if they move at approximately constant rates. Explain how you determined the rates.

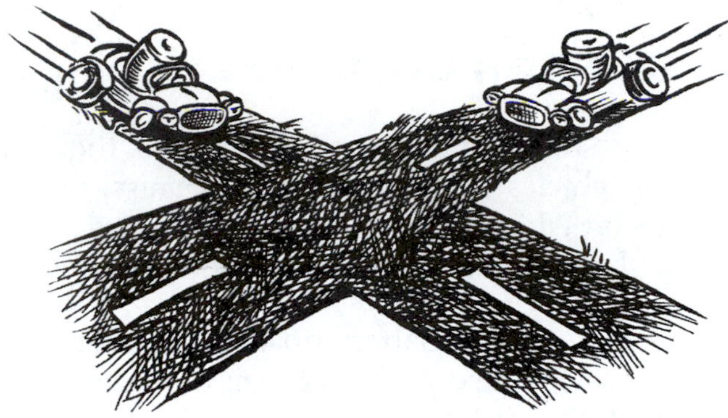

The Stunt. Here's where you get to decide how the stunt will be performed. Remember, you want to create some excitement for the spectators—the vehicles should pass through the intersection as closely as possible without colliding.

2. Describe how you will stage this stunt. Following are some issues that can help you plan the stunt.

   - Where on the intersecting roadway will you place the vehicles?
   - If you are using your own toy vehicles, will you start both vehicles at the same time?
   - If both vehicles do not start at the same time, how do you plan to stagger the starts?

   Include in your description the mathematics supporting your design.

The Proof. Now show your stuff! Are you sure this stunt is going to work as planned? You get at most *two* chances to perform the stunt. If you do not feel confident about your stunt go back over your plans. Remember trial and error is not practical. The stunt drivers would not appreciate guess work!

3. Perform the stunt using either your toy vehicles or the calculator simulation program STUNT. Did the stunt go according to your plans? If not, what do you think explains the differences between what you planned and what actually happened?

SUMMARY

In this activity you put your mathematical ideas into action.

- Using data or two toy cars, you built mathematical models for the motions of two cars.
- Using the models, you planned and performed a near collision.

Since you successfully completed that stunt, you're ready to move on to a more complicated one. Activity 8.3 provides the mathematical background you will need for the next stunt.

HOMEWORK 8.2

1. a) The two-vehicle, near collision stunt that you designed in Activity 8.2 used battery-operated toy cars. Describe the distance-versus-time graphs for battery-operated toy cars.

   b) What do these graphs tell you about the velocities of these toy cars?

2. a) Suppose a race car traveling down a straight highway runs out of gas. Draw a sketch of how you think its distance-versus-time graph would look. Assume that distance is measured from the car's starting position. Be sure to label the units on your axes.

   b) Describe in words what your graph tells you about the car's velocity during its drive.

3. Suppose that a battery-operated toy car moving in front of a motion detector produced the distance-versus-time graph in **Figure 8.13**.

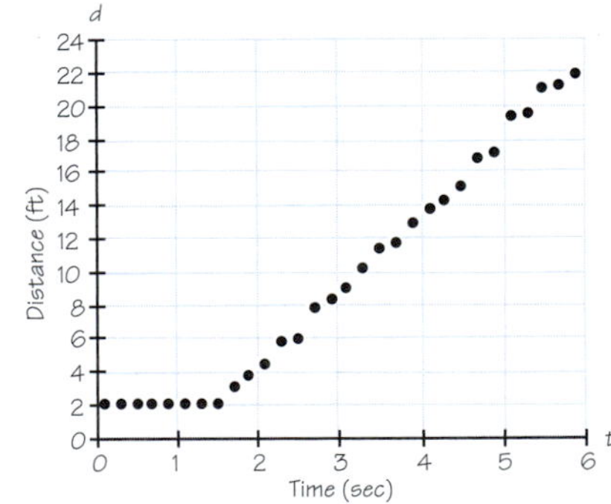

FIGURE 8.13. Distance-versus-time graph for a toy car.

   a) Approximately how far was the car from the motion detector when it began recording data? How far was the car from the motion detector when it stopped recording data?

   b) After the motion detector began recording data, how much time elapsed before the car began moving?

c) Suppose that Nguyen uses a linear model to fit the line in **Figure 8.14** to the data. According to Nguyen's model, what was the approximate velocity of the car? How did you get your answer?

d) Do you think your approximation of the car's velocity based on Nguyen's model is too high, too low, or about right? Explain.

e) Do you think Nguyen's model is a good fit for the data of the car's motion? If not, what would you have done differently?

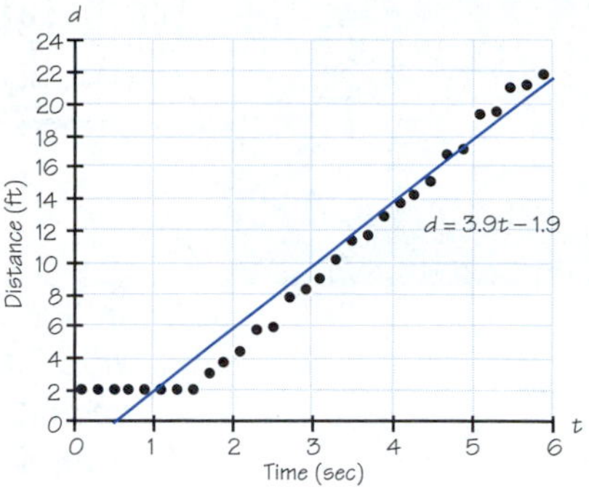

FIGURE 8.14.
Regression line superimposed on scatter plot.

4. A motion detector recorded the motion of a wind-up toy car powered by a spring. The screens in **Figure 8.15** show a distance-versus-time graph for this toy car. The calculator's TRACE feature was used to identify the times, x, in seconds and corresponding distances, y, in feet for three different times.

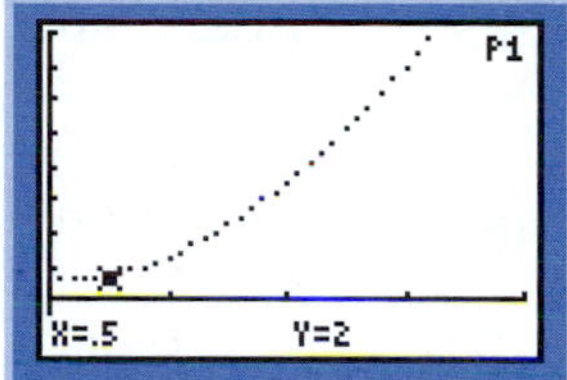

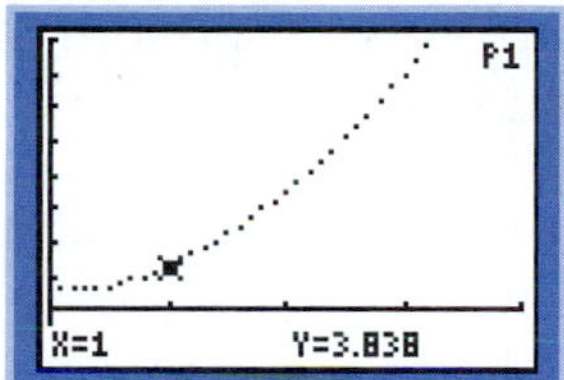

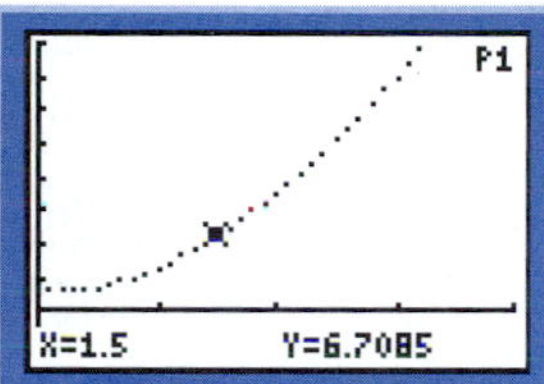

FIGURE 8.15.
Distance-versus-time graph for wind-up car.

a) How can you tell from the first graph whether the car is speeding up, slowing down, or moving at a constant velocity?

b) Approximately what is the average velocity of the car during the first half second that the motion detector records data?

c) What is the average velocity for the time from $x = 0.5$ to $x = 1$?

d) What is the average velocity for the time from $x = 1$ to $x = 1.5$?

e) Based on these average velocities is the car speeding up, slowing down, or traveling at a constant rate?

ACTIVITY

8.3

Walk Like a Parabola

In the first two activities of this chapter, the motion you studied was linear. The graph of this type of motion is a line, and the velocity of the vehicle is constant. Now you will investigate more complex motions. In this activity you will start with a motion that may be represented by a parabola.

Ms. Abbot's math class had a contest to see who could walk the best parabola. Each group planned how their walker would have to walk in order to produce a distance-versus-time graph that looked like a parabola. Questions 1 and 2 describe their experiment.

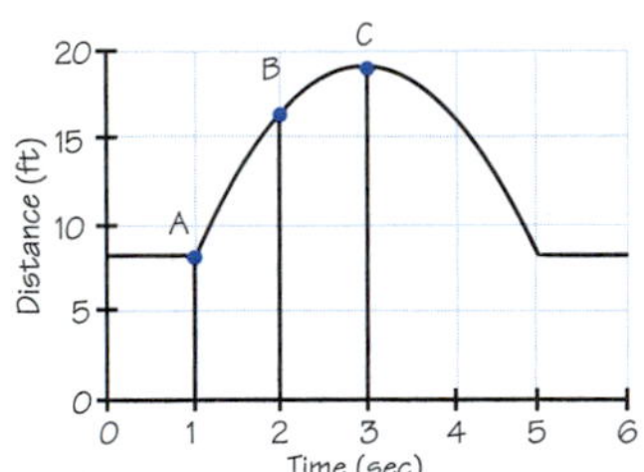

FIGURE 8.16.
Graph of Rosa's walk.

Suppose that Rosa was the walker for her group. The distance-versus-time graph for Rosa's walk is shown in **Figure 8.16.** The coordinates corresponding to points A, B, and C are (1, 8), (2, 16.25), and (3, 19), respectively.

Recall the formula for calculating average velocity from time 1 to time 2:

$$\text{Average velocity} = \frac{\text{distance 2} - \text{distance 1}}{\text{time 2} - \text{time 1}}$$

1. a) Let $t = 1$ be time 1 and $t = 3$ be time 2. What are the corresponding values for distance 1 and distance 2? Use this information to determine Rosa's average velocity from $t = 1$ to $t = 3$.

   b) What was Rosa's average velocity from $t = 1$ to $t = 2$?

   c) What was Rosa's average velocity from $t = 2$ to $t = 3$?

   d) What do the velocities in (b)–(c) tell you about how Rosa's average velocity was changing during the first half of her trip?

2. Rosa's group used quadratic regression to fit a quadratic equation to the data for times between $t = 1$ and $t = 5$ (the time interval over which Rosa was actually walking). They decided that the quadratic model $d = -2.75t^2 + 16.5t - 5.75$ was a good fit for these data.

a) Based on the quadratic model, how far was Rosa from the motion detector when $t = 1.5$?

b) Based on the quadratic model, how far was Rosa from the motion detector when $t = 2$?

c) What was Rosa's average velocity from $t = 1.5$ to $t = 2$?

d) What was Rosa's average velocity from $t = 2$ to $t = 2.5$?

In a car you can determine how fast you are traveling at a particular instant by looking at the speedometer. The velocity at a particular instant is called the **instantaneous velocity.** Since an average velocity is not the same as instantaneous velocity, you cannot use the average velocity formula to find Rosa's velocity at a particular instant. Nevertheless, you can use her average velocity to *estimate* her instantaneous velocity.

3. Explain the difference between instantaneous velocity and average velocity.

4. a) Suppose that Rosa had been hooked up to a speedometer before she began her walk. Approximately how fast would the speedometer indicate she was walking at $t = 2$ seconds? Explain how you determined your answer.

   b) Approximate her instantaneous velocity at $t = 2$ by calculating her average velocity over a small interval containing $t = 2$, such as the interval from $t = 1.9$ to $t = 2.1$. Is your estimate consistent with your answer to (a)?

   c) Use the model determined by Rosa's group to estimate Rosa's velocity at $t = 3$.

   d) Use the model to estimate Rosa's velocity at $t = 4$.

   e) One of the velocities you calculated in this question should be negative. What is the meaning of a negative velocity in the terms of Rosa's walk?

5. Now it is your turn to think about how to "walk like a parabola." Imagine that a motion detector has been set up to record distance-time data as you walk in a straight line in front of it. (Keep in mind that the detector will not be able to track your motion if you are more than 25 feet away.) Your challenge is to walk so that your distance-versus-time graph is the right side of a parabola that opens downward and has vertex at $t = 0$.

a) For your imagined walk, how far are you from the motion detector when it begins recording data?

b) Do you walk toward or away from the detector?

c) Do you walk at a steady pace, start slow and speed up, or start fast and slow down?

6. Mary Lou was the designated walker for her group. She tried to walk like the parabola described in question 5. The rest of her group tracked her with a motion detector. A portion of her distance-time data appears in **Figure 8.17**.

Time (sec)	0	0.5	1.0	1.5	2.0	2.5	3.0	3.5	4.0	4.5	5.0	5.5	6.0
Distance (ft)	13.5	13.4	13.2	12.7	12.1	11.2	10.3	9.1	7.7	6.2	4.6	2.7	0.6

FIGURE 8.17.
Partial distance-time data from Mary Lou's walk.

a) Use the data from Mary Lou's walk or collect your own data from a "walk like a parabola" experiment. Make a sketch of the distance-versus-time graph produced by the walk.

CHECK THIS!

Warning: If you try to walk like a parabola, it may take more than one walk before your group is satisfied that the walker has produced a satisfactory distance-versus-time parabola.

b) Using the data in your calculator, find a quadratic model that describes the relationship between distance and time for the walk data you used in (a).

c) Graph your quadratic model and your walk data in the same calculator screen. Is your model a good fit for the data? Explain.

d) Based on your model from (b), approximate the instantaneous velocities corresponding to $t = 1, 2, 3, 4$, and 5. Record the velocities in a table like **Figure 8.18**. (Remember you can approximate the velocity at $t = 1$ by determining the average velocity over a small interval containing $t = 1$.)

Time, *t* (sec)	Velocity, *v* (ft/sec)
1	
2	
3	
4	
5	

FIGURE 8.18.
Table with times and corresponding velocities.

e) Plot the data from your table. Label the vertical axis velocity and the horizontal axis time. Connect your points with a smooth curve or line. Describe the shape of your graph.

7. a) Alex was the designated walker for his group. He did his best to walk like a parabola. Alex's group used the equation $d = 20 - 0.41t^2$ to describe his walk. Complete a table like the one in **Figure 8.19** using this model.

Time, *t* (sec)	Velocity, *v* (ft/sec)
1	
2	
3	
4	
5	

FIGURE 8.19.

b) Plot the velocity-versus-time graph using the data from your table. Connect your points with a smooth curve or line. Describe the shape of your graph.

c) Find an equation that describes the pattern of your velocity-time data.

SUMMARY

In this activity you investigated motions represented by parabolas.

- You used average velocity to estimate instantaneous velocity.
- You used quadratic regression to find quadratic models of some motions.

This activity gave you some insights you can use for your next stunt. In Activity 8.4 you will plan and test another stunt.

HOMEWORK

8.3

More on Parabolas

Suppose a stunt calls for the hero to step off a rooftop the instant a pickup truck reaches a white line painted across the road. To design such a stunt, you need to calculate specific details about the hero's fall. For example, you need to know how long it would take for her to land on the ground.

1. a) Imagine the hero steps backwards off a rooftop 25 feet above the ground and falls vertically to the ground. Sketch a possible height-versus-time graph for the falling hero. Add appropriate scales and labels to your axes.

   b) Explain why the hero will fall as you have described in your graph.

   c) How might you collect data to decide whether your graph is a reasonable description of the hero's fall?

2. Jackie and Jermaine drew the graphs in **Figures 8.20** and **8.21** as part of their answers to question 1.

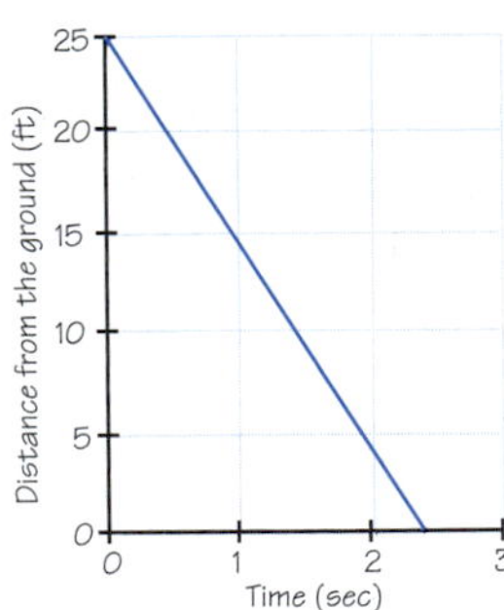

FIGURE 8.20.
Jackie's graph.

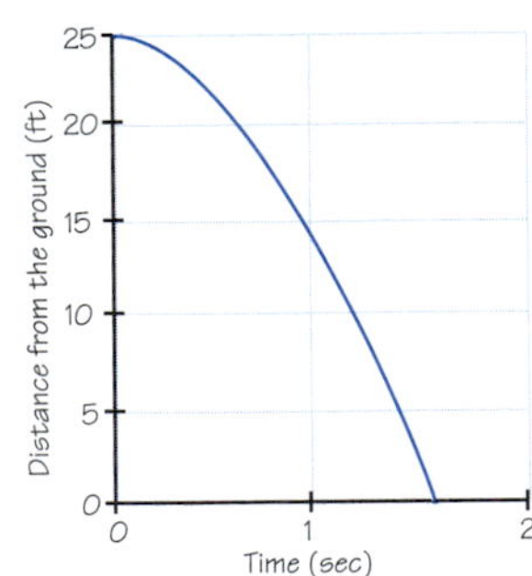

FIGURE 8.21.
Jermaine's graph.

   a) Suppose these graphs represent the motion of a walker in front of a motion detector. Write instructions to a walker so that the distance-versus-time graph would look as much like Jackie's graph as possible.

   b) Write instructions to a walker so that the distance-versus-time graph would look as much like Jermaine's graph as possible.

In Activity 8.4 you will find out whose graph, Jackie's or Jermaine's, better describes the fall of a person from a 25-foot rooftop.

3. a) Rashida was recorded by the motion detector as she jogged. Some of her data for one-second intervals are shown in **Figure 8.22.** Plot the distance-versus-time data for Rashida's jog. Be sure to add labels and scales to your axes.

Time (sec)	Distance from Sensor (ft)	Velocity (ft/sec)
0	2	
1	3.15	
2	6.8	
3	12.95	
4	21.6	
5	32.75	
6	46.4	
7	62.55	
8	81.2	

FIGURE 8.22. Distance-time data for Rashida's jog.

b) Describe the shape of the graph of Rashida's jog.

c) Did Rashida jog at a steady pace, speed up, or slow down? How can you tell from your graph?

d) How was Rashida's velocity changing during her jog?

e) Complete the velocity column in Figure 8.22. To estimate Rashida's velocity at a particular instant, calculate her average velocity during a time interval centered at that time. For example, use the average velocity from $t = 2$ to $t = 4$ to approximate Rashida's velocity at $t = 3$. (Using this method, you will not be able to determine Rashida's velocity at $t = 0$ or at $t = 8$.)

f) Plot velocity-time data for Rashida's jog. Describe the shape of the graph.

g) Based on your answer to (f), what type of model would describe Rashida's distance from the motion detector in terms of time? Explain.

h) Find an equation for Rashida's distance from the motion detector in terms of time. (Use the regression feature on your calculator.)

4. Consider the quadratic function $y = -x^2 + 4x + 4$. The graph of this function is a parabola.

   a) Find the coordinates of the vertex of this parabola.

   b) Find approximate values for the parabola's x-intercepts.

   c) Sketch a graph of this parabola. Label the vertex with the coordinates that you determined in (a).

   d) Modify the equation $y = -x^2 + 4x + 4$ to produce a parabola that looks exactly the same as the one you sketched in (c), except that the y-coordinate of the vertex is three units higher.

CHECK THIS!

Recall from Chapter 4 "Testing 1, 2, 3" that the graph of any quadratic function $y = ax^2 + bx + c$ (where $a \neq 0$) is a parabola.

- To find the parabola's vertex, use the formula $x = \frac{-b}{2a}$ to determine the x-coordinate. Then use this x-value to find the corresponding y-value.
- If you want to find the parabola's x-intercepts, solve the equation

$$ax^2 + bx + c = 0.$$

You can use the quadratic formula to solve this equation

$$x = \frac{-b \pm \sqrt{b^2 - 4ac}}{2a}.$$

ACTIVITY

8.4

The Hero's Fall

In the last activity you studied more complex motions, and you learned how some motions can be modeled by a quadratic equation. In this activity you will put this new knowledge to work. You will investigate how to model the motion of a falling object so that you can plan a hero's stunt.

FALLING BODIES

To design a hero's fall off a building, you need to make accurate predictions of falling objects. You can't perform experiments from the rooftop of your school in order to learn how the objects fall. However, you can analyze data produced by dropping a book and tracking its fall with a motion detector.

Imagine the following experiment: Drop a book from a height of four to six feet and use a motion detector to collect data about its motion. You can use one of the data sets provided or use a motion detector to collect your own data to answer the following questions.

1. a) Make a rough sketch of the distance-versus-time graph of the data. Be sure to add scales and labels to the axes. Explain what each piece of the graph means.

   b) Delete data so that you have only the portion that represents the object's fall. Use your calculator to make a scatter plot of the data. Remove any stray points so that a scatter plot is fairly smooth. Make a quick sketch of this graph.

   c) What type of equation do you think might describe the edited portion of the distance-versus-time data?

   d) Find a model that describes the shape of the edited data. In your model use the variable D to represent the distance between the book and the motion detector. Let T represent the time since the motion detector began collecting data. Be sure to specify your units of measurement.

e) Graph your model and data in the same viewing screen. Does your equation appear to be a reasonable model for these data? (If not, select another type of equation and try again.)

In question 1, you most likely used a quadratic model for the relationship between D and T. (If not, fit a quadratic model to the edited data. Base your answers to the following question on your quadratic model.)

CHECK THIS!

Recall that D represents the distance between the book and the motion detector. When the book hits the floor, the corresponding value of D is not the same as the book's height above the floor; the motion detector has thickness! You need to take into account the height of the motion detector.

2. a) Since the motion detector has thickness, you need to adjust the measurements for this factor. The motion detector used to collect the data in Handout 8.1 was about 11/8 inches. If you used your own motion detector you will need to measure its thickness. Convert the motion detector's thickness to the same units used to measure the book's distance from the motion detector.

   b) Use your results from question 2(a) to adjust your quadratic model so that it describes the relationship between the height of the book from the floor, h, and T.

   c) Graph your model from (b).

   d) Your graph for (c) should be a parabola. Approximate the coordinates of its vertex to four decimal places. Interpret the meaning of the vertex in the terms of the falling book.

   e) Explain the meaning of the coordinates of the vertex in terms of the falling book.

3. a) The model you developed in question 2 describes the relationship between the book's height, h, and the time since the motion detector began recording data, T. The variable T is not the most appropriate independent variable for modeling the book's motion. A better choice would be t, the time since the book was released. Express the relationship between T and t by writing a formula $T =$ ______________.

   b) Replace T in your quadratic equation by the expression for T that you got in (a). The result will be a model that describes the relationship between h and t. Rewrite your equation so that it contains no parentheses, and combine terms.

c) Use your model from (b) to determine how long it took for the book to hit the ground.

d) Approximate the velocity of the book the instant it hit the ground.

4. a) Your model from question 3(b) should have the form $h = at^2 + b$. (If, however, you have a non-zero t-term, the coefficient should be small.) What does the value of b in your formula represent?

   b) Compare your coefficient of t^2 to the coefficients determined by other groups. For a falling book, approximately what value should you use for a?

The value of a that you obtained should be close to –16. This value is related to acceleration of a falling object due to gravity, which is the constant $g = -32$ ft/sec². Physicists have discovered that the height of a falling object is given by the equation

$$h = \frac{1}{2}gt^2 + h_0$$

where h_0 is the object's height at $t = 0$.

ACCELERATION

The acceleration of a moving object is the rate of change of its velocity with respect to time. Recall that when you step on the gas pedal of a car, you say the vehicle is accelerating because its velocity is increasing. The units of acceleration usually are ft/sec² or m/sec².

DESIGNING THE HERO'S FALL

Now you have all the information you need to plan the hero's stunt. Let's call the hero Bruiser. Recall that Bruiser will fall off a rooftop into the back of a pickup truck driving parallel to the building. Bruiser will begin the fall the instant the front of the truck reaches a mark in the road. First you will need to find some basic information about the building and the truck.

5. Choose some measurements for the following details of the stunt. (Be as realistic as possible.)

   a) Building height.

   b) Length and height of the truck bed as well as the length and height of the truck's cab.

   c) Truck's velocity when it reaches the signal mark. Convert this velocity into ft/sec.

6. Use the information you found in question 5 to write a model that describes Bruiser's motion during the fall from the building. State any assumptions you make.

7. a) How long will it take Bruiser to reach the height of the cab?

   b) How long will it take Bruiser to reach the height of the back of the truck?

8. a) Where should you place the mark in the road so that Bruiser will land safely in the back of the truck without hitting the cab of the truck? Find the distance between he mark and Bruiser 's drop point. Explain how you determined your answer.

   b) Make a sketch indicating where Bruiser will land in the back of the pickup.

SUMMARY

In this activity you investigated the motion of a falling object in order to design a stunt.

- You discovered that a model for the height h of a falling book has the form $h = at^2 + b$, where t represents the time in seconds of the fall and b represents the height at $t = 0$.

- You used your model to design a fall so that Bruiser would land in the back of the truck safely.

You have investigated motions represented by linear and quadratic models. In the next activity you will study a motion that is modeled by a new type of function.

HOMEWORK

8.4

Do Drop In

1. Jose found that the model $h = -16t^2 + 2t + 6$ did a good job describing the motion of a book tossed into the air. The height, h, was in feet, and time, t, in seconds.

   a) What was the height of the book the instant it was tossed?

   b) What was the book's maximum height?

   c) Use the quadratic formula to determine how long it took for the book to hit the ground. (Note: For Jose's model, the quadratic formula gives two answers. Only one makes sense in the context of a falling book.)

2. Sonia's group performed the book-drop experiment described in question 1, Activity 8.4. Her group used the equation $h = -15.32T^2 + 5.53T + 3.13$, where h is the height above the floor and T is the time since the motion detector began recording data, to model the book's motion. (Sonia's group accounted for the thickness of the motion detector in their model.)

   a) Sketch a graph of the model found by Sonia's group.

   b) How long after the motion detector began recording data did Sonia release the book? Give your answer to four decimal places. How did you determine your answer?

   c) How high was the book the instant Sonia released it?

   d) Suppose the motion detector had not been in the way. How long would it have taken from the time of release until the book hit the floor? Explain how you arrived at your answer.

   e) What is the relationship between T, the time since the motion detector began recording data, and t, the time since the book was released? Express this relationship as a formula $T =$ ________________.

f) Write a model that describes the relationship between h, the book's height above the floor, and t, the time since the book was released. Revise your equation so that it contains no parentheses, and combine terms. (Don't round numbers until the last step.)

3. Heather's group also performed the experiment in Activity 8.4. They decided the equation $h = -15.46t^2 + 4.04$, where h was the book's height and t was the time since the book was released, described the book's motion. Height was measured in feet and time in seconds.

   a) What was the height of the book when it was released?

   b) Sketch a graph of their model.

   c) After the book was released, did it travel at the same velocity throughout its fall? How can you tell from your graph in (b)?

   d) How long did it take for the book to hit the ground?

The rate of change of velocity with respect to time is known as **acceleration**. Acceleration occurs only when a force acts on an object. You can calculate the **average acceleration** from time 1 to time 2 much as you calculate the average velocity. However, for average acceleration use the ratio

$$\frac{\text{velocity 2} - \text{velocity 1}}{\text{time 2} - \text{time 1}}.$$

Velocity has units of distance per unit of time. The units for acceleration are distance per unit of time per unit of time (This is not a misprint!). For example, if distance is in feet and time in seconds, then the units for acceleration are feet per second per second or ft/sec^2.

Time, *t* (sec)	Instantaneous Velocity, *v*, (ft/sec)
0	
0.1	
0.2	
0.3	
0.4	
0.5	

FIGURE 8. 24.
Velocity-time data.

4. a) Make a copy of the table in **Figure 8.24**. Use the model from Heather's group (question 3) to complete the second column. To compute each velocity, calculate the average velocity during the shortest time interval centered at that time. For example, use the average velocity from 0 to 0.2 to approximate the velocity at 0.1. Do not round your answers.

b) Make a scatter plot of the velocity-versus-time data from (a). Describe the shape of your plot.

c) Find an equation that describes the relationship between velocity, v, and time, t.

5. Use your table from question 4 to find the average acceleration of the book over the following intervals.

   a) What is the average acceleration of the book from $t = 0$ to $t = 0.1$? Be sure to include the units as part of your answer.

   b) What is the average acceleration from $t = 0.1$ to $t = 0.2$?

   c) What is the average acceleration from $t = 0.2$ to $t = 0.3$?

   d) Your velocity equation in 4(c) is linear. What is its slope? How does the slope of the velocity equation compare to the average accelerations you computed in (a)–(c)? Explain why that makes sense.

   e) What does it mean when the acceleration is negative?

By this time you should have a good idea about how to model the motion of a falling book. Imagine dropping a book directly over a motion detector. When you release the book, its velocity is 0 ft/sec. The force of gravity causes the book to fall. The acceleration due to the force of gravity causes the book's velocity to change. Because the distance between the book and the motion detector gets smaller over time, the book's velocity is negative. So negative velocity corresponds to falling. Since gravity pulls down, the velocities became more and more negative (the book falls faster), so acceleration is negative.

6. Imagine you are driving a car and traveling down the highway at 65 mph.

   a) Is your acceleration positive, negative, or zero? Explain.

   b) You decide to pass a truck, so you step on the gas pedal. Is your acceleration positive, negative, or zero? What effect does your acceleration have on your velocity?

   c) After passing the truck, you let your velocity return to the 65 mph speed limit. What can you say about your acceleration as you return to 65 mph? What effect does your acceleration have on your velocity?

7. Michon walked in front of the motion detector. Her distance-versus-time graph was a near perfect parabola. Her classmates found that the equation $d = -0.555t^2 + 7t + 2$ did a good job describing her walk.

a) Complete a copy of **Figure 8. 25** using the equation that Michon's class found. Use average velocity and average acceleration to approximate the velocity and acceleration entries.

Time, *t* (sec)	Distance from Motion Detector (ft)	Velocity at Time *t* (ft/sec)	Acceleration at Time *t* (ft/sec)
0		—	—
1			—
2			
3			
4			
5			—
6		—	—

FIGURE 8. 25.
Data from Michon's walk.

b) What does the sign of Michon's velocity tell you about how she walked? In particular, did she walk toward or away from the motion detector?

c) What does her acceleration, particularly the sign of her acceleration, tell you about how she walked?

ACTIVITY 8.5

Spinning Your Wheels

You have studied some complex motions in this chapter. You have investigated some linear motions and some motions that are modeled by quadratic models. In this activity you will examine motions that are modeled by periodic functions.

INVESTIGATING PERIODIC FUNCTIONS

We'll start by looking at the motion of a carousel. The carousel horses move up and down as the carousel spins. The graph in **Figure 8.26** shows the height above the floor of a carousel horse's back hooves over time during a carousel ride.

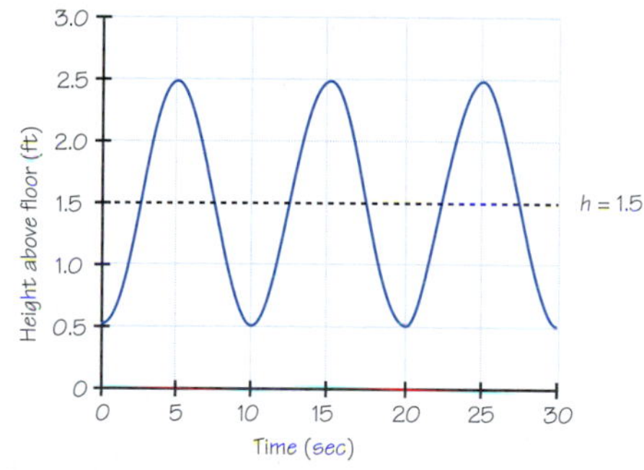

FIGURE 8.26. The height of the hooves of a carousel horse versus time.

The motion of the carousel horse is an oscillating or up-and-down pattern. Any function that has a graph that repeats itself on intervals of equal length is called **periodic**. The line $h = 1.5$ shown on the graph is the center line of the oscillation and is called the **axis of oscillation**.

1. a) The shortest horizontal interval corresponding to the basic repeating shape is called the **period**. Determine the period for the motion graphed in Figure 8.26.

   b) What is the maximum height of the back hooves? What is the minimum height?

   c) The **amplitude** is half of the vertical length of the basic repeating shape or half of the difference between the maximum and minimum y-values of the graph. Use the formula $\frac{\text{maximum} - \text{minimum}}{2}$ to calculate the amplitude.

Now you will investigate the motion of a ferris-wheel ride. Unfortunately you can't fit a full-sized ferris wheel into your classroom to study this motion. Instead of an actual ferris wheel you'll collect data from a bicycle wheel (or some other round object).

2. a) What is the diameter of your wheel?

   b) What is its circumference?

Set up the experiment as follows:

- Lay a tape measure on the floor in front of the bicycle wheel. Fasten the tape measure to the floor with masking tape. (Later you will roll your wheel alongside the tape measure.)
- Place another piece of tape perpendicular to the tape measure, to clearly mark the zero reading on the tape measure. Roll the wheel up to and directly on top of this marker. The tire should sit on the masking tape so that its axle should be centered directly above the tape.
- Place a dot on the tire (a sticker or piece of tape will do) at the point where the tire sits on the masking tape.

3. Roll the wheel forward beside the tape measure. Stop at regular intervals. When you stop, measure and record two things: the height of the dot above the floor and the distance the wheel has traveled. (See **Figure 8.27**.) Record your data in Figure 1, Handout 8.2. (See Handout 8.2 for additional instructions.)

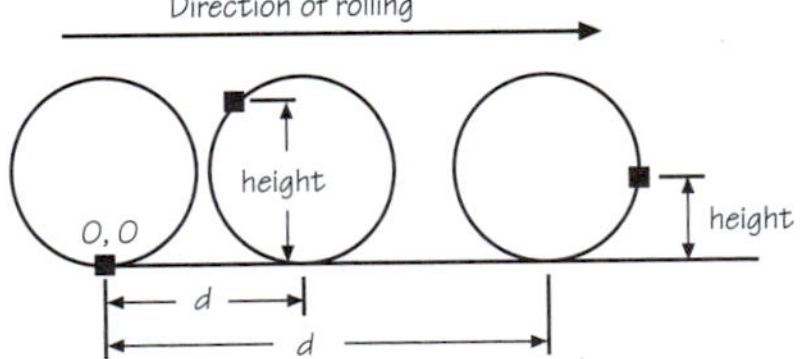

FIGURE 8.27. Measuring the distance the wheel rolls and height of the dot.

GRAPHING THE MOTION

4. a) Plot your data on a sheet of graph paper. Draw the vertical axis near the center of your paper. Because you want to predict height, height should be the response variable (or the dependent variable). This means that height should be on the vertical axis.

b) On your plot from (a), sketch a smooth curve (no corners) that you think best represents the relationship between height and distance.

5. a) Your graph from question 4 should appear periodic. (If not, recheck your measurements.) What is the period of this graph?

CHECK THIS!

Be sure to save your data and your graph for use in question 9, Activity 8.6. At that time you will determine the model for this motion.

b) What is the amplitude?

c) Explain how you could have predicted the period and amplitude before you started your experiment.

d) What is the axis of oscillation?

6. a) Complete the pattern in your graph for negative distance-values by extending your graph to the left.

b) In this situation, what do negative values for distance mean? How would you gather data corresponding to this situation?

Use the extended version of your graph from (a) to answer the remaining questions.

c) For what distances is the dot on the wheel at the top of the wheel? How can you get this answer from your graph?

d) For what distances is the dot horizontally level with the wheel's axle? How can you get this answer from your graph?

SUMMARY

In this activity you investigated another type of motion that can be represented by a periodic graph.

- You learned that a function that has a graph that repeats itself on intervals of equal length is called a **periodic function**.
- The shortest horizontal interval corresponding to the basic repeating shape is called the **period**.
- The **amplitude** is half the difference of the maximum and minimum values.

In the next activity you will write the equations of periodic functions, and use them to model some circular motions.

HOMEWORK 8.5

Round and Round You Go

1. Suppose you pedal your bicycle so the pedals revolve once a second. Each pedal is 5 inches from the ground at its lowest and 18 inches from the ground at its highest. Suppose that you start with the right pedal in its lowest position.

   a) Copy the table in **Figure 8.28** and complete the height entries.

   b) Plot points that show the height above the ground of the right pedal every 0.25 seconds for the time from 0 to 4 seconds.

   c) Use your plotted data from (b) to draw a graph that you think best represents the relationship between height and time.

   d) Determine the period of this graph. What is the meaning of the period in terms of the revolving bicycle pedal?

Time *t* (sec)	Height of Right Pedal (in)
0	
0.25	
0.5	
0.75	
1	
1.25	
1.5	
1.75	
2	
2.25	
2.5	
2.75	
3	

FIGURE 8.28.
Height-time data for a bicycle pedal.

   e) What is the approximate amplitude of this graph?

2. **Figure 8. 29** shows graphs of three periodic functions. Each function's graph repeats some basic shape over and over.

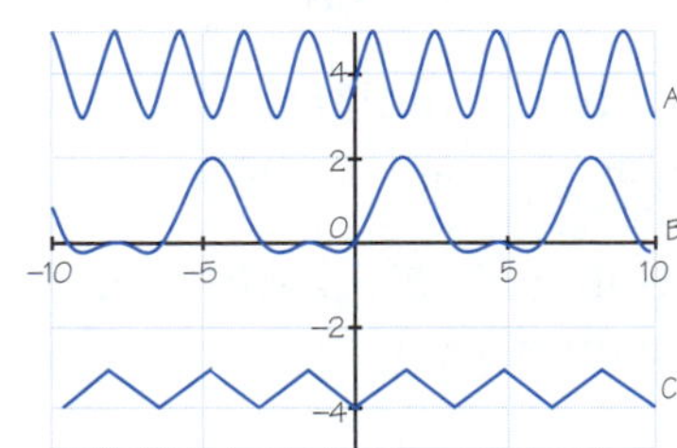

FIGURE 8.29.
Graphs of three periodic functions.

a) Determine the approximate period for each of these graphs.

b) What is the amplitude for each of these graphs?

3. a) Imagine reworking Activity 8.5 using a wheel that has a one-foot radius (or two-foot diameter). How far would you have to roll the wheel for the dot on the tire to make one complete turn? Give both an exact answer (involving π) and a decimal approximation.

   b) How far would you have to roll the wheel for the dot on the tire to move from the bottom of the wheel to the top of the wheel for the first time? And then the second time? Give both exact and approximate answers.

   c) Draw a smooth curve that you think represents the pattern of height-versus-distance data for this wheel. Complete the graph for distances from 0 to 25 ft.

   d) What is the period and amplitude of the graph that you drew in (c)?

   e) What is the axis of oscillation for the graph that you drew in (c)?

Calculations related to circles of radius 1 (such as in question 2) are usually simpler than calculations related to circles with different radii. Mathematicians often study periodic behavior using a **unit circle**, a circle with a one-unit radius that is centered at (0, 0).

Imagine an ant walking along the edge of a unit circle. Suppose as the ant walks, it pushes a straight blade of grass anchored at the circle's center.

- The ant starts at the point (1, 0).
- The ant can choose to walk either in the positive direction—counterclockwise (as shown in **Figure 8.30**) or in the negative direction—clockwise.

Suppose after walking for a while, the ant stops. The initial position of the blade of grass (on the horizontal axis) and its new position form an angle θ, called a **central angle**. Note also that as the ant moves, the blade of grass sweeps through a portion of the circle's circumference forming an arc.

One convenient way to measure this angle is to specify the length of arc together with a + or a – for the direction counterclockwise or clockwise. This method gives the **radian** measurement of the angle. (See Figure 8.30)

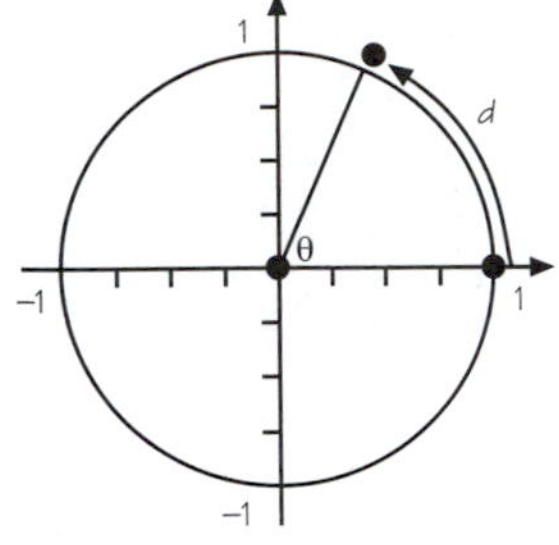

FIGURE 8.30.
The length of the ant's walk (arc length *d*) is the radian measure of angle θ.

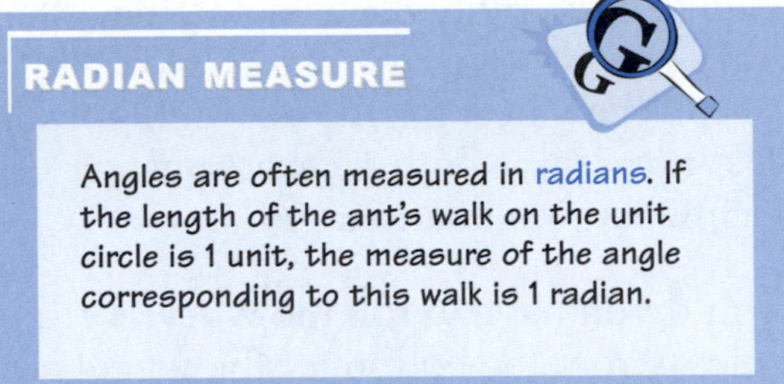

RADIAN MEASURE

Angles are often measured in radians. If the length of the ant's walk on the unit circle is 1 unit, the measure of the angle corresponding to this walk is 1 radian.

4. a) Suppose the ant in Figure 8.30 walks so that the blade of grass it pushes forms an angle with a radian measure of 2π. This means the ant starts its walk at the point (1, 0), walks in the counterclockwise direction, and stops after walking an arc of 2π units. Where is the ant after its walk? How many times has it traveled around the circle?

   b) If the ant walks so that the radian measure of the angle formed by the blade of grass is –4π, how many times has it walked around the circle? Does the ant walk counterclockwise or clockwise?

   c) Suppose that the ant's walk makes an angle of π radians. How far around the unit circle has it walked? In what direction has it walked? Where is it on the circle?

5. a) **Figure 8.31** shows two different walks. (Each arc represents one walk.) During Walk 1, what fraction of the circle did the ant walk?

   b) What is the radian measure for the central angle corresponding to the walk in (a)? How did you get your answer?

   c) During Walk 2, what fraction of the circle did the ant walk?

   d) What is the radian measure for the central angle corresponding to this walk?

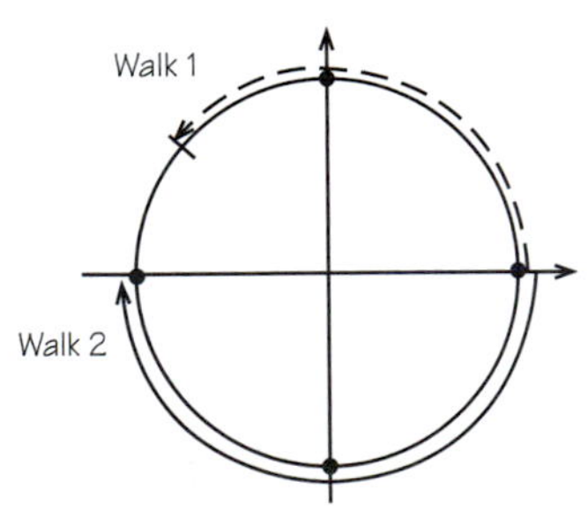

FIGURE 8.31.
Two arcs representing two walks.

6. **Figure 8.32** shows a unit circle superimposed on a coordinate plane. You can use the coordinate plane to estimate the ant's height above the horizontal axis (positive when the ant walks on the top half of the circle and negative when the ant walks on the bottom half of the circle) as it walks around the circle.

FIGURE 8.32. Graph of unit circle.

a) Copy the table in **Figure 8.33**. Use Figure 8.32 to complete the entries in your table.

CHECK THIS!

It may help to convert angles measured in radians to the fraction of a complete turn around a circle. For example, suppose the ant's walk corresponds to a central angle of $\pi/2$ radians. Recall that a central angle of 2π radians corresponds to turning once around the unit circle. Therefore a central angle of 2π radians corresponds to turning $\frac{\frac{\pi}{2}}{2\pi} = \frac{\pi}{2} \cdot \frac{1}{2\pi} = \frac{1}{4}$ of the way around the circle.

Radian Measure of Central Angle	$-\pi$	$-\frac{3\pi}{4}$	$-\frac{\pi}{2}$	$-\frac{\pi}{4}$	0	$\frac{\pi}{4}$	$\frac{\pi}{2}$	$\frac{3\pi}{4}$
Height (Vertical Displacement from Horizontal Axis)								

Radian Measure of Central Angle	π	$\frac{5\pi}{4}$	$\frac{3\pi}{2}$	$\frac{7\pi}{4}$	2π	$\frac{9\pi}{4}$	$\frac{5\pi}{2}$	$\frac{11\pi}{4}$
Height (Vertical Displacement from Horizontal Axis)								

FIGURE 8.33. Height-versus-radian data from a unit circle.

b) Graph the height-versus-radian data from your completed table from (a). Use the scaling shown in **Figure 8.34**.

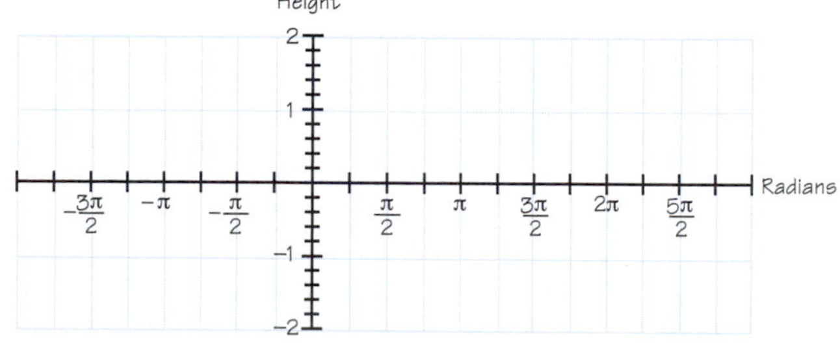

FIGURE 8.34. Axes with scaling used for question 6(b).

c) Draw a smooth curve through your plotted points.

d) What is the period and amplitude of your graph? What is its axis of oscillation?

e) Compare your graph from (b) to the graph that you drew for the one-foot radius wheel in question 2(c). How are your graphs alike and how are they different?

As the ant moves from its starting position, the point (1, 0) on the unit circle, a graph of its height versus its central angle (measured in radians) results in the oscillating graph that you drew in question 6(b). The function that produces this graph is important enough to be given a name. It is called the **sine function**. The equation of the sine function is written $y = \sin(x)$.

ACTIVITY

8.6

Learning Your A, B, C's (and D's)

In Activity 8.5 you performed an experiment and gathered data of a circular motion. You obtained a graph for the motion, but you did not get an equation to represent the motion. In this activity you will experiment with some sine functions to help find a model that is a good fit for the data.

MORE SINES

Your graph of the wheel data in Activity 8.5, question 4(b), had the same shape as a sine curve. The sine function is used to model many real-life phenomena, such as the tides. For your wheel data; the period, amplitude, and axis of oscillation however, were different from the graph of $y = \sin(x)$.

Figure 8.35 shows a graph of the sine function, $y = \sin(x)$, as seen in a calculator screen. The marks on the x-axis are $\pi/2$ units apart. The marks on the y-axis are one unit apart.

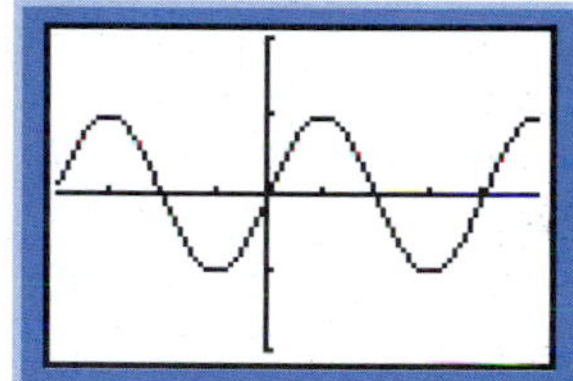

FIGURE 8.35.
Calculator screen showing a graph of the sine function.

1. a) What is the period of $y = \sin(x)$?

   b) What is the amplitude of $y = \sin(x)$?

   c) What is the axis of oscillation?

A function that can be expressed by an equation of the form $y = A\sin(B(x - C)) + D$ is called a **sinusoidal function.** Let's see if a function of this type can be used to describe your wheel data. Before you can use the sine function to model oscillating patterns, you will need to find out what each of the control numbers A, B, C, and D control.

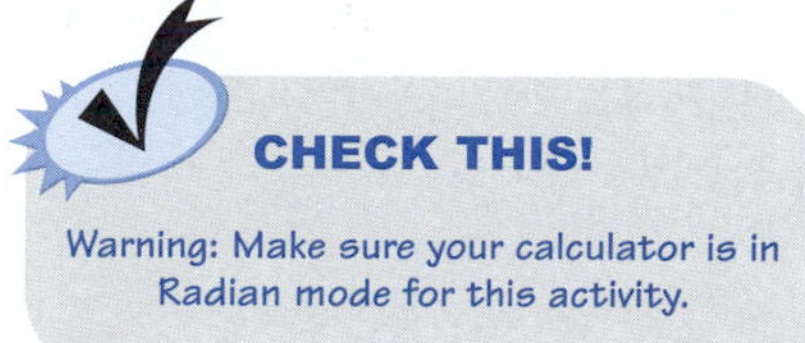

CHECK THIS!

Warning: Make sure your calculator is in Radian mode for this activity.

2. Use your graphing calculator to investigate how changing the constant A affects the graph of the sine function.

   a) Start with the graph of $y = \sin(x)$. Use your calculator's "trig viewing window." You can adjust window settings if you need to later.

b) In the same screen as your graph of $y = \sin(x)$, graph $y = 3\sin(x)$, $y = 0.5\sin(x)$, and $y = -2\sin(x)$. Continue to graph $y = A\sin(x)$ using other values for A until you are sure you know how the value of A affects the graph of the sine function.

c) What effect does the value of A have on the graph if $A > 1$? Include in your summary sketches of graphs that support your findings.

d) What effect does the value of A have on the graph if $0 < A < 1$? Include in your summary sketches of graphs that support your findings.

e) What effect does the value of A have on the graph if A is negative? Include in your summary sketches of graphs that support your findings.

f) When you change the control number A, which of the following, if any, changes: the period, amplitude, or axis of oscillation? Describe the change.

3. Use your graphing calculator to investigate how changing the control number B affects the graph of the sine function. Again begin with the graph of $y = \sin(x)$.

   a) In the same screen as your graph of $y = \sin(x)$, graph $y = \sin(2x)$ and then $y = \sin(0.5x)$. Continue to graph $y = \sin(Bx)$ using other values for B until you are sure you know how the value of B affects the graph of the sine function. Concentrate on B-values greater than zero.

   b) What effect does the value of B have on the graph if $B > 1$? Include in your summary sketches of graphs that support your findings.

   c) What effect does the value of B have on the graph if $0 < B < 1$? Include in your summary sketches of graphs that support your findings.

   d) When you change the value of B, which of the following, if any, changes: the period, amplitude, or axis of oscillation? Describe the change.

4. Next use your graphing calculator to investigate how changing the control number C affects the graph of the sine function. Start with the graph of $y = \sin(x)$.

a) In the same screen as your graph of $y = \sin(x)$, graph $y = \sin(x - C)$ for several choices of C. Be sure to try both negative and positive values for C. Continue to investigate $y = \sin(x - C)$ until you are sure you know how the value of C affects the graph of the sine function.

b) What effect does the value of C have on the graph if $C > 0$? Include in your summary sketches of graphs that support your findings.

c) What effect does the value of C have on the graph if $C < 0$? Include in your summary sketches of graphs that support your findings.

d) When you change the value of C, which of the following, if any, changes: the period, amplitude, or axis of oscillation? Describe the change.

5. Finally, use your graphing calculator to investigate how changing the control number D affects the graph of the sine function. Start with the graph of $y = \sin(x)$.

a) In the same screen as your graph of $y = \sin(x)$, graph $y = \sin(x) + D$ for several choices of D. Be sure to try both negative and positive values for D. Continue to investigate $y = \sin(x) + D$ until you are sure you know how the value of D affects the graph of the sine function.

b) What effect does the value of D have on the graph if $D > 0$? Include in your summary sketches of graphs that support your findings.

c) What effect does the value of D have on the graph if $D < 0$? Include in your summary sketches of graphs that support your findings.

d) When you change the value of D, which of the following, if any, changes: the period, amplitude, or axis of oscillation? Describe the change.

Equations of the form $y = A\sin(B(x - C)) + D$ are used to model many real situations, such as measuring the amount of daylight each day in the year at a certain location. Based on what you have learned from your investigation into $y = A\sin(B(x - C)) + D$, answer the following questions. (If you can't answer these questions return to questions 2–5 and continue your investigation.)

6. a) Which control number, A, B, C, or D, shifts the graph horizontally (left or right)?

   b) Which control number shifts the graph vertically (up or down)?

   c) Which control number affects the graph's amplitude? How is the graph's amplitude related to this number?

   d) Which control number affects the graph's period? How is the graph's period related to this number?

7. a) The period of the graph of $y = \sin(x)$ is 2π. How would you modify this equation so that the period of the graph of the modified equation is 4π?

   b) How would you modify $y = \sin(x)$ so that the period of the graph of the modified equation is 3π?

   c) How would you modify $y = \sin(x)$ so that the period of the graph of the modified equation is 2?

   d) How would you modify $y = \sin(x)$ so that the period of the graph of the modified equation is 10?

8. In (a)–(c) the equation of $y = \sin(x)$ has been changed. Describe how each of the changes to the sine function affects the graph. (In other words, explain how the change affects the amplitude, period, vertical or horizontal position of the sine graph.) Then sketch the graph of the new equation and $y = \sin(x)$ on the same set of axes.

   a) $y = 3\sin(x) + 2$.

   b) $y = 0.5\sin(x - \frac{\pi}{2})$.

   c) $y = \sin(2(x + \frac{\pi}{2}))$.

9. a) Refer to the graph that you drew for question 4, Activity 8.5. Recall this graph showed the relationship between height and distance rolled for your wheel data. Use what you have learned in your investigation to find a sinusoidal model $y = A\sin(B(x - C)) + D$ that could describe this relationship. Explain how you determined each of the control numbers.

b) Enter your wheel data from question 3, Activity 8.5 into your calculator. Make a scatter plot of height versus distance. In the same viewing window graph your model from (a). Does your model do a good job of describing your wheel data?

SUMMARY

In this chapter you developed sinosodial models in the form $y = A\sin(B(x - C) + D)$ to describe the bicycle data you collected in Activity 8.5. You explored control numbers:

- *A* controls the graph's amplitude
- *B* controls the graph's period
- *C* controls the horizontal shift
- *D* controls the axis of oscillation

You explored how changing the value of a control number affects the graph and used what you learned to find a model for the wheel data in Activity 8.5.

HOMEWORK

8.6 Sine Up!

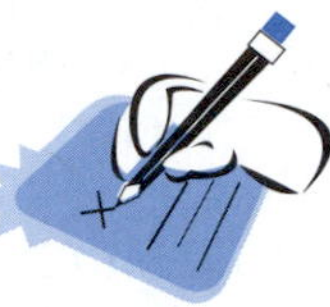

Complete this assignment without using a graphing calculator. After you have completed the assignment, you may use a graphing calculator to check your answers. If you need a refresher on the graph of the sine function, refer to Figure 8.35 in Activity 8.6.

1. a) Sketch the graphs of $y = 3\sin(2x)$ and $y = \sin(x)$ on the same set of axes. Choose a scaling for the x-axis that involves multiples of π. What are the amplitudes and exact periods of these two functions?

   b) Sketch the graphs of $y = 3\sin(0.5x)$ and $y = \sin(x)$ on the same set of axes.

   c) How are the graphs of $y = 3\sin(2x)$ and $y = 3\sin(0.5x)$ the same?

   d) How are the graphs of $y = 3\sin(2x)$ and $y = 3\sin(0.5x)$ different?

2. In the model $y = A\sin(B(x - C)) + D$, the value of C controls the horizontal shift. One way to think of C is that $x = C$ marks the start of the first loop of the graph of the sinusoidal model, $y = A\sin(B(x - C)) + D$.

   a) Graph $y = \sin(x)$ and $y = \sin(x - 1)$ on the same set of axes. What effect does $C = 1$ have on the graph?

   b) Graph $y = 3\sin(0.5x)$ and $y = 3\sin(0.5(x - 2))$ on the same set of axes. What affect does $C = 2$ have on the graph?

   c) Graph $y = 3\sin(0.5x)$ and $y = 3\sin(0.5(x + 2))$ on the same set of axes. What is the value of C in this case? What effect does this value of C have on the graph?

3. In (a) – (c), write an equation of a sinusoidal function having the given amplitude, period, horizontal shift, and axis of oscillation. Then sketch a graph of your function without the aid of a graphing calculator. Scale your x-axis in multiples of π. (After you have sketched your graphs, use your calculator to check that your graphs are correct.)

   a) An amplitude of $3/4$, period of 4π, horizontal shift of $\pi/2$ units to the right, and axis of oscillation $y = 0$.

   b) An amplitude of 1, period of 3π, horizontal shift of $\pi/2$ units to the left, and axis of oscillation $y = 0$.

   c) An amplitude of 2, period of 2π, horizontal shift of π units to the right, and axis of oscillation $y = 1$.

4. Write an equation that describes each of the graphs in **Figures 8.36–8.38**. Check your answers with your calculator.

   a)

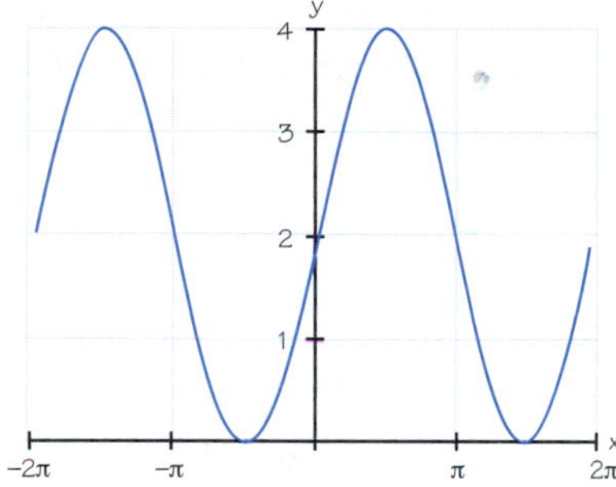

FIGURE 8.36.
Graph of a sinusoidal function.

   b)

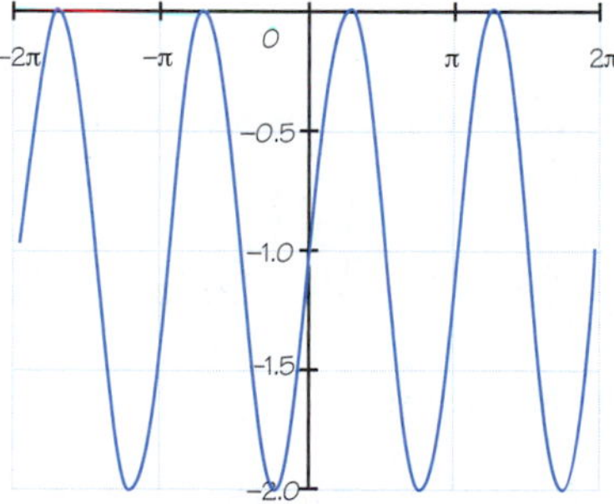

FIGURE 8.37.
Graph of a sinusoidal function.

   c)

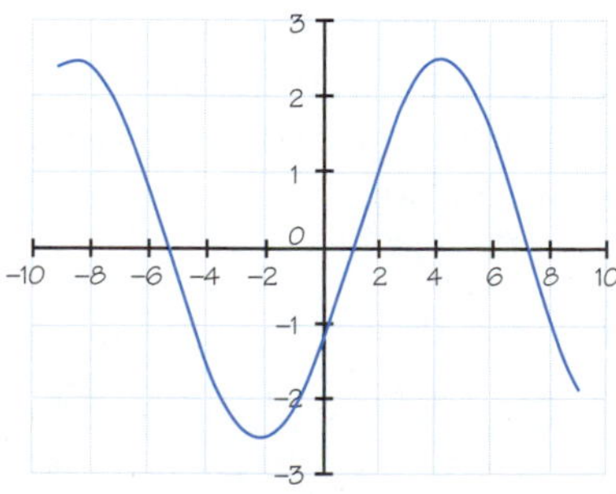

FIGURE 8.38.
Graph of a sinusoidal function.

5. The graph in **Figure 8.39** represents the height above the floor of a carousel-horse's back hooves versus time during a carousel ride. (You first saw this graph in Activity 8.5.)

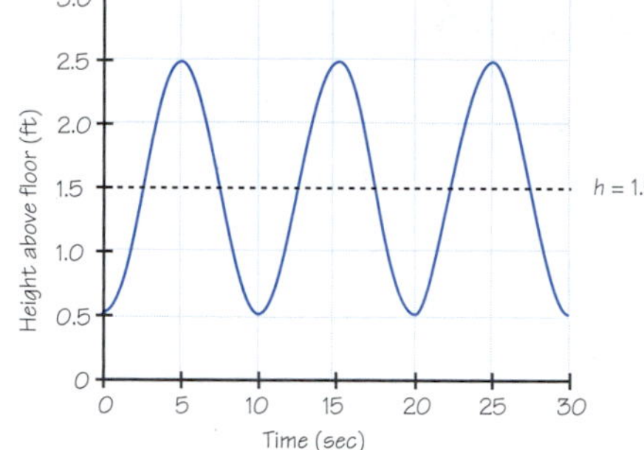

FIGURE 8.39.
The height of the hooves of a carousel horse versus time.

a) What are the period and amplitude of the function graphed in Figure 8.39?

b) Find an equation that describes height versus time. After writing the equation for your model, use your calculator to check that its graph matches the one in Figure 8.39.

6. Suppose Anne is the last rider to board a ferris wheel. The wheel, which sits 3.5 feet above the ground, has a diameter of 29.5 meters. It rotates at a rate of 2.6 revolutions per minute. Assume that the relationship between Anne's height above the ground and time is sinusoidal.

a) What is the period of the graph of the periodic function? (Change time units to seconds.)

b) What is the amplitude of the graph of the periodic function?

c) During the ride what is Anne's maximum height?

d) What is her minimum height?

e) What is the axis of oscillation of the graph of the sinusoidal function?

f) Write a sinusoidal equation that describes Anne's height above the ground over time during the ride. Explain how you determined your model.

Modeling Project A Swinging Function

Pendulums are familiar to everyone. Perhaps your first experience with a simple pendulum was as a child swinging an object from a string.

Pendulums have been used in clocks since Galileo noticed that a chandelier in a cathedral swung at a regular rate.

Pendulum motion is sensitive to gravity. So sensitive, in fact, that slight changes are used to detect gravity variations in the ground that may indicate oil deposits. For example, a pendulum instrument led to the discovery of the Cleveland oilfield in Liberty County, Texas.

Pendulums would not be useful for many tasks unless their motion could be modeled mathematically. In this project you will do just that.

1. Get a piece of string at least five feet long and a weight. A heavy fishing weight is fine. Attach the string to the weight.

   a) Have another member of your group stand about four feet from you and suspend the weight from three feet of string. Use a watch with a second hand to time the pendulum. Pull the string toward you and release it. Observe what happens. How long does it take to return to you after you release it? Repeat, but this time pull the pendulum back even closer to you. Repeat again, but with the pendulum not as close as the first time. Write a summary of your observations.

CHECK THIS!

Your time estimate will be more precise if you let the pendulum swing back and forth three or four times, then divide the total time by that number.

   b) What happens if the string is shorter than three feet? What happens if it is longer than three feet? Repeat the experiment enough times to find out. Describe your findings.

2. For a string length of three feet, describe the (horizontal) distance of the pendulum weight from you over time. Assume that the weight is one foot from you when you release it and the person holding the pendulum is four feet away. Include a rough sketch of a graph of the relationship between distance and time.

3. Develop a mathematical model to predict the pendulum's distance from you at a given time. Use the same pendulum you used in the previous question. Use a graphing calculator to graph your equation and compare it to your sketch in the previous question.

4. a) It is important to be able to adjust mathematical models when circumstances change. Discuss changes in your model if you release the pendulum two feet from you instead of three feet.

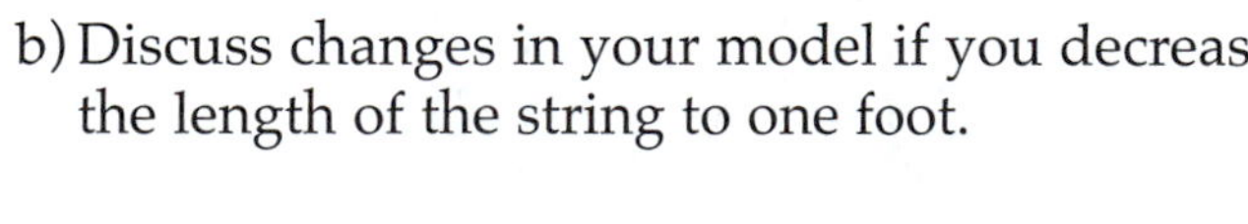

   b) Discuss changes in your model if you decrease the length of the string to one foot.

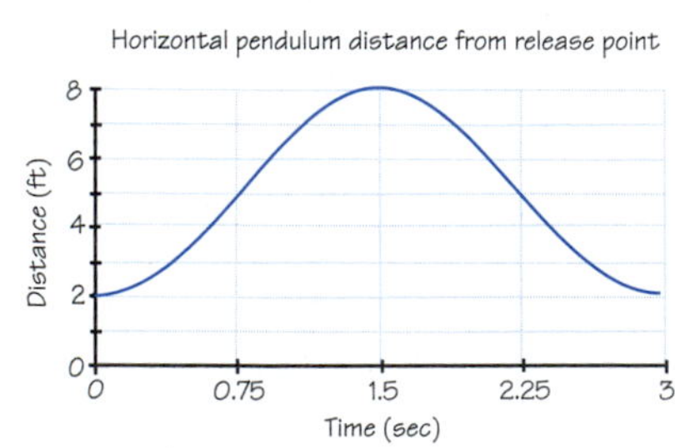

FIGURE 8.40.

5. **Figure 8.40** is a graph from a pendulum experiment. What conclusions can you draw about the pendulum and the experiment?

6. Research pendulum motion. Discuss the implications of your research for your results in the previous questions.

Practice Problems

▪ Review Exercises

1. Tessie was the designated walker for her group. A motion detector recorded her distance from the detector as she walked. A scatter plot of her distance-time data appears in **Figure 8.41**.

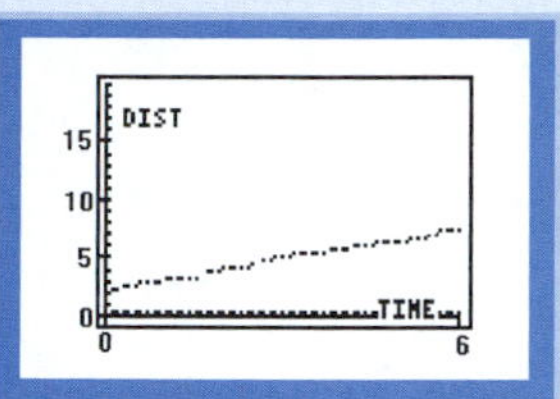

FIGURE 8.41. Distance-versus-time graph for Tessie's walk.

   a) What was the approximate distance between Tessie and the motion detector at the instant the motion detector began taking readings?

   b) How far did Tessie walk during the time the motion detector was reporting readings?

   c) What was Tessie's average velocity over the entire time that the motion detector was taking readings?

   d) Did Tessie walk at a constant rate, speed up, or slow down during her walk? How can you tell from her distance-versus-time graph?

   e) Find a linear model that describes the relationship between Tessie's distance from the motion detector, d, and time, t.

   f) Interpret the values of the d-intercept and slope of your equation in (e) in the context of Tessie's walk.

2. The graphs in **Figure 8.42**, on the right, resulted from students walking in front of a motion detector. In all cases the vertical axis represents the walker's distance (feet) from the motion detector and the horizontal axis represents time (seconds). For each graph, describe the student's motions.

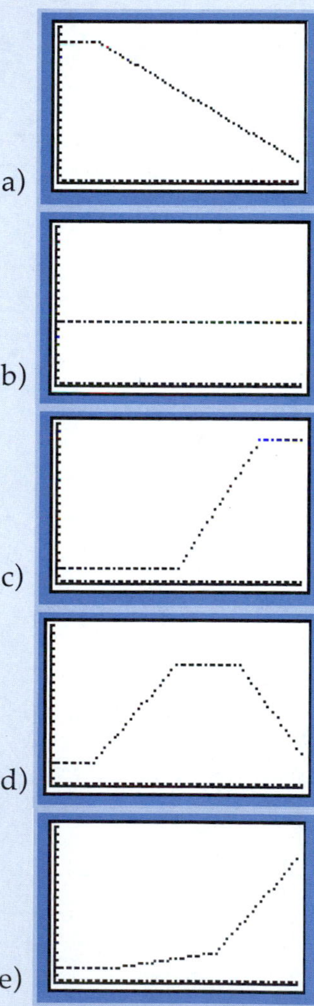

FIGURE 8.42. Graphs recording student walks.

3. Imagine you have set up a motion detector to track the motion of your friend as she walks at a constant velocity *toward* the motion detector. The motion detector records distance-time data for six seconds.

   a) Draw a distance-versus-time graph that might represent your friend's walk.

b) According to your graph in (a), what is your friend's average velocity during her six-second walk? Your answer should be negative. Why?

c) Write an equation that describes your graph from (a).

4. Henry and Andre each felt that he had walked the better line. Data from their walks appear in **Figure 8.43**.

FIGURE 8.43. Data from Henry and Andre's walks.

Elapsed Time (sec)	Henry's Distance from the Motion Detector (ft)	Andre's Distance from the Motion Detector (ft)
0.5	6.0	6.0
1.0	7.5	7.0
1.5	7.9	8.1
2.0	8.8	9.2
2.5	9.5	10.6
3.0	11.0	12.2
3.5	12.5	13.2
4.0	13.5	15.0
4.5	15.5	16.8
5.0	16.5	18.6
5.5	16.8	20.5
6.0	17.4	22.4

a) Enter the data from Figure 8.43 into calculator lists.

b) Make a distance-versus-time graph for Henry's walk. Fit a least squares line to Henry's distance-time data. Graph the least squares equation and Henry's data in the same screen. Did Henry walk a good line?

c) Repeat (b) using Andre's data. Did Andre walk a good line?

d) Assume for the moment that your equations in (b) and (c) are reasonable models for both Henry's and Andre's walks. According to these models, approximately how fast was Henry walking? What about Andre? (Remember to include the units as part of your answer.)

e) Make a residual plot for each model. (To calculate the residuals subtract the predicted distance from the actual distance. Then plot the residuals versus time.) Based on the residual plots, who do you think walked a better line? Explain.

5. Rashawn's group dropped a book over a motion detector. **Figure 8.44** shows the distance-versus-time graph for the falling book.

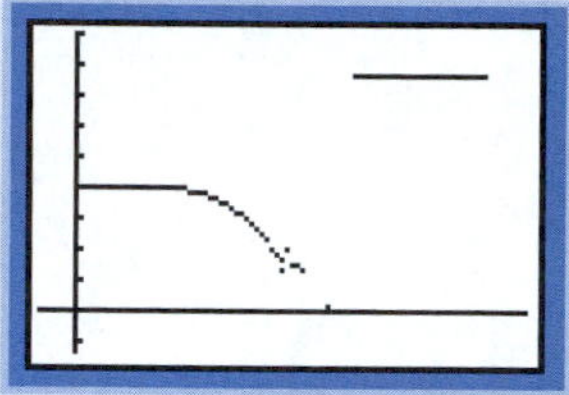

FIGURE 8.44.
Calculator screen showing distance-versus-time graph for a falling book.

a) Interpret the display shown in Figure 8.44.

Rashawn's group edited their data so that only a smooth section corresponding to the book's fall remained. Then they fit a quadratic equation to their edited data. Calculator screens showing a graph of their edited data and the results of a quadratic regression appear in **Figures 8.45 and 8.46.** The variables x and y represent the time (seconds) since the motion detector began recording data and the book's distance (feet) from the motion detector, respectively.

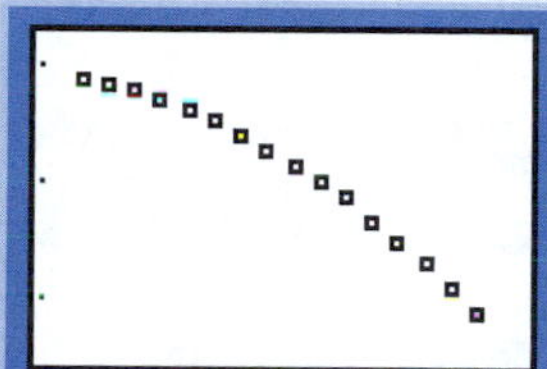

FIGURE 8.45.
Plot of edited data.

FIGURE 8.46.
Quadratic equation fit to edited data.

b) Rashawn's group wanted an equation that expressed the relationship between h, the book's height above the floor, and x. They measured the height of the motion detector and discovered that it was approximately $11/8$ in ≈ 0.1146 ft. What is their equation for the relationship between h and x? (Keep at least four decimal places in your answer.)

c) Your answer for (b) should be a quadratic equation. Its graph is a parabola. What is its vertex? (Use four decimal places in your answer.) Interpret the coordinates of the vertex in the context of the falling book.

d) Use your answers to (b) and (c) to determine an equation describing the relationship between h, the book's height (feet) above the floor, and t, the time (seconds) since the book was released. Explain how you got your answer. (Round the constants in your final answer to two decimal places.)

6. Juanita is driving on the Interstate at 55 mph. She comes up behind a slow car and decides to pass. To pass, she steps on the gas pedal and goes around the slower car. Once around the car she slows back to 55 mph. **Figure 8.47** shows a graph of her velocity versus time for this event.

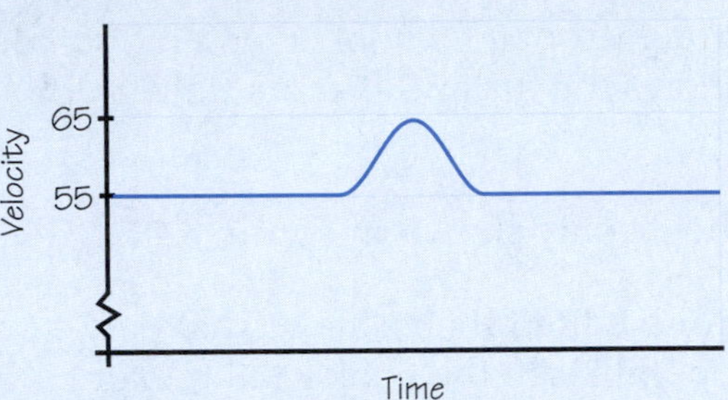

FIGURE 8.47. Graph of velocity versus time.

a) Describe her acceleration during this trip.

b) Sketch a possible graph of her acceleration versus time.

7. Renee attempted to walk away from the motion detector so that her distance-versus-time graph would be a parabola. Her data are repeated in **Figure 8.48**.

Time t (sec)	Distance from Sensor (ft)	Velocity at Time t (ft/sec)	Acceleration (ft/sec^2)
0.0	2.0	—	—
0.5	2.7		—
1.0	3.6		
1.5	4.8		
2.0	6.5		
2.5	8.7		
3.0	11.7		
3.5	15.6		—
4.0	21.0	—	—

FIGURE 8.48. Data from Renee's walk.

a) Make a careful graph of Renee's data. Be sure to scale and label your axes.

b) Based only on your graph, do you think Renee was slowing down or speeding up during her walk? Describe the feature of your graph that helps you decide, and explain your reasoning.

c) Given your answer to (b), was Renee's acceleration positive, negative, or zero? Explain.

d) Complete a table similar to the one in Figure 8.48. To calculate each velocity and acceleration, calculate the average during the shortest time interval centered at that time. For example, use the average velocity from 0 to 1.0 to approximate the velocity at 0.5.

e) Using the information from (d), is Renee's distance-versus-time graph a parabola? Explain your reasoning.

f) Use your calculator to fit a quadratic equation to Renee's data. Plot the data and your equation. Does your equation appear to do a good job in describing Renee's walk?

g) Make a residual plot for your equation from (f). Based on your residual plot, does your quadratic model do a good job describing Renee's distance-time data? Explain.

8. **Figure 8.49** shows portions of distance-time data collected from a falling book.

a) Enter these data into calculator lists. Make a scatter plot of the distance-time data. Does your plot resemble a portion of a parabola?

b) Use quadratic regression to find an equation for the book's fall.

c) What do the coefficients in your model from (b) tell you?

d) Approximate the velocities for $t = 0.16, 0.18, 0.20$, and 0.22. Record the velocities in a table like the one in **Figure 8.50**. Did you use the data in Figure 8.49 or your model to determine the velocities?

e) Use your velocity-time data from (d) to determine an equation for velocity versus time.

f) What was the book's acceleration as it fell?

Time t since book was released (sec)	Height h of book above floor (ft)
0.12	5.0424
0.14	4.9638
0.16	4.8732
0.18	4.7704
0.20	4.6556
0.22	4.5287
0.24	4.3897

FIGURE 8.49.
Book drop data.

Time (sec)	Velocity (ft/sec)
0.14	
0.16	
0.18	
0.20	
0.22	

FIGURE 8.50.
Velocity-time data for falling book.

9. Refer to your model for the falling book in question 8(b).

a) How long was the book in the air before it hit the floor? How do you know?

b) What was the book's velocity when it hit the floor?

c) What was the book's acceleration when it hit the floor?

10. The equation $h = -15.2(t-2)^2 + 40$ models the height of a rock shot straight upwards. Distance is measured in feet and time in seconds. Assume the rock started to move when $t = 0$.

a) How far was the rock from ground level when it was shot? Was it shot from above the ground or below the ground?

b) How high above the ground did the rock travel? Explain how you determined your answer.

c) What was the rock's initial velocity? How did you determine your answer?

d) Complete the velocity table in **Figure 8.51**. To approximate each velocity, calculate the average velocity during the shortest interval centered at that time.

FIGURE 8.51. *Velocity-time data for rock.*

Time (sec)	1.8	1.9	2.0	2.1	2.2
Velocity (ft/sec)					

e) What do the velocities in (d) tell you about the rock's motion? What, if anything, do they tell you about the rock's acceleration?

11. At a tennis match, a video is used to gather information on the height of the ball during a long volley.

a) If you plot the ball's height versus time, will your graph be periodic? If so, how would you determine its period and amplitude? If not, why not?

b) Suppose that the height-versus-time graph for the volley was periodic. What would this graph tell you about how the players were hitting the ball? Do you think it's possible for the players to hit the ball this way?

12. The only carousel owned by the United States government is located in Glen Echo Park, Washington D. C. The carousel is 48 feet in diameter. It turns counterclockwise at a top speed of five revolutions per minute. Suppose you have chosen to ride a horse that is in the outer row. Your horse is about two feet from the edge.

a) Each time the carousel makes a complete turn, how far (in feet) have you ridden?

b) How fast, in feet per minute, are you riding? How fast would this be in miles per hour? (Remember there are 5280 feet in 1 mile.)

c) Imagine that your friend is riding on an inside horse that is approximately six feet from the edge. In terms of miles per hour, how much faster are you moving than your friend?

13. Over a year, the length of daylight (the number of hours from sunrise to sunset) changes every day. **Figure 8.52** shows the length of daylight every 30 days from 12/31/97 to 3/26/99 for Boston, Massachusetts.

Date	12/31	1/30	3/1	3/31	4/30	5/30	6/29	7/29
Day Number	0	30	60	90	120	150	180	210
Length (hours)	9.1	9.9	11.2	12.7	14.0	15.0	15.3	14.6

Date	8/28	9/27	10/27	11/26	12/26	1/25	2/24	3/26
Day Number	240	270	300	330	360	390	420	450
Length (hours)	13.3	11.9	10.6	9.5	9.1	9.7	11.0	12.4

FIGURE 8.52.
Data on length of day.

a) Draw a set of axes similar to **Figure 8.53**. Then plot the data from Figure 8.52.

b) Draw a smooth curve through the points on your graph. Extend your graph to show how you think it should look over a period of two years. Explain why you think it should look this way.

c) What is the period of your graph?

d) What is its amplitude?

e) What is the axis of oscillation?

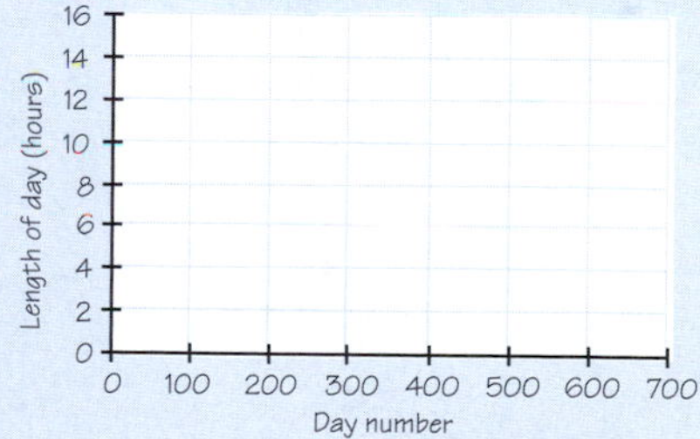

FIGURE 8.53.
Axes for length of day versus day number.

14. Figure 8.52 gives data on length of daylight versus day number for Boston, Massachusetts. A plot of the data indicates that the relationship between length of daylight and day number is sinusoidal.

a) Write an equation that could describe this relationship. Explain how you determined your equation. (Use your calculator to check that your model provides a good fit to the data.)

15. Ariana and Terry are turning a long jump rope for Tracy, who is waiting to jump in. Ariana and Terry turn the rope one time per second. The maximum height is seven feet while at its lowest it just touches the ground. Assume the rope was on the ground moving up and away from Tracy at starting time.

    a) Write an equation expressing the height of the rope (feet) as a function of time (seconds) since Ariana and Terry started turning it. Explain how you arrived at your equation.

    b) Tracy is 5 feet 3 inches tall. When will the rope be higher than Tracy's head? Explain how you determined your answer.

16. The Potomac River flows through Washington, D. C. Tide levels for the Potomac are approximately sinusoidal and range from a height of 5 feet to a low of 1.5 feet when measured against a post at Harbor Place (a small plaza with a landing dock for boats).

    a) Suppose on July 1, high tide is at 6:30 A.M. and low tide is at 12:45 P.M. Write an equation describing the height of the water for every hour after midnight on July 1. Explain how you arrived at your model.

    b) How high will the water be at noon on the fourth of July?

    c) If the owners of Harbor Place need to hire someone to help people docking when the water is below three feet, for about how long every day should they hire someone? Explain how you reached your answer.

17. Suppose you board one of Nauta-Bussink's 55-meter ferris wheels. Once the wheel is up to speed, you rotate one-and-one-half times around the wheel every minute. During your ride your height above the ground oscillates between three meters and 55 meters and can be modeled by the sinusoidal function $h(t) = 26\sin(0.16(t - 10)) + 29$. For this model, height is measured in meters and time in seconds.

    a) On the way up when you reach the height of the axle, what is the approximate rate of change of your height above the ground with respect to time? (Use the average velocity over a small interval to approximate this rate.)

b) On the way down when you reach the height of the axle, what is the approximate rate of change of your height above the ground with respect to time?

c) Compare your answers to (a) and (b). What accounts for the sign difference in your velocities?

18. Write an equation for each of the mystery graphs (**Figures 8.54 and 8.55**). Check your answer with a calculator.

a)

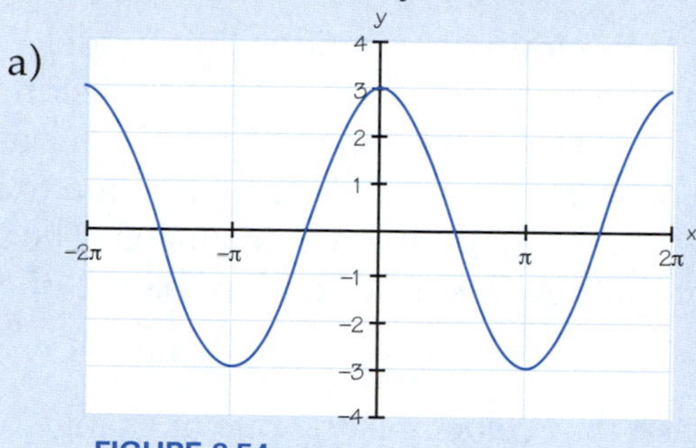

FIGURE 8.54.
Mystery graph 1.

b)

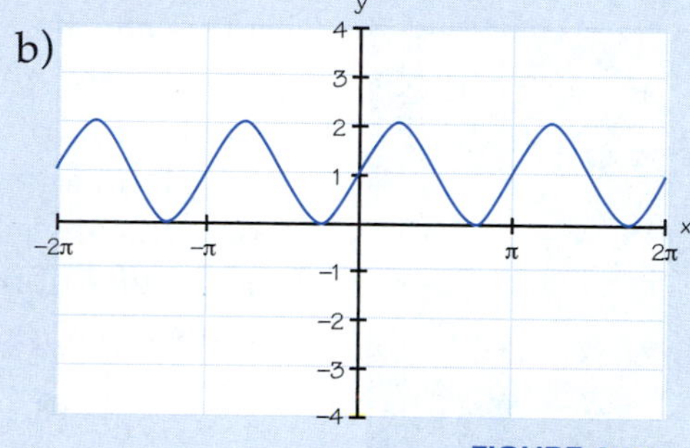

FIGURE 8.55.
Mystery graph 2.

19. For **Figure 8.56–Figure 8.59**, determine an equation for a sinusoidal model that describes the graph. Also state the period and amplitude.

a)

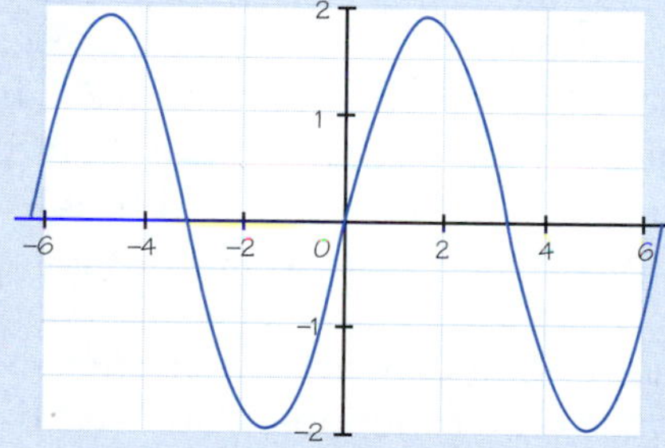

FIGURE 8.56.
Graph of a periodic function.

b)

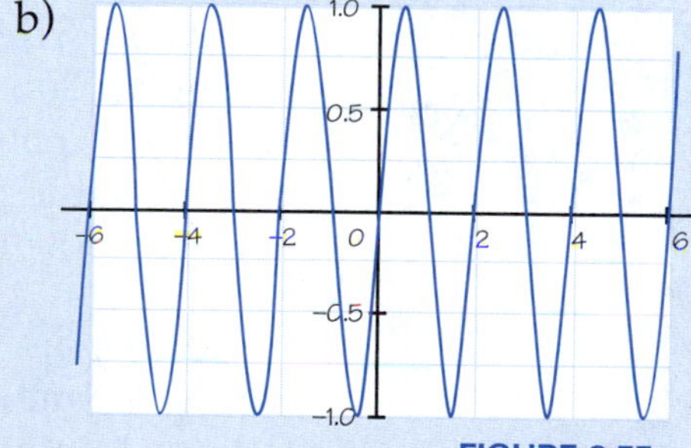

FIGURE 8.57.
Graph of a periodic function.

c)

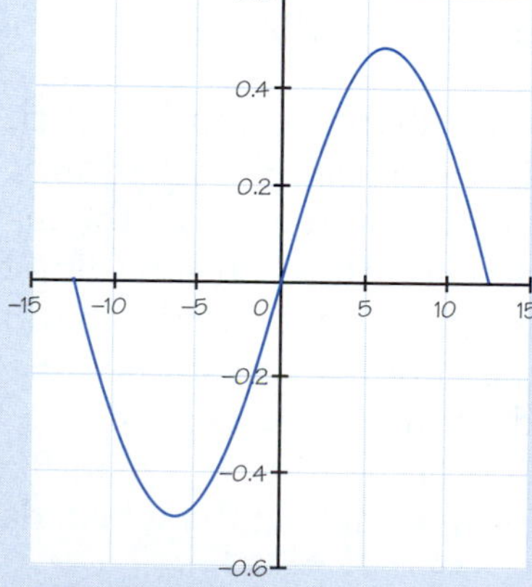

FIGURE 8.58.
Graph of a periodic function.

d)

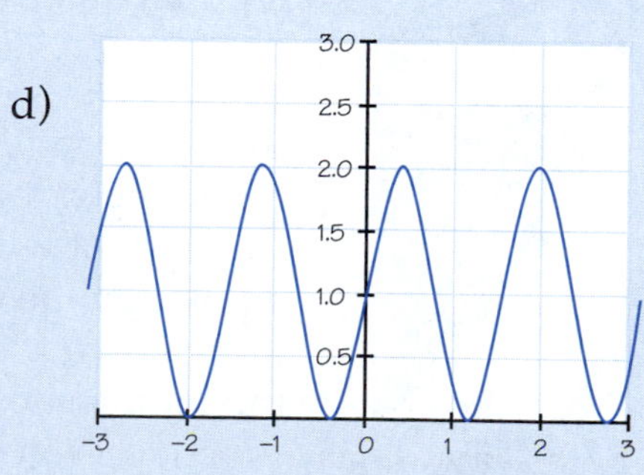

FIGURE 8.59.
Graph of a periodic function.

Describing Motion

REVIEW

Model Development

In this chapter you have determined equations that model various motions. You modeled motion based on time-distance data from motion detectors. Using the models of the motions, you planned and performed some stunts. The model of a motion at a constant velocity was used to plan a near collision. The model of a nonconstant motion was used to plan the fall of a hero from a rooftop.

In Activity 8.1 you learned that velocity is the rate of change of distance with respect to time. You examined some motions that had constant velocity to determine the kind of model for these motions. You discovered that when objects (or walkers) move at a constant rate, their distance-versus-time graphs are linear. In Activity 8.2 you put these concepts to work to plan and test a near collision.

In Activity 8.3 you studied motions that may be represented by parabolas. You learned about instantaneous velocity, which is the velocity of a moving object at a particular instant. You discovered that when the relationship between distance and time is quadratic, the relationship between velocity and time is linear.

In Activity 8.4 you studied falling objects. The motions of falling objects are modeled by quadratic functions. You used data of a falling object to determine a model for the motion using a quadratic model. You entered the data into a calculator and found a quadratic model that was a good fit for the data. You used your results to plan the hero's fall.

In Activity 8.5, you investigated circular motion, such as revolving carousels and bicycle wheels. You performed an experiment with a round object and measured the distances that a point on the object traveled and the corresponding heights. You graphed your data and discovered that a periodic function can model this motion. In Homework 8.5 you studied the sine function, which is a periodic function. In Activity 8.6, you developed sinusoidal models of the form $y = A\sin(B(x - C)) + D$ so that you could determine a model for the wheel data you collected in Activity 8.5.

Mathematics Used

To model motion you must first investigate the velocity of the motion. To calculate the average velocity from time 1 to time 2, use the ratio $\frac{\text{distance 2} - \text{distance 1}}{\text{time 2} - \text{time 1}}$.

To approximate instantaneous velocity at a particular time t_0, calculate the average velocity over a small time interval centered at t_0.

The simplest motion you studied was that of a walker (or car) moving in a straight line at a constant rate. This type of motion was modeled by distance-versus-time equations of the form

$d = vt + d_0$, where d is the distance and t is the elapsed time.

The slope, v, of the graph tells you the object's velocity. The constant term d_0 gives the object's distance when $t = 0$.

The motion of a falling object is more complex because the object's velocity is not constant. A linear model is not an appropriate model for this type of motion. The distance-versus-time graphs for falling objects are parabolas. The velocity-versus-time graphs are lines.

The story of a falling object can be told by its quadratic model: $h = at^2 + h_0$, where h is the height and t is the elapsed time since the object was released. The value of a is half the acceleration and h_0 is the object's height when $t = 0$.

Acceleration is the rate of change of velocity with respect to time. After determining velocities at time 1 and time 2, the average acceleration from time 1 to time 2 may be calculated using the ratio $\frac{\text{velocity 2} - \text{velocity 1}}{\text{time 2} - \text{time 1}}$.

More complex still is motion that oscillates between two extreme values. Circular motions are modeled by functions that are periodic. Periodic functions have graphic patterns that repeat over fixed intervals.

- The period of a periodic function is the shortest horizontal length of the basic repeating shape of its graph.
- The amplitude is half the vertical height (top to bottom) of the basic repeating shape.

The sine function is an example of a periodic function. The sine function is defined as the vertical displacement versus the radian measure of an angle made by a dot rotating around a unit circle. Any function of the form $y = A\sin(B(t - C)) + D$ belongs to the family of sinusoidal functions. The control numbers A, B, C, and D control the graph's amplitude, period, horizontal shift, and axis of oscillation.

When fitting a sinusoidal model to data by hand, you need to use your data to estimate the values of the control numbers A, B, C, and D. Here is one method for estimating the control numbers. Sometimes, however, a variant of this method will provide better results. So don't get locked into a single method.

- To estimate A: Average several of the maximum (peak) values in the data to get an estimate of the function's maximum value. Then do the same for the minimums. Let A be half the difference between the estimated maximum and minimum values.

- To estimate D: Let D be the average of the estimated maximum and minimum values.

- To estimate B: First estimate the period. Select two data points associated with peaks, or maximums, but separated by several cycles of oscillation. Divide the time difference of these two data points by the number of cycles separating these maximums. This gives you an estimate of the period. Let $B = \frac{2\pi}{(\text{estimated period})}$.

- To estimate C: The sine function $y = A\sin(Bt) + D$ should have its first maximum at $t = \frac{(\text{period})}{4}$. Locate the t-value associated with the first maximum in the data. Call it t_0. To approximate the horizontal shift, set $C = t_0 - \frac{(\text{period})}{4}$.

Acceleration: The rate of change of velocity with respect to time.

Amplitude (of a periodic function): Half the vertical height of the basic repeating shape; that is, half of the vertical distance between maximum and minimum values of the periodic graph.

Average acceleration: The rate of change of velocity from time 1 to time 2:

$$\frac{(\text{velocity 2} - \text{velocity 1})}{(\text{time 2} - \text{time 1})}, \text{ or } \frac{(\text{change in velocity})}{(\text{elapsed time})}.$$

Note that force is the cause of acceleration. Typical forces are due to gravity, friction, or the push of a hand.

Average velocity: The rate of change of distance from time 1 to time 2:

$$\frac{(\text{distance 2} - \text{distance 1})}{(\text{time 1} - \text{time 2})}, \text{ or } \frac{(\text{distance traveled})}{(\text{elapsed time})}.$$

Axis of oscillation (of a periodic function): The center of the oscillations, determined as the horizontal line midway between the maximum and minimum values of a periodic graph.

Central angle: An angle whose vertex is the center of a circle.

Instantaneous velocity at time t_0: The rate of change of velocity at the instant $t = t_0$. You can approximate the instantaneous velocity by calculating the average velocity over a small interval centered at $t = t_0$.

Period (of a periodic function): The shortest horizontal length of a basic repeating shape.

Periodic function: Function that repeats itself on intervals of a fixed length (equal to the period).

Radian: The radian measure of a central angle with one side along the positive x-axis is the directed length of an arc of the unit circle that begins at (1, 0). If the angle turns in the counterclockwise direction, the radian measure is positive; if it turns in the clockwise direction, the radian measure is negative. You can extend this definition to any circle by defining its center as the origin and its radius as one unit.

Sine function: The sine function pairs the vertical displacement (the output) of a point with the radian measure (the input) of an angle. The point must lie at the intersection of the unit circle and one side of the angle. Since the angle must be a central angle, its other side lies along the positive x-axis.

Sinusoidal function: Any function that can be expressed in the form $y = A\sin(B(x - C)) + D$, where neither A nor B is zero.

Unit circle: Circle with radius one unit and centered at the origin.

Velocity: The rate of change of distance with respect to time.

9
CHAPTER
Mathematics in the Music
ACTIVITY 9.1
NOW IS THE TIME
ACTIVITY 9.2
MOVING THE MUSIC
ACTIVITY 9.3
HARMONICS SERIOUS
ACTIVITY 9.4
TWELVE TONES
ACTIVITY 9.5
SCALING MAJORS
ACTIVITY 9.6
LET'S JAM
CHAPTER REVIEW

Can you hear the mathematics in the music?

You probably don't think about mathematics when you see 'N Sync performing. In fact pop stars use mathematics to write and play their songs. Although music is an art form, it is also very mathematical. Arithmetic, algebra, geometry, and sinusoidal functions are just a few of the types of mathematics that are used in music.

Musicians and composers often structure their works using mathematics. Musicians keep time by counting the beats. Composers have used special numbers to determine when to repeat a musical phrase in their compositions. In the twentieth century, an entire new form of music was created that is heavily mathematical. In some ways the new mathematics that developed in the nineteenth century made an important contribution to this new form of music. Get out your pens, calculators, and musical instruments. You are about to learn all about the mathematics behind the music.

PREPARATION READING

Music Makes The World Go Round

It is said that music is the universal language. Music is played in many forms all over the world. Instruments vary from the didjeridoo of the Australian aborigines (a hollow tree branch that plays only two tones) to large concert organs. No matter what the sound, people everywhere can appreciate good music. This explains why so many forms of world music are often used in American pop music.

In the Western World we have developed an entire language and notation for music. Let's start with some basic ideas. Look at Handout 9.1 that provides a summary of music notation.

- **Beat:** When you tap your foot or bob up and down to the music, you are feeling the beat.
- **Rhythm:** The duration of tones within a beat and across several beats.
- **Melody:** When you sing along with some music you are singing the melody. It is the series of musical sounds that is the basis for a song.

Musicians have an entire system of note writing to communicate their works. This *notation* uses a staff like the one in **Figure 9.1**.

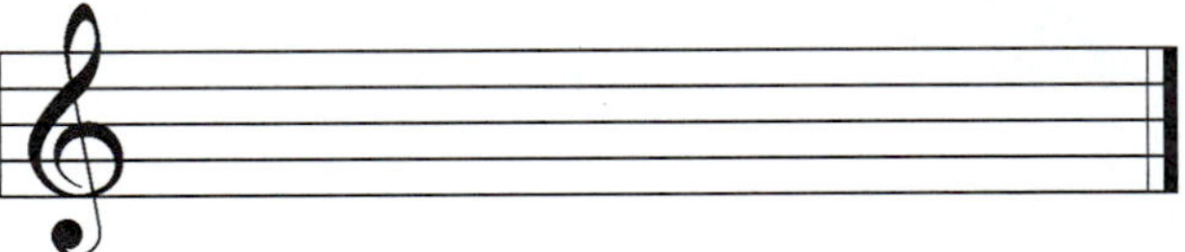

FIGURE 9.1.
A Staff.

To create a melody, musical notes are written on the staff. Here are the symbols for a few of the basic notes.

♫ eighth notes (2 notes)

♩ quarter note

𝅗𝅥 half note

𝅝 whole note

These basic ideas will be used regularly in the following activities. Refer back to this page when you need to. Now on with the show!

ACTIVITY 9.1

Now Is the Time

Music would be disorganized noise if it was not played at the right speed. If the right notes are not played at just the right moment, the result could be a mess. (Think about when you heard a young person trying to learn to play an instrument!) Timing is vital to good musicians. In this activity you will learn about the metrical structure of music and how mathematics is the key to understanding this structure.

THE TIME SIGNATURE

Many musicians consider the beat the most important element of their music. The beat provides the foundation for the melody and harmony. Performers study rhythms within the beat and the stress patterns for various combinations of beats.

A piece of music is divided into units of time called **measures**. The **time signature** of the music tells a musician how many beats are in a measure and what kind of note receives 1 beat. The time signature looks like a fraction without the fraction bar, such as $\substack{3\\4}$. The time signature is written on the staff at the beginning of the work.

FIGURE 9.2. A Measure.

- The number 4 on the bottom tells the kind of note that receives the beat. For $\substack{3\\4}$, 4 means a quarter note receives a beat.
- The top number 3 shows the number of beats in a measure. For $\substack{3\\4}$, 3 means 3 beats per measure.

Time Signature	Meaning
$\substack{4\\4}$	four beats to a measure; the quarter note receives a beat also known as C or common time
$\substack{2\\8}$	two beats to a measure; the eighth note receives a beat
$\substack{2\\2}$	two beats to a measure; the half note receives a beat

Figure 9.3 shows some other time signatures and their meanings.

FIGURE 9.3. Common time signatures.

1. Determine the number of beats per measure and the note that gets a beat in each of the following.

   a) $\frac{2}{4}$

   b) $\frac{6}{8}$

   c) $\frac{5}{4}$

   d) $\frac{3}{2}$

   e) $\frac{5}{8}$

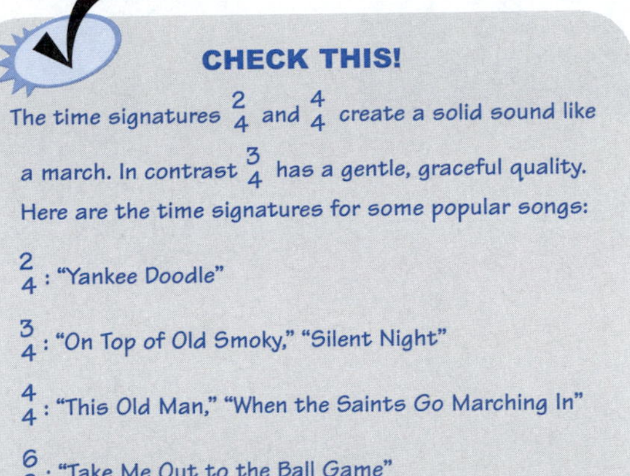
CHECK THIS!

The time signatures $\frac{2}{4}$ and $\frac{4}{4}$ create a solid sound like a march. In contrast $\frac{3}{4}$ has a gentle, graceful quality. Here are the time signatures for some popular songs:

$\frac{2}{4}$: "Yankee Doodle"

$\frac{3}{4}$: "On Top of Old Smoky," "Silent Night"

$\frac{4}{4}$: "This Old Man," "When the Saints Go Marching In"

$\frac{6}{8}$: "Take Me Out to the Ball Game"

Here are the note values of some notes for music written in $\frac{4}{4}$ time.

♫	eighth notes (2 notes)	$\frac{1}{2}$ beat each (single eighth: ♪)
♩	quarter note	1 beat
𝅗𝅥	half note	2 beats
𝅗𝅥.	dotted half note	3 beats
𝅝	whole note	4 beats

2. Write the number of beats for each combination of notes in $\frac{4}{4}$ time.

   a) ♪ ♪ ♪

   b) ♩ ♩ 𝅗𝅥

   c) ♩ 𝅗𝅥 ♪ ♪

   d) ♩ ♪ ♪ ♪ ♪

   e) 𝅗𝅥 ♩ 𝅗𝅥

Written music often includes a **tempo** indicator that tells how fast the beat is played. Sometimes terms, such as *Allegro* (fast) and *Largo* (slow), are used to determine the tempo. Often a specific number of beats per minute is indicated. For example,

♩ = 128 means to play 128 quarter notes per minute.

If the quarter note gets a beat, as it does in $\frac{2}{4}$, $\frac{3}{4}$, or $\frac{4}{4}$ time, 128 beats would be played per minute.

3. Calculate how long each of the following selections of music would last.

   a) 48 measures in $\frac{4}{4}$ time at ♩ = 128. (Hint: There are 4 beats in each of the 48 measures and 128 beats per minute.)

   b) 132 measures in $\frac{2}{4}$ time at ♩ = 120.

   c) 252 measures in $\frac{5}{4}$ time at ♩ = 210.

4. Write an equation that models the time t to play a piece (in minutes) written in $\frac{4}{4}$ time. Let m represent the number of measures, and let s represent the tempo in beats per minute.

5. Write an equation that models the time, t (in minutes), it takes to play a piece where m is the number of measures, b is the tempo in beats per minute, and s is the number of beats per measure.

6. Use the equation from question 5 and your calculator to find to the nearest *second*, how long it would take to play the following selections.

   a) 54 measures in $\frac{4}{4}$ time at ♩ = 120.

   b) 140 measures in $\frac{2}{4}$ time at ♩ = 132.

   c) 204 measures in $\frac{5}{4}$ time at ♩ = 210.

7. Rewrite the equation in question 5 to solve for the time in seconds rather than minutes.

8. Solve the equation in question 7 for m, the number of measures.

9. Manuel is writing a piece of music for a television commercial. The commercial lasts exactly 36 seconds, and the music should cover the entire commercial. He wants to write the piece in $\frac{2}{4}$ at ♩ = 120. How many measures must he write?

10. Another commercial lasts 44 seconds. How many measures are needed in $\frac{3}{4}$ time at ♩ = 180?

11. Solve the equation in question 5 for b, the tempo.

12. Manuel has a contract to write another piece for a radio commercial. The commercial lasts exactly 40 seconds. He has written 32 measures, m, in $\frac{4}{4}$. At what tempo, b, must the music be performed?

13. For a commercial 55 seconds long, at what tempo must a piece be played if it contains 44 measures in $\frac{5}{4}$?

14. How many **seconds** would it take to perform the selection below if the performer follows the notation perfectly?

♩ = 48

15. How many seconds would it take to perform the following piece?

♩ = 96

16. How many seconds would it take to perform this selection? Note that the lines of music joined by the braces, {, are played at the same time rather than one after another.

SUMMARY

In this activity you found out how musicians use mathematics to keep time.

- You learned about time signatures and how a piece of music is divided into measures.
- Using the tempo and the number of beats per measure, you calculated the time it takes to play a piece of music.

In the next activity you will look at the melodic aspect of music.

HOMEWORK

9.1

And The Beat Goes On

Zhia has written four melodic phrases shown in **Figure 9.4**. She believes these phrases may be played in any order except that they must end with phrase B. She plans to use them to write music for several commercials.

FIGURE 9.4.

The term *combination* means any set of possibilities. In mathematics a combination is a set in which order does not matter. In contrast, permutations are sets in which a different order produces a different result.

- Choosing a committee of three from a group of five people is an example of a combination. It does not matter who is chosen first, second, or third. They make up the same committee.
- Selecting the officers of the school council is an example of a permutation. Order is important. It's better to be president of the school council instead of being selected vice president!

1. In creating music from the set of phrases given, does a different order produce the same or a different result?

2. Based on your answer to question 1, are the sets of phrases combinations or permutations?

3. Count the number of beats in each of the four melodic phrases in Figure 9.4. (Hint: There are two ways to count the beats. You can refer to the beat counts for notes on page 532 or you can use the time signature and count the measures.)

Using the letter names for the phrases in Figure 9.4, write the combinations or permutations requested in each commercial situation. In each case use the entire phrase in order.

For example, if a melodic phrase 1 has 8 beats and melodic phrase 2 has 7 beats, there are two possible permutations to create a 15 beat piece of music.

- melodic phrase 1 first with 8 beats then melodic phrase 2 with 7, totaling 15 beats or
- melodic phrase 2 with 7 beats first then melodic phrase 1 with 8, totaling 15 beats

So the different permutations are 1, 2 and 2, 1.

4. Write all the possibilities for a commercial with exactly 43 beats using only phrases B and D.

5. Write all the possibilities for a commercial with exactly 24 beats.

6. Write all the possibilities for a commercial with exactly 25 beats.

7. Write three possibilities for a one-minute commercial at ♩ = 120.

8. Write three possibilities for a 30-second commercial at ♩ = 180.

9. Write one possibility for a 20-second commercial at ♩ = 150.

ACTIVITY 9.2

Moving the Music

In the last activity you were introduced to two important elements of music: duration (time) and pitch (melody). Now you will discover some methods composers use to write music. In this activity you will learn how transformations are used to write music.

TRANSLATIONS

To unify a composition, a composer will repeat a sequence of tones several times during a piece. This sequence will occur at the beginning of the piece, and it may reappear in different forms to create richer, more complex music. The composer may use different variations of the original sequence of tones. These variations can be represented by a geometric transformation.

FIGURE 9.5.

A **translation** is a geometric transformation in which each point of a figure is moved in exactly the same direction by exactly the same amount. See **Figure 9.5**.

The following phrases in the popular folk song "Shoo Fly" can be represented by a translation.

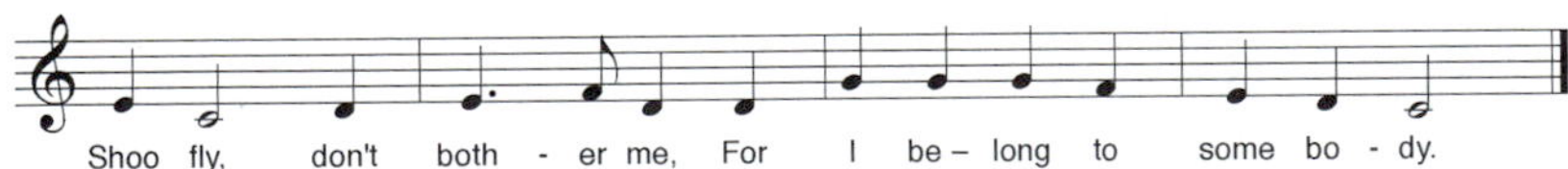

To see the translation we consider measures 1–2, which contain a melodic sequence. We can create a graph of these two measures using the following steps.

I. Replace the staff with a coordinate plane.

II. Think of the note heads (the circular parts) as points on a coordinate plane.

III. Plot the "points" on the coordinate plane according to their position on the staff.

IV. Connect each pair of points with a straight line.

Using these steps we obtain the following graph of measures 1–2.

1. a) Make a copy of **Figure 9.6** and plot the notes that correspond to measures 3–4.

   b) How does the shape of the graph of measures 1–2 compare with the shape of the graph of measures 3–4?

   c) How can you move the graph of measures 1–2 to produce the graph of measures 3–4?

FIGURE 9.6. A graph of measures 1–2 of "Shoo Fly".

The type of variation you found in question 1 is called a **musical sequence.**

REFLECTIONS

Your graph in question 1 showed a translation of the original musical phrase. Another type of musical variation can be represented by a geometric transformation called a reflection (or mirror image). A **reflection** of a figure is obtained by rotating the figure across a line. See **Figure 9.7**. Notice how figure A has been rotated to create figure B. We say that the figure B is a reflection of figure A.

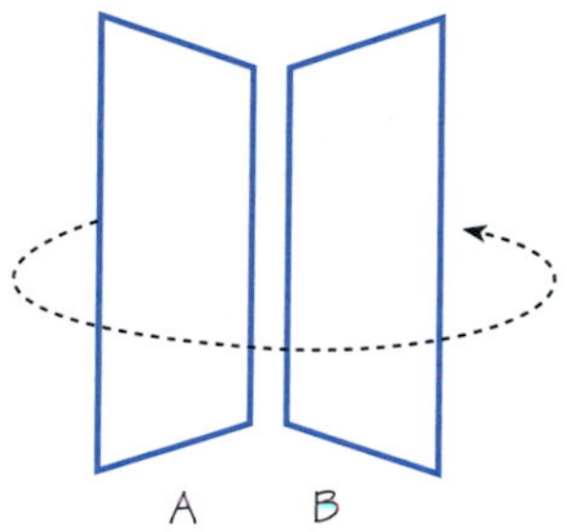

FIGURE 9.7.

Now let's see how a selection can be represented by a reflection. Consider the following folk song.

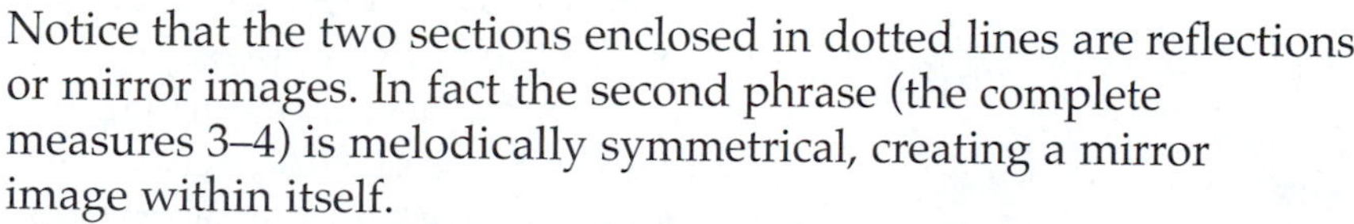

Notice that the two sections enclosed in dotted lines are reflections or mirror images. In fact the second phrase (the complete measures 3–4) is melodically symmetrical, creating a mirror image within itself.

Figure 9.8 shows the graph of the section in measures 3–4 within the dotted lines. Notice that the first and last notes in the graph are connected to form a polygon so that you can see the transformations of this figure.

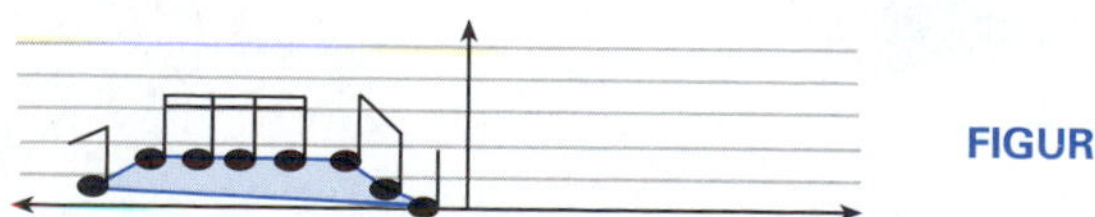

FIGURE 9.8.

2. a) Make a copy of Figure 9.8 and plot the notes in the section of measures 5–6 within the dotted lines. Plot the notes on the right side of the y-axis so that you can see the transformation. Connect the first and last notes to form a polygon.

   b) How does the shape of the polygon you created compare with the shape of the original polygon?

The figure you drew in question 2 should show that the polygons derived from the two sections can be represented by reflection across the y-axis. In music this is called **retrogression**.

Here's another example. Consider this harmonization of "The Gypsy Rover."

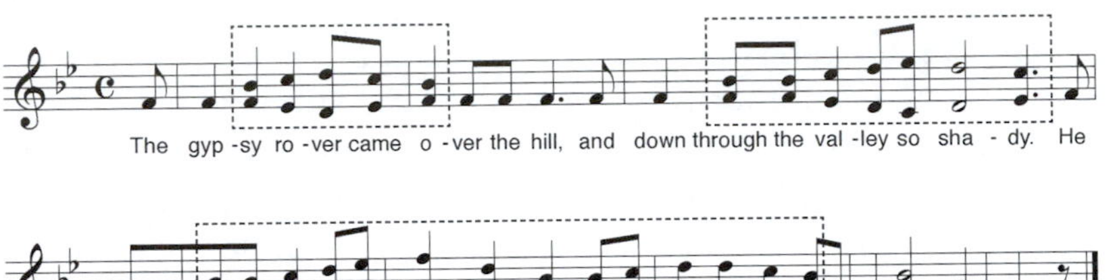

FIGURE 9.9.

Note C indicates $\frac{4}{4}$ time signature.

The mirroring in the melody (top notes) and harmony (bottom notes) is abundant in this piece. Study the sections in dotted rectangles to see the vertical reflection. The top notes of the chords form one polygon, and the bottom notes of the chords form a second polygon. The graphical representation of the first section in dotted lines is shown in **Figure 9.9**.

You can see that the polygons formed by the top notes are a reflection across the x-axis of the bottom notes. Inversion is a musical transformation corresponding to reflection across the x-axis. The melody and harmony (top and bottom) parts illustrate this type of reflection.

ROTATIONS

One more type of musical variation can be represented by a rotation. A **rotation** is a geometric transformation in which a figure is rotated about a point. The same effect can be achieved by reflecting the figure across both axes. See **Figure 9.10**.

FIGURE 9.10. A rotation of a figure about a point.

Consider the folk song "Jim Along Josie."

We will rotate the notes in measures 2–3 to obtain the notes in the dotted lines in the last two measures. To rotate the first phrase, reflect each note across the y-axis and then reflect it again across the x-axis. **Figure 9.11** shows the graph of this rotation.

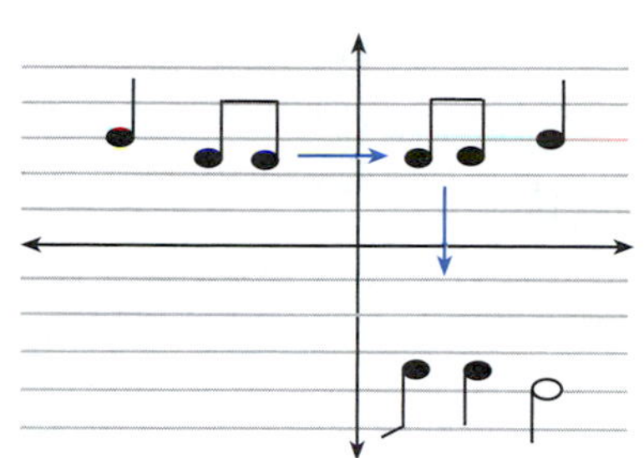

FIGURE 9.11.

The term for this musical variation is **retrograde inversion.** This term is derived from the terms retrogression (reflection about the y-axis) and inversion (reflection about the x-axis).

3. Complete the empty measures in the following composition by drawing the notes. Draw the notes that are produced by each transformation of the given melodic phrase. Retain the note values (eighth note, quarter note, etc.) of the original phrase. Give your song a title.

4. Write your own composition by transforming the given musical phrase. Use any transformation you learned in this activity. Identify each transformation that you use.

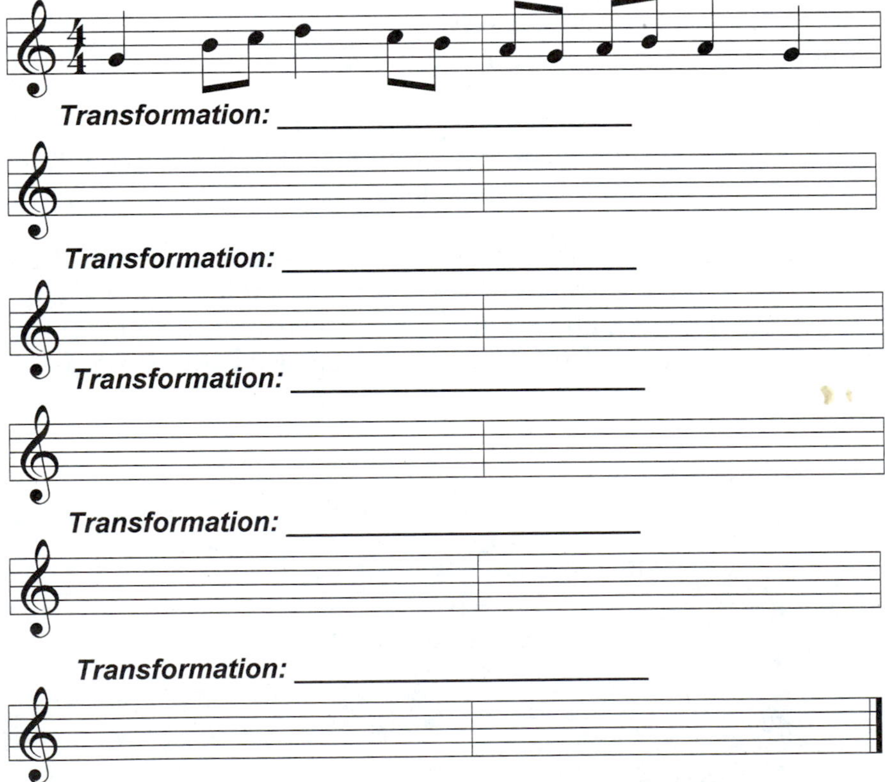

TRANSFORMATIONS AND THE GOLDEN MEAN

Composers often used transformations to write musical variations of a phrase. Occasionally the location of these transformations may be found using special numbers, such as the Golden Mean. For example, works by Wolfgang Amadeus Mozart and Béla Bartók use the Golden Mean to locate musical variations. It is not known whether these occurrences were intentional or accidental. Mozart and Bartók were both fascinated by numbers, and Bartók was also extremely interested in science and nature.

Bartók's *Mikrokosmos* is a six-volume set of piano teaching pieces. Bartók chose the title because the work is a microcosm of his larger works, a "small world" embodying all the elements of his major works. In these works there are many examples of the use of numbers close to the Golden Mean that find the section breaks. Sometimes he begins an inversion or retrogression at these points. Sometimes he returns to the theme at the beginning of the piece.

For example, in a piece with 44 measures, using the proportions of the Golden Mean, a section break may appear at measure 27 (because $44 \cdot 0.618 = 27.192$).

Mikrokosmos Volume 4, Number 99

5. "Hands Crossing" has 23 measures. Using the proportions of the Golden Mean, where would you expect to find the section break (where the musical passage returns to the theme or a variation)?

GOLDEN MEAN

The number 1.6180339887... (the ..., or ellipse, means the number goes on forever) is called the Golden Mean. This number was first calculated by the ancient Greeks using the formula $\frac{\sqrt{5}+1}{2}$ and was used to design their temples. The ratio $\frac{1}{0.618...}$ is called the Golden Ratio because $\frac{1}{0.618...} = 1.6180339887...$.

Very close estimates of this ratio appear often in art, music, architecture, and natural objects, such as flower petals. It is called "Golden" because it is seen so often, and because it seems to be connected with our definition of physical and natural beauty.

SECTION BREAKS USING THE GOLDEN MEAN

Composers often change the structure of their music at specific places in the piece. The point before this change happens is called the section break. If a composer is using the Golden Mean to determine where section breaks will happen, they will multiply the total number of measures in the piece by 0.618.

6. Find and mark the section break for "Hand's Crossing".

7. Which of the following occurs at the section break for "Hand's Crossing"? (Hint: Examine the top staff of measures 1–4 and the first four measures at the break.)

a) inversion

b) return to original theme

c) retrogression

d) sequence

e) retrograde inversion

8. "Melody in the Mist" has 44 measures. Using the proportions of the Golden Mean, where would you expect to find the section break?

9. Find and mark the section break for "Melody in the Mist".

10. Which of the following occurs at the section break for "Melody in the Mist"? (Hint: Examine measures 20–23 and the first four measures at the break. It may not be a perfect transformation.)

a) inversion

b) return to original theme

c) retrogression

d) sequence

e) retrograde inversion

SUMMARY

In this activity you learned how musical variations can be represented by geometric transformations.

- You studied how a musical phrase is transformed into another phrase.
- You used translations, reflections, and rotations to write variations of musical phrases.
- You used the Golden Mean to find section breaks in some works by Béla Bartók.

Now that you have some experience with writing music, in the next activity you will learn how the instruments produce sounds.

HOMEWORK 9.2

Play it Again Sam

1. In the musical selection below, circle and classify all examples of the following transformations of at least four notes: sequence, inversion, retrograde, and retrograde inversion.

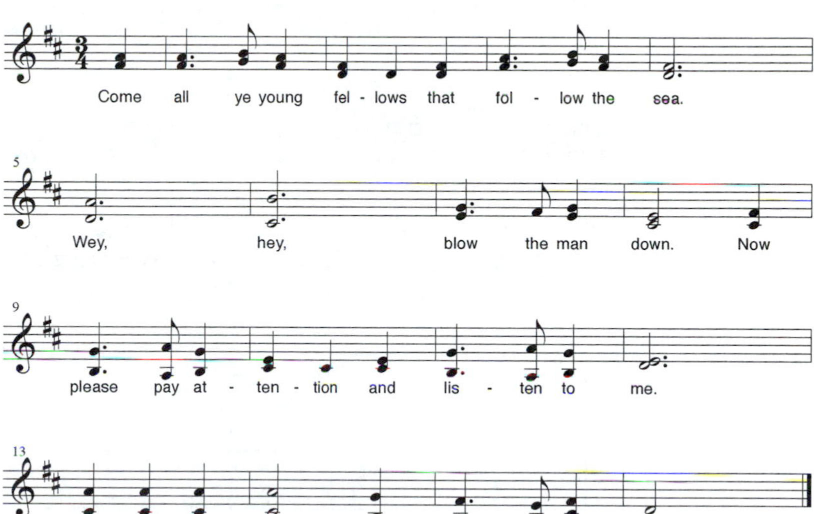

2. "Once Upon a Time…", Number 94 in *Mikrokosmos Volume 3* has numerous time signature changes. Therefore the section break must calculated according to the number of beats in the piece rather than the number of measures. What number tells you the number of beats in a measure?

3. Calculate the number of beats at which you would expect the section break to occur for "Once Upon a Time...".

4. Find and mark the section break.

5. Which of the following occurs at the section break? (Hint: Examine the bottom staff of measures 1–5 and the top staff of the first five measures at the break.)

   a) inversion

   b) return to original theme

   c) retrogression

   d) sequence

   e) retrograde inversion

ACTIVITY

9.3

Harmonics Serious

In Activity 9.2 you learned some of the tricks of the composer's trade. Now you will take a look at how instruments play musical tones. In this activity you will learn about the mathematics of stringed instruments.

HOW SOUND IS MADE

When you pluck a violin or guitar string, the string vibrates back and forth many times. To see this effect try plucking a rubber band once. The blur you see is that rubber band vibrating back and forth many times.

CHECK THIS!

If you pull and release a violin string, the first vibration sends the string through its original rest position. The string continues past the rest position before it returns back to the original position. This motion is called a cycle. The frequency of a vibrating string measures how fast the string is moving. Frequency is measured in cycles per seconds or hertz, a term created in honor of the discoverer of radio waves.

1 hertz (hz) = 1 cycle per second

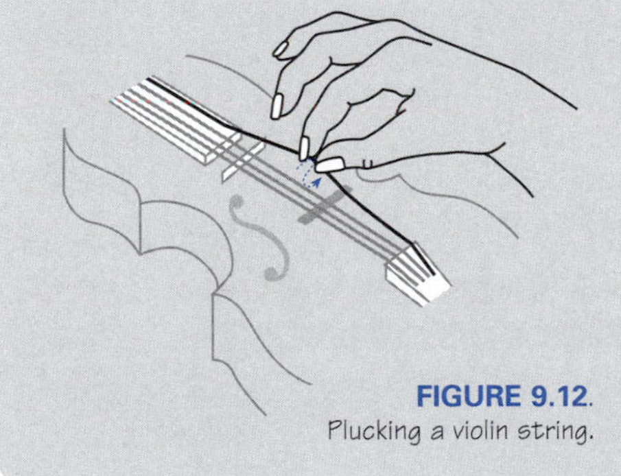

FIGURE 9.12.
Plucking a violin string.

When set into motion, a string vibrates at a rate directly proportional to its length. In addition to vibrating over its entire length, a string simultaneously vibrates over fractional divisions of its length (1/2, 1/3, 1/4, 1/5, and so on) producing higher frequencies that are inversely proportional to the string length divisions. The lowest frequency, corresponding to the primary mode of vibration of the string, is called the **fundamental**. The series of vibrations including the whole length of the string and the fractional divisions is called the **Harmonic Series**. The series of vibrations produces the sound you hear. Physicists have discovered that the frequency of each harmonic after the first harmonic is a multiple of the frequency of the first harmonic.

1. a) Here are the frequencies for the fundamental and the first few higher harmonics of a note. Find the frequencies of the next three tones in the harmonic series. 131, 262, 393, 524, ____, ____, ____

   b) Describe the pattern you found.

2. If the frequency of the fundamental is 220 hz, find the frequencies of the next five tones in the harmonic series.

3. a) The harmonic whose frequency is twice the frequency of the fundamental is called the **second harmonic.** Using this information and your results from questions 1–2, define the third harmonic.

   b) Define the fourth harmonic.

   c) Define the nth harmonic.

4. If the frequency of the fundamental is 330 hz, write the frequency of the third harmonic.

5. a) Given a fundamental of frequency 88 hz, write the frequency of the fourth harmonic.

   b) Write an algebraic expression for the frequency of the fourth harmonic in an harmonic series. Let x represent the frequency of the fundamental, and let y represent the frequency of the fourth harmonic.

   c) Write an algebraic expression for the frequency of the fifth harmonic in a harmonic series. Let x represent the frequency of the fundamental, and let y represent the frequency of the fifth harmonic.

   d) Write an algebraic expression for the frequency of the nth harmonic in a harmonic series. Let x represent the frequency of the fundamental, and let y represent the frequency of the nth harmonic.

   e) In your equation of the frequency of the nth harmonic, what is the independent variable? What is the dependent variable? Explain.

GRAPHING THE HARMONICS

The graph of the fundamental is the familiar sine wave shown in **Figure 9.13**. In this graph the horizontal axis represents the time, and the vertical axis represents the loudness of the tone at that time.

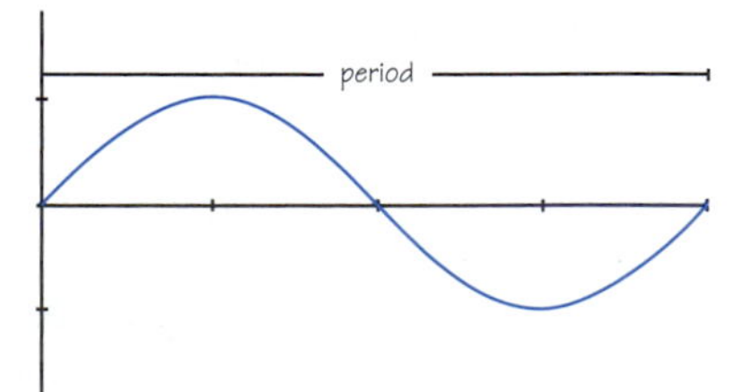

FIGURE 9.13. The graph of a sine function.

6. a) On your graphing calculator set the MODE to degrees. To graph the fundamental A-220 (the A below middle C), use the equation $y = \sin(220x)$. Set the window as follows: Xmin = 0; Xmax = 2; Xscl = 0; Ymin = –2; Ymax = 2; Yscl = 0.

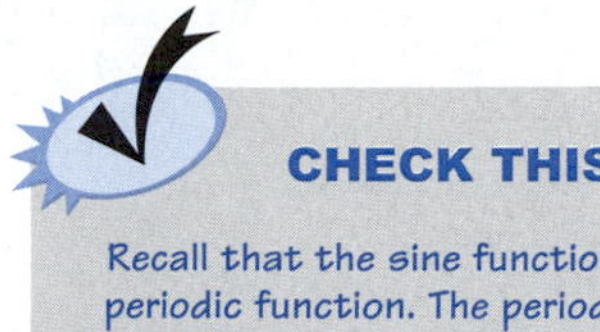

CHECK THIS!

Recall that the sine function is called a periodic function. The period of the sine function is the x-interval in which the graph makes one complete cycle.

   b) Which variable (x or y) represents time on the graph?

   c) Use the TRACE feature to determine the period (how many seconds it takes for one wave to be completed). Remember the period is found by the difference in the x-values for a complete cycle. Round to the nearest hundredth of a second.

You can use the period to find the frequency of this fundamental A-220. There is an inverse relationship between period and the frequency. This relationship is given by $f = \frac{1}{p}$ where p is the period and f is the frequency. Since your calculator MODE is in degrees, multiply by 360°.

UNITS, FREQUENCY, AND PERIOD

If your calculator MODE is in degrees, the relationship between the frequency and period of a harmonic is given by

$f = \frac{1}{p} \cdot 360.$

The units for the period will be seconds, and the units for the frequency is hertz.

7. a) Use your result from question 6 to find the frequency of the fundamental A-220.

   b) If the frequency of the first or fundamental is 220 hz, what is the frequency of the second harmonic?

   c) Graph the equation for the second harmonic along with the graph of the first harmonic on your calculator.

   d) What equation did you use for the second harmonic?

In music, the frequency of a tone determines how high or low the tone sounds, which is called the **pitch**.

Since the frequency of a tone is related to the period, then the frequency also is related to the pitch. You can use the graphs from question 7(c) to investigate the relationship between the pitch and the frequency of a tone.

PITCH

The pitch refers to the highness or lowness of a tone. For example, a low note at the left end of the piano keyboard has a low pitch. A high note at the other end of the keyboard has a high pitch.

8. a) The fundamental A-220 has a lower pitch than the pitch of the second harmonic. Compare the periods of the two graphs in question 7(c).

   b) What can you say about the frequencies of A-220 and the second harmonic?

   c) Use your results from (b) to make a general statement about the relationship between the frequency and the pitch of a tone.

   d) What numbers in the equations of A-220 and the second harmonic indicate the harmonic with the lower pitch?

9. a) Write an equation with a lower pitch and longer period.

   b) Graph the equation in (a) using your calculator. How does the graph show that this tone has a lower pitch than A-220?

   c) Can you see a complete wave with the current WINDOW?

   d) What parameter in the WINDOW must you change to see a complete period?

AMPLITUDE, PERIOD, AND FREQUENCY

The amplitude of the graph of the harmonic shows the loudness of the sound. The greater the amplitude, the louder the sound. Recall that the amplitude of a periodic function is

$$\text{Amplitude} = \frac{1}{2}(\text{maximum value} - \text{minimum value})$$

10. Determine the amplitude of each graph.

a)

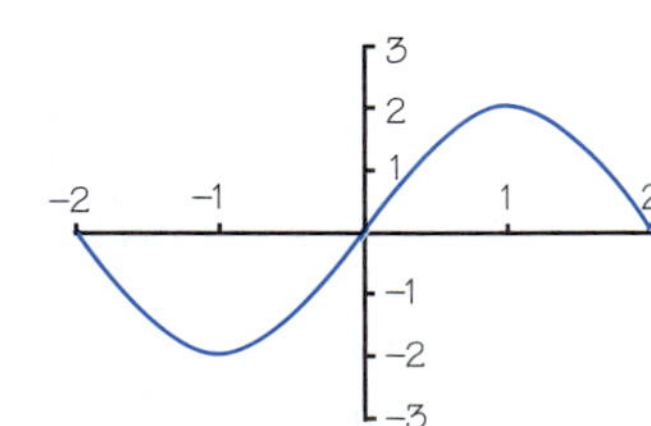

FIGURE 9.14.

b)

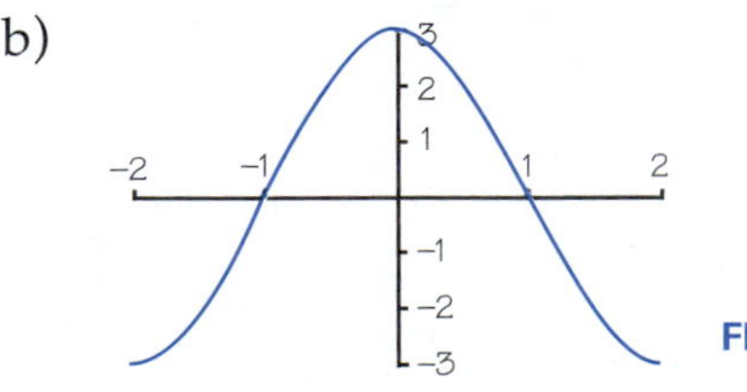

FIGURE 9.15.

c)

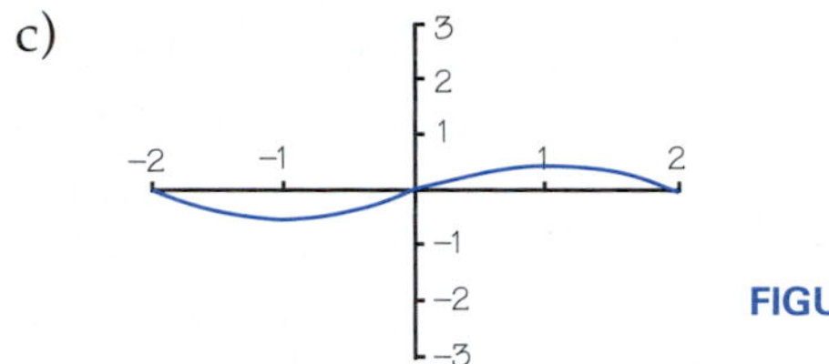

FIGURE 9.16.

11. a) Graph the following equations together using the following WINDOW: Xmin = 0; Xmax = 1.8; Xscl = 0; Ymin = –5; Ymax = 5; Yscl = 0.

$$y = \sin(220x)$$

$$y = 2\sin(220x)$$

b) Suppose these graphs represent two sounds. If the amplitude represents how loud the sound is, compare the amplitudes of the two graphs.

c) Compare the periods of these two graphs.

d) Based on your answer to (c), how do the frequencies of the sounds represented by these graphs compare?

e) Which number in the equation represents the frequencies of the sounds represented by these graphs?

f) Which number in the equations represents the amplitude? Which equation represents the louder sound?

Most musical instruments produce a collection of different harmonics. The unique sounds of instruments, such as flutes, clarinets, and trumpets, are determined in part by their harmonics and their amplitudes. This unique sound is called the **timbre**. This quality is one of the four characteristics of music. The other three characteristics are duration, pitch, and intensity. We considered duration in Activity 9.1 when we studied rhythm and beat. We discussed pitch in this activity when we studied frequency, and we discussed intensity when we studied amplitude.

Complex sound is created by combining harmonics. Mathematically this sound may be modeled by the sum of some sine functions.

12. a) A sound of a flute may be represented by the following composite function:

$$y = 16\sin(440x) + 9\sin(880x) + 3\sin(1320x) + 2.5\sin(1760x) + 1.0\sin(2200x)$$

Graph the function for the flute above, using the following WINDOW:

$X\text{min} = 0$; $X\text{max} = 1.5$; $X\text{scl} = 0$; $Y\text{min} = -25$; $Y\text{max} = 25$; $Y\text{scl} = 0$.

b) Find the amplitude of the flute equation using the TRACE feature or by calculating one of the maximums. Round to the nearest unit.

c) What do the x-coefficients in this equation represent?

d) What is the frequency of the fundamental harmonic for this sound?

e) Which other harmonics are present?

13. Suppose another sound is represented by the equation $y = 16\sin(440x)$. Graph this equation with the flute equation. Compare the amplitudes and frequencies of these sounds.

a) What is the same? Explain what it means.

b) What is different? Explain what it means.

SUMMARY

In this activity you learned about the harmonics.

- You determined the frequencies of higher harmonics based on the frequency of the fundamental.
- You graphed the harmonics and learned the graphs of the harmonics are the graphs of a sine function.

In the next activity you will study an entirely new system of music that was introduced in the twentieth century.

HOMEWORK 9.3

More Harmonies

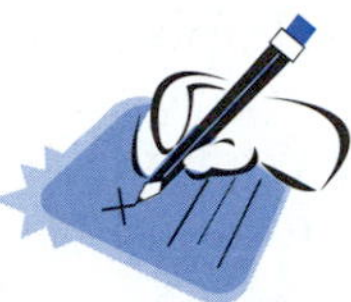

1. If the second harmonic of a note has a frequency of 330 hz, find the fundamental.

2. If the third harmonic of a note has a frequency of 330 hz, find the first and second harmonics.

3. If the fourth harmonic of a note has a frequency of 340 hz, write the frequencies of the first five harmonics in the series.

4. Clarinets produce mostly odd-numbered harmonics. Write the first five frequencies in the series for an instrument producing only odd-numbered harmonics, playing a note with a frequency of 256 hz.

5. Use a bar graph to graph the frequency versus the number of the harmonic for the first five harmonics if the fundamental has a frequency of 220 hz.

6. Use a bar graph to graph the frequency versus the number of the harmonic for the first five harmonics of a fundamental frequency of 220 hz, for an instrument producing only odd-numbered harmonics.

7. a) The normal frequency range for human hearing is 20 to 20,000 hz. Write the equation of a sound your classmates can hear (different from any we have graphed so far).

   b) Sketch its graph giving the WINDOW you used to view it.

8. a) A violin may produce the following composite function:

$$Y = 19\sin(440x) + 9\sin(880x) + 8\sin(1320x) + 9\sin(1760x) + 12.5\sin(2200x) + 10.5\sin(2640x) + 14\sin(3080x) + 11\sin(3520x) + 8\sin(3960x) + 7\sin(4400x) + 5.5\sin(4840x) + 1.0\sin(5280x) + 4.5\sin(5720x) + 4.0\sin(6160x) + 3\sin(6600x).$$

   On your graphing calculator set the MODE to degrees. Graph the equation above. Set Xmin to 0 and Xmax to 2. Determine the appropriate values for Ymin and Ymax. Explain how you chose those values.

b) Graph the equation.

c) What is the frequency of the note the violinist is playing (the fundamental)?

d) Give the first ten frequencies in the harmonic series.

9. List the numbers of the harmonics present in the equation for the violin in question 8 in a table like **Figure 9.17**. Give the amplitude, frequency, and period (rounded to six decimal places) for each harmonic.

FIGURE 9.17.

Harmonic Number	Amplitude	Frequency	Period

10. Create a bar graph showing the amplitude for each harmonic present in the sine wave for the violin in question 8.

ACTIVITY
9.4

Up to now, the techniques you have studied may be used to read and play many different kinds of music, including classical music written before 1900. In this activity you will learn how the methods of the new music of the twentieth century may be understood using some simple arithmetic.

THE NEW MUSIC AND BABBITT SQUARES

Mathematical set theory has been around for many years. However, in the twentieth century musicians began to use this system for composition. The term *set* applies to a collection of pitches. This set is then used in various ways to create music. From this idea came the twelve-tone technique.

When creating or examining a twelve-tone composition, it is helpful to be able to easily see the forty-eight possible forms of the series that are created by the transformations studied earlier. This is done by means of a matrix often called Babbitt Square after Milton Babbitt.

To explore the Babbitt Square, we must first discuss musical half steps. A half step is the distance to the next nearest note. It is helpful to look at the piano keyboard if you are not acquainted with music. The black notes have two possible names, but we will use only one to avoid confusion. We will call the black keys sharps (#), such as A#. See **Figure 9.18.**

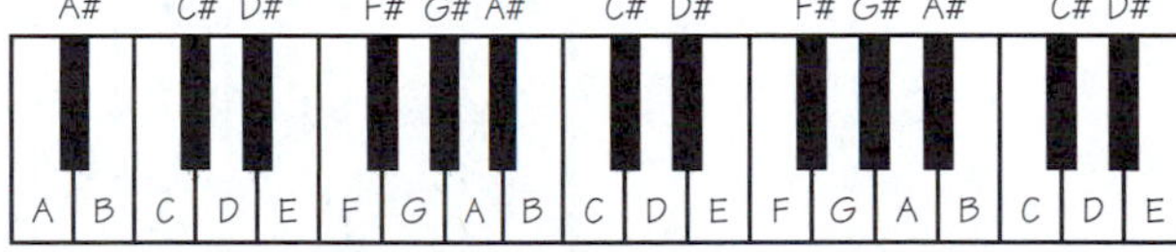

FIGURE 9.18.

For example, A to D is 5 half steps because if we begin on A and count to D, we will count 5 black and white keys.

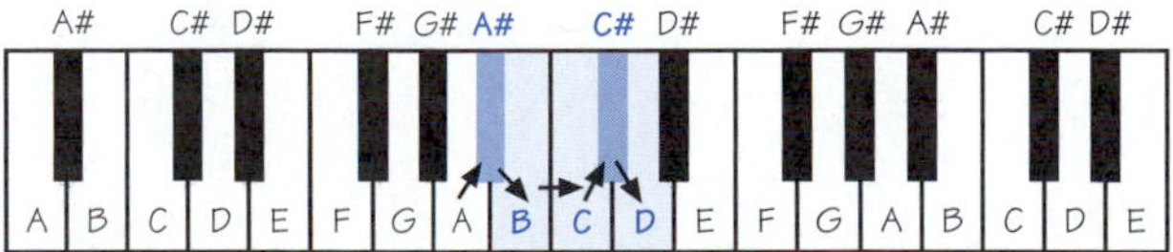

FIGURE 9.19.

In a Babbitt square, you start with a series of notes in the first row (see **Figure 9.20**). Then you use index numbers on each side of the matrix to indicate how the notes are transformed. The numbers indicate the number of half steps. Let's try one to see how this process is done.

Suppose the following tone row is chosen as the original or prime row:

A F# G G# E F B A# D C# C D#

Step 1: Find the number of half steps from the first note (A) to each note. These numbers are written above and below the matrix. For example,

- since the number of half steps from A to A is 0, write 0 above the A in the first column
- since the number of half steps from A to F# is 9, write 9 above the F# in the second column
- since the number of half steps from A to G is 10, write 10 above the G in the third column

Continue this process for every note in the series. Then rewrite these numbers at the *bottom* of the matrix.

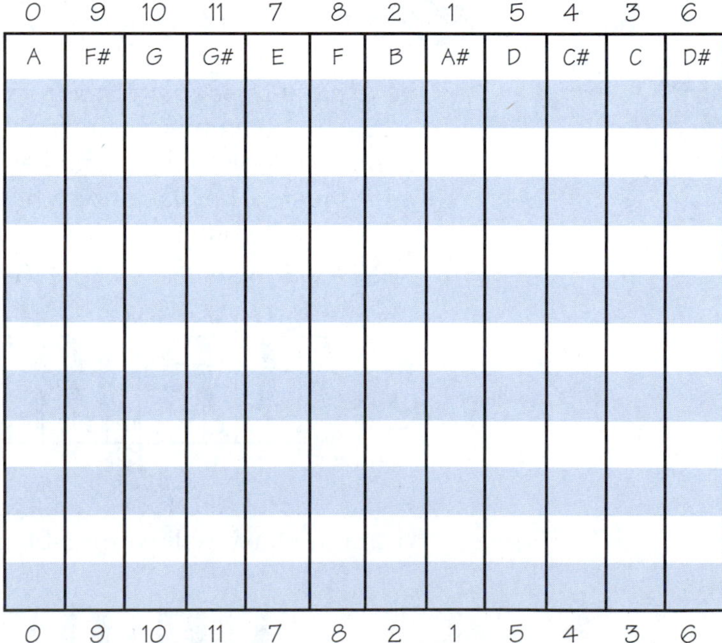

FIGURE 9.20.
Starting a Babbitt Square.

Step 2: Subtract each number from 12 to find the interval for each element in the first column (with the exception of the first number which is still 0 half-steps from itself). These numbers are written to the left and right of the matrix for easy reference.

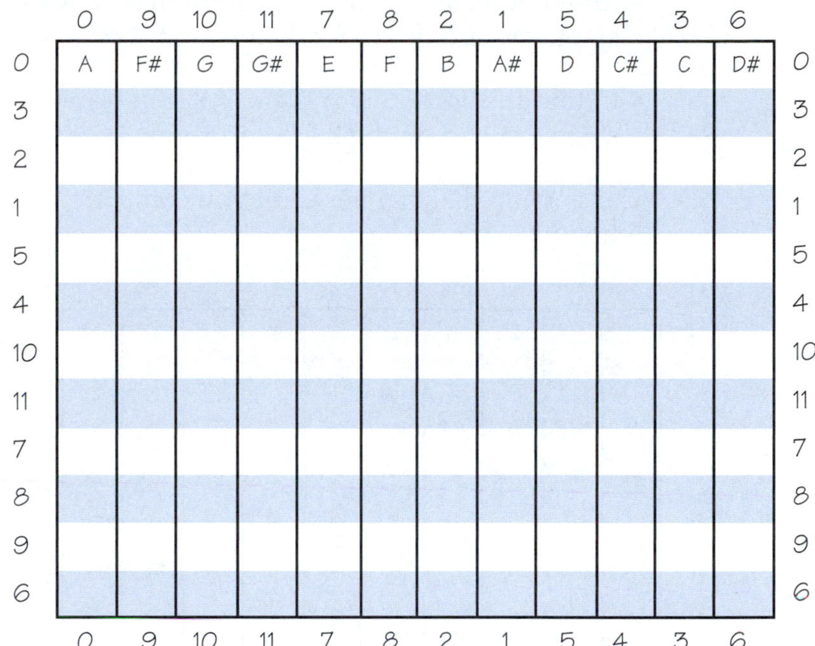

	0	9	10	11	7	8	2	1	5	4	3	6	
0	A	F#	G	G#	E	F	B	A#	D	C#	C	D#	0
3													3
2													2
1													1
5													5
4													4
10													10
11													11
7													7
8													8
9													9
6													6
	0	9	10	11	7	8	2	1	5	4	3	6	

FIGURE 9.21. Step 2 of completing the Babbitt Square.

Step 3: Using the numbers on the side of the matrix, count half steps from A on the piano keyboard to find the notes for the first column.

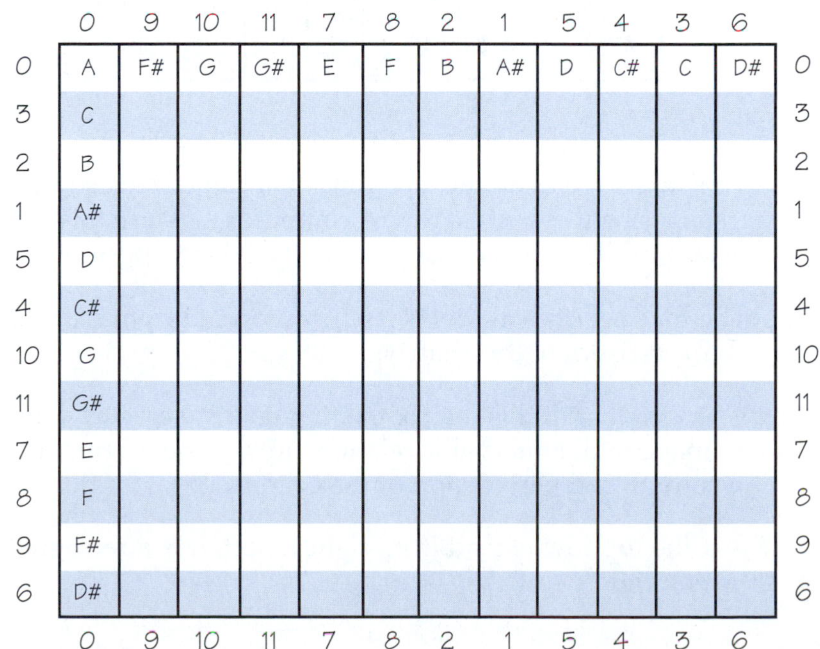

	0	9	10	11	7	8	2	1	5	4	3	6	
0	A	F#	G	G#	E	F	B	A#	D	C#	C	D#	0
3	C												3
2	B												2
1	A#												1
5	D												5
4	C#												4
10	G												10
11	G#												11
7	E												7
8	F												8
9	F#												9
6	D#												6
	0	9	10	11	7	8	2	1	5	4	3	6	

FIGURE 9.22. Step 3 of completing the Babbitt Square.

Step 4: Repeat Step 3 for the other columns. For example,

- to find the entry under F# in column 2, move three half steps forward on the keyboard from F# to A
- to find the entry under G in column 3, move three steps forward on the keyboard from G to A#
- to find the entry under G# in column 4, move three steps forward on the keyboard from G# to B

Continue in this manner until you complete the whole matrix.

	0	9	10	11	7	8	2	1	5	4	3	6	
0	A	F#	G	G#	E	F	B	A#	D	C#	C	D#	0
3	C	A	A#	B	G	G#	D	C#	F	E	D#	F#	3
2	B	G#	A	A#	F#	G	C#	C	E	D#	D	F	2
1	A#	G	G#	A	F	F#	C	B	D#	D	C#	E	1
5	D	B	C	C#	A	A#	E	D#	G	F#	F	G#	5
4	C#	A#	B	C	G#	A	D#	D	F#	F	E	G	4
10	G	E	F	F#	D	D#	A	G#	C	B	A#	C#	10
11	G#	F	F#	G	D#	E	A#	A	C#	C	B	D	11
7	E	C#	D	D#	B	C	F#	F	A	G#	G	A#	7
8	F	D	D#	E	C	C#	G	F#	A#	A	G#	B	8
9	F#	D#	E	F	C#	D	G#	G	B	A#	A	C	9
6	D#	C	C#	D	A#	B	F	E	G#	G	F#	A	6
	0	9	10	11	7	8	2	1	5	4	3	6	

FIGURE 9.23. A completed Babbitt Square.

This matrix gives all the tone rows that can be created by transposing the original row. A composer may use this matrix to determine all possible tone sets allowed in composition of this type.

The tone set in each row is read from left to right. Here are some characteristics of this matrix.

- Notice that in this matrix the first column creates a mirror image of the intervallic relationships of the original row. This column is the **inversion** of the original row.
- If the top row is read from right to left, it is a r**etrogression** of the original row.

- The first column read from bottom to top gives the **retrograde inversion** of the original row.

In the same manner the inversion, retrogression, and retrograde inversion of each tone row (**sequence**) are found in the matrix.

Notice the symmetry in the matrix.

- The set transposed up a level of 6 (the last row of the matrix) is the same as the retrogression of the original row.
- The diagonals contain the same note throughout. These traits will not always occur.
- The diagonal from top to bottom, left to right, will always contain the same element throughout. This characteristic can be used as a checkpoint when completing the Babbitt Square.

To see what some of this music sounds like, listen to a recording of Webern's *Symphony Op. 21*, Berg's *Violin Concerto*, Stravinsky's *In Memorian Dylan Thomas* or *Sonata for Two Pianos*, or Schoenberg's No. 1 of *Three Songs, Op. 48*.

Figure 9.24 is the written form of the first few measures from "Sommermüd." In this piece Schoenberg chose to use the first 8 tones of the tone row in the top staff and tones 9–12 repeated in the piano accompaniment. The first occurrence of the tones is marked for you.

FIGURE 9.24.

1. Make a Babbitt Square for the following tone row:

   G# C# F# B E A D G C F A# D#

   Be sure to check the diagonal for obvious mistakes.

SUMMARY

In this activity you learned about twelve tone music.

- You studied a Babbitt Square and discovered how it is used to find all the variations of a sequence of notes.
- You found the symmetries in a Babbitt Square.

In the next activity you will study the major scales.

T^3—Twelve Tones Transformed

1. Suppose the following tone row is chosen as the original or prime row:

   A F# G G# E F B A# D C# C D#

   Complete the Babbitt Square below for the given tone row. The rows and columns are numbered for easy reference.

FIGURE 9.25.

Column —		1	2	3	4	5	6	7	8	9	10	11	12	
|	Row	0	9	10	11	7	8	2	1	5	4	3	6	
1	0	A	F#	G	G#	E	F	B	A#	D	C#	C	D#	0
2	3	C	A	A#		G		D		F		D#		3
3	2	B	G#		A#	F#		C#		E	D#	D	F	2
4	1		G	G#	A	F	F#	C	B			C#		1
5	5		B		C#	A		E		G	F#		G#	5
6	4	C#		B		G#	A	D#	D	F#	F	E		4
7	10		E		F#	D		A	G#				C#	10
8	11	G#			G	D#	E	A#	A	C#		B		11
9	7		C#	D	D#	B	C		F	A			A#	7
10	8		D	D#		C	C#	G	F#	A#	A	G#	B	8
11	9		D#	E	F					B	A#	A		9
12	6	D#	C	C#	D	A#	B		E	G#		F#	A	6
		0	9	10	11	7	8	2	1	5	4	3	6	

Using your knowledge of musical transformations, find the relationship between the given rows and columns.

2. Row 1 and Column 1
3. Row 1, left to right and Row 1, right to left
4. Row 1 and Column 1, bottom to top
5. Row 1 and Row 5
6. Row 1 and Row 10
7. Column 1, top to bottom and Column 1, bottom to top
8. Row 3, left to right and Row 3, right to left
9. Row 1 and Row 7

ACTIVITY 9.5 Scaling Majors

In Activity 9.4 you learned about twelve-tone composition and how to use a Babbitt Square to find the transformations of a tone set. Now you will consider the major scales that are the basis of traditional western music composition.

THE MAJOR SCALES

A scale is an ascending and descending tonal arrangement of the pitches within an octave. This arrangement varies according to different cultures. For example, the notes from C to C in **Figure 9.26** are often referred to as the C Major scale. Major scales have eight notes. The fifth of a scale is the fifth note of the scale. For example, in a C scale, G is the fifth.

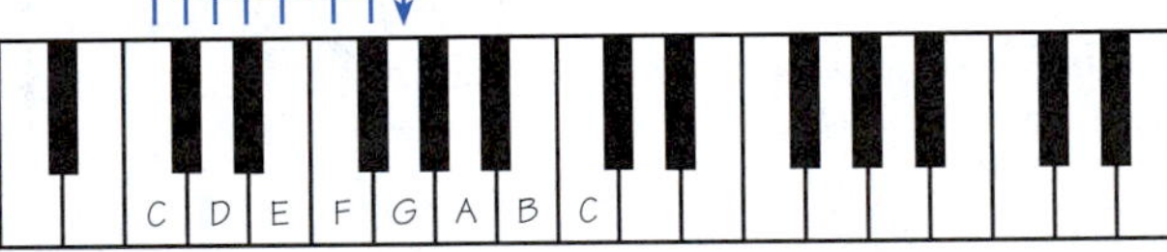

FIGURE 9.26.

1. The interval from C to G on the C scale is a perfect fifth. How many half steps must you move to find the perfect fifth? (Recall that a half step is the distance to the next closest note, black or white.)

Using the ratios for the octave (the interval from C to the next C) and the fifth, Pythagoras constructed a scale that could be used to produce melodies that were considered particularly pleasing to the ear. Using these ratios, we will determine the frequencies for a Pythagorean C major scale. This scale is shown in **Figure 9.27** where N is the frequency of the note C.

Recall from Activity 9.3 the pattern for the harmonic series. If the frequency of the fundamental is N, we know that the frequencies of the first three harmonics are N, $2N$, and $3N$, respectively. Thus we have the following:

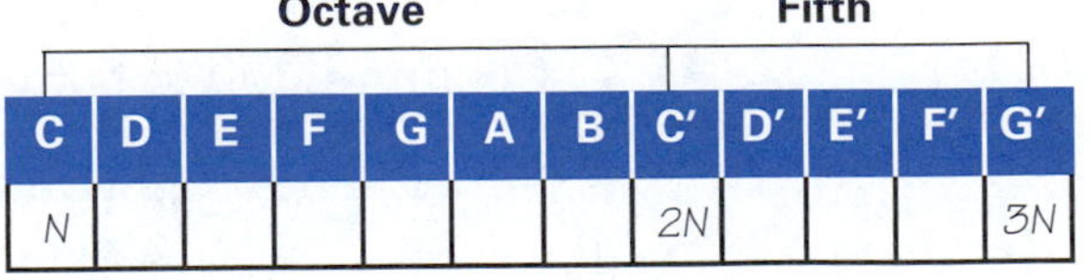

FIGURE 9.27.

Since the ratio of the frequency of G′ to the frequency of C′ is $\frac{3N}{2N} = \frac{3}{2}$, the ratio of the frequency of G to the frequency of C is also $\frac{3}{2}$, and the frequency of G is $\frac{3}{2}N$. Thus we have the following ratios:

- the ratio of a fifth above a given note to the original note is $\frac{3}{2}$
- the ratio of an octave above a given note to the original note is $\frac{2}{1}$

From these two ratios, we can find the frequencies of all the notes of the scale.

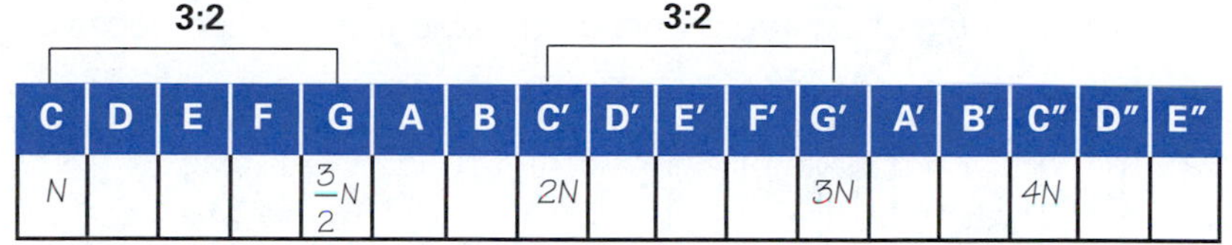

FIGURE 9.28.

2. If G′ has a frequency of 3N, what is the frequency of G if G is an octave lower than G′?

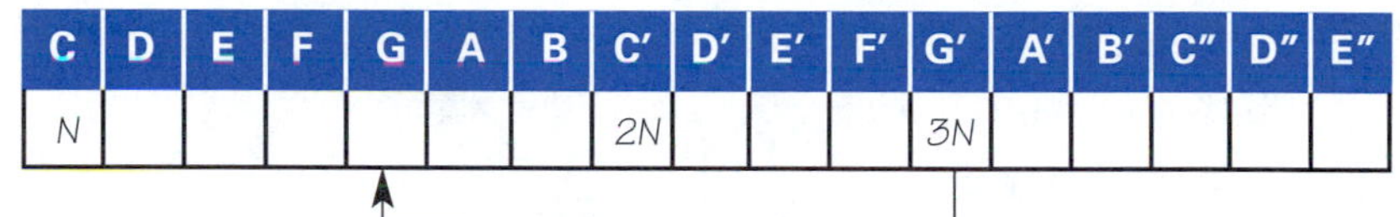

FIGURE 9.29.

3. Make a chart like the one in **Figure 9.29**. Fill in the frequency for G.

4. a) What note is a perfect fifth above G on the keyboard (refer to the keyboard)? What is the frequency of that note? Fill in the frequency on your chart.

   b) Fill in the frequencies for D (an octave lower) and D″ (an octave higher).

   c) Continue moving up a fifth for a new note. Then fill in the frequencies for the octave(s) above and below.

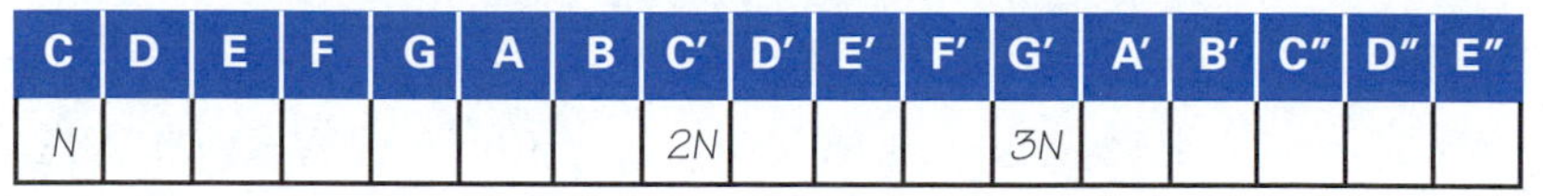

FIGURE 9.30.

5. a) Look at the piano keyboard. Which note is a fifth **below** C′?

   b) If the frequency of a note a fifth **above** has a $\frac{3}{2}N$, what is the frequency of a note a fifth **below**?

6. Complete the table in **Figure 9.31**.

FIGURE 9.31.

C	D	E	F	G	A	B	C′	D′	E′	F′	G′	A′	B′	C″	D″	E″
N				$\frac{3}{2}N$			$2N$				$3N$			$4N$		

7. If the frequency of C is 256 hertz, complete **Figure 9.32**, keeping exact answers and using fractions where necessary.

FIGURE 9.32.

C	D	E	F	G	A	B	C′
256							

8. Using the values from question 7, calculate the following ratios.

   a) $\dfrac{\text{frequency of D}}{\text{frequency of C}}$

   b) $\dfrac{\text{frequency of E}}{\text{frequency of D}}$

   c) $\dfrac{\text{frequency of F}}{\text{frequency of E}}$

   d) $\dfrac{\text{frequency of G}}{\text{frequency of F}}$

   e) $\dfrac{\text{frequency of A}}{\text{frequency of G}}$

   f) $\dfrac{\text{frequency of B}}{\text{frequency of A}}$

   g) $\dfrac{\text{frequency of C}}{\text{frequency of B}}$

The ratio of $\frac{9}{8}$ is called a whole step. The ratio of $\frac{256}{243}$ is called a half step.

A succession of tones that follows the pattern WWHWWWH (where W is a whole step and H is a half step) is called a **major scale**.

SUMMARY

In this activity you calculated the ratios for the frequencies of the major scale devised by Pythagoras.

- You used the ratios for the frequencies of notes at the intervals of a fifth and an octave to find the relationship of the frequencies of all eight tones of a major scale.
- You calculated the actual frequencies of the tones of a major scale beginning on C-256.
- You discovered the ratios of the frequencies of half steps (notes next to each other on the piano keyboard) and whole steps.
- You learned the pattern of half and whole steps that defines a major scale.

In the next activity you will use the Pythagorean ratios to play a musical instrument.

HOMEWORK 9.5

Using the major scale pattern and the ratios calculated in question 8 of Activity 9.5, complete **Figure 9.33** for the frequencies of the G major scale where the frequency of G is *N*.

G	A	B	C	D	E	F#	G′

FIGURE 9.33.

ACTIVITY

9.6 Let's Jam

In Activity 9.5 you calculated the ratios for frequencies of half steps (semi tones) and whole steps. You found the ratio for the half step is $\frac{9}{8}$ and the whole step is $\frac{256}{243}$. The pattern for a major scale is shown in Figure 9.34.

FIGURE 9.34. The pattern for the major scale where T is tonic (beginning note), W is whole step, and H is half step.

Scale Tone	1	2	3	4	5	6	7	8
	T	W	W	H	W	W	W	H

MONOCHORD

An instrument that was popular in the Middle Ages. The instrument is a string stretched over a wooden box with tone markers indicated for finger placement. one hand is used to pluck the string and the other holds the tone marker for the tone you wish to hear.

In this activity we will be using Handout 9.2 to build a simple instrument called a **Monochord**. Since we want an instrument that plays tones correctly we will have to measure the length of the string carefully, and accurately place scale tone markers. Finally we will test our instrument by playing some familiar tunes.

Calculate and mark the major scale tones for your monochord.

1. Measure the length of the string on the monochord from the bridge (nail) to the other end (eye screw). Mark scale tone 1 at the eye.

2. Use the ratios you calculated in Activity 9.5 to calculate placement of the major scale tones to the nearest tenth of a centimeter. C (frequency of N) is the tonic you used in Activity 9.5.

When calculating string lengths, you will be shortening the length to use portions of the entire string length. So rather than dividing by $\frac{9}{8}$ for the next whole step, you will multiply by the reciprocal $\frac{8}{9}$. In the same manner, when marking a half step, you will multiply the string length by the reciprocal of $\frac{256}{243}$, which is $\frac{243}{256}$. For example, if the string length is 22 cm, scale tone 2 will be marked as follows:

$$22 \cdot \frac{8}{9} = \frac{176}{9} \approx 19.6 \text{ cm.}$$

Scale tone 3 will be marked:

$$19.6 \cdot \frac{8}{9} = \frac{157.8}{9} \approx 17.5 \text{ cm.}$$

Remember to use the ratio for the half step for scale tone 4 and scale tone 8.

FIGURE 9.35.

Scale Tone	Interval	Ratio	Process	String Length
1				
2	1	$\frac{8}{9}$	$\_\_\_ \cdot \frac{8}{9}$	
3	1	$\frac{8}{9}$	$\_\_\_ \cdot \frac{8}{9}$	
4	$\frac{1}{2}$	$\frac{243}{256}$	$\_\_\_ \cdot \frac{243}{256}$	
5		$\frac{8}{9}$	$\_\_\_ \cdot \frac{8}{9}$	
6		$\frac{8}{9}$	$\_\_\_ \cdot \frac{8}{9}$	
7		$\frac{8}{9}$	$\_\_\_ \cdot \frac{8}{9}$	
8		$\frac{243}{256}$	$\_\_\_ \cdot \frac{243}{256}$	

3. Measuring from the nail towards the eye screw, mark the scale tone numbers on the masking tape.

4. Use one piece of molding as a bridge placed against the nail. Play the tune below in the same way you would pluck a guitar string: Use the second piece of molding to press down the string on the proper scale tone. Pluck the string with your index finger.

   a) Scale Tones:

   3 2 1 2 3 3 3

   2 2 2 3 5 5

   3 2 1 2 3 3 3

   2 2 3 2 1

   Identify the song.

   b) Scale Tones:

   3 3 3 3 3 3 3 5 1 2 3

   4 4 4 4 4 3 3 3 3 3 2 2 3 2 5

   3 3 3 3 3 3 3 5 1 2 3

   4 4 4 4 4 3 3 33 5 5 4 2 1

   Identify the song.

c) Scale Tones:

535 535 6543234

5 1 111 12345

5224321

Identify the song.

d) Scale Tones:

1155665 4433221

5544332 5544332

1155665 4433221

Identify the song.

e) Scale Tones:

33211 224321

554333 21231

Identify the song.

SUMMARY

In this activity you used Pythagorean ratios to make music.

- You calculated placement of scale tones for a major scale.
- You marked the placement on a stringed instrument.
- You performed several songs based on the major scale.

Modeling Project Be a Composer

Choose one of the following projects:

I. Compose a melody and expand with transformations.

II. Compose a twelve-tone piece.

III. Compose a piece for a 30-second television commercial.

Compose a melody and expand with transformations

1. Compose a simple melody about four measures in length.
2. Choose a time signature. Make sure each measure has the correct number of beats.
3. The melody may be written by randomly choosing pitches and rhythms or by having someone play the notes you think might sound good and changing them until it sounds the way you want it to sound.
4. Use a series of transformations to expand the piece.
5. Give the piece a title. Be sure to write your name on the right at the top of the piece as composer.
6. Write a description of your process and the transformations used to create your composition.
7. If possible have someone play your piece and make a recording or enter it into the computer for playback.

■ Compose a twelve-tone piece

1. Write a tone row consisting of all twelve pitches either by playing and determining the sound you like or by randomly choosing them.
2. Create a Babbitt Square.
3. Choose a time signature. Make sure each measure has the correct number of beats.
4. Use the tone row along with any combination of columns and/or rows to create a melody 16 measures or longer.
5. Give the piece a title. Be sure to write your name on the right at the top of the piece as composer.
6. Write an explanation of how you chose your tone row and which row/columns you used. Did you consider transformations when making your choices?
7. If possible have someone play your piece and make a recording or enter it into the computer for playback.

■ Compose a piece for a 30-second television commercial

1. Design and describe a scenario for the commercial. What are you advertising? What is the mood? Should the piece be fast or slow?
2. Determine the number of measures (16 or more) and the tempo.
3. Calculate the number of beats needed.
4. Choose an appropriate time signature.
5. Create a melody of your own or use part of a childhood tune. Many commercials use parts of classical music that is available for public use.
6. Give the piece a title. Be sure to write your name on the right at the top of the piece as composer or indicate there that you are the arranger of an existing piece. Be sure to write the tempo marking on the left.
7. If possible have someone play your piece and make a recording or enter it into the computer for playback.

Practice Problems

Review Exercises

1. Draw a geometric representation of a musical sequence.

2. Draw a geometric representation of a musical retrogression.

3. Draw a geometric representation of a musical inversion.

4. Which geometric term and which musical term describe the transformation shown below?

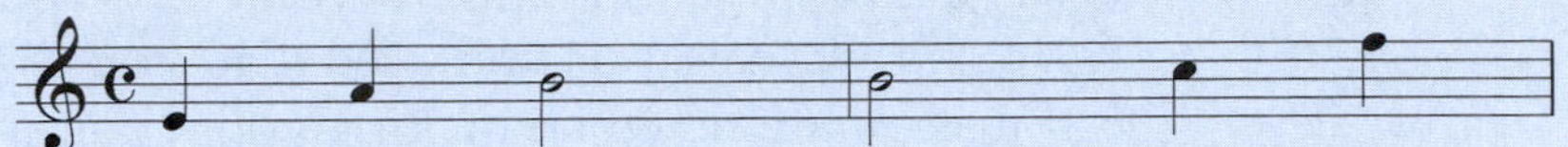

Note: remember common time is $\frac{4}{4}$.

5. Which geometric term and which musical term describe the transformation shown below?

6. Write the inversion of the following phrase.

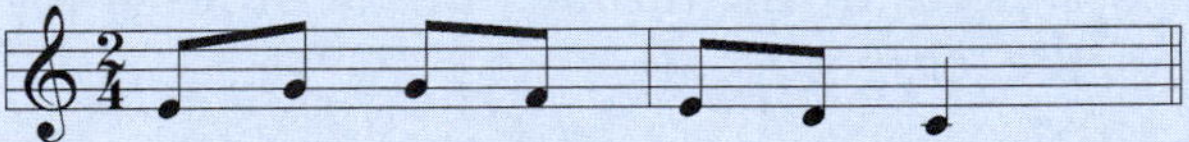

7. Write the retrograde form of the phrase in question 6.

8. A Babbitt square is a/n _______ used in composing twentieth century music.

9. The type of twentieth century music that uses the Babbitt Square is _______________.

10. One characteristic of the Babbitt Square is

    a) diagonals are equal

    b) the first row and last column are alike

    c) the first and last rows are alike

d) the left to right, top to bottom diagonal has a single element throughout

11. Give the retrograde form or retrogression for the following tone row:

A# D E C D# B F A C# F# G G#

12. Calculate how long, to the nearest second, each of the following selections of music would last.

a) 64 measures in $\frac{4}{4}$ time at ♩ = 128.

b) 132 measures in $\frac{2}{4}$ time at ♩ = 140.

c) 304 measures in $\frac{5}{4}$ time at ♩ = 210 (give answer in minutes and seconds).

13. For a commercial 110 seconds long, at what tempo must a piece be played if it contained 88 measures in $\frac{5}{4}$?

14. If A_4 on the piano has a frequency of 440 hz, calculate the frequency for the following notes according to Pythagorean ratios.

a) A_5 (one octave higher)

b) A_6 (two octaves higher)

c) E_4 (one fifth higher)

15. If the frequency is doubled what is the speaking length (the part that sounds) of the string?

a) $\frac{1}{2}$ the length of the original string length

b) twice the length of the original string length

c) $\frac{3}{4}$ the length of the original string length

d) 25 centimeters

16. Given a fundamental of frequency 495 hz, write the next five numbers in the harmonic series.

17. Fill in the chart for the frequencies of the notes of a major scale (TWWHWWWH) using Pythagorean ratios. Graph the frequency versus the number of the term in the series.

Term	Note	Frequency
1	C	512
2	D	
3	E	
4	F	
5	G	
6	A	
7	B	
8	C	

18. Which graph shows the correct correspondence of frequency to pitch?

a)

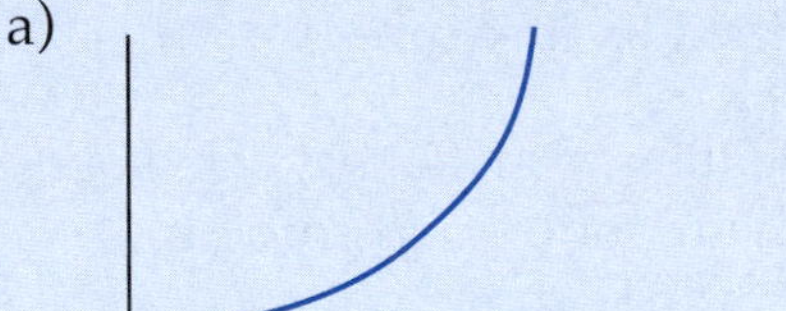

b)

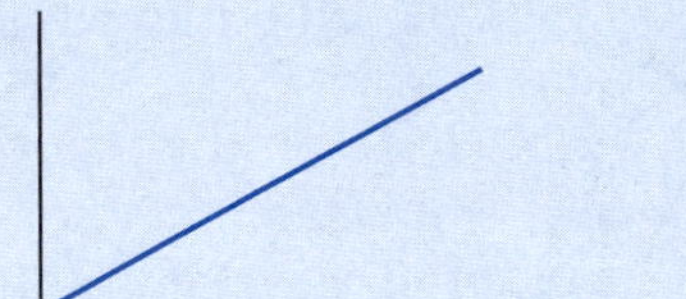

c)

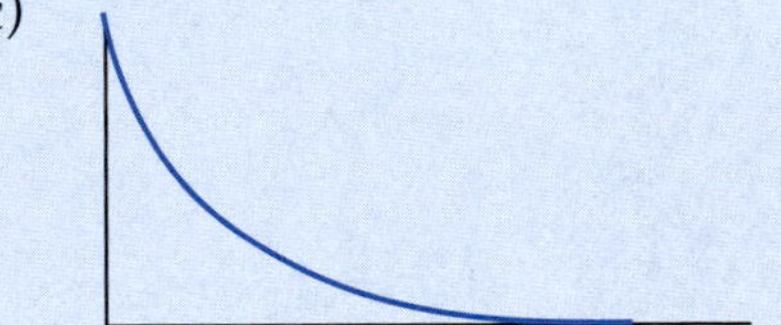

d)

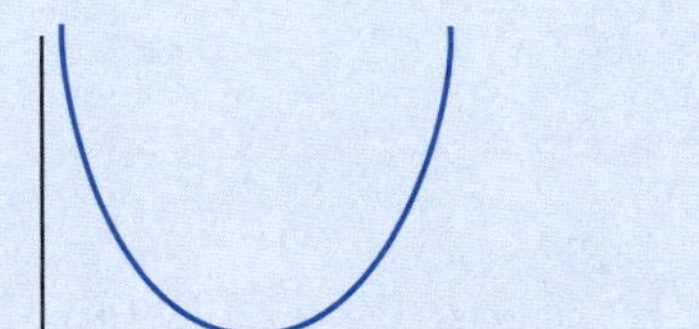

19. If the heart beats 18 times in 15 seconds, how many times will it beat in one minute?

20. If a heart beats 100 times a minute, how many times does it beat in 15 seconds?

21. If a composer creates a section break according to the proportions of the Golden Mean, where would you expect the break to occur in a piece of music with 144 measures?

22. Wave frequency is measured in __________.

23. A sound with a lower pitch has a small _________.

    a) amplitude

    b) frequency

    c) period

    d) wave velocity

24. A sound with a lower pitch has a longer _______.

    a) amplitude

    b) frequency

    c) period

    d) wave velocity

25. What is the amplitude of the sine graph generated by the following functions.

    a) $y = \sin(440x)$

    b) $y = 2\sin(440x)$

    c) $y = 3\sin(220x)$

26. Give the fundamental frequency for the equation $y = 25\sin(110x) + 16\sin(220x) + 9\sin(330x) + 7\sin(440x) + 5\sin(550x) + 2\sin(660x)$.

27. Describe what happens to the frequency of a graph when the period is decreased.

Modeling Music

REVIEW

Model Development

Musicians face several mathematical problems, particularly in the composition of music. Before musicians begin to compose, they study and analyze existing music. Here, mathematics and mathematical thinking are used extensively. Rhythm, tempo, and performance time were discussed in the first activity. The study was limited to a few time signatures and note values. All of these factors play a part in composition. Performance time is especially important when composing music for advertisements, movies, etc. In advertising, companies buy a certain amount of time and the music must be timed exactly to fit. In movie production, music is used to set the scene and accent actions in the movie. Thus music must be perfectly timed to have the intended impact. The use of permutations was introduced to find possible ways of incorporating a phrase or section of music in the homework activity. This exercise set the stage for later development of twelve-tone music, which sets up a matrix to find all the possible permutations for a tone row.

Then we studied ways musicians combine variety and repetition using transformations. These transformations are related to geometric transformations with which you are already familiar. Since knowledge of the harmonic series is important in understanding timbres and sounds of various instruments, we looked at the harmonic series. The Golden Mean was studied in order to analyze the music of composer Béla Bartók. We looked at graphs of sine waves in relation to sound. We completed a Babbitt Square and applied our knowledge of permutations and transformations to a study of twelve-tone music.

We used our knowledge of the harmonics series along with Pythagoras' ratio for the perfect fifth to calculate frequencies for a major scale. From there, the ratios were found for half and whole steps and applied to the pattern for a major scale. Your knowledge of Pythagorean ratios was applied to string lengths of a musical instrument to build an instrument related to the major scale. You then performed songs on the instrument.

▪ Mathematics Used

Composition requires not only a good ear for music but also knowledge of music theory. Music theory involves mathematics and mathematical processes. In this chapter we review a variety of mathematical concepts and processes. We begin by examining time signatures, tempos, and note values, and writing a linear equation relating these variables. The equation for the different variables was solved in order to calculate the needed number. Then combinations and permutations were reviewed in order to use them along with the calculations to create music.

In Activity 9.2 geometric transformations and their relationship to melody patterns was considered. Students not only recognize geometry patterns but also apply them to a musical situation. We also looked at the Golden Mean and some natural or contrived relationships in musical compositions. We not only calculate the point at which a musical device may occur but also identify the device from the work we completed on transformations. In Activity 9.3 we reviewed arithmetic sequences associated with the harmonic series in music, wrote an algebraic expression for the *n*th harmonic in a series, and used bar graphs to represent the harmonics present in given situations. Then we modeled the sound wave on the graphing calculator. We set appropriate viewing windows and explored amplitude and period of a periodic function. In the discussion of twelve-tone music in Activity 9.4, we created a matrix of numbers and used it to examine transformations that occurred as musical devices used to create compositions.

Fractions and ratios were used to calculate frequencies of tones of the major scale and to find fractions relating tones and half step or whole step apart. Then we extended and applied the knowledge of Pythagoras' ratios to find tones for a major scale on a stringed instrument. The same process could be used to find tones for other scales as well. Finally we used the instrument to play and identify well-known tunes. All these mathematical processes are used by musicians to compose and analyze music, and design instruments for various sounds.

Amplitude: Half the vertical distance between maximum and minimum values of a periodic graph.

Babbitt Square: A matrix used to find all combinations for a chosen tone row.

Beat: The basic unit of time or pulsation in a composition.

Béla Bartók: Hungarian twentieth century composer who used transformations and the Golden Mean in his compositions. He authored the *Mikrokosmos*.

Combination: A selection of objects from a set in which order is not important.

Frequency: The number of cycles or completed alternations per unit time of a wave or oscillation.

Fundamental: The tone that is the basis of the harmonic.

Golden Mean: The number 1.6180339887… (this number goes on forever) is called the Golden Mean. The ratio $\frac{1}{0.618...}$ is called the Golden Ratio because

$$\frac{1}{0.618...} = 1.6180339887\ldots$$

Half step: The smallest interval commonly used in music. It is found on the keyboard by moving from one note to the next nearest note, whether black or white.

Harmonic Series: A series of harmonics including the fundamental.

Harmonics: The secondary tones that form a component of every musical sound, though they are not heard distinctly.

Hertz (hz): The unit of frequency, equal to one cycle per second.

Inversion: A musical transformation corresponding to reflection across the x-axis.

Major scale: A scale developed about 1700, whose pattern of whole/half steps after the first note is WWHWWWH (W = whole note, H = half note).

Matrix: A rectangular array of numbers (or notes in the Babbitt Square).

Measure: A group of beats (equal units of time), the first of which is often accented, set off by bar lines.

Mikrokosmos: A six-volume set of 153 short piano pieces by Bartók embodying a "little world" of twentieth century technique and style.

Milton Babbitt: The twentieth century composer for which the Babbitt Square is named.

Monochord: A single string stretched over a long wooden box used widely in the Middle Ages. It is designed for the investigation and demonstration of the basic laws of accoustics, particularly the relationship between lengths of strings and intervals.

Octave: A tone on the eighth degree from a given tone.

Perfect fifth: The interval of notes consisting of seven half steps.

Period: The length of the x-interval required for the graph of one complete cycle before the graph begins to repeat itself.

Permutation: An arrangement or selection of objects from a set when order is important.

Pythagoras: A man who lived about 550 B.C. who devised a scale in which all the tones are derived from the interval of the fifth.

Retrograde inversion: The combination of retrograde motion (retrogression) and inversion. It is represented geometrically by a reflection across both axes.

Retrogression or Retrograde form: The repetition of a short section in reverse or backward motion, beginning with the last note and ending with the first. It is represented geometrically by a reflection across the y-axis.

Rhythm: The duration of tones within a beat and across several beats.

Scale: A group of ascending or descending notes.

Section Break: A point in the music before a change in the structure of the piece. The change in structure could include a different tempo, time signature, or other change to the music.

Sequence: The repetition of a short section at different pitches. Its geometric equivalent is a translation.

Tempo: The rate of speed of a composition of section as indicated by tempo marks.

Timbre: Tone color; the peculiar quality of a tone as sounded by a given instrument or voice. The timbre is determined by the harmonics produced and their amplitude.

Time signature: A sign given at the beginning of a composition to indicate its meter (time). It consists of two figures written like a fraction, the lower figure indicating the note that receives a beat, the upper figure indicating the number of beats in a measure.

Tone row: A series containing the twelve chromatic tones in a succession chosen by the composer.

Twelve-tone music: A twentieth century style of music composed by creating a tone row and using its transformations to create a composition.

Whole step: Two half steps.

Wolfgang Amadeus Mozart: Austrian composer who lived from 1756–1791. He was fascinated with numbers and used transformations and the Golden Mean in his compositions.

Index

Defined terms and their page numbers are in bold.

G

H

I

S

T

U

V

W

X

Y

Z

Acknowledgements

PROJECT LEADERSHIP

Solomon Garfunkel, COMAP, Inc., Lexington, MA
Roland Cheyney, COMAP, Inc., Lexington, MA

LEAD AUTHOR

Jerry Lege, Comap, Inc.,

AUTHORS

Ronald E. Bell II, University of Texas at Austin, TX and MOVES, Inc.
Marsha Davis, Eastern Connecticut State University, Willimantic, CT
Sandra Nite, Sweeny ISD, Sweeny, TX

DEVELOPMENT EDITOR

Tony Palermino, Watertown, MA

EDITORIAL CONSULTANTS

Henry Pollak, Teacher's College, Columbia University, NY
Gary Froelich, COMAP, Inc.

Sections of *Mathematical Models With Applications* were developed from a variety of source materials including COMAP's *Mathematics: Modeling Our World* program. The following authors and curriculum developers helped develop and test these source materials:

Landy Godbold, Westminster Schools, GA
Nancy Crisler, Pattonville Schools, MO
Bruce Grip, Etiwanda High School, CA
Rick Jennings, Yakima School District, WA
David Moore, Purdue University, IN
Dédé de Hahn, The Freudenthal Institute, The Netherlands
Jan de Lange, The Freudenthal Institute, The Netherlands
Anton Rodhardt, The Freudenthal Institute, The Netherlands

FIELD TEST DIRECTOR

Joyce Q. Collett, Clear Creek ISD, League City, TX

FIELD TEST TEACHERS AND EVALUATORS

Shirley Cary, North Shore Senior High School, Houston, TX
Juan Castillo, North Shore Senior High School, Houston, TX
Bill Chamberlain, Texas City High School, Texas City, TX
Christine Coffey, Cuero High School, Cuero, TX
Diane Dauer, Clear Creek ISD, League City, TX
Tracie Dearmond, B.F. Terry High School, Richmond, TX
Lillian Downs, Bryan High School, Bryan, TX
Jeannie Frankie, Lamar ISD, Rosenberg, TX
Stephen Fulbright, Clear Brook High School, Friendswood, TX
Tonya Gerdes, Bryan High School, Bryan, TX
Will Haltom, Clear Creek High School, League City, TX
Marilyn Hicks, Clear Lake High School, Houston, TX
Jacky Martin, Clear Lake High School, Houston, TX
Wendall Molix, Montwood High School, El Paso, TX
Sandra Nite, Sweeny High School, Sweeny, TX
Nancy Paterson, Texas City High School, Texas City, TX
Derek Prather, Sweeny High School, Sweeny, TX
Sara Szymanski, Bryan High School, Bryan, TX
Lana Zimmer, Clear Creek Isd, League City, TX

COMAP STAFF

Laurie Aragón, Rafael Aragón, Lynn Aro, Jan Beebe, Peter Bousquet, Clarice Callahan, Ricky Carter, Roland Cheyney, Nancy Crisler, Kevin Darcy, Michele Doherty, Gary Feldman, Gary Froelich, Solomon Garfunkel, Frank Giordano, Daiva Kiliulis, Sue Martin, Sue Rasala, Sheila Sconiers, George Ward, Gail Wessell, Pauline Wright

EDITORIAL INTERNS

Kesha Gianakos
Michael Gray

INDEX EDITOR

Seth Maislin, Focus Publishing Services, Arlington, MA

CHAPTER OPENER ILLUSTRATIONS

Ashley Van Etten, Narragansett, RI

ILLUSTRATIONS

David Barber Illustrations, Marblehead, MA
Lianne Dunne, Sandwich, MA

PHOTO RESEARCH

Michele Doherty, COMAP, Inc.
Susan Van Etten, Wenham, MA

PHOTO CREDITS

AP/Wide World Photos, New York, NY
68

Vin Catania Studios, Boston, MA
268, 275, 278, 286, 297

Courtesy of Paul Conklin: © Paul Conklin
162, 553

Corbis Images, Tacoma, WA
vii, viii, x, xiii, xiv, xv, 7, 8, 18, 29, 38, 48, 49, 54, 70, 89, 91, 99, 101, 114, 116, 120, 123, 130, 138, 146, 168, 169, 194, 195, 196, 199, 204, 212, 221, 240, 260, 281, 311, 355, 388, 402, 415, 477, 480, 486, 489, 491, 497, 500, 552, 557, 561, 564, 572

Dover Publications, Inc., Mineola, NY
xi, xiii, 259, 265, 274, 283, 291, 297, 304, 394, 398, 408, 570

Courtesy of Dunbar Armored, Inc.
416

EyeWire, Seattle, WA
x, 105, 107, 113, 148, 167, 196, 202, 246, 337, 340, 349, 418, 428, 447, 468, 482, 494, 513, 534, 558

Daiva Kiliulis, COMAP, Inc.
247, 377, 424, 538

Dr. Dexter W. Lawson, Wenham, MA
80

Michele G. Lege, Davis, CA
184

Kathy Osborne, Leominster, MA
356

The Bridgeman Art Library International LTD., New York, NY
260, 261, 271, 279, 281, 299

The Folger Shakespeare Library, Washington, D.C.
261

The Image Bank, Dallas, TX
xi, 274, 280, 284

Three Pounds of Bakon, Los Angeles, CA
47

Susan Van Etten: © Susan Van Etten, Wenham, MA
viii, ix, xii, xiii, 20, 33, 83, 85, 93, 94, 132, 134, 137, 144, 153, 154, 173, 176, 180, 182, 223, 231, 232, 328, 330, 331, 334, 336, 348, 351, 354, 359, 363, 365, 371, 378, 379, 389, 392, 396, 431, 434, 439, 441, 444, 454, 469, 473, 475, 484, 495, 510, 530, 536, 540, 547, 555, 569

Permission to use ***Mikrokosmos*** pieces ***Once Upon A Time...*** (Vol. 4, no. 94); ***Hands Crossing*** (vol. 4, no. 99); and ***Melody in the Mist*** (Vol. 4, No. 107) was generously provided by the Estate of Béla Bartók and Boosey & Hawkes, New York, NY. To obtain sheet music for Mikrokosmos and other works by Béla Bartók visit your music dealer or Boosey & Hawkes at http://www.boosey.com/
544, 545, 547, 548

References

CHAPTER 2

Anscombe, F.J., Graphs in Statistical Analysis, The *American Statistician 27 (1973)*: 17-21.

Behrman, Richard E. ed. 1991. *Textbook of Pediatrics.* Philadelphia, PA: W.B. Saunders Company.

Conroy, Glenn, et.al., 1992. Obituary: Mildred Trotter, Ph.D. (Feb. 2, 1899- Aug. 23, 1991), *American Journal of Physical Anthropology 87*:373-374.

CHAPTER 4

Wright, J.E., and V. 1990. Cowart. *Anabolic Steroids, Altered States.* Carmel, IN: Benchmark Press.

CHAPTER 9

Bartók, Béla, *Mikrokosmos, Once Upon a Time...* Vol. 4, No. 94; *Hands Crossing* Vol. 4, No. 99; and *Melody in the Mist,* Vol. 4, No. 107, (Hawkes and Son (London) Ltd. 1987)